OCCASIONAL PUBLICATIONS OF THE
CAMBRIDGE UNIVERSITY MUSEUM OF ARCHAEOLOGY AND ETHNOLOGY

II

THE PREHISTORIC CULTURES OF THE HORN OF AFRICA

IN THIS SERIES

I The Archaeology of the Santa Elena Peninsula in South-west Ecuador
By G. H. S. BUSHNELL

II The Prehistoric Cultures of the Horn of Africa
By J. D. CLARK

Other volumes are in preparation

PRESENT-DAY TROGLODYTES AND 'ICHTHYOPHAGI' OF THE SOMALI COAST (see pp. 286–7)

THE

PREHISTORIC CULTURES
OF THE
HORN OF AFRICA

An analysis of the
Stone Age Cultural and Climatic Succession
in the
Somalilands and Eastern Parts of Abyssinia

BY

J. D. CLARK

With a foreword by

M. C. BURKITT

This Octagon edition contains
a new preface by the author

OCTAGON BOOKS

A DIVISION OF FARRAR, STRAUS AND GIROUX

New York 1972

Reprinted 1972
by permission of Cambridge University Press

OCTAGON BOOKS
A DIVISION OF FARRAR, STRAUS & GIROUX, INC.
19 Union Square West
New York, N. Y. 10003

LIBRARY OF CONGRESS CATALOG CARD NUMBER: 70-159173

ISBN 0-374-91657-8

Printed in U.S.A. by
NOBLE OFFSET PRINTERS, INC.
New York, N.Y. 10003

CONTENTS

PART III

THE PREHISTORY OF THE SPECIAL AREAS

PART IV

A TENTATIVE CORRELATION OF CULTURES AND CLIMATES

LIST OF PLATES

LIST OF TEXT-FIGURES

PREFACE TO THE OCTAGON EDITION

The past decade has seen a quickening of interest in Africa as the continent where man is most likely to have evolved and where his direct relatives among the primates—the chimpanzee and the gorilla—are still to be found. This heightened interest stems from the wealth of fossil hominid remains, mostly from East Africa and sometimes associated with actual living places, that has come to light in recent years. These hominid remains are of Lower and early Middle Pleistocene age but there are also even earlier finds—from Fort Ternan (Mio-Pliocene) and from Rusinga (Lower Miocene)—that suggest that the Hominid and Pongid lines had separated sometime during the Miocene, perhaps even earlier.

At the same time the prehistory and history of the last 2000 years have also assumed special importance, since it was during this period that the present-day peoples of Africa became differentiated and the events took place which brought about the existing pattern of ethnic distribution. Over large parts of the interior of the continent there is little written documentary record before the beginning of the last century so that it is the evidence from archaeology, together with oral tradition, that will provide one of the major sources for writing the history of the African peoples south of the Sahara.

To this general picture of subsaharan prehistory the Horn of Africa is no exception. Although there are more documentary sources available here, they are, aside from Ethiopia, rarely specific nor are geographic and ethnic identifications possible with any exactness. Except for the Omo basin, which geographically belongs more in the East African Rift region, the record in the Horn starts later than in other parts of east Africa. In addition, the prehistoric cultural sequence cannot, as yet, be connected with historic times since the archaeology still remains to be undertaken at the medieval Arab towns of the Indian Ocean coast and in the interior. Also, while the excavations that have been carried out by the Ethiopian Institute at the pre-Axumite and

Axumite centres, throw much interesting light on the culture of north-eastern Ethiopia from the Vth century, B.C., we still lack the studies of older sites necessary to permit an understanding of the change from hunting and gathering to food production.

The reprinting of this present volume on the Stone Age archaeology of the Horn will serve to make available again the information on this prehistorically little known part of the continent where, except in the Omo valley and at certain sites on the Ethiopian plateau, little new work has been done since the mid fifties. At the same time, the now greatly increased knowledge of the "neolithic" and early Iron Age Archaeological Occurrences in the Sahara, as well as the extensive investigations that have been undertaken in Nubia in the valley of the Upper Nile in connection with the construction of the Aswan High Dam, have provided a background against which the prehistoric industries of the Horn can profitably be restudied and their association with the Sahara and the Nile, with southern Arabia and with east Africa, be more meaningfully assessed and given a new perspective.

The book has been out of print for eight years or more and the field-work on which it is based was carried out between 1941 and 1946. But, in spite of the pace at which African archaeology is developing today, it is not significantly out of date, except in its lack of statistical data and radiocarbon chronology. No attempt has been made to amend the text. However, in the paragraphs that follow, reference is made to such new work as has come to the author's attention since 1951 and to the significance of this and of current thought on terminology and other matters as they affect the present volume.

SOMALIA

Since 1946 very little new work has been done in Somalia (formerly Italian and British Somaliland). A further surface collection was made at Jesomma in the Webi Shebeli valley and amplifies knowledge of the "Magosio-Doian" but to the author's knowledge this has not yet been published.

A small, unpublished excavation in a rockshelter at Gan Libah

(9°52′: 44°48′) in northern Somalia gave evidence of occupation and cultural material to a depth of 1.52 m. (5 feet). The resulting collection is not large and has been examined by the writer. Ostrich eggshell (some burnt) and bone from food waste occur at all levels, though decreasing towards the base. The raw material used for artifacts is predominantly chert with a small percentage of quartz. The waste comprises blades, bladelets and flakes struck from one- or two-directional cores. Also present are one flat, discoid core and a flake struck from a radially prepared core of this type. The 61-91 cm. (2-3 ft) level yielded the upper half of a lanceolate point of Doian type. This had been bifacially worked, probably by some form of punch technique, and it is of interest that the specimen, unlike the remainder of the collection, shows a glossy patina which suggests possible heat treatment of the stone before retouching. The 91 cm.-1.22 m. (3-4 ft) level yielded a blade retouched by *écaillé* flaking down the length of one edge, one micro-blade core used as a scraper and the disc core and radially prepared flake referred to above. The only shaped tool in the 1.22-1.52 m. (4-5 ft) level is an end scraper on a blade. Small as it is, this assemblage suggests affinities with the Doian and "Magosio-Doian" industry of southern Somalia, rather than with the "Somaliland Wilton" of the north. As such, it is of interest as it represents the most northerly site at which Doian forms (the lanceolate and blade with *écaillé* trimming) have been found and it perhaps provides also a link with the advanced foliate point aggregates from Dire Dawa in the Danakil Rift.

Another small collection from a rockshelter at Abdul Kadir, near Borama, is more closely comparable to a "Somaliland Wilton" assemblage. It comprises micro-blades and cores together with one well-made lunate in obsidian and one typical example of a struck, small Levallois core and quartzite. There is possibly some depth of deposit at this site, which may contain a sequence comparable to that from the Porc Epic cave at Dire Dawa. It is of interest that obsidian, quartz and quartzite were used for artifacts as well as the usual chert. The obsidian must have been imported since the nearest source is in the Danakil Rift some miles distant to the west.

The so-called "Galla Graves", that are often numerous round watering areas in the Horn, have been divided into two main types—small, rough cairns of stone covering a burial chamber constructed of wooden poles (Series A) and more carefully built, larger cairns with no evidence of a burial chamber (Series B). Two of the smaller (Series A) cairns near Gan Libah were investigated by I. M. Lewis and were found to contain respectively two and three burial of males. Wood forming a support from one of the burial chambers, has been dated to 100 ±150 years B.P. thus showing that some of these "Galla Graves" are of Somali origin and were being constructed as late as, if not later than 250 years ago (Lewis, 1961). Larger (Series B) type cairns at Mandera and Wajir, south of the Juba river, had earlier yielded pottery and a copper bracelet in association with burials but no proper investigation has yet been undertaken of any of these structures which are traditionally associated with the Galla people. The age of the oldest of the cairns, therefore, still remains to be determined.

ETHIOPIA

In Ethiopia since 1946 several important investigations have been undertaken. In 1967 and 1968 work by interdisciplinary teams from Chicago, Paris and Nairobi on the Lower Pleistocene Omo beds in the southwest, north of Lake Randolph, has resulted in the discovery of further hominid remains of Australopithecine type associated with an extremely rich fossil fauna, within sediments dated by the potassium argon method to between 1.8 and 3.7 million years ago (Howell, In Press).

Nowhere in the Omo beds are there any cultural remains but at Melka Kontoure on the Abyssinian Plateau c.50 Km. southwest of Addis Ababa, an important stratified cultural sequence has been investigated by Bailloud (Bailloud, 1965(a)) and currently by Chavaillon (Chavaillon & Taieb, 1968). This sequence is preserved in the sediments of a small lake that formed behind a volcanically produced dam. It comprises stratified, waterlaid sands, silts and clays intercalated with tuffs and land surfaces that were occupied by groups of hunters.

The sequence starts with an artifact assemblage of Developed Oldowan type, followed by several Acheulian occurrences of evolved form. These in turn are overlain by beds and surfaces with "Middle Stone Age" and "Later Stone Age" Archaeological Occurrences. Indeed, it has been claimed that Melka Kontoure provides an undisturbed sequence that may extend back to the Lower Pleistocene but, on the other hand, the absence of any major stratigraphic unconformities within it makes it more probable that the beginning of the succession is not older than the end of the Middle or the beginning of the Upper Pleistocene. In any case this site has shown clearly that the Ethiopian Plateau had been successfully occupied by the makers of the Acheulian Industrial Complex prior to, or at the beginning of, the Upper Pleistocene and, apart from the Omo river sequence, Melka Kontoure now stands out as the most important stratified site in Ethiopia. When the descriptions of the Archaeological Occurrences from the different horizons become available they will constitute the first detailed analysis of unselected artifact aggregates from that country.

Analysis of the pollens preserved in these sediments gives also the first indication of the nature of the vegetation pattern on the Ethiopian Plateau contemporary with a "Developed Oldowan" tool kit. This is characterised by much grass pollen together with arboreal pollens of subtropical and montane species (Bonnefille, 1968) suggesting a cooler and open environment.

Most of the remaining fieldwork in Ethiopia has been confined to Axumite and related sites and so lies outside the purview of this book. However, two publications of relevant interest must be noted. Recently an Acheulian-type handaxe has been found near Dallol in the Danakil Rift, in association with corals and shell beds relating to a marine transgression that turned this low-lying trough into a gulf of the Red Sea. The find, the sediments and their elevation suggest that this marine episode should be dated no later, certainly, than the beginning of the Upper Pleistocene and it may possibly be found to correlate, as we had previously suggested (pp. 120-121), with the last episode of worldwide higher sea level that took place prior to the general lowering during the Last Glacial. This handaxe, together with the presence of obsidian artifacts of "Later Stone Age" type associated with later formations and the possibilities for chronometric dating of the corals, holds out high promise for a long and well dated cultural sequence in this little known part of the Horn (Faure & Roubet, 1968). This discovery, like that of Melka Kontoure, points to the need for systematic surveys, both of the Rift and of the Abyssinian plateau, since it is now certain that many other sites are bound to exist there and, in fact, the cultural sequence could be as old as that in Kenya and northern Tanzania.

Sites of small village settlements have also been recorded from the northwestern part of the plateau near Agordat in the province of Eritrea (Arkell, 1954) but at present these are known only from surface collections made over twenty years ago. However, from the presence of lugged stone axes, palettes, mace heads, pottery, a possible ox figurine and other traits, including rare bronze or copper artifacts, Arkell suggests that these settlements may have had cultural connections with the C-Group people of the Upper Nile and so may date to between c.2000 and 1300 B.C.

The presence of settled communities of mixed farmers on the Ethiopian Plateau in the second or even the third millennium B.C. is not unexpected in view of the evidence from Deir-el-Bahari—and the small collection from Quiha near Macalle (p.324), as yet undated, may perhaps be further confirmation. So far, little attention has been paid to the possibility that domestic sheep and goats may have been present in the Horn prior to the introduction of cattle. If this was the case, then the presence of the sheep and goat would date to a time before that of Hatshepsut's expedition to Punt in 1496 B.C. since the bas-reliefs show that cattle were already present in the northern part of the Horn by that time. Sheep/goat bones and teeth have been provisionally identified by Graziosi from the upper levels of the Gure "Makeke" at Bur Eibe (in Stratum III associated with ox molars) and from the Eil rock-shelter in the Nogal Gorge. In the former they are associated with the later Doian and in the latter with an, as yet, unnamed occurrence based on the use of small blades and large scraper forms (Graziosi, 1940, pp.33,61). It is of interest that goat has been provisionally identified also from one of the south Arabian "neolithic" hunting sites in the Rub-al-Khali desert. These are, in part at least, contemporaneous with the Saharan "neolithic" sites, the latter dating mostly between the fourth and the second millennium B.C. (Smith, 1962, p.22; de Bayle des Hermens, 1965).

Besides emmer wheat, barley and sorghum, there are at least three other important local domesticates, which must have an early origin, in use in Ethiopia today—*ensete* (*Ensete edule*), *tef* (*Eragostris*) and finger millet (*Eleusine*). The initial attempts at domesticating these can hardly have taken place as late as the first millennium B.C. and the experimental period, which resulted in the development of the first domestic strains, is likely to date back at least to the second or third millennium (Clark, 1967, pp.613-5), comparable in time with the beginning of plant cultivation in west Africa. Indeed, it is possible that the domestic plant that is represented on Egyptian Predynastic pottery of the fourth millennium from Naqada is the *Ensete edule* (Simoons, 1965). A programme of excavation of selected sites can be expected to document the development of crop cultivation and animal domestication in Ethiopia and to show the extent to which this was influenced from Egypt, the Sudan or southern Arabia.

It is to be expected that stone continued to be used for certain purposes for some time after metal was introduced to the Horn and stone-using agriculturalists may, therefore, have existed contemporaneously with the southern Arabian, pre-Axumite centres (5th century B.C. to 1st century A.D.) and even with Axumite culture 1st-10th century A.D.) on the Ethiopian Plateau. Generally, metal would replace stone, firstly by trade and later by local manufacture, those groups in the north and east adjacent to the Red Sea coast being the first to adjust to a bronze- and iron-using economy.

In this connection it is of interest that Anfray (Anfray, 1968. p.353) reports finding at the bottom of the sequences at Matara and Adoulis, a stone industry based on the use of obsidian, which he believes was contemporary with the beginnings of pre-Axumite culture. The finding of obsidian flakes ("knives") and *lames écaillées* in one of the Axumite buildings at Matara, dated provisionally to the sixth century A.D. and of polished stone axes and stone mortars of Yeha type (some of which recall the "Neolithic Stone Bowls" of the East African Rift) in another structure at the same site, also suggests that stone was still being made use of for some purposes at that late date (Anfray & Annequin, 1965).

ROCK ART

Very pertinent here to determining whether, and by how long, food production preceded the coming of metallurgy in the Horn is a study of the subject matter of the prehistoric rock art.

Three important new localities with paintings have been reported by Graziosi (Graziosi, 1964) in the Akkele Guzai region, notably in the districts of Senafe and Adi Ugri and near the port of Karora on the Sudan border. This art depicts pastoralists and there is very little evidence of hunting. The subjects are cattle, humans and, more rarely, antelopes and ?goats. Again we are reminded of the eastern Saharan art, notably some of that from the Gebel 'Uweinat region. What are believed to be the oldest paintings and are certainly the most naturalistic and carefully executed (from Ba'atti Sollum [Mai Aini]) do not have human figures associated. However, warriors with long triangular bodies, armed with spears and carrying oval and circular shields, are found with what Professor Graziosi considers is probably a later stage. These are also found at several rockshelters, notably Zeban Ona Libanos, where they are associated with cattle, including a milking scene, and a musician playing a zither-like instrument.

Similar human figures with spears are found in the Adi Qanza rockshelter, though they are less skillfully drawn here and humped (?zebu) cattle are shown together with the long horned, humpless breed. There are no camels represented in any of this art. If the spear and shield carrying warriors in the Saharan art, sometimes associated with the two horsed chariots, can be dated to between c.600 and 300 B.C., as has been suggested (Law, 1967), then the similar styles and subject matter in the Ethiopian art could date back to this time. The older, more naturalistic styles with cattle might, therefore, be the work of pre-metal age "neolithic" pastoralists.

In 1962-3 G. Bailloud carried out a study of the rock art sites in the region of Harar and Dire Dawa and discovered six new painted sites. He distinguishes two main styles: an earlier "Ourso style" and a later "Laga Oda style" (Bailloud, 1965(b), pp.9-11). Bailloud associates this art with Archaeological Occurrences of "Wilton type" and believes that their makers were cattle herders. The sculptured relief carvings at Chabbè in Sidamo depicting humpless cattle with sinuous, filiform horns similar to those so well known at the painted sites in the Harar and Adi Ugri districts, now extend the known distribution of pastoral art styles some five hundred kilometres south and west to the north end of Lake Abaya (Anfray, 1968, p.351).

A further site with engravings has been reported from a fissure in gypsum rock at Godhardunneh in northern Somalia, eight miles west of El Afwein between Burao and Erigavo. These are pecked engravings belonging to the later of the engraving styles known from Somalia. They depict camels with enlarged humps, lions or leopards and male human figures. It has been suggested by Lewis, who examined and reported on the site, that the engravings may date to a time prior to the introduction of Islam in the 9th or 10th century A.D. (Lewis, 1958).

There are apparent similarities between the Ethiopian rock art styles and one or more of the hunting-pastoral rock art groups of the pre-literate period of central Arabia recently made known by Anati (Anati, 1968). These include the triangular treatment of the upper part of the human body; clothed female figures depicted with hour-glass bodies; leaf-shaped, metal spears; shields; daggers (worn mostly horizontally in a belt); the sometimes "thread-like" way of representing the horns and the "segmented" style of depicting the bodies of cattle. Although there are also many differences, the general similarities remain, in particular with the styles depicting Anati's "Oval Headed People". These were hunters and pastoralists who are believed to have moved into and settled in central Arabia about the end of the fourth millennium, B.C. There they remained until they were replaced, towards the end of the second or the beginning of the first millennium, B.C. by peoples introducing new artistic styles and Semitic script. Anati suggests that the "Round Headed People" were negroids who, following the arrival of the Semitic immigrants, then moved further south into the southern parts of the Arabian peninsula. If this were so there is no reason why some of these people may not also have crossed the Red Sea into Ethiopia and the eastern desert, certainly the chronology would not conflict with this view. It is possible, therefore, that the later stages may be associated with the immigrants from southern Arabia who had established themselves in Eritrea by the fifth century, B.C.

Unfortunately, although the broad, twofold division of the rock art of the Horn into an earlier stage with long horned, humpless cattle and a later stage with spear-carrying herdsmen, humped cattle and camels still holds good; and several sub-styles can now be distinguished, we still lack, as yet, satisfactory means of dating any of them. While a late first or early second millennium, B.C. date for the earliest surviving art styles seems permissible, the possibility should be borne in mind that some, even of the earlier art, might be as late as pre-Axumite times. The curious, rectangular body shape of many of the cattle paintings at the Eritrea sites and the method of rendering the interior of the body by a series of parallel lines or segments, might, perhaps, have an origin or proto-type in the bronze, open-work *marques d'identité* found in the pre-Axumite shaft graves at Yeha and dating between the third century B.C. and the first century A.D. (Anfray, 1963, Plate CLIV (a)-(d)). One of these also is possibly a stylised human figure (*op. cit.* Plate CLIV (e)) which, though not identical, nevertheless calls to mind the stylised humans in the painted shelters of Sollum Ba'atti and Ba'atti Abba Keisi, near Mai Aini (Graziosi, 1964).

RECENT WORK IN RELATED REGIONS AND ITS BEARING ON THE PREHISTORIC INDUSTRIES OF THE HORN.

More precise knowledge of the nature and composition of later Pleistocene and "neolithic" Archaeological Occurrences in surrounding regions provides an opportunity for understanding better any possible connections these may have with the equivalent, presumed contemporary, lithic aggregates in the Horn. The relationship of the Doian to the Saharan and southern Arabian "neolithic" industries, for example, will not, of course, be determined on the nature of the attributes of the lithic component alone, but the evidence now exists for making a more exact and meaningful comparison between these aggregates as an essential basis for study of the occupation sites themselves.

In the example just cited there appears to be no close comparability though they possess some significant tool forms in common. The triangular, hollow based and foliate arrowheads, trihedral rods and "slugs", bladelets with invasive *écaillé* retouch along one edge and the transverse arrowhead (sometimes with Heluan retouch) found with the Doian industry and related Archaeological Occurrences, are now seen to be very widely distributed in the Sahara from *Ténéré* (Adrar Bous) (Hugot, 1962, pp.154-63,333-48) and Tibesti (Wanyanga Kebir) (Arkell, 1964, pp.6-8) to the Egyptian oases such as Kharga (Beduin microlithis) (Caton Thompson & Gardner, 1952, pp.158-64).

There are, however, significant lithic elements at the Saharan "neolithic" sites that are missing from Somalia. These comprise among others ground and polished stone axes, the percussion flaked adze forms, palettes, saddle querns and the tanged arrowheads. If, therefore, diffusion has had a hand in the spread of a hunting and pastoral economy from the Sahara to the Horn, the desert and tropical savanna through which the traits would have to be passed were sufficient of a filter to have let through only those that proved significant for the hunting populations of the Horn.

A general, though probably more superficial resemblance can be seen also between the Doian and the "neolithic" aggregates from the southwest corner of the Rub-al-Khali desert in southern Arabia (Zeuner, 1954; Field, 1958, 1960(a) & (b), 1961; Smith, 1962). One of the Rub-al-Khali sites has been dated to 5090 ±200 years (3,140 B.C.) and there are several dates of between 4000 and 2500 B.C. from "neolithic" sites in the Sahara (de Bayle des Hermens, 1965).

The long, rod-like, foliate points, both unifacial and bifacial, the triangular, bifaced forms with bevelled base, the occasional larger, leaf-shaped or ovate examples, the fan-shaped scrapers and the blades with invasive *écaillé* retouch are all part of these hunting camps in the Arabian desert; there is a similar absence of ground and polished stone tools. The Arabian collections differ from the Doian in that the commonest arrowhead form appears to be absent and the percentages in which the shaped tools occur appear to be very different, though these might be explained in terms of activity differences.

In northeastern Uganda, further excavations by Posnansky at Magosi yielded an aggregate with some lithic forms that are reminiscent of certain of those found with the Doian—notably one bifacial foliate and bladelets with flat, marginal, *écaillé* flaking on the dorsal face. So far as I am aware, these forms do not occur on the Kenya plateau or other east African sites. Well-made pottery is also associated, as with the later Doian in the Horn, though that from the southern Somalia sites is too fragmentary for significant comparisons to be made.

A charcoal sample from just above the level with the lanceolate point in the Magosi II shelter gave a radiocarbon date of 6080 $\pm$130 (4130 B.C.) (SR-64; Sheppard and Swart, 1966) and may help to suggest a date for the Doian of Somalia though the only really sure way is, of course, by dating carbon samples from the Somalia sites themselves.

The Uganda occurrence suggests the possibility that people possessing a similar hunting and, later, a hunting and herding economy may have occupied the zone of arid country separating the Ethiopian from the East African Plateaux, stretching from southern Somalia throughout the Northern Frontier District of Kenya into northeastern Uganda and the southern Sudan.

The question may be asked whether these common elements could be indications of cultural diffusion pointing to the existence of a widespread hunting and herding economy in the more arid parts of northern Africa, the Horn, East Africa and southern Arabia. If this indeed existed the indisputable evidence still needs to be assembled, just as any inferred relationship between the populations must be firmly based on skeletal data.

At one time it was fashionable to explain typological similarities as well as changes in lithic assemblages in terms of population movement. Today, however, there is more general appreciation among archaeologists of the complexity of the various factors influencing culture change and comparability and a corresponding caution in selecting any one, or combination, of these factors as the explanation.

Desert and steppe country invite mobility but culture traits can be transmitted without there having necessarily been any accompanying change in population. There are several hypotheses that might explain the existence of the Doian with its Saharan type artifacts in the southern part of the Horn. It could be the result of diffusion from the south-eastern Sahara, either via Danakil or through the savanna south of the Bar-el-Ghazal, northeastern Uganda and the Northern Frontier District of Kenya—or from both these directions. Again, in view of the similarities with the Rub-al-Khali Archaeological Occurrences, it could reflect influences from south Arabia across the straits of Bab-el-Mandeb. Either of these possibilities could have come about without any significant change in the human population. On the other hand, either might also have been accompanied by actual population migration, the result of the continued desiccation of the desert regions. The crossing of the Red Sea probably did not present too difficult an obstacle by this time, as a "Later Stone Age", "Wilton"-type assemblage with lunates and round scrapers in obsidian has now been described from the Dahlac Islands off the Eritrean coast, all the raw material having to be transported to the islands from the mainland (Blanc, 1955). Such initial movement, if, indeed, it did take place, need not have amounted to any significant number of individuals, provided the spread and increase of any new ethnic elements were favoured by social and cultural selection as would be expected to be the case with any group bringing a superior technology.

Within the Horn the local "Magosian" in the south is the immediate precursor of the Doian while in the north and centre the antecedents lie more with the Hargeisan. Certainly there is no sudden change in technology or in the basic composition of the occurrences between the

"Magosian" and the Doian, either in the Bur Eibe rockshelters or at Bur Hakaba. If substantiated from other stratified sites, this continuity in the general composition of the lithic component suggests that some, at least, of the aboriginal inhabitants successfully completed the economic change from hunting and gathering to hunting and herding when the opportunity was presented to them.

The probability that the Hargeisan Archaeological Occurrences in northern Somalia, based upon the punched blade, are part of a related diffusion of late Upper Palaeolithic blade technology that was spread widely from northeast Africa and the Levant between 15,000 and 10,000 B.C. now receives support from the recent discovery of blade and burin industries of late Pleistocene age in the Nile valley in Nubia (Smith, 1966).

The excellent descriptions and illustrations of the well dated cultural sequence in the Nile valley in Sudanese Nubia (Wendorf, 1968) now provide a much more certain basis on which to compare the industries of the Horn when more complete, stratified collections become available from that region. Such comparison as can be attempted at present serves to confirm the existence of common technological elements in the industries of the two regions during the first part of the Later Pleistocene.

TERMINOLOGY AND NOMENCLATURE

The main shortcoming of this volume, as has been said, lies in the lack of statistical data on which to make detailed comparisons, both between the aggregates from within the Horn and with those from regions further afield. This is especially so in regard to the unmodified waste of which, for various reasons, only selected samples, believed to be representative, were kept. In a number of cases it should be possible, however, to obtain the relevant data from the original collections housed in the University Museum of Archaeology and Anthropology at Cambridge, England. In the case of the so-called "Levalloisian", however, the aggregates are probably inadequate to show the full range of forms and attributes since they are mostly known from small collections only; in this case, further fieldwork is the only answer.

The changes that have come about recently in concepts and approaches to terminology and nomenclature clearly require some modification of the terminology for the Horn set up in this volume. In the light of the recommendations for clarifying African archaeological terminology made in 1965 at the Wenner-Gren Foundation Symposium "Systematic investigation of the African Later Tertiary and Quaternary" (Bishop and Clark, 1967) and subsequently endorsed by the VIth Pan-African Congress on Prehistory and Quaternary Studies, Dakar, 1967, it will eventually be necessary to find alternative terms with which to describe some of the industrial components of Somalia.

The "Acheulio-Levalloisian" can now better be seen as one "phase" of Upper Acheulian in which the Levallois technique is well represented. A second, perhaps earlier "phase", in which this technique appears to be absent, is represented by the assemblages from Upper and Lower Sheik. Since the term "Levallois" is now reserved to describe a special method of flake production, new terms will be necessary by which to refer to those industries here called lower and upper "Levalloisian". The largest assemblage of the former comes from the tug at Garoe and of the latter from Jesomma. Probably neither of them, however, is sufficiently complete or representative to justify its use as a basis for redefinition.

The assemblages of the "Somaliland Stillbay" from Hargeisa will probably provide an adequate basis for definition but the name "Hergeisan" is pre-empted by the terminal Pleistocene occurrences based on the punched blade.

Similar redefinition is necessary for the "Somaliland Magosian" and the "Somaliland Wilton" and, since the use of these names never implied anything other than the broadest typological comparability, new definitions and new terms will be required. So far as redefinition of the "Magosian" is concerned the assemblages from southern Somalia and Alablah are only barely adequate to provide the basis. Redefinition of the "Wilton" is possible using the aggregates from Mandera and the Webi Shebeli sites.

It should be stressed, however, that no purpose would be served by substituting a new local name for one already in the literature without having first justified this by redefinition on the basis of statistical data either obtained from a restudy of the original collections or from further fieldwork.

Future prehistoric research in the Horn may, therefore, be expected to concentrate on assembling larger and more representative collections as a basis for redefinition and closer comparison with those from other parts of the continent. Of particular interest are the precise age, distribution and composition of the Hargeisan blade occurrences as well as of the Doian with its "neolithic" elements. Excavation of stratified occupation sites in the open and the recovery of spatial distribution patterns will provide evidence of changing complexities of human behaviour, even though the deposits rarely preserve faunal remains that are older than the terminal Pleistocene. Similarly, the sequence of events and the changes in behaviour that led from hunting and gathering to the pastoral and the full farming ways of life need to be documented by a systematic programme of excavation, the results of which can then be correlated with those from the mediaeval urban centres and so with the written and oral word.

J. Desmond Clark.
Berkeley.

January, 1969.

REFERENCES

Anati, E. 1968. Rock Art in Central Arabia: Vol. I The "Oval headed" People of Arabia.
Institut Orientaliste, Bibliothèque du Muséon, 50, Louvain.

Anfray, F. 1963. Une campagne de fouilles à Yéha (Fév.-Mars, 1960).
Annales d'Ethiopie, V. pp.171-232.

1968. Aspects de l'Archéologie Ethiopienne.
Journal of African History, 9(3) pp. 345-366.

Anfray, F. & G. Annequin. 1965. Matara: deuxième, troisième et quatrième campagnes de fouilles.
Annales d'Ethiopie, VI. pp.49-142.

Arkell, A. J. 1954. Four occupation sites at Agordat.
Kush, 2. pp.33-62.

1964. Wanyanga. London. Oxford University Press.

Bailloud, G. 1965 (a). Les gisements paléolithiques de Melka-Kontouré.
Cahiers de l'Institut éthiopien d'archéologie, I. p.37.

1965 (b). In Anfray, F. Chronique d'Archéologie (1960-1964) avec des notes de G. Annequin, G. Bailloud et R. Schneider.
Annales d'Ethiopie, VI, pp.9-11.

Bishop, W. W. & J. D. Clark, Eds. 1967. Background to evolution in Africa.
University of Chicago Press.

Blanc, A.C. 1955. L'industrie sur obsidienne des Iles Dahlac (Mer Rouge).
In Balout, L. *Ed.* Actes du Congrès Pan-africain de Préhistoire, IIe Session, Alger, 1952. Paris.

Bonnefille, R. 1968. Contribution à l'étude de la flore d'un niveau pleistocène de la haute vallée de l'Aouache (Ethiopie).
C. R. Acad. Sci. Paris. 266. pp.1229-1231. (Série D).

Caton Thompson, G. & E. W. Gardner. 1952. The Kharga Oasis in Prehistory.
London.

Chavaillon, J. & M. Taieb. 1968. Stratigraphie du Quaternaire de Melka Kontouré (vallée de l'Aouache, Ethiopie): premiers résultats.
C. R. Acad. Sci. Paris. 266. pp.1210-1212. (Série D).

Clark, J. D. 1967. The problem of Neolithic culture in subsaharan Africa.
In Bishop, W. W. & J. D. Clark, *Eds.* Background to Evolution in Africa.
University of Chicago Press. pp. 601-624.

de Bayle des Hermens 1965. Datages par la méthode du C14 des formations géologiques et des gisements préhistoriques de l'Afrique du nord et du Sahara.
Libyca, 13. pp.367-380.

Faure, H. & C. Roubet. 1968. Découverte d'un biface acheulien dans des calcaires marins du golfe pleistocène de l'Afar (Mer Rouge) (Ethiopie).
C. R. Acad. Sci. Paris. 267. pp.18-21. (Série D).

Field, H. 1958. Stone implements from the Rub al Khali, southern Arabia.
MAN, 58 (121) pp.93-4.

1960 (a). Stone implements from the Rub al Khali.
MAN, 60 (30) pp.25-6.

1960 (b). Carbon 14 date for a "Neolithic" site in the Rub al Khali.
MAN, 60 (214) p.172.

1961. Palaeolithic implements from the Rub al Khali.
MAN, 61 (9) pp.22-3.

Graziosi, P. 1940. L'Etá della Pietra in Somalia.
Centro di Studi Coloniali: Univ. degli Studi di Firenze.

1964. New discoveries of rock paintings in Ethiopia, Parts I and II.
Antiquity, 38. pp.91-98 and 187-190.

Howell, F. C. In Press. New investigations in the lower Omo basin, Ethiopia.
Actes du 6e Congrès Panafricain de Préhistoire et de l'Etude du Quaternaire, Dakar (1967).

Hugot, H. J. 1962. Premiers aperçus sur la préhistoire du Ténéré du Tefassasset. pp.149-78.
In H. J. Hugot, *Ed.* Missions Berliet—Ténéré—Tchad. Paris.

Law, R. C. C. 1967. The Garamantes and trans-Saharan enterprise in classical times
Journal of African History, 8 (2) pp.181-200.

Lewis, I. M. 1958. The Godhardunneh Cave decorations of northeastern Somaliland.
MAN, 58 (234) pp.178-9.

1961. The so-called "Galla Graves" of northern Somaliland.
MAN, 61 (132) pp.103-6.

Posnansky, M. & G. H. Cole. 1963. Recent excavations at Magosi, Uganda: a preliminary report.
MAN, 63 (133) pp.104-6.

Sheppard, J. G. & E. R. Swart. 1966. Rhodesian radiocarbon dates.
Radiocarbon (8). p.423.

Simoons, F J. 1965. Some questions on the economic prehistory of Ethiopia.
Journal of African History VI (1) pp.1-13.

Smith, P. E. L. 1962. Two "Neolithic" collections from Saudi Arabia.
MAN, 62 (17) pp.21-22.

1966. The late Palaeolithic of northeast Africa in the light of recent research.
American Anthropologist, 68. Pt. 2 (2) pp.326-55.

Wendorf, F. 1968. The prehistory of Nubia. 2 vols.
Southern Methodist University Press.
Zeuner, F. A. 1954 "Neolithic" sites from the Rub-al-Khali, southern Arabia.
MAN, 54 (209). pp.133-6.

FOREWORD

Only those who have trained distinguished students know what a source of pride and satisfaction these become to their former teachers. I cannot therefore over-stress what a pleasure it is to me to write this preface for Dr Desmond Clark.

Dr Clark is the Curator of the Rhodes-Livingstone Museum of Northern Rhodesia and he has already written a valuable work on the Zambesi gravels in the vicinity of Livingstone, so important indeed that, presented as the major part of his thesis it largely gained for him the Ph.D. degree of the University of Cambridge last year. And now, again, here he is producing another equally valuable monograph. During the recent war Clark was sent overland to Somaliland where he was stationed for more than two years. He had to travel up and down the country and thus had a unique opportunity to study the prehistory of this part of the world. Lying in the so-called Horn of Africa, Somaliland borders Kenya to the north. The latter region was already fairly well understood from the point of view of its prehistory thanks to the work of another old pupil of the Cambridge School, Dr L. S. B. Leakey. But while rich stone industries were known to occur in Somaliland little or no systematic excavation or study had been undertaken. Dr Clark thus brought to bear a trained mind experienced in field-work on an almost virgin area and the present volume is the result. All too seldom did the hazards of war bring people and places so fortuitously together!

I do not imagine that Dr Clark would claim that the final word with regard to the details of his subject has been said. But I think one can, without any fear of contradiction, suggest that the present volume goes far to elucidate the early story of the area and will form the indispensable basis for all further studies. The prehistory of the world is becoming ever more complicated as fresh regions are explored and new evidence has to be fitted into the jig-saw picture of the whole. Indeed, the problem is even more troublesome than a jig-saw which at least is all in one plane. It is more like trying to put together one of those Chinese puzzles in which pieces of wood fretted out in all directions ultimately fit together to form, say, a solid cube. The simple cultural sequence invented by Gabriel de Mortillet is now seen to have only a local application. It is becoming more and more difficult to form even the most generalized theories—and Africa is one of the most 'problemiferous' of continents. But this is a challenge to all prehistorians and the sort of work Dr Desmond Clark has done both in Northern Rhodesia and now in Somaliland supplies us with precious and reliable evidence about the industrial and cultural successions in these regions. It will be for our successors to fit the pieces into their proper places and, drawing the final generalized conclusions, to set out clearly the history of the whole great world during the fascinating period of the Stone Ages.

M. C. BURKITT

UNIVERSITY MUSEUM OF ARCHAEOLOGY AND ETHNOLOGY
CAMBRIDGE

October 1951

PREFACE

Although the Horn of Africa was one of the first regions of the African continent outside Egypt to furnish tools of Early Man yet it has remained largely neglected until recent years and its richness has been somewhat eclipsed by the systematic discoveries in East and South Africa and Rhodesia. Collections from Somaliland have found their way into a number of European and African museums but with the exception of excavations by Teilhard de Chardin, the Abbé Breuil and P. Wernert in the cave of Porc Epic at Dire Dawa in 1930 and 1933, and by P. Graziosi at Bur Eibe and in the Nogal valley in 1935 all the data collected and described up to the present have consisted of surface finds the provenance even of which was sometimes but imperfectly known. Previous workers have therefore had to rely on typological evidence and physical condition for estimates of the age and the cultural associations of the specimens they found. While therefore typologically the framework of the cultural succession was already foreseen our own work was directed towards providing wherever possible the necessary stratigraphical evidence on which to base the cultural and chronological succession for the Stone Age cultures in the Horn.

It has been thought advisable at the risk of becoming tedious to give full typological descriptions with tool lists and stratigraphical dating evidence for each culture. In a better known region this would probably be unnecessary as these particulars can usually be found in a fairly complete literature. This is not the case with the Horn, however, where, with rare exceptions, the literature is sparse and confined usually to brief notes and often incomplete descriptions of surface collections. All the evidence has been set out in full therefore so that readers may judge for themselves whether the conclusions that have been reached are indeed justified. This book remains, however, very much of a 'pioneer' work and must not be considered as an exhaustive treatise on the prehistory of the Horn. The cultures still are but imperfectly known and much of the detail remains to be elucidated and confirmed.

The field-work on which this book is based was undertaken in the Somalilands and parts of Abyssinia between 1941 and 1946 in the course of some two and a half years spent at various times in those territories during service with the East Africa

Command. The time that it was possible to spend in any one area was, particularly in the north, determined essentially by military necessity so that it was not always possible to obtain from a site evidence of as conclusive a nature as we should have liked. Sometimes it was unfortunately necessary to leave some problem quite unsolved while many important areas could not be examined at all. In spite of the rather interrupted nature of our work important evidence was nevertheless recovered and has enabled us to determine with some degree of certainty the main outlines of the geological and cultural succession during a large part of the Stone Age in several areas of the Somali plateau.

In 1941 parts of western British Somaliland, the Danakil desert, and the Reserved Areas were studied, and the first area was again visited in 1943 under more congenial conditions when it was possible to check several of the results obtained from our previous field-work, as well as to undertake work on new sites. In 1944 during service with the Military Administration in Somalia the country between the Webi Shebeli and the Juba rivers was studied and time was granted in which to conduct excavations at one open station and one cave site in that area. At the end of 1945 and the beginning of 1946 the Administration agreed to a preliminary archaeological survey being undertaken in certain areas not previously worked—notably the lower Nogal valley, the Obbia hinterland, and the Ogaden, in order that the northern sites might be linked with those found in the south. The cost of this survey was borne in the main by the Somalia Administration, and my best thanks are due to not a few of the officers of this administration for their help, but in particular to the late Brigadier D. H. Wickham without whose assistance this survey could not have been undertaken.

A distribution map of recorded sites will be found at Pl. 29, and a topographical map faces this page.

All measurements of height by the writer are barometric measurements. Two sets of readings were made wherever possible and the findings were checked against known heights and compensated for temperature variation. The diurnal variation was often considerable and in the Nogal valley and in the eastern Mudugh (Mudug) it was sometimes as much as 100 ft. or more.

The spelling of proper names has presented a considerable difficulty. In some cases as many as three or four different spellings for a single place-name

General Map of the Horn of Africa

NOTE

This topographical map has been taken from a map published by the National Geographic Society, Washington, and is based upon a combination of British, French and Italian maps. For this reason not a few of the spellings of place names and topographical features differ from the spellings which have been adopted in the text and Pl. 29. This discrepancy was not considered sufficient justification, however, for completely redrawing the map, which is reproduced here to show the topographical features.

exist according to whether the name appears in British, French or Italian publications or maps. In the case of well known place-names the British spelling most commonly used at the time that the writer was in the Horn has been employed, but where these differ from the spellings used in Hunt and Viney's *Gazetteer of British Somaliland and Grazing Areas* (1946) the latter spelling is given in brackets. In the case of sites and places of minor importance visited by the writer, and of which the positions are clearly fixed though they may not always appear on the maps, the name has been spelt according to the R.G. II System. In the case of minor sites recorded by Italian, French or German authors and which are not marked on any map, it has been found necessary to retain the spelling used in the original publication.

I am most grateful to a number of colleagues and specialists who have helped me in a number of ways with their advice and specialist reports. The late Miss D. M. A. Bate examined and reported on the faunal remains recovered from excavations in Somaliland (Appendix F) and I am deeply grateful to her for the help that she gave me. Dr L. R. Cox has kindly examined and identified the molluscan fauna from raised beaches on the east coast (Appendix G), and the Director of the Royal Botanic Gardens, Kew, has provided the report on the charcoals (Appendix H).

I am very greatly obliged to Dr W. A. Macfadyen for the great care that he has taken to read through the geological part of this book and for the helpful and detailed criticism and suggestions that he has made. It is indeed a privilege and a comfort to have one who knows the geology of the Somalilands as he does to check these chapters especially as I felt at times in the field the lack of geological assistance very acutely. To Professor F. E. Zeuner also for his determination of soil and rock samples and suggestions as to their value as indicators of past climatic change I am most deeply grateful. Also to Dr A. F. Hallimond for his determination of an iron slag from the Biede rock-shelter I must express my thanks.

Count Björn von Rosen most kindly made me a gift of a copy of his very fine book *Berget och Solen* with his description and photographs of the important Harar group of painted rock-shelters. I am very grateful to him for allowing me to reproduce several of his photographs and for his kindness in providing me with prints to facilitate this.

I must also record my thanks to the Trustees of the Rhodes-Livingstone Institute and Museum for financial assistance towards the transport of the collections.

I am especially obliged to Mr M. C. Burkitt for his Foreword and have been very greatly helped by the constant advice and suggestions that he has so kindly given during the time that this book was being written. His kindness and help have been invaluable and without them it is likely that this work would never have been started.

I have also profited very greatly by discussions with the Abbé Breuil, Professor C. van Riet Lowe, Dr L. S. B. Leakey and Dr C. B. M. McBurney and other friends and colleagues and I wish to record my thanks to them for the help that they have given me.

The drawings of the stone implements, indispensable in a work of this nature, I owe to my wife and my most grateful thanks are due to her for these illustrations and also for the constant help that she has given me in correcting proofs, checking statistics and in many other ways.

I have been very fortunate in having been able to examine all the important collections that have been made from the Horn with the exception of those in Italy, and I am greatly obliged to the following for allowing me to examine and to make full use of the collections that they control. To Mr Harper Kelley for permission to examine the collections from Somaliland and Abyssinia in the Musée de l'Homme in Paris. To the Abbé Breuil and Professors H. Vallois and R. Vaufrey for their kindness in letting me examine the collections in the Institut de Paléontologie Humaine where the material from French Somaliland and the bulk of the Porc Epic collection is housed. The collections in the University Museum of Archaeology and Ethnology have been examined by courtesy of Mr M. C. Burkitt and Drs T. T. Paterson and G. H. S. Bushnell. Mr E. M. M. Alexander of the British and Medieval Antiquities Department at the British Museum placed at my disposal the collections made by Mr Seton-Karr, and Dr L. S. B. Leakey made available the collections, many of them formed by British military personnel during the course of, and after, the East African Campaign, which are housed in the Coryndon Memorial Museum at Nairobi.

Captain H. B. Gilliland has also put at my disposal the collections made by him in the eastern part of British Somaliland and now housed in the

Archaeological Survey at Johannesburg, and I am grateful to Professor C. van Riet Lowe for permission to examine these.

The bulk of my collections have been given to the University Museum of Archaeology and Ethnology at Cambridge, but comparative collections have been given to the Institut de Paléontologie Humaine at Paris, to the Archaeological Survey at Johannesburg, South Africa, and to the Rhodes-Livingstone Museum at Livingstone, Northern Rhodesia.

J. D. C.

September 1951

PART I

INTRODUCTION

CHAPTER 1

PHYSICAL GEOGRAPHY

PHYSIOGRAPHICAL FEATURES

The territory of the Somalis forms part of what is known as the Somali Plateau, one of two vast structurally and morphologically similar areas into which the Horn of Africa may be geologically divided. These two areas, the Abyssinian High Plateau and the Somali Plateau, have been separated the one from the other by faulting and earth movement on an immense scale. The structure of the Somali Plateau is to a great extent determined by faulting along three main trends—the East African trend (N.E.–S.W.), the Red Sea trend (N.W.–S.E.), and the Gulf of Aden trend (E.–W.). The first has caused the development of the Abyssinian and Danakil Rifts, which form part of the major Rift Valley system stretching from the Palestinian lakes in the north to Lake Nyasa and the lower Zambesi in the south. Work undertaken by Nilsson[1] in the Abyssinian section of the Rift Valley in the basin of Lakes Zwai and Shala (5900 to 4900 ft.) has shown that the last major period of faulting in this section of the Rift is to be assigned to a time later than the deposition of Kamasia-type sediments, which in Kenya and northern Tanganyika have been dated to the Middle Pleistocene, and have yielded an evolutionary series of the Chelles-Acheul Culture.[2] These beds have been faulted and occur on the escarpment in steps rising up to the mountain ranges which flank both sides of the Rift. The highest beach of the Zwai-Shala basin has been worked out in these Kamasian beds.[3] The Kamasian sediments are exposed again in the valley of the Awash (Hawash) river in open position to the Danakil desert and the Red Sea, no doubt deposited in an arm of the latter.[4] If therefore we can accept the contemporaneity of these Kamasia-type sediments with dated Kamasian deposits in East Africa, then an appreciable part of the present rifting in the Abyssinian section of the Rift Valley is of end-Middle Pleistocene age.

The Danakil Rift (3300 ft. to −400 ft.) is a roughly funnel-shaped area bordered on the west by the massive escarpment of the Ethiopian plateau and on the south-east by the scarp of the Somali Plateau. The southern part of the Danakil Rift, the Aussa desert, lies between the Ethiopian plateau and the

[1] Nilsson, E. (1940). [See Bibliography, p. 369.]

[2] (i) Leakey, L. S. B. (1936). (ii) Solomon, J. D. (1931).

[3] (i) Negri, G. (1913). (ii) Nilsson, E. (1940, p. 39).

[4] Nilsson, E. (1940, p. 38).

Danakil Horst, a triangular mountain ridge (average 3300 ft.) bordered on the north and east by the Red Sea. The Aussa desert is made up of stony delta deposits, clays and silts, some of which have yielded human artifacts.[1]

The Somali Plateau lies to the south-east of its Abyssinian counterpart and, as can be seen from the inset map (Pl. 29), its western edge is determined by the fault-scarp, the plateau here falling steeply almost precipitously to the floor of the Rift. This feature of the scarp is maintained in the north-west where, at the junction of the East African and Gulf of Aden fault trends,[2] the high mountain walls bend round to the east, past the Danakil or Aussa trough, and along the shores of the Gulf of Aden, the plateau in places approaching, particularly to the east of Berbera, almost to the coast itself.[3]

The highest parts of the plateau, apart from the strip of highland marking the edge of the Gulf of Aden fault-scarp, lie close to its western edge, where the Harar and Arussi-Bale mountains reach average heights of 7000 ft., with occasional peaks rising above this to as much as 10,000 ft. or more.[4] From this mountainous region, now within Abyssinian territory and inhabited for the most part by the agricultural Galla, the plateau slopes very gradually to the east-south-east to the Indian Ocean, indicating uplift of the plateau along its western side and gentle tilting about an axis at right angles to this slope.[5] The initiation of the last earth movement along this axis has been deduced by Nilsson to have taken place during the Last Interpluvial, and to have been completed before the end of the first part of the Last Pluvial.[6] These western highlands form the major watershed of the plateau.

Within the Somalilands the plateau falls naturally into three parts divided by the three great drainage basins of the Nogal, the Webi Shebeli, and the Juba

[1] Garesleh Tug, see p. 116.

[2] Gregory, J. W. (1921, p. 31).

[3] The following Admiralty Charts were examined with a view to ascertaining whether any submarine faults could be recognized on either the north or east coasts. East Coast Charts, nos. 597, 100*a*, 671, 848, 670. North Coast Charts, nos. 6B, 919, 100*b*. It was apparent that at a distance varying from 1 to 4 miles from the coast the sea-bed sloped fairly steeply down from 60 to 120 fathoms on an average along the north coast. Whether this is the result of submarine faulting or represents merely the normal continental-slope zone could not be determined. A definite and deep fault is, however, clearly discernible in the Gulf of Tajura.

[4] Mount Kaka 13,232 ft.; Chillalo Mountains 11,961 ft.

[5] (i) Nilsson, E. (1940). (ii) Dainelli, Prof. G. (1936, p. 97).

[6] Nilsson, E. (1940, p. 42). That some degree of early tilting along this axis must have existed prior to the formation of the major river systems is evidenced by the general south-east trend of the latter. Stefanini, Dainelli and others attribute this initial tilting to the fact that the plateau, composed of successive deposits of sedimentary rocks laid down for the most part in direct stratigraphical succession following three major periods of marine submergence and subsequent uplift, emerged evenly but very gently inclined towards the south-east, the strata becoming progressively later in age from south-west to north-east.

rivers, the directions of flow being towards the south-east. In the north the plateau of British Somaliland slopes gently from its northern edge, a strip of high country known generally as the Ogo, to the south, and east to the Nogal depression and the great plains of the Haud.[1] This mountain escarpment which may be considered as an extension of the Harar high plateau, attains average heights of 6000 ft., with a maximum of 7900 ft. at Shimbir Beris in Erigavo District and forms the main watershed for the streams and rivers flowing north into the Gulf of Aden, and south-east into the Indian Ocean; it has been affected by tilting on the south-east axis of a similar nature to that in the west. Over the northern edge of the Gulf of Aden fault-scarp the terrain falls often precipitously and abruptly to the coastal plain or to the sea itself. Approximately two-thirds of the way going eastwards along this coast, however, the scarp is broken by a broad gap of less elevated country (600–1600 ft.) known as the Karin Gap, formed by northward-throwing faults, and through which at one time the Darror (or Daror) depression appears to have been drained into the Gulf of Aden.[2] A somewhat similar break occurs in the scarp further to the west near Las Dureh, known as the Asseh Gap. The eastern corner of the plateau lying between the Gulf of Aden and the Nogal is known as the Mijertein and is one of sharp relief characterized by steep, flat-topped, often isolated hills and plateaux, of which the desolate and waterless Sol Haud (or Sawl Haud) is an example, and numerous deeply entrenched streams or 'tugs' to give them their Somali name, which today are almost invariably dry for the greater part of the year and contain water for a few hours only after rain: they wind through steep gorges and gullies to the sea.[3] These drainage channels have been carved out of anhydrite beds limited to the north and south by hills of overlying limestones. The central plateau, lying between the Nogal and the Webi Shebeli, consists of a vast undulating plain known as the Haud, waterless for the most part but storing water for three or four weeks during the rains in a number of natural reservoirs known as 'ballehs'.[4] The soil of this plain is of a bright red sandy nature. Much of it is probably derived from the disintegration of the underlying limestones and sandstone,[5] but some is undoubtedly of aeolian origin

[1] Average altitude for the Ogo is between 5000–6000 ft. Average altitude for the southern Haud is 1000 ft.

[2] (i) Hunt, J. A. (1944). (ii) Barrington Brown, C. (1931*a*, p. 259).

[3] Such seasonally dry river courses will be referred to by their Somali (Isaak-Darod) name of 'tug'. In the south various terms are used: 'bohol' (Rahanwein), 'bug' (southern Darod: Hawiyia), etc.

'Webi' means a seasonally flowing river, e.g. the Webi Shebeli, Webi Gestro.

[4] The 'ballehs' or shallow pans, sometimes artificially enlarged, are of paramount importance to the Somali tribes who rigidly guard their rights over these water-holes. In the south the term 'balleh' is replaced by 'saha'.

[5] Stefanini, G. (1936, p. 83).

and covers cultural material in the neighbourhood of some of the 'ballehs'.[1] In the eastern part of the central region the red Haud sand gives way to monotonous exposures of the underlying Eocene limestone and gypsum beds, relieved in places by solution valleys and sinks. This vast plain slopes very gently from the Abyssinian boundary towards the coastal plain and the Indian Ocean. In the west, however, near the territorial boundary with Abyssinia are found the parallel depressions containing the valleys of the Tugs Gerer (or Jerer) and Fafan, running south-east towards the Webi Shebeli. They probably run directly down the dip slope though it is possible that they may have been developed along old fault lines connected with the south-east Red Sea trend as may be also in part the upper reaches of the Dakata. It is unfortunate that with the exception of the lower courses of the Fafan no opportunity was found of examining these valleys. In the south-west in the bend of the Webi Shebeli, the plateau is dissected by deep gorges incised by 'tugs' that drain into the parent river. These 'tugs' have deepened their beds in geologically recent times and the time at which this rejuvenation took place can be ascribed to the latter half of the Last Inter-Pluvial and initiation of the Last or Gamblian Pluvial when earth movement was causing uplift and tilting of the Somali plateau.[2]

The southern Somali plateau, between the Webi Shebeli and the Juba rivers again comprises a gently undulating limestone surface with occasional isolated low hills rising above it, of inselberg form.[3] Water is comparatively common over this part of the plateau and is found in shallow wells. Here also natural ponds form small lakes and contain water for as much as two or three months after the rains. Two of these 'sahas' as they are here called are often perennial and old tufa, and surface limestone deposits with artifacts have been noted in their basins as also in those of several others.[4] The junction of the plateau and the coastal plain where the crystalline rocks are exposed is often imperceptible especially to the north of the Webi Shebeli in the Mudugh (or Mudug), but sometimes, notably in the region of Isha Baidoa, the junction is effected by a short steep scarp, remnant of the old coast zone in the south.

With the exception of the narrow belt of high country bordering the Gulf of Aden fault and low hills of inselberg form there are no ranges of any size on the plateau within the boundaries of the Somalilands. The plateau here presents the character of a gently warped and folded peneplain surface and is probably of mid-Tertiary Age. Only in the country between the Webi Gestro and the Webi Shebeli, at El Carre, do the outliers of the high country within the

[1] See pp. 208, 217, 218. [2] See pp. 78–79.
[3] Bur Eschunle, Bur Gheider Idda, Cal-ie-Corar, and the Ted Surface.
[4] Ted, Burdo; see also El Dubbo, Saha Gurral, Sijjia, Morogavi, etc.

Abyssinian boundary obtrude and form the foothills of the Arussi-Bale mountains.

Of the principal rivers, all of which flow into or towards the Indian Ocean, the most important is the Juba. Fed by three large tributaries, the Webi Gestro or Web, the Ganale Doria, and the Daua, all of which have their source in the Arussi-Bale range, the Juba has entrenched itself to a depth of 1000 ft. below the plateau. In its lower reaches the river meanders over the wide coastal plain (having here a width of some 200 miles), supporting a thick riverine forest fringe, and reaches the sea near Kismayu.

The Webi Shebeli, also perennial throughout the greater part of its course, rises in the western part of the Ghedeb basin close to the source of the Juba, near the western edge of the plateau,[1] and flows south-east and south for the first part of its course through a deeply incised gorge and then, near Imi, opens out into a wide valley of mature profile. The river flows through this valley, flanked by clearly defined scarps and pediments on both sides, as far as Afgoi where, only fifteen miles from the sea, its course is deflected by coastal sand dunes and it meanders south-west, running parallel to the coast until it finally disappears into the plain at Ballei yer (0° 40′ N., 43° 4′ E.).

The Nogal is a short deeply entrenched river, or more correctly 'tug', now flowing only during the rains. It may be divided, however, into two separate physiographic areas which comprise the short but deep gorge of the lower Nogal reaching the sea at Eil, and the upper Nogal, a wide flat depression draining a considerable area of broken country flanked on the north and south by steepish scarps which descend by a series of steps of predominantly structural origin to the plain itself; this wide and arid plain is dissected near the scarp by a number of 'tugs', which eventually lose themselves in the sands of the central plain.

The Darror (or Daror), between the Nogal and the north coast, is a smaller counterpart of the Nogal and reaches the sea near Ras Hafun. With the possible inclusion of the Issutugan and Raguda Tugs, which flow north-west into the Gulf of Aden, this is the only other large river of note.

Distinct from the plateau is the coastal plain which to the south, in Jubaland and southern Somalia, is low-lying, the average altitude is less than 1000 ft., and reaches for a considerable distance inland up the valley of the Juba nearly to Bardera, and thence across the lower Webi Shebeli to the 'Bur country'. These 'burs' are typical granite inselbergs which rise steeply from the surrounding plain.[2] The highest of these, Bur Eibe, reaches to a height of 1000 ft. and is an impressive sight from the Baidoa scarp. Proceeding northwards the coastal

[1] Abruzzi, Duc d' (1932). [2] Aloisi, P. & de Angelis, A. M. (1938).

plain becomes gradually more constricted by the encroachment of the plateau until from just north of Obbia in central Somalia, to Alula, near Cape Guardafui it occurs only as narrow isolated plains separated by long stretches of the plateau, which reach to the coast itself. Restricted patches of this coastal plain occur also along the northern coast, where it is known as the Guban to the Somalis,[1] but it rarely attains to more than thirty miles in width and is often, as for example near Berbera and Benda Ziada, divided into two parts by a line of rocky hills, usually faulted, which run parallel to the coast and the Aden Rift scarp. Only at the western end of the Gulf does the coastal plain open out once more between Bulhar and Zeila before joining the Aussa plains of southern Danakil.

The chief feature of this plain on the east coast is a line of consolidated fossil dunes, on an average from 200 to 400 ft. high but rising at Gowein, inland from Obbia, to a thousand feet or more, and having a width of from one to eighteen miles.[2] These fossil dunes run parallel to the coast and form an effective barrier between it and the inland plain beyond. The latter is composed for the most part of sandy alluvial soil, often attaining considerable depth, but has no conspicuous features except in the south where the monotony is broken by numerous small isolated 'burs'.

SOLID GEOLOGY

Brief mention may now be made of the underlying solid geology as this has controlled in varying degree the morphology of the area and the distribution and availability of man's raw materials.

The crystalline rocks, granites, gneisses and schists, which form the Archaean basement complex on which the geological superstructure is built up, are nearly everywhere overlain by more recent sedimentary deposits. They are, however, found exposed by the faulting and uplift of the plateau at points along the northern seaboard within the Aden Rift fault, near the base of the scarp, as for example between Hargeisa and Bulhar, at Mandera between Hargeisa and Berbera, at the foot of the Al hills, at Ras Hantara, and on the Harar plateau, and again in the south where the coastal peneplain has eroded the overlying sedimentaries exposing these basement complex rocks which are often covered by only a few feet of residual and wind-blown sand, while the harder and more resistant elements rise here and there to form the characteristic 'bur' between the Middle Webi Shebeli and Bardera on the Juba river.[3] In these areas, while chert is still the predominant raw material of prehistoric man, quartz (occurring

[1] Swayne, H. G. C. (1895).

[2] (i) Stefanini, G. (1936, p. 85). (ii) Artini, E. (1915).

[3] (i) Stefanini, G. (1936). (ii) Barrington Brown, C. (1931). (iii) Macfadyen, W. A. (1933).

in large concentrations within the granite and syenites) is often extensively used. The earliest sedimentary beds, ascribed to the Trias and Lias, are found only in the middle course of the Juba valley in the area of Lugh Ferrandi (where they are overlain by the Jurassic series). Grouped together under the term 'Lugh Series' they include sandstones, gypsiferous and salt-bearing marls.[1]

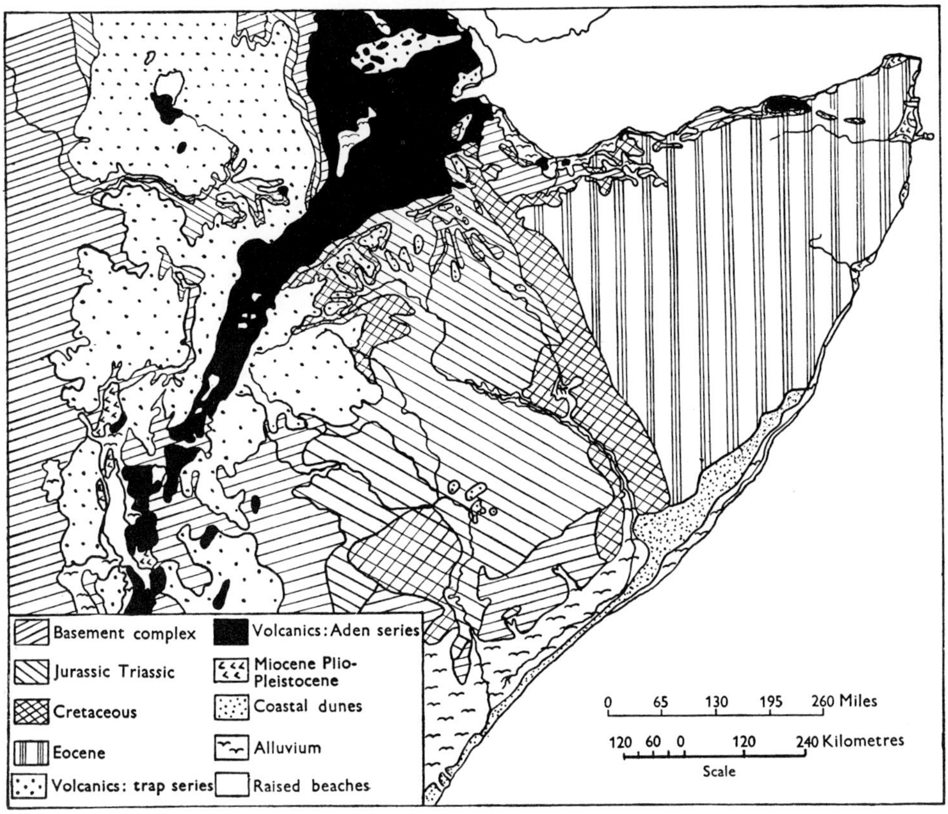

Text-fig. 1. Modified geological map of the Horn.

The main portion of the southern plateau is formed of Jurassic limestones. These comprise a variety of fine-grained fossiliferous limestones, red to grey in colour intercalated with marls and chert-bearing limestone. These beds are fairly rich in fossils and are consequently well-dated. The Jurassic overlie the basement complex rocks also at points along the Gulf of Aden Rift fault and here comprise sandstones, grits and limestones; they are dated by contained fossils mainly to the Upper Jurassic.[2] Rocks of this age also outcrop in the upper

[1] Stefanini, G. (1929). The name denotes its geographical position and lithological nature; the age is not well proven, the fossils (fish teeth) being rare and not typical, neither is the stratigraphy clearly established. In this connexion, see p. 84.

[2] (i) Stefanini, G. (1936). (ii) Barrington Brown, C. (1931). (iii) Macfadyen, W. A. (1933).

reaches of the Webi Shebeli and in the region of the Harar plateau both on its south side (northern Ogaden) and north side (at the foot of the rift near Dire Dawa).[1]

These Jurassic rocks are followed by various Cretaceous deposits along the middle Webi Shebeli (predominantly the right bank) from Missarole to the Shaveli country, into the Ogaden and west again to Arussi where there outcrops a white or ochreous limestone with a rich fauna (Upper Cretaceous: Giglei limestone). On the left bank of the middle Webi the Cretaceous is represented by grey and purple sandstones passing into grey, yellow and red quartzites which contain ferruginous nodules (Jesomma sandstones). These Jesomma sandstones are correlated with the Nubian sandstones and with rocks of similar age in Abyssinia, Eritrea and the Hadhramaut. The Marehan sandstones south of the Juba and in the Northern Frontier District of Kenya are considered to be of the same age.[2] Fossiliferous Cretaceous beds outcrop again along the fault-scarps in the Northern Frontier District, where limestones, some argillaceous, and sandstones, sometimes ferruginous, overlie, often conformably, the Jurassic sediments.[3] The Cretaceous series is seen again on the northern plateau in the area of Hargeisa and in the Marodijeh (or Marodijehh) Tug where the quartzite from these beds is almost invariably the material used for the larger stone implements. Chert and chalcedony nodules (highly coloured) and quartz pebbles from the Cretaceous beds are characteristic and were extensively used, as was also the silicified sandstone found in the middle Webi Shebeli.

The central and northern parts of the plateau are built up of Tertiary sediments, predominantly of Eocene age, which comprise creamy limestones and intercalated anhydrite beds, the former yielding a characteristic honey and grey coloured chert, and the latter light brown and black cherts, which form a very satisfactory medium for the production of implements and throughout the north were universally used from Levalloisian times onwards.[4] On the southern side of the upper Nogal valley where the black-banded cherts outcrop within the

[1] Dreyfuss, M. (1931).

[2] Stefanini, G. (1936, pp. 33, 41 and 46).

[3] (i) Barrington Brown, C. (1931, p. 263). (ii) Reck, H. & Dietrich, W. O. (1923). (iii) Stefanini, G. (1925).

[4] Burkitt, M. C. & Barrington Brown, C. (1931, p. 164). Barrington Brown describes the raw material used on the north-eastern boundary of the Protectorate as follows: 'The material used is mostly a chert, of a yellowish, honey colour, which patinates to either a white or black tint. This chert occurs abundantly as large concretions in the Eocene limestones which form the greater part of the elevated plateau (at 2000 to 3500 ft.) constituting a great part of the interior of British Somaliland. A thick bed of gypsum and anhydrite outcrops over considerable areas on the plateau: no flakes were found on these areas. It is noticeable that white quartz from veins which outcrop in slates at the foot of the scarp is utilized to some extent on the coastal plain.'

gypsiferous series are factory sites covering many acres in extent which were mined continuously throughout Upper Pleistocene and Recent times.[1]

To the Tertiary, but perhaps in part also to the Upper Cretaceous, belong the volcanic rocks, predominantly basalts and liparitic tuffs. Basalt lavas (traps) thought to be of Miocene Age, form the mountain ranges of the western scarp which also form the watershed between the basins of the Abyssinian, or Galla

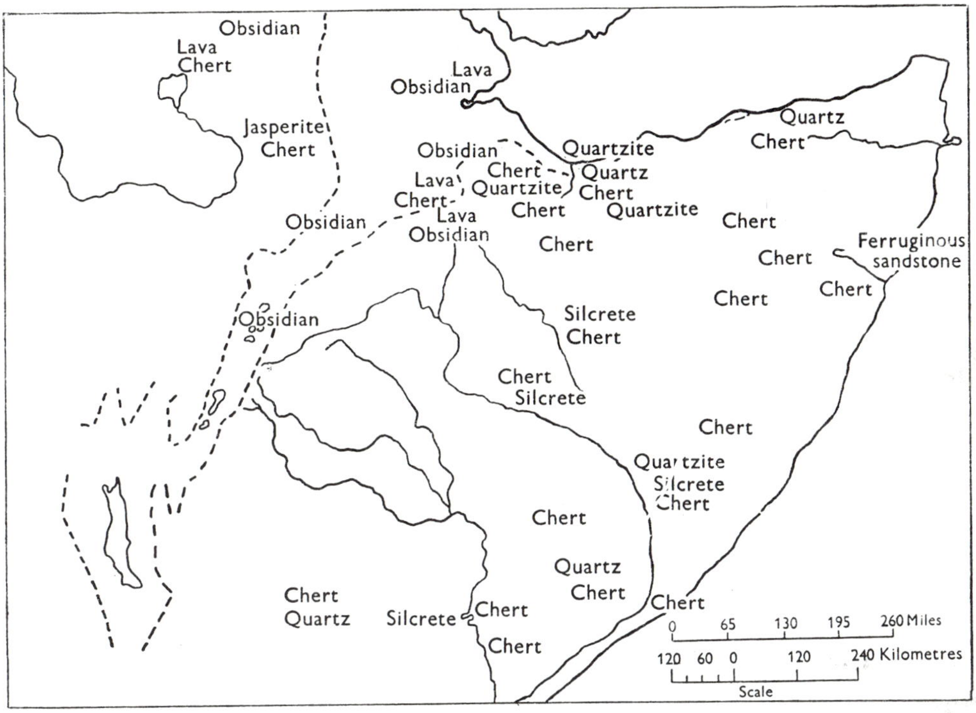

Text-fig. 2. Distribution map of raw materials used during the Stone Age in the Horn.

Lakes, in the Rift to the west, and the Webi Shebeli and Juba to the east. The lavas are widely distributed in Bale, and in the country of the Arussi, Sidama and Boran. With the exception, however, of isolated intrusions of lavas and tuffs in the valleys of the middle Shebeli and middle Juba there are no volcanic rocks in the eastern parts of the plateau proper.[2]

In the north, however, in the Danakil and Gulf of Aden Rifts, lavas of two distinct ages are commonly found and are ascribed to the Aden Series.[3] Examples of the earlier are seen forming nearly flat or tilted tablelands south of Bulhar,[4]

[1] See Bihen/Donkukoq area, p. 292. [2] Aloisi, P. & de Angelis, A. M. (1938, p. 45).
[3] The Aden Series. For more detailed description, see section on southern Danakil Rift, p. 107 *et seq.*
[4] Macfadyen, W. A. (1933).

again in the Mount Mabla area of French Somaliland,[1] and also along the edge of the escarpment south and east of Dire Dawa.[2] These are probably contemporary with the plateau traps and basalts mentioned above. More recent acid lavas and rhyolites are extensively found in Danakil and at points along the Gulf of Aden coast. They have often a markedly fresh appearance and were examined by the writer in the Dire Dawa-Aisha section of the Aussa plain.[3] Isolated occurrences are found also along the Gulf of Aden coast, notably at Siyara and Benda Ziada.[4] Obsidian occurs in the Danakil Rift and in the north-west of the British Somali plateau where this material was extensively used by the makers of the Late Stone Age Cultures: it is common also on the Harar plateau, in the upper courses of the Webi Shebeli and is occasionally met with at sites along the edge of the plateau bordering the Gulf of Aden fault-scarp. Text-figs. 1 and 2 have been drawn to show the main geological features of the Somali Plateau and the predominant raw material used by prehistoric man in each of the main areas studied: it indicates that with few exceptions man made use of the local material ready to hand and clearly no difficulty was experienced in adapting this to his needs.

POPULATION, RAINFALL, WATER SUPPLY AND VEGETATION

If we are able to accept as true the portrayal of the inhabitants of the Land of Punt on the bas-reliefs in the Temple of Deir-el-Bahari,[5] then the Horn of Africa was already inhabited in early historic times by both Hamitic and negroid peoples. Today we find that the negroid peoples have been, with very few exceptions, merged with the dominant Hamitic element, or else restricted to the valleys of the principal rivers. The mountains and highlands on the west are inhabited by the now agricultural Galla and the pastoral Boran, while in the central plains, and in fact throughout the Horn proper, the dominant type is the Somali. The Galla seem to have controlled the whole of the Horn of Africa since at least the first few centuries of the Christian era, and during the sixteenth century spread inland to the south and west but subsequently were displaced by the Somalis pressing down from the north, and the main group of the Galla were forced inland to their present habitat.

All Somalis are fundamentally pastoralists of migratory habit, owning sheep, goats and camels, following the rains in search of pasture for their animals, and

[1] Teilhard de Chardin, P., Lamare, P., Dreyfuss, M., Lacroix, A. & Basc, E. (1930*b*).

[2] Arsandaux, J. (1905).

[3] See pp. 115–20.

[4] Macfadyen, W. A. (1933).

[5] Naville, E. (1898), Pt. III, Pl. LXIX–LXXVI.

living upon their stock in times of drought; their movements are controlled largely by the existence of the 'permanent water' on which they concentrate during the dry season. Stock raiding is an essential element in their culture, and possessing considerable powers of endurance in their impoverished environment they are capable of travelling long distances with the minimum of food and water. The agricultural communities are predominantly negroid and are concentrated in the valleys of the Juba and Webi Shebeli; their origin, however, is obscure. Most of the country between the Juba and Shebeli is inhabited by the semi-agricultural, millet-growing, Rahanwein and Digil Somalis, amongst whom strong negroid admixture can be seen. The comparatively pure northern and central Somali tribes, comprising the Darod, Dir and Hawiya confederacies, have an 'outcast' tribe distributed amongst them known as the Midgans, who were specialists in hunting.[1] Free hunting groups of purer type exist among the southern tribes in the country between the Shebeli and Juba and beyond into the Northern Frontier District of Kenya. The general term for these groups is Bon or Ribi, and they are for the most part a small-statured, slender-boned people of negroid extraction who hunt with the use of dogs and nets in conjunction with the bow and poisoned arrow; these groups own very little stock, and may represent the remnant of the original population, though they are now of very mixed descent.

The heaviest rainfall in the Horn, known as the 'Gu', is brought by the south-west monsoon, which blows in the south from March to May and continues as the 'Hagai' from June to August. In the north the rain falls from April to June. This monsoon is characterized by a hot and dry wind known as the 'Karif' which blows from May to October; the average rainfall for the whole of the Horn is probably little more than 10–15 in., if as much. The north-east monsoon carries the 'Dair', or 'little rains', and blows from October to December. In the highlands within the Abyssinian border, as also on the Isha Baidoa/Hoddur plateau, the rainfall is greatest with this monsoon, the latter area enjoying from 18 to 20 in., and even 25 in. according to recent records.[2] The greatest factor influencing rainfall, however, is altitude, and the mountain escarpment edging the Gulf of Aden Rift carries one of the heaviest rainfalls in the Somalilands.[3]

The vegetation found in the more elevated parts of the country is almost identical with that of the Abyssinian highlands, but as the country falls to the plateau proper and the coastal plain the temperature rises, the rainfall decreases,

[1] Two other 'outcast tribes' also exist, the Yibirs (sorcerers) and Tomals (blacksmiths). See Hunt, J. A. (1943–4).

[2] Dainelli, Prof. G. (1936, p. 140).

[3] Hunt, J. A. (1944).

and there is a rapid transformation from an evergreen montane type to a deciduous semi-desert type of wide extent, with enclaves of thicker growth in the south and in the river valleys. On the coastal plains and in the north-eastern parts of the plateau true desert conditions exist with only a sparse vegetation of gum and incense bush.

The game is entirely of open grassland form and is still commonly met with though it is not as plentiful as it once was according to the accounts of early hunters and travellers.[1]

Evidence for past climatic fluctuation from a heavier rainfall to that which persists today has been deduced by Gilliland from a study of vegetation types in eastern British Somaliland. This is based in the main on the survival in specially favoured sites of vegetation types of the heavier rainfall phase of climate (termed a post-climatic vegetation).[2]

In the Horn water must have been the most important factor in controlling the migration and settlement of prehistoric man, just as it is for the present-day pastoral Hamites who inhabit this area. When water was comparatively plentiful, as it must have been during a period of increased rainfall, man would have been enabled to spread over the plateau finding water, perhaps an annual supply, at the pans and sink holes which are such a common feature of this area. When the climate was drier settlement must have been restricted more or less to the immediate vicinity of the larger river valleys or to those places where escarpments or gorges still permitted spring water to seep out upon the lower slopes. For example the sites of the Somaliland Stillbay Culture which existed during the dry period at the end of the Upper Pleistocene, are confined very largely to the valleys of the bigger rivers or the large pans and springs, but the occupation sites of the Doian and Somaliland Wilton peoples on the other hand, who lived at the time of the 'Neolithic Wet Phase' in the Sahara and North Africa, are extensively distributed throughout the plateau, where many of the smaller pans bear evidence of occupation at this time. During this time also many of the smaller tugs, which are now perpetually dry must have contained standing pools of water, which, besides simplifying the movement of these Late Stone Age peoples, would have ensured a plentiful supply of game being always available. Such, for example, must have been the shallow tug at Aisha in the southern Danakil Rift, where was found evidence of small encampments of Somaliland Wilton people. Macfadyen has informed the writer that one such old water channel called the Bokh, now dry, runs from about Odweina, near Eik, south of Bohotleh and on to Domo. This may connect up with the depression at

[1] (i) Revoil, G. (1882). (ii) James, P. L. (1888). (iii) Swayne, H. G. C. (1895).
[2] Gilliland, H. B. (1947, p. 43).

Badwein, on to Geriban and so to the east coast. The tufas of the pans and dried-up springs bear evidence of the fact that they once supported a settlement by prehistoric man, which would no longer be possible today. Around many of the granite 'burs' (inselbergs), on the coastal plain of southern Somalia, are found 'fossil' rock cisterns, similar to that at Magosi in northern Uganda,[1] which have become filled with sediment and sometimes contain cultural material. The present-day inhabitants of the region still make use of these cisterns not a few of which they have re-excavated.

SUMMARY

It will be seen therefore that the Somali plateau consists of a huge, gently inclined, landmass tilted towards the south-east. This vast and undulating surface is relieved only occasionally by low residual hills or ridges, which indicate a surface of peneplanation. Only on its western side (and to a lesser extent also along the northern edge) is the plateau relieved by high mountain ranges of volcanic and other rocks which line the edge of the Abyssinian rift. Approximately 1000 ft. below this flat peneplain surface have been eroded the wide valleys of the principal river systems—the Juba, the Webi Shebeli, and the Nogal—which were coincident with the formation of the coastal plain, and therefore presumably postdate the initial period of faulting in the Rift Valley.

Such then are the main physiographical features, but before passing to consider their interpretation in the light of the Quaternary and Recent succession it will be necessary to review briefly the works of previous investigators in the field of prehistory in the Somalilands.

[1] Wayland, E. J. & Burkitt, M. C. (1932).

CHAPTER 2

THE WORK OF PREVIOUS INVESTIGATORS

SOMALIA (ITALIAN SOMALILAND)[1]

The earliest collections of which we have record were made by Revoil, who between 1882 and 1888 picked up artifacts on the coast and plateau of the northern Mijertein; he was often lax in recording the localities of his finds but published a short description and an illustrated plate, which included drawings of a Middle Stone Age type of point from the Angora valley, a small discoidal scraper also of Middle Stone Age type, and a small blade.[2] Some ten years after the first implements were found, the Robecchi-Brichetti Expedition recovered in 1891 some flake tools from the district of Obbia, particularly from Warandi,[3] which, from the description given, exhibit a Levalloisian or faceted platform technique.[4]

In 1901 Dr Brumpt, the naturalist of the du Bourg de Bozas Expedition, made a large collection of implements which were examined and described by Breuil & Kelley in 1936. These finds were well documented and enabled them to make out a list of some forty sites.[5] A brief record of the more important sites is given below. Sites are particularly common on the upper Webi Shebeli and in the Dakata valley between Segag and the confluence with the Webi Shebeli. Some sixteen sites in all are recorded from here of which the most important are Orahaut (155), yielding Stillbay and Wilton; Segag (153), Levalloisian and Stillbay; the Dakata/Webi confluence (159), Stillbay; a little farther south on the left bank of the Webi (160–2), Levalloisian, Stillbay and probably Wilton; the Dagahbur/Webi confluence (163), one possible Acheulio-Levalloisian pointed hand-axe, Levalloisian and possibly Stillbay; south and again on the left bank of the Webi (164–5), Levalloisian, Stillbay and Wilton; Oksa (166), Levalloisian and Stillbay; Dabrin (169), Levalloisian.

A second group of sites is situated between the Webi Gestro and the Webi Mana, tributaries of the Juba (170–7). Notable are Kadjoua (176), Levalloisian and Stillbay; Robabouta (173), one pointed Acheulio-Levalloisian implement;

[1] Including the Ogaden returned to Abyssinia on 23 September 1948.

[2] Revoil, G. (1882, pp. 286–7).

[3] Figures in parentheses refer to sites recorded on the Main Distribution Map (Pl. 29).

[4] Robecchi-Brichetti, L. (1899). (Note by Giglioli.)

[5] Breuil, H. & Kelly, H. (1936).

Abbé (174), in the Mana valley, Wilton; Goba (177), one doubtful 'Neolithic' *tranchet.*

Further south on the upper Webi Shebeli at Sheik Hussein (178) the Expedition found Levalloisian implements and near Roukatcha Wells (219), Levalloisian and probably Magosian.

Finally, there is a group of sites on the Harar plateau. Kouloubï (226); Tchalinko (225), in the Ahmar mountains; Dubasso (228); Kourkourron (227); and Warka (224) where is reported 'a rock-shelter with a layer of grey slime under one of grey clay containing worked flints and a fauna similar to that of today'.

There is also a group of sites whose exact position is uncertain. These are mostly situated in the valleys of the Arba, Awash and Shoa rivers and have yielded obsidian tools with a double patina. Breuil & Kelly attribute them to the Magosian and early Wilton cultures on typological grounds.[1]

Briefly, therefore, the industries represented by the du Bourg de Bozas collection may be summarized as follows:

(*a*) Acheulio-Levalloisian.
(*b*) Levalloisian, Somaliland Stillbay and Magosian.
(*c*) Somaliland Wilton.

In 1913 the Stefanini-Paoli Geological and Zoological Expedition collected a series of twenty-nine stone implements from various parts of southern Somalia which were figured by Puccioni.[2] Typologically they can be referred for the most part to the Levalloiso-Stillbay Complex and in some instances to the Somaliland Magosian and Doian. They come from sites at Washaga Guran (102), four specimens, Middle and Late Stone Age; Bur Erimu, three specimens, Doian; and El Ure (103), eleven specimens, Stillbay and Doian; Lugh Ferrandi (100), three specimens, Stillbay and Doian; Guretca (Curetca) (99), one specimen, Stillbay; Allengo (98), three specimens, Doian; and Salagle (94), one core, Levalloisian. The greatest number of implements was provided by El Ure, notable among which is a good example of a small Doian backed blade.

In 1924 Professors Stefanini and Puccioni in a joint geological and anthropological expedition collected a number of artifacts which were later described and illustrated by Puccioni.[3]

[1] Of the lesser sites, Chekifton and Iddi yielded Stillbay- and Magosian-type material; Fidadaedi (Fidadedih), ?Wilton; Dallehallé, Magosian and Wilton; Djilohanno, ?Wilton; Karsaalek, Stillbay and Wilton; Salout (Sullul), Stillbay and ?Wilton; Guiner (Ghimir), Magosian; Habaoule and Rahaïtou, atypical material.

[2] Puccioni, N. (1919, p. 166).

[3] Puccioni, N. (1936, p. 107).

They were derived from three main areas:

(1) The middle Webi Shebeli where three sites were located: Sheik Gure near Bugda Acable (83), fourteen specimens including Levalloisian, Stillbay and Magosio-Doian material, one good example of a subtriangular Doian point requires special mention; Bur Rari, nine specimens, Levalloisian and possibly Doian; Bur Goura,[1] one Levallois-type point.

(2) The country inland from Obbia in the Mudugh: here seven sites were found. Durgale (109), yielding three Levalloiso-Stillbay implements; Adad (107), eleven specimens, Levallois-type and Late Stone Age; Afgudut (47), six specimens including two disc cores, a blade and an end-scraper; Adani (110), four Late Stone Age blades, one with marginal retouch; Warandi (111), which yielded thirteen specimens of Levalloisian type comprising some of the earliest implements from the Mudugh. They include bifaced choppers, large and small flakes with prepared platforms, large blades, disc cores and one Upper Levallois-type unifaced point. They were recovered from 'terraced' deposits of gravels and scree on the slopes of the low hill or 'bur' at Warandi, near the wells of that name. Ilbehla (108) yielded one Levallois-type point only, while from Obbia (49) came one atypical flake probably derived from the strandlooper site located here by the writer. Puccioni assigns the material from Warandi to the Nanyukian (Kenya Fauresmith) but his illustrations suggest rather a Levalloisian age and this is confirmed in some measure by Graziosi.

(3) Northern Somalia: in the upper Nogal valley and the Mijertein, Puccioni found a greater number of implements than in the two previous areas. The sites occurred in two main areas—along the coast between Eil and Hordio (Hafun), and in the upper Darror (or Daror) near the territorial boundary with British Somaliland. The sites may be listed as follows:

The lower Darror (or Daror) (133), twenty-three specimens of Levalloiso-Stillbay and Somaliland Wilton character.

Between Martissor Dunti and Scerer Hor (131), twelve specimens, predominantly of Late Stone Age type, but including three examples comparable to the Somaliland Stillbay.

Badi Ahmud (129), thirty-two specimens all fairly typical Somaliland Wilton material with a predominance of small blades.

Between Warassimoghe and Bender Beila (128), seven specimens, including four Levallois-type blades and flakes and two small Late Stone Age blades.

Between Bender Beila and Bihen (127), ten specimens of Levalloiso-Stillbay affinities.

Bihen (126), thirty specimens, including Levalloiso-Stillbay unifaced points, scrapers and disc cores, probably to be assigned to the Somaliland Stillbay; and small blades and end-scrapers of Somaliland Wilton facies.

[1] The writer was unable to locate these two sites when visiting the area in 1946.

Between Bihen and Ariri (125), fourteen specimens, of Stillbay and Late Stone Age types.

Between Bakbah and Bio Ado (124), nineteen specimens again of Stillbay and Late Stone Age types.

The Doldol plateau (123), four specimens, typologically Middle Stone Age.

The upper Nogal (119), fourteen specimens, Middle and Late Stone Age.

Taleh (136), four specimens, two backed blades and two unworked blades of Late Stone Age type.

The plateau between the Nogal and Darror (or Daror), seven specimens, Middle and Late Stone Age.

Between Golol and Buran, eleven specimens, predominantly of Late Stone Age form.

The upper Darror (or Daror) (134), twenty-six specimens, Middle and Late Stone Age.

Ras Filuc (135), one finely retouched blade probably to be assigned to the Late Stone Age.

Cubo (130), near Gardo, one large, atypical blade.

In 1927, during a brief visit to the Mijertein, Professor Cipriani found in the Hafun peninsula (132) a series of implements which comprise one Levallois-type point, a hollow scraper and disc core; these should probably be referred to the late Middle Stone Age or Somaliland Stillbay.[1]

The material so far described consists of surface finds, mainly flakes, points and scrapers of late Levalloiso-Stillbay type, together with a smaller percentage of diminutive blade industries which can be assigned to the Late Stone Age. The earlier cultures, with the exception of the Warandi material (Puccioni) and the isolated finds from the upper Webi Shebeli (du Bourg de Bozas), are not represented due probably to the fact that they are, unlike much of the later Stone Age material of Somaliland, not usually found on the surface, but in sealed deposits.

In 1935 Graziosi & Puccioni visited Somalia with a view to conducting stratigraphical research in order to establish a systematic chronology. Excavations were made at a rock shelter at Bur Eibe in southern Somalia, and in the valley of the lower Nogal, and surface stations in different parts of northern and southern Somalia were studied. The results have been published by Graziosi.[2]

Of the principal surface sites examined in the Juba valley, that at Bun Mess (95) yielded Late Stone Age (Magosian and Doian) implements and some few Middle Stone Age flakes. An industry of microlithic blades and small tools was found at the foot of Bur Mun (96). Graziosi attributes it in part to the Kenya Wilton, but it is probably of local Doian affinities. Lugh Ferrandi (100) has also yielded microliths amongst which are some typical straight, blunted-backed blades and scrapers, together with a few flakes of Middle Stone Age type. At Bur Eibe (18) Graziosi found a rich site yielding 'artifacts of different facies, some approaching

[1] Cipriani, L. (1932, p. 17). [2] Graziosi, P. (1938*a*, 1940).

the Mousterian and others made with a Solutrean technique—a combination similar in some cases to the Stillbay of South Africa'.

Graziosi's rock-shelter at Bur Eibe[1] yielded him an industry in the upper levels which he was unable to compare in detail with that from any other site or area. 'It comprises [he says] an assortment of artifacts, mostly microlithic, a number with a characteristic Solutrean-Neolithic retouch, including small limaces of most careful workmanship, minute and delicately retouched points, small scrapers, fine gravettes, etc., some of flint and others of very transparent quartz.'[2] This industry he described under the name 'Eibian'.[3] The lower levels yielded Levallois-type flakes associated with 'feuilles de laurier' and seemed to him to indicate the presence of the Somaliland Stillbay Culture.

In the Mudugh (or Mudug), between Galkayu and Garoe (117, 118), Graziosi recovered atypical flakes and artifacts of Levalloiso-Stillbay types, while from Garoe (54) he found a surface site yielding a microlithic blade industry of Hargeisan or more probably Somaliland Wilton facies.[4] His other sites in the Mijertein comprise the following: Deavani (120), and Trogo Dait (121), in the upper Nogal, producing flakes of the normal Levalloiso-Stillbay facies; Eil, near the mouth of the Nogal (56), which yielded Stillbay flakes, as also did the neighbouring valley of the Gulule to the south (122). Graziosi also partially excavated a rock shelter in the Nogal Gorge, near Eil, and at a depth of more than 2 m. found 'alternate layers of hearth deposit containing fauna and a crude industry, which did not change typologically throughout the occupation level'. Its rather crude facies disposes him to regard it as Mesolithic.[5]

Graziosi obtained two surface collections from southern Somalia sent to him by Dr G. Bernadetti. The first from the foot of Bur Dinsor (97) included small blades, some showing secondary retouch, a few scrapers and some poorly made crescents. These should presumably be assigned to the ubiquitous Doian Culture of the south. The second collection, from the Gurerka at Marta-Arba, from the Gehe Kin Tug, and elsewhere near Lugh Ferrandi, includes Middle and Late Stone Age forms. Graziosi also records a collection of 'Levalloiso-Mousterian' flakes and implements made by Major Giorrio in 1933 during the advance of the Italian troops into the Ogaden along the line Sassabannah-Dagahbur-

[1] This is established as the Gure Makeke—the largest shelter situated on the east side of the Bur near Berdale the village of the Sultan of the Eile.

[2] See pp. 203 and 226 for description of the Somaliland Magosian and Doian Cultures.

[3] Graziosi, P. (1940).

[4] In spite of careful search this site was not found when the writer visited Garoe in January 1946.

[5] The writer was not able to identify this shelter with certainty, but it would seem that the industry found by Graziosi is identical with the strandlooping form of the Wilton Culture recovered from the rock-shelter at Biede at the mouth of the Nogal. For description of this, see p. 282.

Jigjiga. Sites of particular mention are the wells at Ubitale (113), Wardere (sometimes Warder) (63),[1] and Jigjiga (4).

To conclude the discoveries of Italian workers mention must be made of the collections formed by Dr Tavani during his Geological Expedition in 1937 and described by A. C. Blanc and Tavani.[2] The collection consists for the most part of Levalloiso-Stillbay material, but some Late Stone Age implements are also included. The sites from which they were discovered are given by Tavani as—a river bank near Guriagale, near Ferfer (42), hills near Lugh (100). Bulo Burti (86), Mustahil (106), Wejit (28), hills north-east of Bur Bitale and the upper slopes of the Shebeli valley between Bugda Acable and Tijieglo (105). Bio Ado near Walwal (115); the country between Beletwein and Tijieglo; Wadi Kor Degheisto; Ubitale (113); the neighbourhood of Mahaddeiwein (104); a hill, 59 km. east of Gorrahai (112); the neighbourhood of Dolo (101); Walwal (114); the neighbourhood of Gabredarre (65); and east of the Gabredarre-Jigjiga road, 15 km. from Gabredarre. The most important sites are those at Bio Ado from which come well-made diminutive points of yellow chert, as well as implements showing typical pressure-flaking, and Guriagale, which yielded small pressure-flaked tools, and well-made straight-backed blades and a heavily patinated basalt grindstone. These tools are probably closely related to the Somaliland Magosian-Doian industries from Jesomma (85) and other sites in the middle Shebeli valley.[3] Gabredarre yielded a large number of Levalloiso-Stillbay implements, probably similar to those found in this area in the upper calcretes of the Tug Fafan.[4]

In addition to the usual Levalloiso-Stillbay material, therefore, Graziosi's and Tavani's discoveries at Bur Eibe and Guriagale provided strong evidence of the presence of an ill-defined but apparently new cultural element in southern Somalia during the Late Stone Age, which is characterized by pressure-flaked points, straight-backed microliths, and limaces.

In 1942 Captain A. Phillips of the British Military Administration in Somalia collected a considerable number of specimens from Lugh Ferrandi (100) which, while considerably weathered due to long surface exposure, are typologically to be assigned to the Somaliland Magosian and Doian Cultures.[5]

[1] A small, unannotated collection from well excavations at Wardere is contained in the Garesa Museum, Mogadishu. This material, which is in chert, is fresh and appears to have been derived from a light grey clay covered by surface kunkar or tufa within the deepish pan in which the Wardere wells are situated. It includes faceted flakes and a good example of a developed Levallois or early Stillbay unifaced point.

[2] Graziosi, P., on Blanc, A. C. & Tavani, G. (1938). [3] See p. 356 and 358–9. [4] See p. 175–9.

[5] For a record of collections made by British personnel during and after the East African campaign, see pp. 32–41.

BRITISH SOMALILAND

For many years the British Protectorate has yielded numbers of well made implements from surface sites. Shortly after 1890 MacConkey found artifacts in the broken country between the fault-scarp and the coastal plain, between Zeila and Berbera (144); further specimens were collected by him to the south of Bulhar. His discoveries were studied by Jousseaume who refers to sites at El Anod, south of Bulhar (142) and on the Tug Darar.[1] From the published description and illustrations, MacConkey's material may be referred in the main to the Levalloiso-Stillbay Culture complex, but includes also a percentage of Late Stone Age blades and scrapers.

In the course of numerous journeys in the country from 1897 onwards, Seton-Karr made a large collection of stone implements from sites between Burao and the sea. The Tug Issutugan (143), about 60 miles south-west of Berbera, has yielded some of the most interesting material and numbers of specimens from here have been distributed to European, African and American museums.[2] From the Issutugan Seton-Karr recovered many implements in a perfect state of preservation, including a good series of chert, quartzite and diorite hand-axes of a developed Chelles-Acheul type which Breuil is inclined to identify with the Micoquian of Europe: in addition to the hand-axes the Issutugan sites also yielded Levalloisian and Somaliland Stillbay material.[3]

Seton-Karr's sites were situated apparently near the head-waters of tributary tugs of the Issutugan along the lower slopes of a line of low hills running parallel with the bed of the main tug. These hills were probably composed of Cretaceous sedimentaries as they consisted of inclined sandstones interbedded with quartzite bands. Smith[4] gives the name of this site as Jalelo[5] situated 68 miles south-west

[1] Jousseaume, F. (1895). Perhaps Daror (9° 43′ N., 44° 14′ E.).

[2] Mr Seton-Karr distributed specimens to over fifty different museums, but as very few of these collections contain any data as to their provenance and associations they are now of very little value except typologically. The more important collections are to be found in the British Museum, the University Museum of Archaeology and Ethnology, Cambridge; the Pitt-Rivers Museum, Oxford; the Musée de L'Homme, Paris; the Musée de St Germain, Paris; the Cape Town, Nairobi, and Bulawayo Museums.

[3] (i) Seton-Karr, H. W. (1896, p. 271), (1898, p. 93). (ii) Evans, Sir J. (1896). (iii) de Morgan, J. (1921). (iv) Breuil, H. (1931). (v) British Museum (1926), Pl. XI, p. 182.

A typical series of *coups-de-poing* from the site are illustrated in Pl. XI of the *British Museum Stone Age Guide*. While one may regret the lack of information accompanying Seton-Karr's collections yet they yielded the first and, with one exception, that of Sheik (or Sheikh), some of the most finished examples of hand-axes yet recovered from the country. For discussion of this material, see pp. 160 *et seq.*

[4] Smith, R. A. (1937), pp. 93–4 and illustrations 935–48.

[5] Perhaps Jalelo (9° 48′ N., 44° 17′ E.).

of Berbera. The implements lay in profusion on the surface presumably eroded by sheet and gully erosion from alluvial beds not described. Seton-Karr recognized, however, that these implements, which were usually in mint condition and made predominantly of quartzite (the raw material ready to hand), were being eroded from a factory site. His specimens while they included many examples which were undoubtedly roughouts, as would be expected from a factory site, included also finely finished examples of Micoque-type: pointed, strangulated and subtriangular forms as well as ovate, cordiform and limande examples. Two typical cleavers occur in the Sturge collection. There is an almost complete absence of utilized flakes that might have been associated but this is probably accounted for by the fact that only the more recognizable specimens (i.e. the hand-axes and cleavers) were collected.

The two collections in the British Museum contain some of the best material collected by Seton-Karr. The hand-axes are typical of those found at Sheik, Hargeisa and other sites in the north-west, and comprise forms that are typologically late Acheulian, Micoquian and Acheulio-Levalloisian. Not a few of the hand-axes and both the cleavers in the collections are made on flakes, but there is no indication that the prepared core was present. Two scrapers on flakes and one large utilized flake-blade, all with plain, inclined striking-platforms may also belong. In addition to the hand-axe material the British Museum collections contain some typical Upper Levallois specimens, the provenance of which, however, is unknown.

In 1897 Count Wickenburg found *coups-de-poing* at Goban (?=the Guban), in approximately the same area as Seton-Karr's finds: these were published by Paulitschke who illustrates and describes *coups-de-poing* and flake tools referable to the Early and Middle Stone Age.[1]

In 1924 the Stefanini-Puccioni Mission visited Taleh (136), and Buran (137), close to the north-eastern boundary and collected backed blades of Chatelperron type and some flakes of chert: they should be referred to the Late Stone Age, to the Hargeisan and Somaliland Wilton Cultures.[2]

While serving as geologist on the Anglo-Italian Boundary Commission in 1929–30, Barrington Brown made a collection of artifacts from several surface sites in the north-eastern part of the Protectorate between the Nogal valley and the Gulf of Aden: the most important of these sites are situated at Hormo (139) and Buran (137) on the northern edge of the Sol Haud. These collections have been described briefly by Burkitt,[3] who divides the material into four series, using difference of patina as the basis. It is of interest to note how closely such

[1] Paulitschke, P. (1898, p. 15).
[2] Puccioni, N. (1936, p. 107).
[3] Burkitt, M. C. & Barrington Brown, C. (1931, p. 164).

differences can be correlated in sufficiently representative collections with typological differences and so linked with the material found later *in situ*.

Burkitt divides the material into four series: (*a*) and (*b*) implements with a deep patina of black, mahogany, white or cream colour of Mousterian or Levalloisian type; (*c*) later types, probably Late Palaeolithic, with a mottled cream or pinkish patina, often glossy; (*d*) flakes and crude scrapers of later date and often unpatinated. The eleven figured implements include typical Levallois prepared cores, points and a faceted flake in addition to end-scrapers, two straight, blunted-backed blades and one transverse, concave angle-burin, characteristic of the Hargeisan Culture. The latter are of importance as all, with the exception of the angle-burin, were collected from what must have been a homogeneous home and factory site on a terrace north of the tug bed one mile north-north-west of the Triangulation Station 2422 at Hormo (139). The material from the Hormo site is consistently fresh in appearance, with pinkish white glossy patina (compare with material from the type site at Hargeisa) and comprises the following artifacts:

(i) Cores. One developed, subtriangular, prepared core; one bi-polar blade core.
(ii) Flakes. Two pointed flakes with faceted platforms and parallel flaking on the upper surface; one thin, radially flaked example with prepared butt.
(iii) Long blades of microlithic type with unprepared striking-platforms.
(iv) Single and double end-scrapers on blades.
(v) Straight-backed blades, mainly large (microliths possibly not collected), some with characteristic reduction of bulb by reverse retouch.

Specimens from this site and from a similar site at Hobat (138) are figured at Pl. 21, figs. 15–33, and should be compared with examples from the type site at Hargeisa. The remainder of the material collected by Barrington Brown is listed in Appendix A, pp. 333–4.

In 1931 A. T. Curle collected numerous specimens from the western districts on the plateau west of Hargeisa and again to the east between Hargeisa and Gerbileh (?=Gebile). The collections from the latter area are now in the University Museum at Cambridge and come from the Teisa (Tessah) Tug (195), and the Dubburro Hills; Arabsiyo (196); Gebile Tug (197), and Ijara Tug (198). Curle also records notes of sites at Badewanak (or Bedr Wanak), Adadleh, Borad, on the Ununuf Plain, Sik, Shilmaleh, Daba Gumbur, Au Berkadle (or Au Bakadleh), Haleya and Hargeisa, but the collections contain no implements from these sites. The specimens are washouts collected on eroded areas of red alluvium overlying gravels in the upper courses of tributaries of the Marodijeh (or Marodijehh) and Marodile Tugs. Details of this collection will be found in Appendix A. Curle's sites contain a high proportion of Lower Somaliland Stillbay and Developed or Upper Levalloisian material, the former being distinguished by its superior technique, end-products and physical condition. It is usually unrolled and has not been appreciably patinated. The assemblages include typical prepared cores, and faceted flakes, bifacial and unifacial limande points and tools showing Emireh reduction of the bulb and butt. The upper or typical Somaliland Stillbay is well seen at the Teisa Tug, the majority of tool types being represented. Hargeisan implements are also included in the collections from this tug.

Curle's material is comparable in every respect with that recorded by the writer from the later deposits in the Hargeisa valley near Hargeisa and thus adds valuable corroborative evidence for the wide distribution on the north-western plateau of the culture sequence established at Hargeisa.

John Parkinson in the course of geological work in the north-western districts collected from fifteen sites yielding for the most part surface material. Although the Somali name of the site is recorded the map reference is not given, so that with few exceptions it is not possible to pin-point these sites; they have accordingly been omitted from the distribution map. Three sites in particular are worthy of note. The Marodijeh (or Marodijehh) Tug—at Hargeisa where from the level of the tug bed are derived unpatinated and unrolled Somaliland Stillbay specimens including one flake with faceted butt, one utilized blade or flake-blade, one utilized pointed flake with converging flake scars on upper surface, and one faceted flake with converging scars showing secondary pressure flaking down one edge. Sharab Ad (perhaps ? Gharab Ad) Tug where was found a very good example of a Developed Levalloisian flake tool—a typical thin flake with finely faceted oblique butt, 25 ft. below the surface of the alluvium of the tug wall. Aror,[1] from which the majority of specimens were collected is a surface site yielding weathered and patinated specimens, several showing a double patina. The material may be divided into two main groups—the first comprising, blades, blade-cores, one microlith and one straight-backed blade, is referable to the Late Stone Age, probably by reason of the fresh and unpatinated condition to the Somaliland Wilton. The second group, to which the majority of the specimens belong, is usually weathered and patinated white to grey in colour, and includes crude discoidal prepared cores and rough flakes with unprepared or poorly faceted platforms. Secondary work is usually confined to steepish notching and nibbling of the edges of the flake, which is often snapped

[1] Two places of this name occur in British Somaliland, at 9°14′ N., 43°50′ E. and at 10°53′ N., 48°22′ E.

across. A more definite and recognizable tool type is represented by fourteen crude 'points' formed by the intersection of the rounded point of two opposing *encoches*. One broken specimen was probably bifacial while the remainder are unifaced examples, the secondary work forming the notches being directed from the main flake surface: the flakes are approximately subtriangular in cross-section. A variation on this type is provided by three faceted flakes in which the opposing *encoches* do not intersect at the point but form a waisted tool, the end of the flake opposite the bulb remaining blunt and flat. A number of other sites yielding crude material of this nature are known—notably Damero, between Burao and Erigavo on the western plateau, and Hararile (55) in the Nogal valley: both are surface sites.

It is of interest that the material from Group 2 at Aror, from Hararile, and from the later series at Damero shows much in common with Caton-Thompson's oasis development of her epi-Levalloisian industries of Egypt—which she has termed Khargan.[1] Whether these Somaliland sites really represent a late and degenerate form of the Levalloisian cannot as yet be proved owing to the absence of stratigraphical evidence on which to base a relative date for the industries. The evidence from Hararile, such as it is, suggests that the industry here is fairly late in the Stone Age sequence, but it must be remembered that Levalloisian material subjected to long surface exposure and weathering exhibits just such characteristics of broken flakes and steep 'nibbled' retouch as are present on the greater number of specimens from these three sites. While therefore we may postulate the presence of epi-Levalloisian industries in the Horn of Africa, further corroborative evidence is required before this can be proved to be a certainty.

FRENCH SOMALILAND AND THE SOUTHERN DANAKIL RIFT

The first scientific report of archaeological material from this area we owe to Teilhard de Chardin[2] who, in the company of H. de Monfried, visited sites in French Somaliland, and the country round Dire Dawa and Harar, between November 1928 and January 1929.

The most important area examined by Teilhard was in the coastal region of French Somaliland between Obok (150) and Mount Mabla (149). Here he noted evidence for two periods of marine transgression, the second of which he correlated with extensive gravel spreads descending from Mount Mabla. Stone implements of rhyolite were very numerous and those contemporary with the

[1] Caton-Thompson, G. (1946).

[2] Teilhard de Chardin, P. (1930*a*).

gravels were heavily patinated and abraded. The implements lying on the Obok coral became progressively smaller and less patinated as one approached the sea. Many tools show signs of later re-utilization, for the same purpose as the original specimen was first made, and indicate that the same industry persisted over a long period. Tools include Levallois-type points, pressure-flaked points, blades and scrapers of different kinds, while hand-axes are associated with the earlier, more heavily patinated specimens.

On the north-western edge of the Somali Plateau around Harar (208) Teilhard found 'Mousterian-type' points and scrapers in obsidian and chert which he considered were derived in part from the thick deposits of laterite that cover much of the Harar plateau. *In situ* in the sections exposed in tributary tugs of the Errer, notably at Djildgiga implements were found even in the lower levels of the Pleistocene alluvium and include 'Mousterian-type' points, small scrapers, choppers and axes, made from crudely flaked basalt pebbles.

Finally, de Monfried & Teilhard excavated a trial trench in a painted, limestone rock-shelter a few kilometres from Dire Dawa which was named the Porc-Epic cave (152). The shelter was about 20 m. in depth and width, but a fall of rock had partially blocked the entrance; the walls were covered with stalactites. Under 30 cm. of ash containing pottery, a further loose deposit was discovered, 1·70 m. thick, which yielded numerous implements of chert and obsidian, the predominant tools being small triangular points, either worked on the edges only of one side, or all over the one side, or worked over both faces of the tool. No pottery was found with the industry, and animal bone had almost entirely decomposed and formed a thick layer of phosphate of lime on the floor of the cave.[1] Teilhard considered that the workmanship of the points suggests that they are fairly recent in date but that the nature of the deposit in which they were found points rather to their being contemporary with the Upper Palaeolithic of Europe. The stylized paintings on the walls of the cave were studied by Breuil & Wernert and will be discussed later.[2]

Teilhard de Chardin again visited this area in 1930 in company with Breuil & Wernert, and further excavations were carried out in the Porc-Epic cave[3] (see p. 208). Their findings in French Somaliland[4] have been summarized below in some detail as they provide valuable corroborative evidence for our own sites on the Gulf of Aden coast and in the Danakil Rift.

[1] Compare the sequence at Porc-Epic with that from the Gure Warbei at Bur Eibe (p. 230), and from Bur Hakaba (p. 236).

[2] See p. 295.

[3] Breuil, H., Teilhard de Chardin, P. & Wernart, P. (1951).

[4] Teilhard de Chardin, P., Breuil, H. & Wernert, P. (1940, pp. 497–522).

OBOK AREA (149, 150)

(i) *Stratigraphy*[1]

The area studied consists of a coastal strip some 12 km. in width from east to west between Obok and the basalt and rhyolite mountain named Mount Mabla. Two roughly concentric zones are recognizable—the raised coral bench to the east comprising the immediate coastal strip, and an area of extensive gravel spreads to the west centred round Mount Mabla.

A. The raised coral bench is deeply dissected by steep-sided tugs and the surface, with the exception of an occasional pebble, is quite bare. Three series of deposits could be recognized corresponding to three successive movements of the coast.

(*a*) At the base occurs an old red coral, which was formed apparently in relatively deep water: this deposit consists of sands and decomposed coral. It was not seen along the coast of the Gulf of Tajura. Teilhard considers it as being possibly earlier than the post-rhyolitic basalts of Mount Mabla.

The emergence of the bench was followed by two periods of marine transgression.

(*b*) Above the old red coral is found a younger coral, better preserved than the former, with a pronounced unconformity between the two. It comprises sands and corals alternating with thick detrital gravels and beds of small rhyolitic gravel. This deposit is correlated with the raised corals of the Tajura coast which are always later in date than the Upper Basalts.

(*c*) The topmost beds which are localized in shallow depressions in the younger coral are considered to indicate the presence of old swamps or lagoons, and this is borne out by the molluscan fauna, notable among which is a large Cerithid.

This sequence shows therefore, first, the gradual emergence of the shoreline, then the sudden appearance of gravels in the sediments and finally a lagoon phase.

B. The gravel zone is well preserved, thick and extensive, and extends from Mount Mabla towards Obok in the east, and the Gulf of Tajura in the south. These, in the latter area, where the earliest spreads probably alternate with the Later or Upper Basalts, are certainly of Pleistocene Age. Two raised surfaces are recognizable:

(*a*) *Upper gravels*

Deeply incised by tugs, form a flat bench approximately 60 m. above the bed of the Latela Tug; the writers correlate these gravels with the younger coral.

(*b*) *Lower gravels*

These are relatively better developed and are found at approximately 30 m. below the upper gravels.

[1] (i) Teilhard de Chardin, P. *et al.* (1930*b*). (ii) Teilhard de Chardin, P. (1932).

The Latela Tug appears to have developed its drainage after the deposition of the lower gravels, that is to say after the tectonic movement which caused the drying up of the lagoons nearer the coast. This tug contains evidence of two low terraces which have yielded a few derived implements from the earlier Mabla gravels.

(ii) *Culture sequence*

The Mabla gravels were found to be very rich in cores, flakes and utilized flakes, the predominant raw material being rhyolite pebbles. All are surface finds, however, so that the classification is necessarily based on physical condition and patination.

The authors identified by this means eight distinct series which they relate to the physiographical development of the region.

(*a*) *Upper gravels*

Series I. A few clearly rolled specimens would appear to prove conclusively the existence of an industry contemporary with the gravels. With the exception of a single biface tool this industry comprises crude flakes and heavy blades the great majority with inclined unfaceted striking-platform (out of nineteen specimens only one shows preparation of the platform). Cores are of various types but details are not given. Illustrations are given of one Levallois-type flake, the butt of a hand-axe, a large flake and a primitive form of *coup-de-poing* which from the illustration is not unlike some of the subtriangular prepared cores found with the Acheulio-Levallois Culture.

Series II. Tools of this series are unrolled but weathered. The assemblage is entirely lacking in *coups-de-poing*; the unprepared, inclined platform found on flakes of Series I is no longer found, and striking platforms now show clear evidence of preparation and a faceting technique. Cores are prepared and both Levallois and Mousterian types occur. Blades are common while the predominant tools are scrapers and crude points.

Series III. Also found in an unrolled but patinated condition resting on the upper gravel terrace. The industry is now essentially Levalloisian with typical cores and faceted flakes: tools include scrapers, borers and burins. The authors also include, however, both unifaced and bifaced hand-axes which they state to be of developed Acheulian type. The three figured specimens bear little resemblance to the advanced forms of *coup-de-poing* found further to the east in British Somaliland, but this may perhaps be explained by the possibility of the Mabla specimens having been derived from a factory or mining site and being, therefore, unfinished examples.

(*b*) *Lower gravels*

Series IV. Specimens are unrolled and comprise rare *coups-de-poing*, Levallois-type cores and numerous flakes with retouched platforms; blades are not so common as in the previous series. The industry is still clearly of Levalloisian type.

Series V. Hand-axes now appear more common and this series has yielded the largest number. Of the two examples illustrated one is limande and the other pyriform in shape. Stone technique appears to have been the method employed for their manufacture. Blades are no longer found but flakes are still predominantly of Levallois type and prepared cores are common.

(*c*) *Industries of the coral bench*

Series VI, to which the greater number of specimens collected are assigned, is essentially a surface industry which the authors considered to have continued over a considerable period of time. It is found lying on the coral in the Obok area; its lower limit is, therefore, formed by the drying up of the swamps and lagoons of the coast while its upper is marked by the hut circles observed on the higher slopes and at Djebel Djinn. Only one example of a biface was recovered. Cores consist of Levallois-type specimens, pyramidal cores for the production of blades and triangular flakes. Striking-platforms are still predominantly faceted but light blades occur and end-of-blade scrapers, burins and the bifaced laurel-leaf point are now found. The industry is assigned, therefore, to the Middle Stone Age and can be dated to a period subsequent to the final emergence of the shoreline and to the deposition of the lower or 30 m. gravels. Certain of the small points found lying on the coral rocks near Obok are compared with similar tools found associated with the industry in the lower cave earth at the Porc-Epic rock-shelter near Dire Dawa.

The two later series described below were found in association with hut circles and are assigned to the Late Stone Age.

Series VII. With one exception, heavy axe elements no longer occur. The commonest form of core is a blade-producing pyramidal type, but small and polyhedral cores are also represented. Flakes with unprepared striking-platform are predominant but a few showing preparation are included. Blades are numerous and tools include large end-of-blade scrapers, convex side-scrapers, borers, etc.; burins were not found. One example of a pick is associated with the industry.

Series VIII. Cores are the same as those of Series VII as are also the striking platforms. Blades now have a microlithic tendency and burins are common. The usual type of scraper occurs and a form of hollow side-scraper is found. The authors note also a micro-biface and a fragment of a laurel-leaf point.

DJEBEL DJINN (151)

From the surface of this isolated, flat-topped mountain Breuil & Wernert recovered numerous artifacts referable to Series VI of Mount Mabla as well as implements and factory debris associated with hut circles. In addition to the usual circle of stones there was found a rectangular outline of stones which apparently showed indications of a 'mihrab' and so possibly claims recognition as a mosque: no microlithic debris was found associated with this latter. Only the later series, VI, VII and VIII, are represented at Djebel Djinn.

THE SHORELINE OF THE GULF OF TAJURA (147)

Everywhere along this coast were found industries comparable to those described from Mount Mabla. The peninsula of Kor Ali (148) was particularly rich; other sites are situated at Ras Kiro (145), Ambouli, Baie de l'Etoile, and Debeleba Bay. Specimens included rolled flakes (series I of Mount Mabla), associated with numerous cores, discs and flakes of rhyolite, having a fresh appearance but sometimes showing evidence of lime incrustation. Mixed with the above is also a microlithic industry in obsidian and chert. The main industry from this area is considered by the authors to postdate series V but to antedate series VI from Mount Mabla. It is characterized by weathered Levalloiso-Mousterian flakes, a pick, blades, angle-burins, end-scrapers and laurel-leaf points. Series I, V, VI, VII and VIII from Mount Mabla are also recognized as occurring at these sites.

LAKE ASSAL (146)

On the neck of land separating the Lake from the Gulf of Tajura, Wernert found triangular-sectioned picks which had a fresh appearance: it is suggested that they may have been used for extracting the salt found in quantity in the Lake Assal basin and are, therefore, probably quite late in the Stone Age sequence. On this neck is found a deposit of approximately 2 m. of gravels on the eroded surface of which were found a large number of cores, discs and flakes. They are usually heavily weathered while some show indications of abrasion by water action and are presumably, therefore, older than the picks.

SUMMARY

To summarize briefly the above evidence we find first, a primitive industry of large flakes, occasional Levallois-type specimens and *coups-de-poing* contemporary with the 60 m. gravels from Mount Mabla and with the younger coral of the Obok area which appears to represent a marine transgression. To the decline of this phase are assigned industries of definite Levallois affinities but including crude *coups-de-poing*, perhaps a factory or mining phase of the Levalloisian. A second phase of gravel deposition from Mount Mabla, at 30 m., is correlated with the final phase of marine transgression when lagoons were formed at the coast. The industries now encountered show more developed Levallois affinities. Later in date is an industry associated with hut circles. The Lake Assal evidence will be considered when discussing the geological evidence from that area. Comparison should be made with the sequence discovered at Berbera in British Somaliland and at the Garesleh Tug in Aussa, approximately 30 miles east of Dire Dawa.

COLLECTIONS MADE BY BRITISH MILITARY PERSONNEL, 1941–46

During and after the East African campaign several important collections of archaeological material were made by British military personnel, other than the writer, while serving in the Somalilands, the Reserved Areas and Abyssinia. Seven of these collections are housed in the Coryndon Museum, Nairobi, one in the Archaeological Survey, Johannesburg, and one in the University Museum at Cambridge. A summary of the contents of the collections is given below, but for details of site lists reference should be made to Appendix A.

(*a*) SOMALIA

(i) *B. A. Phillips Collection.* See p. 21 and Appendix A.

LUGH FERRANDI (100). Late Stone Age (Doian) surface material. See also under Brit. Som., p. 36.

(ii) *H. B. Gilliland Collection.* See Appendix A.

SIRRAU (or SIRAU) BALLEH (199). Late Stone Age surface site, see Appendix A. See also under Brit. Som., p. 38. A pan situated in the Central Haud. The specimens are made from coloured cherts derived from the Cretaceous sandstones and Eocene limestones. All are much weathered due to long surface exposure, trampling by stock and possible immersion in the 'balleh'; this has resulted in the production of many small broken fragments of blades and flakes the majority of which show secondary 'nibbling' of the edges due in the main to these natural causes. Specimens are predominantly microlithic and should probably be referred to a local variant of the Late Stone Age—Somaliland Magosian and Doian[1] Cultures.

(iii) *P. E. Glover Collection.* See Appendix A, also under Brit. Som., p. 36.

(*a*) SIRRAU (or SIRAU) BALLEH (199). Late Stone Age surface site, see Appendix A. See also under Brit. Som., p. 36. The material collected by Glover from this site differs in no way from that found by Gilliland and summarized above. It should be considered complementary to the industry from Alablah Balleh.

(*b*) BARDERA (90). Late Stone Age surface site. See Appendix A. Specimens from surface 'gravel' at Staging Post, three miles north of Bardera on the left bank of the Juba river in southern Somalia. An indeterminate collection of Late Stone Age factory debris of Somaliland Magosian and later facies. Raw material—chert and quartz pebbles.

(*c*) ALABLAH BALLEH (206), Wardere (or Warder) District. See Appendix A. An important and comprehensive collection from the western Haud; the majority of the specimens were derived from pits dug for root investigations through red Haud sand approximately

[1] See remarks under Aror on p. 25, however, in regard to the possible persistence into late times of local epi-Levalloisian industries using snapped blades and steep secondary retouch. See also under Alablah Balleh, p. 217.

1 m. deep and about 200 yd. from the edge of the balleh on the west side. Specimens are said to have occurred singly or in groups with no apparent sequence at varying depths below soil, from the surface to 59 cm. They are fresh and for the most part unpatinated except by glossing and surface staining by iron oxides from the red sand in which they are contained. This sand if it can be correlated with the red sands of the eastern Haud and Mudugh (or Mudug), was first blown into these shallow pans during the dry period marking the end of the Upper Pleistocene. Specimens include large and microlithic flakes and blades; micro-cores, notably the bipolar blade core; small discoidal and subtriangular prepared cores and the small faceted flakes with radial flaking struck from them; redirecting flakes; utilized flakes and blades; straight and convex microlithic backed blades; crescents; *lames écaillées*, or fabricators; polyhedral and gouge-burins;[1] scrapers, for the most part rare (small end- and disc-scrapers); one lanceolate-type limace point pressure-flaked on the upper surface; bifacial points, subtriangular and limande in form. The whole assemblage is closely comparable with the Somaliland Magosian and derivative industry from southern Somalia and the Webi Shebeli, and the stratigraphical horizon indicates that it may with certainty be referred to that culture complex. Examples are illustrated at Pl. 21, figs. 1–14.

In addition to the above this 'balleh' yielded a few derived specimens of Levalloisian affinities from the root excavations, and slightly weathered and waterworn Magosian specimens from the eroded surface on and near the balleh. They are said to be very numerous indeed here, and there is little doubt that they are derived from the red sand where this is eroding near the edges of the balleh.

(*d*) WARDERE (WARDER) (63). See Appendix A. One flake of honey-coloured chert with faceted platform, found on the surface in the grounds of the Petrol Point. Late Stone Age.

(iv) *P. Jackson Collection*

DAGAHBUR (116). See Appendix A. A small series collected at Dagahbur from erosion gullies in the valley of the Tug Gerer (Jerer), is referable to Somaliland Stillbay and Magosian industries. Of particular note, however, is one rolled specimen of an Acheulio-Levallois limande hand-axe covered with a calcareous concretion.

(*b*) BRITISH SOMALILAND

(i) *J. C. Trevor Collection*. See Appendix A

The collections comprise surface specimens from the following localities, mainly in the western half of the territory:

SHEIK (SHEIKH) (141). A small mixed series comprising some Somaliland Stillbay implements and factory debris together with a few specimens typologically referable to a Hargeisan industry.

GEDEIS (232). Implements and factory debris in typical white patina of Somaliland Stillbay form, but including also a small series of Hargeisan material.

[1] It is often difficult to distinguish between gouge-burins of this type and diminutive 'cores' flaked from both sides of an edge. Whether these latter are merely discarded cores is open to doubt, and the writer is of the opinion that they represent finished artifacts of burin type.

MANJASSEH (MANJA ASSEH) (233). Small series of discoidal prepared cores and faceted flakes of developed form; probably derived from a Somaliland Stillbay factory site.

DARBRUK (DARBURUK) (218). Stillbay and Levalloisian forms predominate; broken unifaced and bifacial points, scrapers, prepared cores and faceted flakes.

TUG ARGAN GAP (217). Typical small series of late Middle Stone Age material—predominantly Somaliland Stillbay.

BORAMA (5). Material is mainly quartzite and comprises four specimens which are probably Levalloisian and may have been derived from the Younger Gravels. Remaining specimens suggest a Somaliland Stillbay or later date.

DUBATO (230). Comprises one of the most important collections made by Trevor and consists predominantly of a typical and representative Somaliland Stillbay series. The specimens consistently show a thickish white patina with occasional calcareous concretions adhering to them. They were recovered presumably from surface exposures of red alluvium on the slopes of the Dubato valley for similar specimens were noted by the writer in similar circumstances when he visited the area in 1945. In no way did the deposits from which these implements were eroding differ from those found in the Marodijeh valley at Hargeisa further to the west. The Dubato specimens reflect the later or upper division of the Somaliland Stillbay Culture, but a few larger faceted flakes should perhaps be assigned on typological grounds to the Levalloisian, and two or three unpatinated microlithic specimens to the local variant of the Wilton Culture. The series comprises a number of flakes with faceted butts, small and well-worked prepared cores both discoidal, oval and subtriangular forms, scrapers and numbers of late forms of Stillbay unifaced and bifaced points—the latter containing some very finely made leaf-shaped, lanceolate and cordiform examples, and a few backed blades and burins. Fuller details will be found in Appendix A. The whole assemblage comprises perhaps the most important corroborative evidence outside the Hargeisa valley for the development of the Somaliland Stillbay Culture on the northern plateau.

UPPER SHEIK (SHEIKH) (141). Unfortunately the specimens from this site are all from the surface, but typologically and in view of the sequence now established, they are divisible into three main groups probably derived from three different horizons. They were recovered from extensive exposures of alluvium and gravel near the headwaters of the northern tributaries of the Tug Der.

(i) Acheulio-Levalloisian material (predominantly chert and some quartzite) consisting of small side-scrapers on flakes with unprepared striking-platforms and diminutive limande and ovate *coups-de-poing*; tortoise cores; a polyhedral stone ball; typical broad Levallois flakes (several small triangular-sectioned unifaced points showing resolved flaking of the edges may also belong perhaps); several typical U-shaped cleavers; pointed Micoque-type hand-axes, and ovates (one finely made specimen showing Z-twist). The majority of the pointed hand-axes have the butts unflaked and thick, while in the case of the ovates the butt is usually carefully trimmed and rounded; limande and pear-shaped specimens also occur as well as small subtriangular examples—a variant on the pointed Micoque-type. Like the Acheulio-Levalloisian material from the type site at Hargeisa the majority of implements were probably made from large flakes struck from prepared cores; the same type of flake tool occurs. Generally speaking, however, the implements from Sheik, made as they are from large nodules of chert from the Eocene limestones and not from the more refractory quartzite of the Cretaceous series, comprise some of the finest and most carefully made specimens the writer has yet seen from the Somalilands and are comparable to Seton-Karr's collections from the Issutugan Tug. In Trevor's collection from Sheik, as well as in that made by McInnes and described below[1] the bifacial elements predominate over the purely flake tools and, while the smaller discoidal core is present the large circular prepared cores are not represented. This is probably explained by the fact that it was in the main the more finished and recognizable specimens that were collected, particularly when the limiting factor of storage space is considered, and these sites if re-examined might be found to yield the missing elements. It must, however, be borne in mind that the bifacial element from Sheik is certainly of finer workmanship than that from Hargeisa which may not be entirely explained by difference in material.[2] It is quite possible, therefore, that the Sheik and Issutugan material, if indeed all the specimens included in these (?) Acheulio-Levalloisian series are homogeneous, represents an earlier phase of culture than is represented at Hargeisa with a predominance of well-made bifacial elements and less emphasis on the Levalloisian, and may in fact represent an Upper Somaliland Acheulian. The hand-axes and cleaver illustrated in Plates 3 and 4 are typical of the finest examples yet recovered from the Horn of Africa.

(ii) In addition to the Acheulio-Levalloisian series are a few specimens of Somaliland Stillbay and Developed Levalloisian points and faceted flakes that differ in no essential degree from other material attributable to these cultures.

(iii) Five unpatinated specimens should perhaps be referred to a Somaliland Wilton industry.

DAMERO. Specimens come from a low hill said to be situated beside the road between Burao and Erigavo.[3] They are typical of material found in great abundance in and around the edges of the Nogal depression. Usually factory sites, they are found littering the slopes round the outcrops of chert in this region. What are presumably the earliest specimens show signs of heavy weathering, long exposure on the surface, and the characteristic glossy dark brown or chestnut patina. They comprise nodules and truncated flakes, retouched, apparently haphazardly, by steep-angled flaking and notching of the edges. They probably indicate an early Levalloisian assemblage but weathering and mechanical flaking have made determination difficult.[4] Many of these glossy brown specimens have been reflaked at a later date and show a greyish brown patina. More recognizable forms are found and include

[1] See p. 36.

[2] Quartzite and silcrete were also used at Sheik, and more rarely quartz, limestone, schist, and porphyry: implements in these latter materials reflect in varying degree the more refractory nature of the rocks.

[3] A more exact position for this site cannot unfortunately be given. (? = Damero 8° 20′ N., 47° 24′ E.)

[4] They bear a superficial resemblance to certain primitive tools made on flakes and nodules from Rhodesia and termed the Hope Fountain Culture, but the resemblance is superficial only and no closer connexion is implied. See Clark, J. D. (1950, p. 79).

a discoidal prepared core and several flakes which, although having unfaceted butts, suggest more definitely a Levallois or proto-Levallois technique. This site has also yielded certain notched or eared artifacts which cannot as yet be associated with any definite industry or culture, but are certainly more recent than the Levalloisian material from this site. It may represent a factory or mining site of the Levalloisian.

ERIGAVO (140). With the exception of the notched implements the same forms are represented here as occurred at Damero, with the inclusion of definitely associable Levallois flakes: there is also an insignificant Late Stone Age element present.

(ii) *B. A. Phillips Collection*

Comprises Levalloisian surface material from south-west of Hargeisa (exact provenance not indicated) comparable to that from Aror and Damero. Notable is one notched or crescentic artifact similar to those from Damero and other sites.[1]

Similar material was collected by Phillips from Sheik and El Af (? = El Afwein).

(iii) *D. McInnes Collection.* See also under Reserved Areas and Abyssinia, p. 39.

UPPER SHEIK (141). Surface material comparable to that collected by Trevor. The Acheulio-Levalloisian series includes two well made pointed and strangulated Micoque-type hand-axes, pear-shaped, ovate and subtriangular examples; two round stone balls or bolas and several ovoid to spherical 'trimming' stones of Levallois type should probably be referred on typological grounds to a pure Levalloisian industry as the 'trimming' stone is a type tool of this culture complex.[2] From Manjasseh (Manja Asseh) comes a fine example of an Acheulio-Levalloisian ovate.

(iv) *P. E. Glover Collection*

This collection was made by Major P. E. Glover during 1944–5. It comprises, for the most part, surface specimens from the western half of British Somaliland, but of particular note and importance is the Magosian industry derived from root excavations at Alablah Balleh and summarized on pp. 32–3. Details of the Glover Collection will be found in Appendix A.

GERBAKELE (209). Recovered from the surface at the entrance to the third rock-shelter in Gerbakele Gully[3] are two *bec-de-flute* burins, two end-scrapers, a *lame écaillée*, a microlithic backed blade, a microlithic core, and one sherd of well-fired black pottery: the latter is probably later in date than the stone implements. Typologically referable to the Somaliland Wilton Culture.

SIRRAU (SIRAU) BALLEH. See p. 32.

HARGEISA/WARDERE (WARDER) ROAD 45 MILES FROM HARGEISA (200). Three specimens only, comprising a microlithic blade core, a microlithic double-backed blade or rod of Magosian type, and an eared-scraper.

GED DOBO, 13 miles from Manjasseh (202). Specimens collected at random on the surface on the old Oil Company's road include bifaced points referable to the Somaliland Stillbay

[1] See p. 294. [2] Examples are figured in Pl. 12, no. 11 and Pl. 14, no. 15.

[3] For description and illustration of paintings at Gerbakele and other sites discovered by Major P. E. Glover, see p. 301 and Text-fig. 35. Also Burkitt, M. C. & Glover, P. E. (1946).

and a unifaced 'backed point' typical of later Magosio/Hargeisan forms, as well as *coups-de-poing* and a cleaver of Acheulio-Levalloisian affinities.

LAS ANOD (181). Specimens assignable to a Levalloisian and a later series and comparable to the earlier and later groups recovered by Gilliland.

BARDERA (90). See p. 32.

DUDUBO (DUDUB) COFFEE SHOP (201). Between Hargeisa and Burao. Several typical Levalloisian flakes of developed type as well as specimens typologically assigned to the Somaliland Stillbay and Hargeisan Cultures.

DUKOKOLOL YERO (213). A small surface collection from the mouths of rock-shelters. Specimens are of Hargeisan and Somaliland Wilton affinities but include one typical large radially flaked Levallois flake in silicified sandstone, slightly abraded.

HALEYA (203), 5 miles east of Hargeisa. Here Glover recovered from *in situ* in the 'boulder bed' exposed in the wall of the Marodijeh Tug one subtriangular hand-axe made from a large flake, two crude prepared cores and one considerably abraded, faceted flake with thick section. This horizon is probably to be correlated with the Older Gravels II at Hargeisa, and the contained implements to the upper division of the Acheulio-Levalloisian.

SKUTAR (204). Specimens were found on a 'boulder bed' exposed by recent erosion in the Tug Der, about 20 miles north of Burao. Amongst other specimens may be mentioned a Micoque-type hand-axe, a U-cleaver, Levallois-type flakes, long blade-like flakes, and discoidal prepared cores apparently derived from the gravel on which they were found and typologically referable to the Acheulio-Levalloisian.

WARDERE (63). See p. 33.

JIFA URE (or JIFO URI) (205). Material was collected at the foot of the granite hill named Jifa Ure, 20 miles from Borama on the Hargeisa/Borama road. The material is of interest in that it is obsidian and was presumably transported to this site from a distance. The specimens have suffered due to heavy weathering on the surface and abrasion by natural causes. While the majority of the specimens are waste products from a microlithic factory site, recognizable implements include *lames écaillées*, microlithic cores, two broken straight-backed microliths, thumb-nail scrapers, mostly small, one small end-scraper, one crude backed blade, a well made small bifaced point of Magosian type, and one weathered fragment of micaceous pottery. The assemblage shows definite Somaliland Wilton characteristics, but the small Magosian-like point suggests an industry of Hargeisan/Wilton transition.

(v) *H. Lambert Collection*

This material comprises a large collection of factory debris and a small percentage of artifacts from a single home and factory site situate to the north of Hargeisa town (3) on the left bank of the tug. The specimens are washouts found lying on the surface on the side slopes of a small tributary tug: the bulk of the specimens can be referred on typological and physical grounds to the Hargeisan Culture together with a small number of specimens derived from earlier industries. The Hargeisan material includes a predominance of blades and long flakes showing longitudinal flaking, as well as blade cores. Implements include a *bec-de-flute* burin and angle-burins, end-of-blade scrapers, backed blades, typical triangular-sectioned unifaced points, and a number of flakes with faceted platforms. The assemblage forms useful corroborative evidence for the type sites on the right bank of the main tug.

(vi) *Gilliland Collection*

Details and discussion of this collection will be found in Appendix A, so that a summary only of the sites and the affinities of the material yielded by them is given here. With one exception all Gilliland's sites are situated in the eastern part of the Protectorate and the collections were made in the course of a plant and pasture survey in that area between 1944 and 1945. They are of particular interest in that they connect the better known sites in the western half of the country with those recorded by the Italians in the northern Mijertein and with the writer's own research in the eastern part of the Nogal valley from Bihen to Eil. The material used is almost exclusively chert from the Eocene formations. The most important site is the Governor's camp area: Las Anod (181). Las Anod is situated on the edge of the southern scarp of the Nogal depression on the gypsum-anhydrite beds of Eocene age which are exposed in the upper end of this valley. The material from this site is derived from scree at the base of two hill slopes, one at the side of the District Commissioner's house and the other about one mile to the north-east. That at the former site appears to be eroding from beds of secondary gypsum which are being cut into by many small erosion gullies broken irregularly by ledges on to which the implements have weathered out. From the description and photographs of the site it would seem that the deposit is a scree cemented in places by gypsum from which factory debris and a few implements showing a pink and grey patina are derived. No thickness of deposit was noted by Gilliland and there would appear to be no connexion with alluvial beds of the Las Anod tug. On the evidence of typology and *état physique* the material can be divided into two groups: the first group includes long blade-like flakes, or large true blades; broad flakes, some with faceted butts, a crude ovate 'hand-axe' or chopper, two roughouts for *coups-de-poing* or possibly for reduction into limande-shaped points, a good example of an oval prepared core, Levallois-type unifaced points, flakes struck from prepared cores and from which the bulb and platform have been intentionally reduced, and one limande-shaped bifaced point. This latter specimen, which shows a greater degree of patination by iron oxides than the other implements, should perhaps be excluded, but the writer believes, in view of the evidence from other sites (Jesomma, Hargeisa, etc.) that it should be included as an integral part of the assemblage. The collection is probably a factory and home site and is comparable with the Developed Levalloisian industries from the gypsiferous gravels and sands of Garoe Tug (54). Specimens are figured at Pl. 16.

The second group is not necessarily a homogeneous assemblage but is undoubtedly later in date than the first: it may be referred to the Somaliland Stillbay and later cultures. Specimens certainly referable to the Somaliland Stillbay come from Bohotleh (183), while of Gilliland's other sites those at Kirit (182), Merkanwein (or Marqanweina) Balleh at Bohotleh (183), Raguda Tug (185) and possibly Sirrau (Sirau) Balleh (199) yielded specimens of prepared cores, Levallois-type flakes, points and scrapers with resolved and pressure-flaked retouch of Somaliland Stillbay/Magosian facies. The greater part of the collection must be referred to the Late Stone Age, however, and comprises series from Kirit (182), El Dur Elan (184), Raguda Tug (185); Hudin/Ainabo road (186); Bur Khansa (187); Halin (188); Erigavo (140); Haded: Balad Ad (the western part, Balad Ad, of the Haded Plains) (189); Shimbir Beris (190); Las Anod (181); Hudin/Halin road (192); Sirrau (Sirau) Balleh (199); specimens from these sites are distinguishable by their predominantly microlithic flavour, blades, burins, and large discoidal and end-scrapers.

(*c*) RESERVED AREAS AND ABYSSINIA

(i) *D. McInnes Collection*

JIGJIGA (4). Weathered surface specimens comprising small subtriangular unifaced and bifaced points, faceted flakes and diminutive prepared cores, the secondary work on these specimens is crude and of a resolved nature—true pressure-flaking does not appear to be present. The implements resemble the pseudo-Stillbay industries from the end-Kamasian Pluvial lake or swamp beds on the Kinangop plateau of the Kenya Rift Valley.[1] The evidence, however, is lacking whereby we may claim any definite connexion other than that of typological similarity and it is more likely that any culture connexion if it is to be inferred, should be made with the end of the Stillbay sequence or with the Magosian as is found at sites in the Webi Shebeli valley.

MEGA. Although lying outside the area at present under consideration mention must be made of the collection of small and typologically late ovate and limande *coups-de-poing* made by McInnes from Mega comparable to some of the Sheik specimens of either Upper Acheulian or Acheulio-Levalloisian date. The distribution of the Acheulio-Levalloisian industries suggests that this culture is very widespread in the Horn of Africa, occurring as it does from Gondar in the north[2] to Mega in the south of the Abyssinian plateau, to Hargeisa, Sheik, Dagahbur and the upper Webi Shebeli in the north and west of the Somali plateau.

(ii) *W. B. Fagg Collection*

A small but valuable collection from the surface and talus of two rock shelters situate in Jurassic limestone formations near the foot of the Cercer escarpment which forms the south-eastern wall of the Danakil Rift.

GROTTE DE PORC-EPIC—NEAR DIRE DAWA (152).[3] The talus from this cave contained several flakes with faceted butts probably to be assigned to a Stillbay industry. The bulk of the material, however, must be referred to the Late Stone Age, and besides specimens typical of the Somaliland Wilton contains examples that exhibit Magosian features. Among

[1] Leakey, L. S. B. (1936, p. 52).

[2] Clark, J. D. (1945*a*).

[3] For summary of previous work in this cave, see T. de Chardin (1930*a*) and T. de Chardin, Breuil & Wernert (1940), pp. 497–522, and for record of paintings, Breuil, H. (1934), p. 31; also Breuil, H. *et al.* (1951).

the Wilton assemblage may be included a *lame écaillée*, thumbnail scraper, two pieces of haematite showing striation due to grinding or rubbing, a crescent, a backed blade, a large hollow scraper and end-scrapers on blades. There also occur small subtriangular bifaced points (in several the retouch is of a fine resolved nature, and is confined to the edges of the tool): they are characteristically Magosian. Also included is one well-fired potsherd showing a fine red paste: it is possible that this is later in date than the majority of the implements listed above.[1]

LAGO ODA (221)[2] situate approximately 50 miles to the south-west of Dire Dawa. The specimens are derived from the talus at the entrance to the shelter and, although a small series, include microlithic blade cores and some blades, three crescents, two small crude disc-scrapers on a core and a nodule, and a gouge-burin.[3] They can with confidence be assigned to a Somaliland Wilton industry.

(iii) *F. Moysey Collection*

Although outside the immediate area with which this book deals, the collections made by Lieut.-Col. Moysey from the Abyssinian plateau must be briefly mentioned here as later it will be necessary to refer to them when discussing the affinities of certain of the Somaliland Stone Age Cultures. Further details of Moysey's finds (which are housed in the Coryndon Museum at Nairobi) will be found in Appendix A.

The most valuable site is a rock-shelter situated at Gorgora at the north end of Lake Tana in Tigré Province. Here Moysey excavated and recovered a stratified sequence of industries which he described jointly with L. S. B. Leakey.[4] The lower half of the deposits yielded a Stillbay industry in which a gradual development and decline could be clearly traced. The characteristic tool of the industry at the height of its development is a finely made bifaced point markedly subtriangular in form. The upper half of the deposit yielded a poorly represented Magosian industry which in turn was superseded by a Wilton industry with pottery.

An industry from a rock-shelter at Quiha near the edge of the eastern or Red Sea fault-scarp also contains an interesting series referable on account of the presence of pottery, and the typology of the stone industry, to the Late Stone Age. The assemblage is predominantly microlithic, but contains large utilized blades, burins and disc-scrapers reminiscent of the Hargeisan Culture of the northern Somali plateau and of the Elmenteitan Culture of Kenya.

The remainder of Moysey's sites yielded for the most part late industries of Hargeisa/Wilton cultural affinities and the impression is given that the 'blade and burin' influence is much greater and probably earlier in Abyssinia than in the Somalilands where its presence can be shown to date to the later part of the prehistoric sequence. However, two sites

[1] For records of pottery associated with stone implements, see pp. 224 and 222 (Hargeisa: Mandera); at Bur Eibe and Bur Hakaba pottery definitely lies above the Late Stone Age levels.

[2] Lago Oda Cave also contains paintings which were photographed by W. B. Fagg. Details of these will be found on p. 297.

[3] See p. 33, n. 1.

[4] Leakey, L. S. B. & Moysey, F. (1943, p. 196).

(Wadelai Delanta and Amba Alagi) have yielded flakes with faceted platforms and small discoidal prepared cores which indicate the presence of industries of Levalloiso-Stillbay affinities.

(iv) In 1941 the writer had occasion to examine and collect from two areas in Abyssinia. One on the edge of the Dabat plateau to the east of Gondar which yielded an industry showing very close affinities to the Kenya Fauresmith and the Acheulio-Levalloisian[1] and a second site at Yavello[2] on the southern plateau where two small rock-shelters yielded cultural material—the first contained a Stillbay industry overlain by one of Wilton affinities, while the second yielded only a degenerate Wilton and some much-faded paintings of a schematic nature which were possibly associated.

(v) Graziosi[3] records certain additional finds from Abyssinia but in most instances they are isolated, unrelated surface specimens classifiable only on a typological basis.[4]

SUMMARY

From this record of evidence of previous investigators it can be seen that while the actual material is plentiful enough, the necessary stratigraphical evidence is, in the majority of cases, lacking, by means of which the cultural sequence could be fitted into the geological and climatic framework, so that any attempt at establishing the culture sequence must be based primarily on typology and *état physique*. When examined, however, in the light of the geological evidence from our own investigations, the majority of these surface collections can be placed in their right context and provide valuable corroborative material particularly as to cultural distribution.

It is evident that the earliest material from the Somali plateau includes the hand-axes and associated tools collected by Seton-Karr from the Issutugan, by Trevor & McInnes from Sheik, and by Glover from Hargeisa, and is comparable to the Acheulio-Levalloisian from Hargeisa and the Kenya (or Abyssinian) Fauresmith from Gondar. The absence of Lower Palaeolithic implements from southern and central Somalia is perhaps worthy of note. This negative evidence cannot as yet be taken to imply that the Acheulio-Levalloisian did not stretch

[1] Clark, J. D. (1945*a*).

[2] Clark, J. D. (1945*b*).

[3] Graziosi, P. (1938).

[4] Blanc, A. C. (1938), describes a series of finds from Moggio, sent to him by an Italian soldier. Specimens include 'points and scrapers of Mousterian type and also Upper Palaeolithic and Mesolithic implements such as gravette blades and crescents'. Blanc has classified them as Stillbay, Magosian and Wilton.

In 1938 a member of the Dainelli Expedition found a greatly weathered biface at Gondar. In the same year Professor A. Fantoli informed Graziosi that he had identified prehistoric sites at Axum, Addis Ababa, Bisciofta, Cleritta, Adamitullo, Lake Bisciaff, Isha Baidoa (Somalia) and other areas including Moggio and Harar. He stated that the majority of his finds were made from obsidian.

into these areas but rather that deposits likely to yield this evidence have been insufficiently examined and indeed the isolated finds from Dagahbur and the upper Webi Shebeli suggest that this culture will be found in these areas also.

Finds attributable to the Middle Stone Age are particularly plentiful and Levalloisian and Stillbay material are equally common throughout the plateau and coastal plain. In certain cases specimens can be accurately assigned to Levalloisian or Stillbay industries, but in the case of many of the Middle Stone Age surface collections a more exact classification is often impossible owing to the very slow development and essentially conservative nature of the long Levalloisian complex in our region, and in fact (as is evidenced by the factory debris) the Somaliland Stillbay is essentially a Levalloisian derivative which has adapted and developed controlled flaking and flaking by pressure.[1] The differences between the factory debris (to which so much of the Middle Stone Age material from those surface collections belongs) which is associated with the Levalloisian and the Stillbay is often very slight, therefore.

The Late Stone Age is well developed and finds of this period are, somewhat naturally, much more numerous than those of previous periods. Industries of Magosian type can be shown to be well established with distinct regional differences in the north and south: in the south and centre in particular it is well represented. Some local divergence in the succeeding cultures is apparent. Widespread in the north and west is a regional form of the Wilton Culture, while more restricted but more highly developed are certain industries of the south and centre, with small pressure-flaked points and limaces which cannot be identified with the Wilton complex.

[1] It is obvious that many of the bifaced and unifaced implements associated with the Stillbay and later cultures were not produced as a result of flaking by pressure but exhibit characteristics strongly suggestive of a controlled wood or bone technique; experiments by the writer in different raw materials confirm this view.

PART II

THE GENERAL GEOLOGY OF THE SPECIAL AREAS AND STRATIGRAPHY

CHAPTER 3

INTRODUCTION. WESTERN BRITISH SOMALILAND AND THE NOGAL VALLEY

INTRODUCTION

By reason of the wide extent of the area this book attempts to cover it has been decided to submit first the geological and stratigraphical evidence from each of the main areas studied, and after correlating these to pass in later chapters to a consideration of the contained cultural evidence. The geological, climatic and cultural sequence in the Horn of Africa is then compared with the sequence established in other parts of the African continent.

Six main areas will be described:

A. Western British Somaliland—including the north-western part of the plateau proper, the broken country between the plateau and the coastal plain in the western half of the Protectorate, and the northern Haud plain.
B. The Nogal valley—including in addition to the valley itself that part of the plateau lying within the eastern half of the British Protectorate, and the southern Mijertein.
C. The Webi Shebeli valley—including the evidence from the northern tributaries in the Ogaden, the 'Bur country' of the south-eastern coastal plain, and the southern plateau, between the Webi Shebeli and Juba rivers.
D. The Danakil rift.
E. The north coast.
F. The east coast—excluding the 'Bur country' of southern Somalia.

WESTERN BRITISH SOMALILAND

I. HIGH-LEVEL 'PLATEAU BEDS'

The earliest superficial deposits which rest on the solid geology of the western plateau are represented by pebble beds and conglomerates which occur for the most part on the higher parts of the plateau surface, and have been deeply dissected by the present river systems so that they now stand out as high-level 'plateau beds' above the existing valleys. They do not appear to have been faulted but there is evidence to suggest that they have been affected by warping and tilting of the plateau surface.

It is clear that these beds were at one time widespread as is shown by the extent of their present distribution although subsequent erosion has removed them entirely from considerable areas of the plateau. They are partly the result of deposition by water action and partly of disintegration of the underlying sedimentary rocks. They have been examined by the writer in the area between Aware (Awareh), on the south-western border between Somalia and the British Protectorate, and Hargeisa, between Hargeisa and Borama and again in the Laferug area near the foot of the Gulf of Aden fault scarp. In the first two areas they consist for the most part of well-rounded pebbles of Eocene limestone and chert—presumably derived from the limestone beds—about 3–6 in. in diameter set in a matrix of sandy calcareous cement of marl-like consistency in the lower levels: either red or yellow in colour. The surface of these beds is almost universally consolidated and is formed by a hard kunkar (kankar) bed or secondary limestone[1] of varying thickness (average 2–3 ft.) which passes gradually into the underlying less consolidated levels. The surface of the kunkar as well as the pebbles contained within it often show a red to brown ferruginous patination. Elsewhere, notably in the Borama area, the kunkar beds attain greater thickness and the pebbles and boulders are set in a hard calcareous cement. Numerous exposures of these beds occur in borrow-pits dug for road metal and between Hargeisa and Aware (Awareḥ) ridges of these beds are separated by wide shallow valleys filled with sandy reddish brown alluvium.[2] For about 10 miles north of Laferug the pebble beds, which lie unconformably at approximately 500 m. on the surface of the peneplain developed in the crystalline rocks of the Basement Complex, are often unconsolidated and rest on a flat surface elevated above the valleys of the minor tugs which are crossed by the main Berbera/Hargeisa road: these beds may, however, be later in date than those on the plateau. Similar deposits of pebbles, at approximately the same altitude, were noted on the south flank of the valleys of the Harirad Tug between Borama and Zeila.

Macfadyen records pebble deposits of a similar nature, notably in the Bihendula/Dagah Shabel area, and describes them under the term 'Older

[1] The terms 'calcrete', 'surface or secondary limestone', 'kunkar or kankar', and 'tufa' have been used to describe various kinds of calcium carbonate deposits commonly met with in the Horn. With the exception of tufa all these are considered to have been formed by the evaporation of ground water bringing to the surface, or concentrating at a particular level, carbonates derived from the underlying rocks. Their formation is taken to indicate a rainfall requirement of probably not more than 15 in. a year. The term 'tufa' has been reserved for those porous, calcareous deposits which are clearly the result of percolating water—from springs or seepage zones.

[2] Dr Macfadyen informs me that these Older Boulder Beds also occur capping the Nubian sandstones on the north of the Marodijeh Tug at Hargeisa.

Boulder Beds'.[1] They consist mainly of Auradu (Lower Eocene) limestone boulders and pebbles with a proportion of Archaean rocks. The boulders are commonly of from fist to head size and predominate over finer material: all are well-rounded. In places they are set in a sandy cement and elsewhere the whole deposit grades into hard red sandy marls and shaly sands which are reworked Nubian sandstone and Shabel Beds. They stand at altitudes of 500–550 m. at Dagah Shabel, while at Daghani they occur up to 594 m. and one patch lies undisturbed across the Dagah Shabel fault.

Parkinson[2] also notes certain high-level coarse conglomerates of a similar nature at several places in the Gadabursi country, where they lie at various levels above the present surface of weak and spasmodic erosion. They are also recorded by Stefanini as covering a large part of the northern Haud.[3]

These boulder beds and conglomerates, which are presumed to be the result of widespread erosion and weathering of the Eocene limestones and later rocks, are considered to be indicative of the long persistence of humid or pluvial conditions. No artificially fractured stone has so far been recovered from within these beds, although implements of all ages overlie them. This fact, coupled with their high-level position antedating the present drainage system, leads to the conclusion that they cannot be later than Lower Pleistocene and are probably of end-Tertiary Age.

II. RIVER VALLEY AND OTHER ALLUVIAL DEPOSITS

(1) HARGEISA AREA—MARODIJEH (MARODIJEHH) TUG

The deposits in this valley were examined at a number of points west of Hargeisa for a distance of six to seven miles upstream on the right bank.

The Marodijeh Tug, which becomes the Issutugan in its lower course, has cut through the inclined and gently folded plateau surface at Hargeisa a broad, comparatively shallow, valley of mature profile bounded on the south by a short scarp with well developed pediment. (Pl. 30, fig. 1.) The upper part of this scarp is formed by the Auradu and Allahkadjid facies of Lower Eocene limestones which overlie the Cretaceous sandstones in this area, and in the lower half of the scarp the soft sandstones and quartzite bands of the Upper Cretaceous are clearly visible. The river has entrenched itself some 370 ft. within its valley, the floor of which rests on sandstones of the Lower Cretaceous system. Text-fig. 3 shows a section across the valley at Hargeisa.

This initial period of valley cutting was undoubtedly the longest and most important erosional feature in the valley's physiographical development. Sub-

[1] Macfadyen, W. A. (1932, 1933). [2] Parkinson, J. (1932, p. 651).
[3] Stefanini, G. (1936).

sequent to this first and major erosional cycle the history is one of aggradations alternating with erosional phases which gradually diminish in intensity and which fail to erode the valley to its former base-level. During this initial period of erosion a valley was formed some 400 ft. below the surface of the plateau.

Pebble beds and breccias. At isolated points flanking the southern scarp between Haraf and Haleya and also on the northern scarp above Hargeisa occur eroded patches of boulder and pebble beds of a similar nature to the 'plateau pebble beds', except that some ironstone and quartzite pebbles from the Cretaceous sandstones are present in the valley beds: they are covered and cemented by a thick deposit of surface limestone or kunkar. No archaeological material has so far been recovered from them, but no good opportunity occurred of carrying out any detailed examination. From their position high on the flanks of the valley at altitudes of 175–210 ft. they clearly antedate the development of the 75 ft. terrace: at this time also were probably accumulated certain

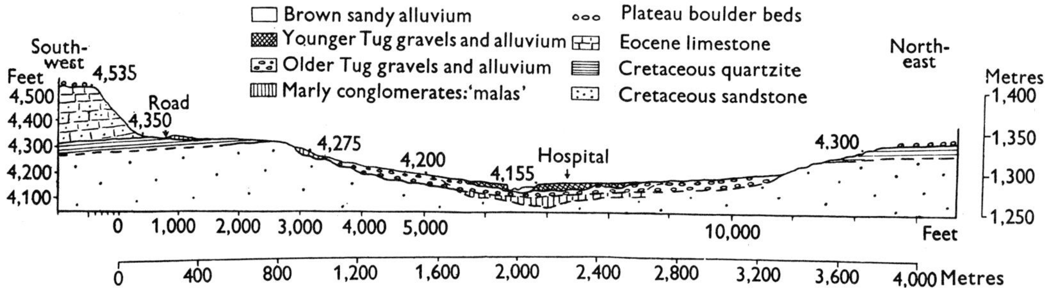

Text-fig. 3. Section through Marodijeh Tug, Hargeisa.

breccia deposits found at similar altitudes some 5 miles east of Hargeisa. They would appear to date to a major period of aggradation following upon the phase of initial valley cutting, and the same or similar deposits form the lowest infilling of the valley.

I am indebted to Dr W. A. Macfadyen for further details of these deposits, which he describes under the term 'Malas' (the local Somali name). Their thickness is proved in drilled wells to range between 14 and 35 ft. Poorly exposed, they consist of well-consolidated whitish marls and clays, with sandy and gravelly material and some Lower Eocene limestone boulders. These beds underlie the tug alluvium where they form the lowest aggradation deposits and overlie (or pass downwards through reworked beds of) the Nubian sandstone. The boulders contained within the 'Malas' are patinated with a light brownish glaze. Similar glaze was found on the gravels of the Webi Shebeli 15–30 m. terrace and on the 60 m. sheet gravels and boulder beds of the upper Nogal, where it was considered to be closely similar to a desert varnish.

The Older Gravels. Subsequent to the deposition of the 'Malas' beds there followed a further denudation phase, when the bed of the tug was cut to below its present channel, which was followed by the aggradation of gravels and finer sediments forming a terrace some 75 ft. on an average above the present river level, though these deposits extend to as much as 50 ft. above and below the 75 ft. level. Sections through the deposits are clearly seen in tributary tugs on the right bank which run through steep gorges into the main river and three of these in particular yielded quantities of artifacts *in situ*.

The deposits of the 75 ft. terrace may be divided into two parts. The lower part comprises a basal gravel (G. 1), coarse and unsorted and composed of subangular pebbles and boulders predominantly of limestone from the Eocene beds and clearly deposited by torrent streams under conditions of intermittent rainfall. At the two main sites where bed rock was visible this gravel rested unconformably on the decomposed and altered surface of the sandstone.

At the site H. 12 as much as 15 ft. and more of decomposed rock underlay the gravels (which are here cemented by a surface limestone or kunkar) before solid rock was reached, and the bluish broken-down sandstone is fissured and replaced by a brown-grey marl with numerous partially formed pisolites of calcium carbonate. This decomposition of the bed rock suggests that a longish period of time probably elapsed between its exposure and the deposition of the overlying sediments. The G. 1 gravels are usually found cemented by kunkar and, at site H. 1, this is exposed in the bed of the tributary tug and is of rock hardness. This gravel contains implements of a Lower Acheulio-Levalloisian industry.

Overlying the basal gravel (G. 1) is a brown to red sandy alluvium sterile of implements and deposited presumably under more arid conditions than those which permitted of the aggradation of the torrent gravels.

The lower alluvium is in turn overlain by a second deposit of torrent-bedded gravels, with little or no evidence of sorting and bedding, but is interstratified with coarse grits in which were found at site H. 12 several living floors of the Upper Acheulio-Levalloisian. These upper gravels and grits (G. 2) show evidence of having been deposited in mounds or banks which are considered to indicate conditions of infrequent rainfall and rapid run off. They are sometimes, as at site H. 8, overlain by a further deposit of red-brown alluvium or else, as at H. 1 and H. 12, by an unsorted detrital bed. The upper (G. 2) gravel and alluvial clays overlie unconformably the lower (G. 1) deposits and small shallow gullies filled with the upper gravel have cut into or entirely eroded the lower alluvium: there is evidence at site H. 12 near the European cemetery that the upper gravels cut more deeply into the underlying G. 1 gravels before passing from sight under the bed of the tug.

Overlying the upper deposit of gravels (G.2), or alluvium if this is present, is a detrital bed, represented by rubble and derived components of the gravels and taken to be indicative of a former land surface. It contains some derived Acheulio-Levallois tools and is dated by the unrolled Developed or Upper Levallois industry found within it to the second half of the succeeding phase of increased rainfall; on its surface and in the overlying red sand occur sealed factory and home sites of the Somaliland Stillbay Culture. This old land surface was probably formed and remained exposed during the whole of the latter half of the Last or Gamblian Pluvial.

The Younger Gravels. The onset of wetter conditions subsequent to the deposition of the Older Alluvium (G.2) is evidenced by a further denudation phase during which the Older Gravels were eroded in large measure from the centre of the valley and the river was again cut down to below the thalweg. This channel was not eroded to bed rock but rests on the 'Malas' and on the denuded Older Gravels. An aggradation phase follows and resulted in the deposition of a coarse gritty gravel (G.3) overlain by a thick deposit of red to brown sandy alluvium subdivided into two phases by the presence of grit lenses. The component parts of this gravel were largely derived from the eroded Older Gravels; rare derived Acheulio-Levallois and early Levallois tools are found. Implements of Developed Levallois type also occur, for the most part unrolled, while at the junction of the gravel and the alluvium occur a few flakes and implements of the Lower Somaliland Stillbay Culture. Much of this alluvium, certainly in the upper levels, appears to be of aeolian origin, and is contemporary with the red to buff sand on the upper slopes of the valley, beneath and in which occur habitation sites of the Somaliland Stillbay. Two good sections of these Younger Gravel beds are seen in Hargeisa itself. The first, where the upper half of the deposits is exposed is at site H.6, 500 yd. beyond Hargeisa hospital going towards the Somali town and on the left-hand side of the road. Here washouts and borrow pits have exposed sections of the red alluvium containing occasional unrolled flakes of a Stillbay or Upper Levallois facies and rare abraded Levallois artifacts. Again near the tug bank in the hospital grounds a well was sunk and has exposed the following section through the lower beds: at the top a black alluvium to 3 ft. underlain by a band of gravel 3–4 in. thick, red alluvium and gravel, separated by grit lenses from a lower gravel, the base of which was not seen: depth of lower beds was some 15 ft.

Post-Pluvial deposits. Following upon the phase of deposition of the Younger Gravels and of the apparently aeolian sand, there is evidence for a further semi-arid or wet phase of lesser intensity when channels were eroded by the tug through the red alluvium, which, however, failed to excavate the bed to its previous depth; small steep-sided tributary gullies were also cut through the earlier sediments at this time as is evidenced, for example, by the 5 ft. bench in the tributary tug at site H. 8. Within the channel cut by the main tug was laid down a thin gravel of small pebbles overlain by a sandy grey alluvium with occasional black to grey clay bands. A good section was seen at the beginning of 1944 at the brick pits at Hargeisa about ¼ mile east of the old Camel Corps camp on the right bank of the Marodijeh. The section showed the eroded surface of the red alluvium, with numerous kunkar nodules at the unconformity, dipping steeply towards the north where a deepish channel had been eroded through it. Within the channel was a deposit of black alluvium 10 ft. thick which, at one

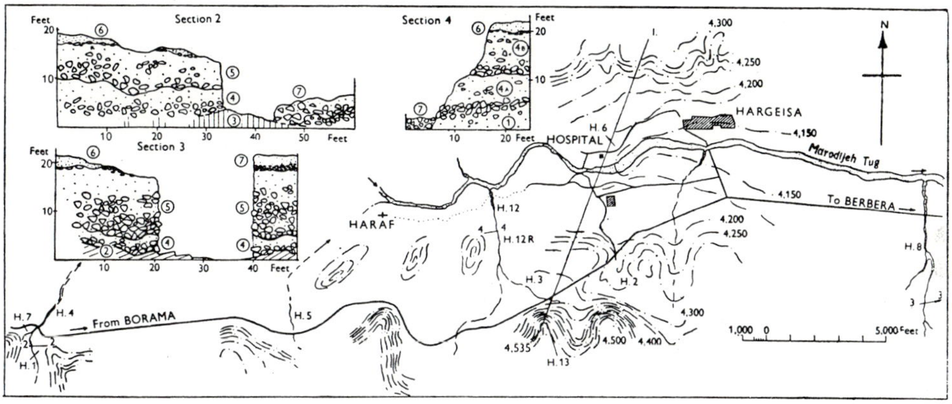

Text-fig. 4. Map and Sections, Marodijeh Tug, Hargeisa.

Key to Sections: Marodijeh Tug, Hargeisa.

1. Decomposed Cretaceous sandstone.
2. Quartzite (Cretaceous).
3. Marly conglomerates ('Malas' beds).
4A, 4B. Older Tug Gravels and Alluvium.
5. Younger Tug Gravels and Alluvium.
6. Land rubble and aeolian/alluvial sand.
7. Re-deposited gravel and gritty alluvium.

point near the north edge of the section, was overlain by a black clay some 5 ft. thick. The Hargeisan Culture is probably contemporary with the beginning of this period, but it dates also in part to the last half of the preceding arid phase: several habitation sites had been located on the south (H. 2, H. 8, H. 12 R) and one on the north bank (Lambert Collection).

The sequence is completed in this area by the present cycle of weak and spasmodic erosion during which the tug and tributary gullies were eroded to their existing contours. The Somaliland Wilton Culture, which is the natural derivative of the Hargeisan, is dated to the decline of the previous phase and to the present period. The detailed sequence at the various sites studied is given below while the more important sections are illustrated at Text-fig. 4; the position of the sites being indicated in the sketch-map.

Sites H. 1, 7 *and* 4. These sites are situated about 9½ miles west of Hargeisa on the road to Jigjiga where this crosses a small tributary tug which has exposed considerable sections of gravels and alluvium on both banks (Text-fig. 4).

The artifacts are derived from the red sandy alluvium (Younger Alluvium) resting on the detrital bed and Younger Gravels: H. 1 is situated a few yards to the left of the road on the right bank of the tug. Quantities of artifacts and factory debris of the Somaliland Stillbay are being eroded from the red sand and many were recovered from *in situ* where they lay partially exposed by sheet erosion. These Stillbay artifacts made from chert and quartzite show an almost consistently thick white patina with matt surface and it would seem that they lay exposed on the surface for a considerable period of time before being covered by a further deposit of red alluvium.[1] No clearly defined fossil horizon or floor is visible and the artifacts come from various depths within the upper part of the alluvium. Apparently coming from a slightly higher level are a few implements again with white patina but showing for the most part a glossy instead of a matt finish. They are considered to indicate the presence of an industry of Hargeisan facies. The uppermost part of the alluvium is altered to a fine grey sand due probably to the presence of humic matter. Within this grey alluvium are found working floors of the Somaliland Wilton Culture with a number of microlithic blades and small cores, but very few finished tools.

Site H. 4 is situated on the same side of the tug as H. 1 and some ¼ mile to the right of the road. Here sheet exposures of the red alluvium have washed out from the upper levels numbers of finely made points and other artifacts which together with those from the previous site can, on typological grounds, be referred to an upper division of the Somaliland Stillbay.

Site H. 7. Just over ¼ mile on the far side of the drift on the south side of the road and the left bank of the tug are extensive exposures of the red alluvium overlying the Younger Gravels. Large numbers of Stillbay artifacts were recovered from the alluvium in the main from the lower half immediately above the gravels and are less developed than finished implements from the upper levels. They are accordingly considered to represent an earlier cultural stage and have been described under the term Lower Somaliland Stillbay. Associated with these artifacts was a very sparse fauna.[2]

The grey uppermost part of the alluvium contains several working floors of

[1] It would seem that this bed, containing as it appears to do a high percentage of aeolian sand, was affected by sufficiently strong wind action to cause shifting of the sand which resulted in the exposure, and reburial of the old surface and the polishing of some of the artifacts found within it.

[2] See Appendix F.

the Hargeisan Culture, one of which is seen in the photograph (Pl. 32, fig. 3). The gravels at these sites have yielded a few specimens usually showing evidence of abrasion by water action and referable to a Developed or Upper Levalloisian stage of culture, together with one derived example of an Acheulio-Levalloisian *coup-de-poing* made from a large prepared lava flake.

Site H.2. The situation of this site can be seen from the map. Here the red to buff alluvium has been eroded exposing numerous implements belonging to what is typologically a derivative industry of the Hargeisan. Particularly noteworthy are certain rare examples of hollow-based point. The buff surface deposit at this site has yielded a few Somaliland Wilton crescents, in addition to the characteristic Hargeisan forms, one fragment of pottery and a nodule of haematite with signs of rubbing or polishing. (Pl. 30, fig. 2.)

Site H.3. A similar site to that just described, it has yielded some fine examples of the Upper Somaliland Stillbay *in situ.*

Site H.5. A small tributary tug has exposed a similar sequence to that recorded from previous sites: artifacts are not as prolific as at other sites.

Site H.6. This is the site previously referred to near Hargeisa hospital.[1]

Site H.8. (Pl. 31, fig. 2.) An important site where exposures of the Older Gravels are well seen for over a mile up this tributary tug.

The basal gravel (G.1) overlying disintegrated bed rock has yielded *in situ* a small but valuable series of Acheulio-Levalloisian artifacts. The grey to brown kunkarous alluvium overlying it is sterile and has been considerably eroded or else removed entirely by the overlying beds. The overlying gravel contains no recognizable implements in an unrolled condition, but a few heavily abraded flakes are probably derived from the lower bed while some are comparable to certain crude forms found with the Lower Levalloisian of Somaliland. In places there is evidence of a gravel and much denuded alluvium interposed between the two gravels and it is probable that this represents the remains of the G.2 gravel seen better at site H.12. Time did not permit of a very detailed examination so that it is not possible to state for certain whether this indeed represents the second aggradation phase (G.2) or merely denotes temporary change of course of the tug and seasonal variation in the deposition of the Older Gravels. The alluvium overlying the upper gravel is red to brown in colour with nodules of kunkar in places, and this gravel is considered to represent the first depositional phase of the Younger Gravels (G.3). The alluvium is sterile except in its uppermost levels and at the junction with the overlying bed where numbers of Stillbay artifacts of upper facies and typical thick white patina are being eroded. A characteristic of the Stillbay and Levalloisian artifacts from the

[1] See p. 50.

majority of open station sites of this nature in the Somalilands is the presence of concretions of lime adhering to them. Locally overlying the red alluvium and Stillbay living floors is a loose gravel of small components which fills shallow gullies in the underlying bed and slopes steeply towards the bed of the tug. No implements were found within it, but it is overlain by a grey to dark brown alluvium of fine texture which covers in places small living floors attributed to the Somaliland Wilton Culture. An aquatic origin for this alluvium is suggested.

*Site H.*12. (Pl. 31, fig. 1.) This is found in the lower reaches of a small tributary tug draining into the Marodijeh on the right bank, the source of which is just to the north of the Hargeisa/Jigjiga Road at the Customs Post, and the junction with the main tug a short distance upstream from the European cemetery.

The Cretaceous sandstone outcrops in the bed of the stream at two points in this tug while at one place a residual spur protected by a capping of quartzite has been dissected by the present channel and the Older Gravels can be seen in sections banked against it, and cemented by surface limestone.

The normal sequence found here comprises: (i) the G.1 gravels resting on bed rock and containing a good series of Acheulio-Levalloisian artifacts, the majority of which are fresh or only slightly abraded; (ii) a brown alluvium sterile of artifacts with pellets of concretionary kunkar formed *in situ*; (iii) bands of pebbles and boulders interbedded with coarse and fine grit bands (G.2) and alluvium in which are living floors of the upper division of the Acheulio-Levalloisian; (iv) sometimes a thin deposit of brown alluvium overlies these grits, but in the majority of sections the grit bands are overlain directly by (v) a rubble bed of derived material representing a fossil land surface as in it are found rare derived Acheulio-Levalloisian artifacts as well as fresh and unrolled tools typical of the Developed or Upper Levalloisian; resting on this bed are living floors of the Somaliland Stillbay; (vi) overlying this pebble bed is a sandy grey alluvium of fine texture. In the lower part of this, are found some implements of Stillbay facies distributed sporadically throughout the area examined. In addition, however, there was found, at one place (H.12 R) in particular, living floors with an assemblage of artifacts comprising blades, blade-cores, end-scrapers, burins, straight-backed blades and microliths. Notably there was here an almost total absence of typical Stillbay forms, and there is little doubt that these living floors represent a single homogeneous industry—a primarily blade and burin assemblage—which throws important light upon the small groups of typologically similar artifacts which had been found at other sites in north Somaliland, and which had up till then formed puzzling groups of unassociated tools. A collection from a surface site on the left bank of the main tug at Hargeisa yielded to Lambert a large number of blades and waste flakes as well as a few implements identical with those from our own site H. 12 R; it was clear, however, from an examination of the material from Lambert's site that it was primarily a workshop site incorporating a percentage of derived artifacts. Artifacts in every way identical with those from Hargeisa were collected by Barrington Brown from Hormo on the banks of a tug at Buran in the eastern half of the Protectorate; and a representative series are figured at Pl. 21, nos. 15–29. We are forced to recognize, therefore, that these blade and burin industries are indicative of an invasion by new technical and typological forms of country that had previously been a province of the Levalloiso-Mousterian Culture complex. The H. 12 R site at Hargeisa is situated some 50 yards from the edge of the tug wall, where sheet and gully erosion is beginning to remove the alluvial deposits. Implements and flakes were found *in situ* and eroding from the red to grey upper alluvium which here attains a depth of (on an average) not more than 2–3 ft. and rests on the detrital bed with associated Stillbay and earlier artifacts. Implements are found in the alluvium from the base almost to the top but the greater number were found in the lowest levels immediately above the 'gravel'. Associated with the blades and scrapers should be included a few Magosian-type points. They are commonest at the south end of the site, but may in part be derived from a Late Stillbay site found here. At one place a small cache of ostrich egg-shell was found which probably formed part of an egg-shell container as several of the pieces fit together.

Two further sites at Hargeisa must be mentioned. The first, situated on the old aerodrome near the right bank of the Marodijeh, has yielded three crude forms of scraper derived from the surface of the grey-brown alluvium (G. 2): these tools bear little resemblance to the Somaliland Wilton scraper, but must, if they are indeed to be included in the Stone Age succession, be relatively late. The second, named Fly Hill after its Somali name, is the hill immediately to the south of the main Hargeisa/Mogadishu road and forms a prominent bluff on the southern scarp. The scarp here shows a good section through the Eocene and Upper Cretaceous beds. Below a hard and resistant capping of Auradu limestone the softer Allahkadjid series has been undercut primarily as a result of wind action and from here a steep scree and talus falls to the foot of the bluff; the scree in the lower levels may be cemented by surface limestone or kunkar. The soft undercut yellow clayey sandstones or marls yielded nodules of chert which have been extensively used as raw material by Middle and Late Stone Age man. Factory debris litters the slope, the earlier artifacts, always patinated white, being found for the most part in the lower half of the slope and at the foot. The Late Stone Age debris is always fresh and found usually in the upper part of the scree; a small overlying floor was found on the flat surface of the Auradu limestone cap on the top of the bluff. It would seem, therefore, that the source of the undermining of the upper half of this bluff may have been caused by man extracting the chert, which in the late Middle and Late Stone Age times was considered a more satisfactory raw material for the production of delicate and specialized tools than the quartzite which occurs in quantity in the lower part of the valley.

(2) BORAMA DISTRICT

The solid geology here shows evidence of considerable earth movement and tilting, the outcome no doubt of the major rifting which developed the Gulf of Aden and East Africa fault trends. Steep often V-sided valleys, evidence of an early and powerful period of erosion, dissect the plateau with its high-level pebble beds, and drain northwards to the coastal plain and the Gulf of Aden. These valleys have been incised deeply within the crystalline rocks of the Basement Complex and the later sedimentaries further to the north. A short examination of the deposits in the small tug at Borama was made some 4 miles upsteam from the town (Text-fig. 5). Here the steep-sided V-shaped profile of the valley has cut through, on the right bank, tilted limestones probably of Jurassic age which overlie sandstones. These beds, some 30 ft. thick, rest on metamorphized schists of the Basement Complex. At one place at a height of 35–45 ft. above the bed of the tug which was visible at this point were found banks of well-rounded quartz pebbles, sometimes in a calcareous matrix, which may in part have been derived from the base of the sandstones: similar quartz pebbles rest on the surface of the tilted limestones, however, and while quartz is the predominant rock, occasional pebbles of (?) diorite and limestone also occur. They represent the first phase of aggradation in the morphology of the valley but were sterile of artifacts. Following upon this aggradation phase was one of continued and deep valley cutting when the stream entrenched itself a further 40 ft. or more within the Basement Schists; the altered bed rock seen in section in the bed of the tug seems to indicate weathering of the rocks due to a time-lag between the deepening of the valley and the deposition of the alluvial deposits which fill it.[1] These are clearly divisible into two phases. An earlier (G. 1) in which some 8 ft. of angular, ill-sorted gravel resting on bed rock is overlain by some 10–12 ft. of red-brown alluvium, and a later (? G. 3) angular gravel some 5 ft. thick is overlain by approximately 10 ft. of brown to grey alluvial sands. The junction of the two depositional series is accompanied by calcification which has resulted in the formation of sheet kunkar in several places. This capping of

[1] Parkinson, J. (1932, p. 651). Confirms the altered nature of the hornblende schists in this area.

kunkar has to a certain extent protected the underlying beds from subsequent erosion and a bench has formed at this level. The third erosional and depositional phase is not seen in this section but at Borama itself in a section at the back of

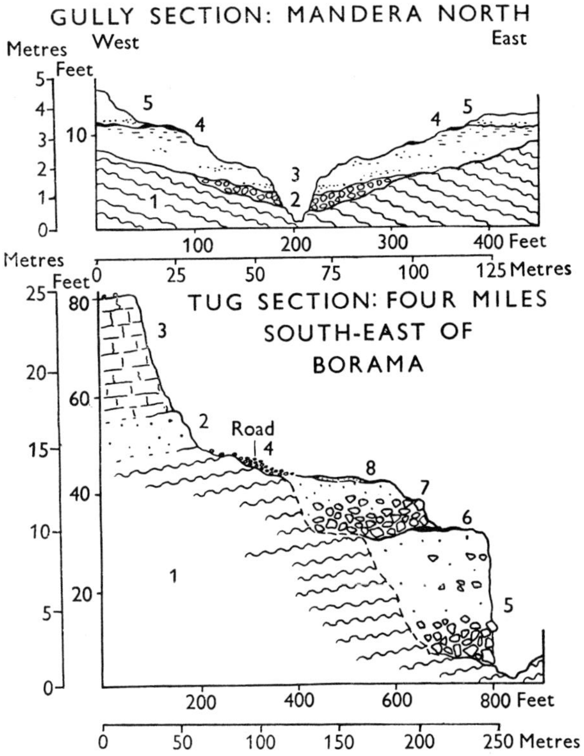

Text-fig. 5. Sections, Mandera North and Borama.

Key to Sections.

Gully Section: Mandera North

5. Aeolian/alluvial sand.
4. Kunkar and land rubble.
3. Younger Tug Alluvium.
2. Older (?) Tug Gravels.
1. Granite gneiss.

Tug Section: four miles south-east of Borama

8. Land rubble and aeolian/alluvial sand.
7. Younger Tug Gravels.
6. Kunkar.
5. Older Tug Gravels and Alluvium.
4. Quartz gravels.
3. Eocene limestone (?).
2. Cretaceous sandstone.
1. Hornblende schist.

the hospital there is evidence in the tug of a minor depositional phase resulting in the aggradation of a low terrace some 3–8 ft. thick, composed of loosely compacted gravels overlain by a brown to grey sandy alluvium. This was followed by the present cycle when the tugs were eroded by gullying to their present depth. At Borama no time was available in which to examine the basal gravels, but the upper deposits yielded Stillbay and later artifacts. The Stillbay implements came from the base of the brown-red alluvium (8) resting on the gravel, while implements comparable with the Hargeisan industries and with glossy patina were derived from the upper levels of the alluvium.

In the course of a journey from Borama to Aisha an opportunity occurred of confirming that the same sequence of alluvial deposits was to be found in the tugs of the north-western part of the district: notably in the valley of the Harirad Tug[1] between Borama and Abdul Kadir. Here a wide deep valley has been eroded in the Jurassic and earlier sedimentary rocks. The highest river valley deposits comprise scattered patches of quartz pebbles stained to a chestnut brown. Below these is the main Pleistocene terrace some 30–50 ft. above the bed of the tug and below this again is a further terrace 5–10 ft. above the thalweg. No opportunity occurred of examining these beds for artifacts.

(3) MANDERA AND DISTRICT

Mandera is situated at about 2200 ft. at the foot of the main Gulf of Aden fault-scarp on the much dissected peneplain surface of which the Berbera coastal plain is the continuation. This peneplain cuts across steeply dipping Archaean rocks, and a similar surface is widely developed in East Africa: it is now usually considered to be Pliocene in age.[2] Rivers have incised valleys—usually with youthful profile—within these crystalline rocks, and a good example of the topographical features of the peneplain between the coastal plain proper bounded by the faulted sedimentary rocks of the Dubar hills, and the foot-hills of the fault-scarp at Mandera is seen between mile 20 on the Berbera/Hargeisa road and Laferug. On the right of the road, travelling from Berbera, the peneplain is bounded by a scarp composed of some 300 ft. of dipping schists which have been planed, and possibly correlates with an earlier peneplain surface recorded by Dixey from the base of the Abyssinian plateau above the Eritrean plain.[3] These planed schists are overlain by some 40 ft. of dark coloured lavas often vesicular in structure.[4] On the left of the road the Pliocene peneplain stretches as far as the eye can see to the east only relieved by occasional residual masses, forming isolated hills or 'burs' which rise abruptly from the plain. In places also the surface of the plain is overlain by pebble beds of varying thickness,[5] which are anterior to the present drainage system, as they are dissected by numerous small steep-sided ravines or tugs. At mile 22 a gorge some 20–30 ft.

[1] Given to me as Dur Dur Ad Tug by Somali informant.

[2] Dixey, F. (1946, pp. 339–88).

[3] Dixey, F. (1946, p. 354).

[4] Macfadyen, W. A. (1933, p. 13).

[5] See p. 46.

deep has been cut through the schists and has been filled by one, sometimes two deposits of angular gravels, banded grits, and brown alluvium. In the tug in question the lower gravel rests on residual deposits resulting apparently from weathering and disintegration of the schists and gneiss; it comprises sometimes as much as 10 ft. of laminated white calcareous alluvium and kunkar containing angular fragments of rock at the base forming a typical calcrete; from this veins of kunkar descend vertically into the underlying rock. The alluvial deposits in the tugs are best seen in the tug at Laferug, the Hinweina (H'enweina). Here the same scarp of lava overlying Archaean rocks bounds the tug on its left and west bank. The surface of this today waterless plateau, which is covered by residual lava boulders and pebbles, has yielded a sparse Late Stone Age factory debris which contributes to the evidence that the Late Stone Age was contemporary with a period of increased rainfall. Within the Hinweina are found two series of deposits of gravels, false-bedded sands and brown alluvium, the upper of which is sometimes overlain by a detrital gravel. Overlying these alluvial deposits are compact dune sands, buff to yellow in colour, possibly evidence of an arid phase. The dunes were subsequently partially eroded and much of the sand leached and redeposited in the bed of the tug. Further wind action during the south-west monsoon (the Karif) at the present day has formed shifting dune sands which encroach upon the road. For the most part these tugs are sterile of implements, but a few Middle and Late Stone Age specimens have been recovered from the surface and eroded probably from the brown alluvium of the upper series.

A good collection of artifacts was recovered from a small tributary gully of the main tug at Mandera (Text-fig. 5; Pl. 32, fig. 4). This commences immediately to the north of Mandera Camp and was examined between the camp and the bakery, approximately ½ mile to the north. The sequence of deposits from the top downwards is as follows. A red predominantly aeolian alluvium grey at the surface due to the presence of humic matter, yielding from the top 6 in. unpatinated Somaliland Wilton artifacts and from the deeper levels a white-patinated Hargeisan-type industry together with a few upper Stillbay points, these latter probably from the base of the alluvium. Apparently in association with the microlithic blades were two pieces of pottery, one with an incrustation of lime similar to that found on several of the implements. This red alluvium usually rests on a detrital bed of pebbles and finer 'gravel' forming in places a typical desert pavement and indicative of a buried land surface, but near the bakery this 'gravel' attains a depth of several feet and is probably deposited by torrent action; from the surface come a few weathered, and possibly rolled Levalloisian artifacts but, with the exception of a worked nodule of chert forming a crude point, the lower levels are sterile. Elsewhere in the exposures, however, notably near the head of the gully this 'pavement' is absent and the red alluvium rests directly on the underlying bed. The red alluvium is deepest near the camp and has been almost entirely eroded near the bakery: it is made up of sand grains finest near the surface but with no evidence of bedding, and pisolitic kunkar at the base. The greater part of the sand suggests deposition by aeolian action but near the camp the lower levels contain a certain percentage of grit which may be indicative of more humid conditions. Underlying the red alluvium, and the detrital bed where this is present, is a grey alluvium with definite evidence of even bedding and containing a high percentage of clay in the upper levels as well as kunkar tending to form vertical pipes. The lower levels are composed of grits and false-bedded fine gravels with an increasing proportion of kunkar. The base is often completely cemented by kunkar showing horizontal piping and consists of a coarse unsorted gravel which rests directly upon the eroded granite rock: near the bakery a greater depth of this basal gravel is seen. The grey alluvium and basal gravel yielded no artifacts, it possibly represents the Older Gravels: the Younger Gravels and Alluvium are represented by the red alluvium.

(4) DAGAH SHABEL/BIHENDULA AREA

Although the writer had no opportunity of visiting this area a brief summary of the sequence found by Macfadyen is given here.[1]

Subsequent to the deposition of the Daban conglomerate which forms the top of the thick Daban fresh-water series, posthumous faulting in the Gulf of Aden trend took place along the Dagah Shabel fault and affected the Daban Series.[2] Next were deposited the Older Boulder Beds standing some 100 ft. above the tug beds, the material being derived from the high ground to the

[1] Macfadyen, W. A. (1932, p. 433).

[2] Macfadyen divides the Daban series, the occurrence of which is apparently limited to the country fringing but within the Gulf of Aden rift, into three phases. The Upper and Middle Daban beds are a fresh-water series of shales and conglomerates with boulders and pebbles of Auradu limestone (Lower Eocene) 1945 m. thick in the section at Bio (Biyo) Gora. Elsewhere it includes flint bands and a silicified band with a species of fresh-water gastropod, ostracod marls and fossil wood; the thickness is said to be at least 270 m. The lower beds of the series are intercalated with fossiliferous marine Middle Eocene beds near the base. During the formation of the Daban series a fault cliff existed south of the main Dagah Shabel fault. The upper series which are considered to represent fresh-water lake beds are dated only by means of the fossil wood—*Capparidoxylon geinitzi* Schenk—and this may be of Lower Oligocene age, but from the extensive time range of such fossils such a date must remain provisional only. Macfadyen considers the apparently extraordinary thickness of these lake beds to be of special note, and difficult to account for satisfactorily; but finds it equally difficult to believe that the outcrops could be merely superficial and arranged *en échelon* in descending steps. A comparison with the Lake Yaya deposits discovered and named by Nilsson in the Gojam on the Abyssinian plateau is not without interest. Lake Yaya is considered by Nilsson to date probably to the Lower Pleistocene.

south. There was then established on this gravel-covered area a drainage system independent of the underlying solid geology and during what must have been a long wet period, several tugs flowing north cut channels and gorges across the strike and against the dip of the underlying beds. Within the valleys so formed there were next deposited a younger series of gravels mainly composed of pebbles of Archaean igneous rocks and Auradu limestone and locally largely made up of reworked Cretaceous beds. At Dagah Shabel these Younger Gravels fill the tug to a height of some 30 ft. above the present bed. A later phase of erosion removed the Younger Gravels from the centre of the valley and the photograph reproduced by Macfadyen appears to show a further low terrace a few feet only above the bed of the tug.

(5) NORTHERN HAUD BETWEEN HARGEISA AND AWARE (AWAREH)

The northern half of this plateau country comprises long stretches of surface pebble beds resting on Eocene limestones and cemented by a thick deposit of surface limestone or kunkar. Three wide and shallow valleys cross this country from west to east filled with a red-brown sandy alluvium of unknown depth. Parkinson[1] records that a borehole put down 50 miles S.E. by S. of Hargeisa passed through 112 ft. of alluvial clay before reaching the white and brown marls and limestone from which it was presumably derived. It contained, in this bore, no pebbles, and its deposition was concluded to be of a lacustrine nature. These plains, Erankasare, Khansaha Harbodleh, and Khansaha Au, are now bare of watercourses with the exception of a few shallow gullies which carry the sparse run-off from the valley sides and lose it in the sands of the central plain. The conformation of these valleys or depressions suggests perhaps some tilting and warping of the plateau surface from west to east. What evidence of drainage exists today is towards the east and the Nogal depression, but it would seem that the gradual desiccation of the Horn of Africa, which occurred at the end of Pleistocene times, did not permit of sufficient rejuvenation of those valleys and the establishment of permanent or semi-permanent rivers. It is suggested that they may represent the silted valleys of rivers once flowing towards the west or north-west, but which have suffered from a reversal of the drainage system.

Near the southern edge of the Khansaha Au plain were recovered *in situ* from the top of the red alluvium several Magosian and Stillbay-type artifacts which help to date the uppermost series of these beds to the period of deposition of the red alluvium and Younger Gravels in the tugs. South of Aware commences the bright red sandy soil of the Haud proper.

[1] Parkinson, J. (1932, p. 651).

(6) JIGJIGA

This town is situated in the Marar plain below the Marda pass through which runs the main line of communication between Harar and the east. The hills bounding the Marar plain on the west are composed of faulted and folded limestones and sandstones with local volcanic intrusions. On the slopes of these hills and also on the surface of the flat grass-covered plain at Jigjiga itself diminutive faceted flakes, discoidal prepared cores and subtriangular unifaced points can be picked up. They show varying degrees of patination from light brown to mottled colouring and are usually weathered due to surface exposure. On a typological basis they compare favourably with the Magosian industries further to the south and east. While, however, the resolved 'stone' retouch is clearly seen, the finer tools associated with the Magosian do not seem to occur but their absence may in fact be explained by the nature of the site itself. There is a certain resemblance to the 'pseudo-Stillbay' from the Kinangop beds in the Kenya Rift valley, but the superficial nature of the brown alluvium from which these tools appear to be derived does not substantiate a date as early for the Jigjiga artifacts.

SUMMARY

Correlating the evidence from north-western Somaliland, therefore, the following sequence of events and deposits is broadly established:

(1) Deposition over a long period of time of plateau pebble beds, which antedate the present drainage system. Previous to this period, faulting in the Gulf of Aden Rift trend had already taken place.

(2) A major erosional cycle and the cutting of deeply incised valleys possibly accompanied by earth movement and tilting.

(3) The aggradation of marly pebble beds and breccias, capped by a thick deposit of surface limestone within the valleys, the 'Malas'.

(4) Erosion of an extensive nature to below present tug bed level followed by the deposition of one, sometimes two, torrent-bedded gravels and grey-brown alluvium (the Older Gravels) containing Acheulio-Levalloisian implements; these deposits were calcified during a semi-arid to arid phase.

(5) Renewed erosion followed by the deposition of gravels and red alluvium (the Younger Gravels) containing Developed Levalloisian tools in the lower levels and Somaliland Stillbay artifacts at the top of the upper red alluvium, which is overlain by some aeolian sand.

(6) A minor phase of erosion and deposition of a coarse low-level gravel and alluvium.

(7) Existing minor erosional phase.

THE NOGAL VALLEY

The Nogal sequence may be divided into two contemporary series of events, that which is evidenced in the upper Nogal, and that found in the gorge of the lower course of this tug.

I. THE UPPER NOGAL AND MIJERTEIN

(1) HIGH-LEVEL 'PLATEAU BEDS'

The plateau to the south of the upper Nogal between Badwein and Donkukoq is characterized by flat, very gently undulating country relieved by an occasional line of low hills, or by shallow valleys of senile profile forming plains of red sandy alluvium. The general direction of these valleys is north-west/south-east and such gully erosion as now exists indicates a south-east direction of flow towards the Badwein depression and Geriban. The rock which outcrops in this area is Eocene limestone as well also as marls and gypsum of the anhydrite series, though for the most part these latter are obscured by the limestones, immediately to the south of the valley, but are exposed within the valley itself and again to the south in the region of Badwein in the northern Haud. Overlying these rocks on certain parts of the plateau are pebble beds, particularly nodules of chert, which are derived presumably from the Eocene Taleh Series, and represent a residual deposit left after the erosion and decomposition of these sedimentary rocks. They should be equated in all probability with the plateau pebble beds noted in north-western Somaliland. Similar deposits are recorded be Stefanini[1] and Parkinson from certain parts of the high plateau of the Mijertein and are ascribed by them to a similar origin. Stefanini states that these residuals quickly undergo a characteristic process of alteration and weathering so that the nodules appear red or dark brown due to ferruginization, and while, as a rule, they do not show much sign of transportation they have frequently been fractured and split into flakes by thermal action: they also provided an almost inexhaustible supply of raw material for prehistoric man. Stefanini has observed these beds in the coastal areas of the eastern Mijertein, in Nobir, Abob, Harar, Hor and Deh—that is to say over most of the plateau between the Nogal and the Darror (Daror): they are absent only in the river valleys. Patches of similar pebble beds occur to the north and west of Dubo near Gardo, in some parts of the Karkar, on the north side of the Darror, and on the high plateau between the Darror and Giaheel. Stefanini also observes that it is thought that a considerable part of the Haud is covered by these formations

[1] Stefanini, G. (1936, p. 83).

underlying the superficial covering of bright red sandy soil. The present writer confirms this observation in so far as the well areas in the Haud are concerned, where deep sections can be observed in the wells themselves. The nodules and pebbles usually show some evidence of rolling, however, particularly in the southern Haud.

In the neighbourhood of Burao these residual and aeolian sands and sandy soils attain a depth of 20–25 m. before the limestones are reached, but they thin out from north to south until at Gherlogubi in the Haud the calcretes and pebble beds again come to the surface. These pebble beds are closely analogous to the high-level beds of a similar nature from the eastern Mudugh (Mudug). In the Karin Gap, through the Gulf of Aden fault-scarp, these beds obscure the underlying geology and are considered by Hunt to have been formed in that position at an altitude of approximately 200–500 m. by the draining into the the Gulf of Aden of the Darror depression through the fault gap. Similar deposits are seen again in the Dudub gorge to the west, which drains the El Dur Elan plains northwards to Ankor and it would seem, therefore, that uplift and tilting away from the rift must have accounted for the later development of these rivers. These plateau pebble beds were not seen in the eastern and lower part of the Nogal and it is presumed that the erosion that has taken place in this part of the plateau has removed them.

(2) VALLEY DEPOSITS

Dissecting this widely developed and evenly tilted peneplain surface in a south-east direction is the Nogal valley. Exactly how this vast depression came into being is uncertain but it seems likely that much of it represents a solution basin in folded strata, and is probably of considerable age, the Eocene limestones and anhydrite series lending themselves readily to erosion of this nature, and in fact it would seem that the overlying limestones have been eroded in fairly recent geological times. The Nogal, as also the Darror, shows an interesting reversal of the normal valley form in that it is wider towards its source than at its mouth, the reason for which is probably to be sought in earth movement and river capture: a similar state of affairs was noted by Caton-Thompson & Gardner in the Wadi Hadhramaut in southern Arabia[1]—in fact in some ways these two river systems seem to show close parallels.

The Nogal, the upper part of which is fed by the Tug Der, is lowered some 1000 ft. below the plateau, the descent to the valley floor being made by a series of steps, which appear for the most part to be benches of structural origin (Pl. 33, fig. 1). No faulting has been observed in the valley but Macfadyen

[1] Caton-Thompson, G. & Gardner, E. W. (1939).

records a single instance of local anticlinal folding of an extremely shallow nature at Yagareh on the southern scarp in the middle Nogal, and that the southern margin of the Bur Anod and Shileh Madu-Dagah Gurgur ranges, which bound the valley proper on the north side, has been faulted.[1] Profiles of the southern half of the valley between Donkukoq and Sinugif and Donkukoq–Bihen–Garoe are shown in Text-fig. 6. The pediment at the foot of the scarp on the south side, between Bihen and Donkukoq, is marked in places by marly conglomerates and breccias, and in one instance near Bihen, by thick tufa beds which have as yet yielded no artifacts or fossil remains of any kind. These breccias and marly pebble beds are thought to be of Lower (perhaps in part, of Middle) Pleistocene Age, on account of the position in which they were found to occur, and their similarity to the valley conglomerates and breccias of the valleys in north-western British Somaliland. For the most part, however, the scarps are denuded of superficial

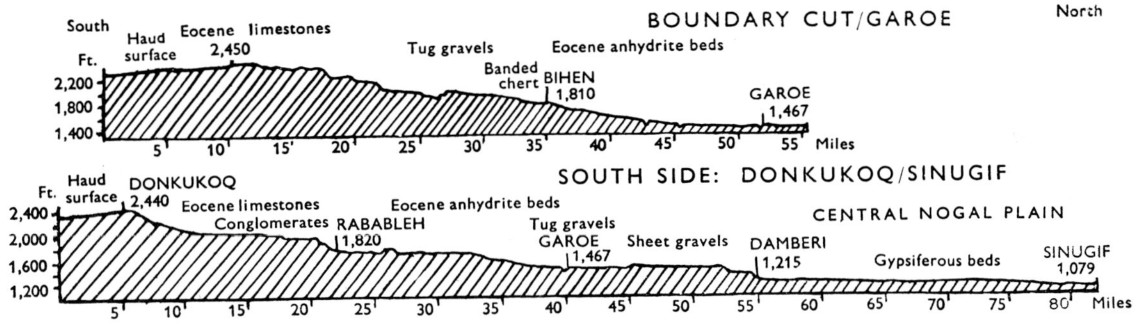

Text-fig. 6. Profile sections, Nogal valley.

deposits, with the exception of screes edging the gypsum residuals—hills and spurs exposing bands of the grey and black chert which comprised the medium for 98 per cent of the implements found in the valley. These screes which stretch for miles along the foot of the scarp are for the most part one vast accumulation of factory debris from workshop sites. Near Donkukoq on the very edge of the valley, exposures in erosion gullies eating into the scarp have laid bare an interesting series of deposits containing some cultural material complementary to the sequence seen in the main tug at Garoe. In one such gully on the left of the road 3 miles north of the tribal retainer post at Donkukoq the following sequence was seen. At the base was found some 4 ft. of typical angular torrent-bedded gravel in a calcareous matrix with lenses and pockets of coarse grits (the base of this deposit was not seen) within which were found abraded Levalloisian tools of early form. This gravel is overlain by a red sandy alluvium some 3 ft. to 3 ft. 6 in. in thickness. From the surface of the gravel and

[1] Macfadyen, W. A. (1933, pp. 9 and 38).

the base of the alluvium were recovered a few typical Levalloisian artifacts of developed form. Overlying this alluvium was a second alluvium of similar colour and composition some 3 ft. thick. The junction between the two alluvia was almost imperceptible but was marked by an occasional pebble and a certain difference in weathering of the surface of the lower deposit. This is considered to indicate a fossil land surface and from it was recovered fresh debitage of late Middle Stone Age form; unlike the underlying artifacts which show a white patina these specimens show glossing but otherwise are unpatinated. Overlying this second red alluvium was a land-surface deposit of pebbles and nodules forming a closely packed 'pavement' some 6 in. thick. The whole of this sequence had been cut through by the present channel.

The scree deposits and their contained artifacts are described under the section on the archaeology of the area on p. 292. Between Rabableh and Garoe and near the centre of the valley, between Garoe and Callis, are flat-topped ridges and low limestone plateaux covered by coarse sheet gravels and boulders (notably near Damberi) deposited apparently under fluviatile conditions. They lie approximately 200–250 ft. above the level of the central plain, and show a ferruginous to black patina with typical desert varnish. In their lower levels they appear to be cemented by an intractable kunkar. In the hasty examination of them that was possible no archaeological material was found to be associated other than Late Stone Age surface scatterings. That their deposition must have taken place during a considerably wetter climate than persists today seems certain: they are tentatively referred to the Great, or Kamasian Pluvial, but they may well be earlier and possibly should be correlated with the high-level gravels of the Dagah Shabel area in north-west Somaliland. These coarse gravels and boulder beds lie between 50–200 ft. above the present subsidiary tugs with their contained datable deposits, in similar manner to the boulder beds of the Dagah Shabel area in the north-west. Since their deposition the central plain has been lowered some 150–200 ft. and gypsiferous beds[1] are alternately exposed over long monotonous stretches or obscured by alluvial fans and aeolian sands, the former consolidated by secondary deposition of gypsum salts (Pl. 33, fig. 2). Whether these gypsiferous deposits represented in fact the Eocene anhydrite series or whether they were in part secondary could not be determined in the

[1] These anhydrite and gypsiferous beds are shown on Italian geological maps as being of Eocene age and intercalated with creamy limestones of the same age; it is possible, however, that some of the gypsum deposits of the central plain are later, but the absence of exposures renders confirmation difficult. Some of the beds resemble closely gypsum and selenite deposits in the middle Webi Shebeli and Juba valleys. Here the latest gypsum beds in the latter, at Lugh, were found by Dr F. Dixey to be banked against the Jurassic limestones and earlier Triassic anhydrite beds and do not underlie the Jurassic limestones as suggested by Italian geologists.

time available, but it is suggested that some at least may well prove to be of a secondary nature.

The tug at Garoe near the south-east corner of British Somaliland and the Mijertein shows good exposures for a distance of some 10 miles upstream and 1 mile downstream. Further sections are seen in tributary streams at right angles to the main river. Some 20 ft. or more of deposits are exposed.

Only in one place in the main tug, approximately 1 mile downstream of the fort, was bed rock seen: here occurred an outcrop of typical reddened gypsum rock. The gypsum outcrops again on the flanks of the valley in the subsidiary tugs. Immediately overlying bed rock is a typical river gravel of well-rounded pebbles (chert and limestone), which shows evidence in some places of even bedding, while in others pockets and lenses of unsorted gravels are intercalated with red sands and grits consolidated by gypsum salts. Such conditions suggest deposition during seasonal rains under semi-arid climatic conditions. There is evidence in some places, particularly in the upper reaches of the tributary tugs, for two phases of gravel deposition at this time, but the second phase was not always present.

Overlying the basal gravel and alluvium is an almost identical sequence of torrent gravel—this time showing no indication of even bedding—overlain by consolidated, red alluvial sands. There is usually no clearly defined stratigraphical break between the lower alluvium and the upper gravels and it would seem that deposition of the latter followed closely upon the accumulation of the alluvium. This upper gravel is deposited more as long drawn out lenses or pockets on the surface of the lower alluvium, but in a few places the gravel is absent and the upper alluvium rests directly upon and passes almost imperceptibly into the lower alluvium. The total thickness of these beds is some 18–20 ft.—of which some 1–3 ft. is basal gravel, 3–5 ft. lower alluvium, 6 in. to 1 ft. 6 in. upper gravel and 6 ft. upper alluvium. The surface of this upper alluvium is always well defined and undulating with in places deepish erosion gullies cutting through it. The surface layers are always well consolidated by kunkar and gypsum salts, forming either a tough kunkarous capping some 2 ft. thick, or, where later deposits cover this surface, being concentrated round vertical cracks or fissures running down into the alluvium for as much as 3 ft. or more, and giving to the deposit the impression of pillared or columnar structure. This horizon clearly represents a fossil land surface and the sequence below it must be considered as representing a single climatic phase subdivisible into two parts. Overlying these red consolidated gypsum gravels is a second gravel and sand some 3–5 ft. thick, which rests disconformably upon the red alluvium. Before its deposition gully erosion had carved out steep V-sided channels in the underlying beds to

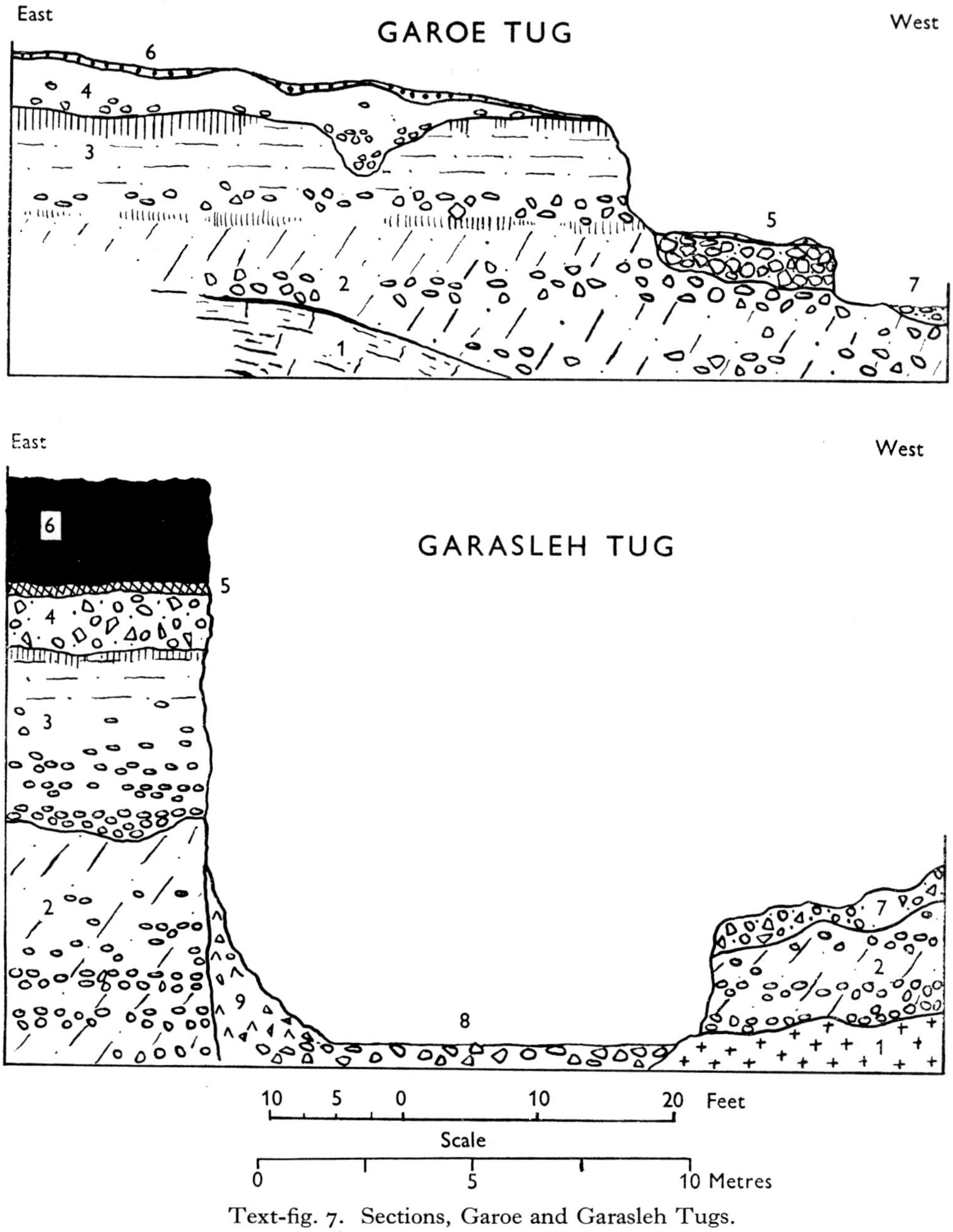

Text-fig. 7. Sections, Garoe and Garasleh Tugs.

Key to Sections

Garoe Tug: upper Nogal

7. Present erosion phase and resorted gravel.
6. Land rubble forming 'pavement'.
5. 5–8 ft. bench and torrent gravels }
4. Grey gritty alluvium }
3. Upper Tug Gravels: Upper Levalloisian.
2. Lower Tug Gravels: Lower Levalloisian.
1. Gypsum.

Garasleh Tug: southern Danakil

9. Scree.
8. Resorted gravel in tug bed.
7. Eluvial gravel.
6. Basalt (?) lava.
5. Red 'sandstone', i.e. baked surface of deposit 4.
4. Detrital bed representing a former land surface.
3. Red-banded gravels and stiff alluvium: Upper Tug Gravels.
2. Red-banded gravels and stiff alluvium: Lower Tug Gravels.
1. Volcanic tuff.

the level of a bench clearly seen in the main stream some 5–8 ft. above the bed of the present tug. At the base is a coarse gritty gravel, showing little or no sorting, and always loosely cemented. It is grey in colour and is overlain by a grey to buff alluvium. This alluvium is very fine textured and much of it is presumably of aeolian origin. Except in the gullies this gravel is always found overlying the red gravels at an average height of 15 ft. above the tug bed. Only in the gullies does the thickness increase to some 6–10 ft. of gritty, buff alluvium with nests of boulders and finer gravels. Overlying this is a deposit of closely compacted pebbles and surface wash such as is commonly met with under desert conditions. The bench mentioned above is a clearly defined feature of the tug. It is not always present but is usually found on the outside of a bend in the river's course—the bench is eroded in the red alluvium and gravels usually to the level of the top of the lower gravels, and resting upon it is some 2 ft. or more of quite unsorted and unconsolidated boulder gravel. In some places wind-blown sands and scree from the tug bank have covered the gravel. During this minor erosion phase the tug failed to erode the earlier sediments down to its original base-level. The present cycle has deepened the tug some 5 ft. below this bench, but in one place only has it succeeded in reaching bed rock. Text-fig. 7 shows a typical section across this tug 1 mile upstream from Garoe fort.

Within the two red gravels (Pl. 33, fig. 3) and the lower red alluvium, are found typical Levalloisian artifacts in some profusion, evenly distributed through the lower part of the deposit: the upper red alluvium was apparently sterile. Both rolled and unrolled specimens occur—the former usually in the gravels and the latter in the intercalated alluvium. The industry shows little development but for the most part the larger artifacts are found in the lower alluvium and gravel. Within the loose gritty gravels are found occasional derived artifacts as well as a few Somaliland Stillbay specimens. The surface wash yields only Late and late Middle Stone Age and indeterminate, presumably derived, debitage of Levalloisian facies.

The middle reaches of the Nogal are clearly defined and constricted by the escarpment hills, and are 'fed' by a number of large tributaries, now for the most part dry, but which have incised in their middle and lower courses, steep U-shaped valleys some 200–300 ft. or more below the plateau surface in the case of the lower, southern, tributaries, i.e. the Tug Hagob. The Tug Babani has eroded a valley 200 ft. deep with a pronounced terrace at 100 ft., while the valley of the Tug Tieg falls in approximately $2\frac{1}{2}$ miles from the plateau surface at 1630 ft. to a terrace at 1305 ft. and the tug bed at 1250 ft. South-east of Callis in the centre of the Nogal plain the road to the lower Nogal at Eil descends over anhydrite beds from 1109 ft. at Callis to 1040 ft. at the foot of the scarp 5 miles distant to the south-east, and from there climbs in 6 miles to 1370 ft. to the top of the flanking limestone hills. The tilt of the central plain at this point is therefore 11·4 ft. per mile towards the south-east.[1] The tops of the hills are sterile of prehistoric sites but some of the tributary river valleys have yielded sparse material, though the time spent in examination was strictly limited. In the Tug Babani, just below the road crossing, the following section was exposed. At the foot a grey sandy clay, the base of which is for the most part unexposed, but in one place, underlying the clay, was noted a gravel, which rested on a shelf of limestone. The maximum thickness of these beds was here 3 ft. Resting disconformably on the grey clay was a red unsorted gravel some 2 ft. thick overlain conformably by 4 ft. of red sandy alluvium with grit lenses. The surface of the red alluvium, where uneroded, was covered by a surface scree of pebbles and talus to a depth of 1 ft. Opposite, on the right bank, is a lower aggradation terrace with resorted gravels, which occur again in the tug bed. It will be noted that after the deposition of the grey clay and gravel the river for the most part failed to re-excavate its bed to its original base-level. The grey gravel was sterile but Levalloisian artifacts are eroding from the red gravels and alluvium, and Late Stone Age debris strewed the surface of the pebble wash. Within these valleys are numerous outcrops of banded grey chert which was extensively worked through the Middle and Late Stone Age, and numerous surface sites occurred between the Babani and Tieg tugs.

The sequence from the upper Nogal is found to be, therefore:

(i) Deposition of residual, but in part alluvial pebble beds on the plateau surface during a long and wet period; perhaps contemporary in part with (ii).

(ii) Major erosional phase when the Nogal valley and the main tributaries were cut; probably in stages over a long period.

(iii) Aggradation of conglomerates, breccias and marls on the valley slopes, affected by calcification.

[1] The inclination of the plateau is gradual for practically three-quarters of the distance between Garoe and the coast. For example corrected barometric readings give heights of 1460 ft. for Garoe; 1100 ft. for Callis (54 miles east); 950 ft. for Hararile 5 miles west of the Tug Babani (82 miles east); 940 ft. for the plateau surface above the Tug Hagob (104 miles east). From here near Bur Wein, where was found a scree containing typical Levalloisian debris, to the Tug Dubet (116 miles east of Garoe), the plateau falls in 12 miles from an altitude of 940 ft. to one of 590 ft. From there on the inclination is again a gradual one, heights of 520 ft. being registered at the Eil aerodrome (124 miles east) and 510 ft. at the top of the gorge at Eil village (128 miles east). The Nogal winds through the bottom of this gorge, here some 360 ft. deep, to the sea $3\frac{1}{2}$ miles distant. At the Nogal mouth above the fishing village of Biede the cliffs marking the limit of the plateau are some 200 ft. high and fall precipitously to the narrow coastal strip. The average slope of the plateau is therefore approximately 9·6 ft. in the mile.

(iv) Deposition of pebble and boulder sheet gravels at heights of 50–200 ft. above the level of the existing tugs which are cut through them. Perhaps in part contemporary with (iii).

(v) Erosional phase when the channels of the subsidiary tugs were cut through gypsum and the anhydrite beds of the main valley.

(vi) Deposition of two, perhaps three gravels and alluvia—red in colour and cemented by kunkar and gypsum salts, containing Levalloisian artifacts. (The Babani and Garoe evidence suggests the possibility of a depositional cycle subsequent to (iv) and antedating (v).)

(vii) A further subsidiary erosion phase (during which the tugs failed to excavate their beds to the previous local base-level) followed by the deposition of torrent gravels and grey sandy alluvium predominantly aeolian.

(viii) Erosion to present contours coincident with occupation by Late Stone Age man.

II. THE LOWER NOGAL

This is an important area as it provided one of the few links between the river terraces and the coastal deposits.

For approximately 60–70 miles before reaching the sea at Biede, the Nogal is confined within a gorge nearly 400 ft. deep, known locally as the Eil Gorge, around the base of which are found springs which provide almost the only fresh water in the whole of the Mijertein. In its lower reaches the gorge is cut through marine limestone, marls and sandstones assigned by Stefanini to an Oligocene or Miocene age.[1] The sides of the gorge abound in small rock-shelters, a number of which were occupied during Late Stone Age times. Many are at the foot of the gorge and of easy access, but a few are situated high up in the walls of the gorge and are often difficult to reach; they contain deposits of occupational midden material of varying depths. The percolating waters from

[1] The age of these beds is not well established, but their similarity to the Dubar and upper Daban (Dagah Shabel) beds of western British Somaliland, and the Hafun series at the mouth of the Darror, which are dated by contained fossils to the Oligocene and Miocene, suggests a similar age for these marine beds at the mouth of the Nogal. It is of interest to note that the Eil and Hafun beds appear as delta-shaped formations extending up the valleys of the Nogal and Darror respectively (Stefanini considers that they may be found at intervals along the north-east coast). On the north coast the Dubar beds lie within the Gulf of Aden Rift: this would seem to prove that their age is less than that of the faulting which determines the northern limit of the plateau. Since these series of beds were laid down the plateau has been elevated some 900 ft. and more on the north and 600 ft. on the east coast.

Stefanini, G. (1936, p. 71). Macfadyen, W. A. (1933).

the springs situated on the left bank have resulted in the formation of terraced tufa beds containing cultural material. These spring waters combine to form a running stream which, at the mouth of the Nogal, widens into an estuary some 200 yd. across, the exit of which is blocked by a sand bar, the water percolating through the sand over the beach to the sea. Text-fig. 8 shows a section through the gorge upstream of the village of Eil, which was found to be typical of the sequence throughout the area examined. This extended some 5 miles upstream of Eil to Got Ad and beyond, and downstream for $3\frac{1}{2}$ miles to the estuary mouth at Biede.

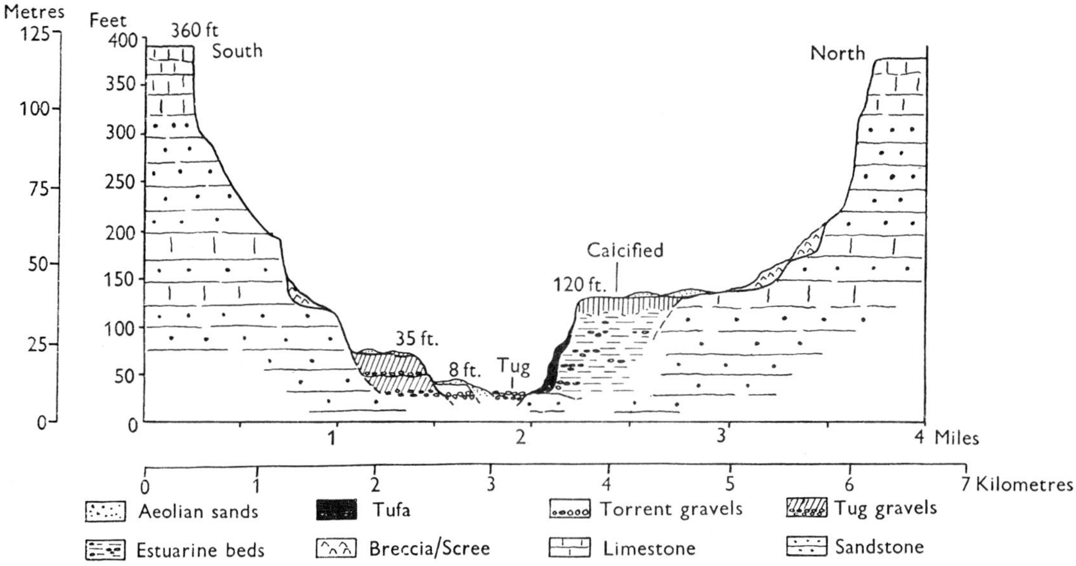

Text-fig. 8. Section through Nogal Gorge at Eil.

No residual or other pebble beds were noticed on the plateau, which at Eil stands at an elevation of between 300–400 ft. above the sea, rising gradually from approximately 200 ft. above Biede to 420 ft. when $3\frac{1}{2}$ miles inland.

The cutting of the Nogal Gorge must have taken place over a long period of time during a major erosional phase and can presumably be equated with a similar event in the upper courses of the valley, and in fact the distribution of the marine deposits at the mouth seems to suggest that the valley may even have existed in an early form prior to the marine invasion which resulted in the deposition of these beds. A similar situation exists in the valley of the Darror. Whether or not a valley existed at such an early time (unless of course the deposits are of later age than has been previously determined) it is certain that the initial or major cutting of the valley is very ancient and must have taken place at least not later than the Lower Pleistocene.

Within the gorge a bench has been cut into the soft limestones and sandstones, level with which are the earliest deposits within the gorge itself: these comprise evenly and false-bedded, fine and coarse gravels, comprising flat, well-rolled pebbles of quartz, sandstone, limestone and chert, sands and marls. The whole, particularly in its lower levels, has been ferruginized and presents a buff to orange appearance. Only in one place, at Got Ad, 5 miles upstream from Eil, was the base of this deposit seen exposed in section in the left bank of the tug (Pl. 33, fig. 4). Here were seen some 20 ft. of ferruginous false-bedded gravels, sands and grits resting on an outcrop of sandstone of the Hafun Series. For the most part, however, the base of the deposit is situated below present tug level. The upper 5 ft. everywhere contains many boulders and large pebbles and has been cemented into a very hard, secondary, conglomeratic limestone with, in places, what appears to be partially formed nodular chert *in situ*.[1] In the area examined the top of this deposit lies at heights of 170 ft., 4 miles upstream of Eil; 118 ft. at Eil; 80 ft. at a bench 2 miles downstream from Eil; and 55 ft. at the estuary mouth. The average height for the deposit and bench is therefore 75–100 ft. (referred to as the 75–100 ft. level), the fall being some 15·3 ft. per mile. The nature of the deposit suggests that it was laid down during a long wet cycle and that it is probably an estuarine deposit of fluvio-marine origin. The gorge in the region of Eil is approximately 1 mile wide narrowing to less than $\frac{1}{4}$ mile nearer the sea and at one time the valley was filled by these estuarine beds; they are now confined practically entirely to the left bank of the tug, which hugs the original right wall of the gorge, and they are practically continuous throughout the area examined, all subsequent erosion having taken place on the right bank. These estuarine beds of the 75–100 ft. terrace are unfortunately sterile of human artifacts or fauna, but at one site round the base of an outcrop of silicified, ferruginous sandstone near the side of the gorge 1 mile north-west of Eil were found Acheulio-Levalloisian artifacts resting on a surface of these estuarine beds and eroding from a matrix of hard secondary limestone. The succeeding deposits in the tug proper contain Levalloisian artifacts. It may be suggested therefore that the aggradation of this deposit was contemporary with a period of high sea-level (during which time the river was aggrading its bed).[2]

[1] This deposit, which, to walk over, one would consider at first to be a Tertiary limestone, is typical of many of the secondary limestones in the Horn. These secondary limestones or sheet kunkars are of rock-like consistency on the surface, but pass down into uncemented alluvial or aeolian deposits which can sometimes be dated by the artifacts contained within them.

[2] On the coast itself no high-level beach was seen, other than the terrace at 55 ft. against which is banked the 38 ft. raised beach; but as the sea came right up to the cliffs at this time the evidence has probably been lost with the erosion of the soft Miocene beds.

The aggradation of these gravels and sands, and the calcification of their surface, was followed by a further period of erosion during which the tug cut for itself a deep channel through the 75–100 ft. terrace deposits, a channel which was subsequently filled by two aggradations of gravels and grey sandy silts. These are best seen at Eil and upstream on the right bank between the village and Got Ad (Pl. 34, fig. 1). The base of these beds is for the most part in a buried channel below the present bed of the tug, but where exposures are seen the basal deposit is a gravel of well-rounded pebbles (derived presumably from the deposits of the 75–100 ft. terrace) set in a matrix of grey sand and grit often cemented by calcium carbonate. For the most part this basal gravel is evenly bedded and passes gradually up into interbedded fine gravels, sands and clays. At Got Ad there is evidence of a slight disconformity marked by a line of partially formed calcrete or kunkar. Above this is found a second gravel occurring as thin sheets which again passes up into grey sandy silts and clays. The surface of this deposit is cemented to a laminated calcrete or tufa on an average 3 ft. thick, containing some angular and rounded pebbles and representing presumably an old land surface. The depth of this deposit is some 18 ft.

On the right bank of the gorge almost opposite Eil village gully erosion has exposed the following section:

(i) 3–4 in. of surface wash with scattered Late Stone Age factory debris.

(ii) An irregular and eroded land surface dissected by small gullies.

(iii) Calcrete or tufa to 3 ft. passing gradually down into grey calcareous sand with lenses of gravels with some angular components and typical Somaliland Stillbay artifacts *in situ*.

(iv) Slight disconformity; grey sandy clay partially consolidated and interbedded with gravels containing occasional Levalloisian implements, some approximating closely to Stillbay form and technique, thickness 3 ft. 6 in. to 4 ft.

(v) Resorted boulder gravel in bed of gully (Pl. 34, fig. 4).

The height at which the grey gravels and silts once stood in the narrow part of the gorge between Eil and the sea is clearly seen at a bench some 2 miles upstream from the mouth. This bench here stands at a height of 40 ft. below the uppermost bench (estuarine beds), and 40 ft. above the present tug bed. Later resorted torrent gravels, overlain by red alluvium bedded with grits and sands passing up into red and orange wind-blown dunes, form a terrace some 10 ft. above the bed of the stream (Pl. 34, fig. 2).

The calcrete or tufa and upper part of the grey silts have yielded typical examples of the Somaliland Stillbay Culture, while from the gravels and lower half of the alluvium were recorded rare, abraded and fresh examples of typical Levalloisian tools.

Contemporary with the aggradation of these grey silts was the formation of tufa round the edges of fossil springs situated in the upper half of the estuarine beds; these were found to occur at several points between Got Ad and the Nogal estuary (Pl. 34, fig. 3). It is from these springs, some of which are still active, that the Nogal estuary is filled. These springs at Eil and upstream were all found to be active, while those downstream are now fossil, their position being marked only by the tufa beds. Tufa is still forming round such springs as are still active, and its growth appeared to be fairly rapid. It was found that these lime-saturated spring waters had resulted in some places in the calcification of the grey silts. Those springs which are still active appear to have been continuously so from the time the grey silts were deposited, and have resulted in the layered tufas down the slope of the estuarine beds to tug level. The tufas contain rare Levalloisian and Stillbay and at the top some Late Stone Age artifacts and factory debris.[1] There was little evidence of stratification within the tufas. This was considered to be due in the main to the continual growth of the tufa causing perpetual shifting of the spring waters. Also we were unable to excavate sections in these beds due to lack of time and to the usually very hard nature of the deposit, which necessitated our relying on such natural sections as were already exposed.

The calcification of the surface of these grey gravels and silts and the first phase of tufa formation at the springs, was followed by deepish gully erosion by tributary streams coming down from the sides of the gorge, and further deepening of the river bed in the main stream. Erosion has cut deeply into the calcrete and grey silts and has deposited a coarse unsorted boulder gravel, some 3 ft. thick at the side of the gorge, everywhere showing signs of deposition by seasonal torrents. This gravel is overlain conformably by 4 ft. or more of red sandy alluvium and the whole is covered by a hill wash and sometimes by unconsolidated dune sands.[2]

In the main stream these torrent gravels rest on and are banked against the

[1] Stefanini collected from these tufas the following fossils: *Melanoides tuberculatus* and several diatoms, *Epithemia argus*, *Synedra affinis* and *Coscinodiscus lineatus*; while those at Gunumale yielded, besides casts of fossil twigs, *Navicula rostellata*, *Nitzschia obtusa*, *Fragilaria bituminosa* and *Trachelomonas volvocina*. Stefanini, G. (1936, p. 87).

[2] Opposite the village of Eil a good section in a side gully shows: (i) surface wash with some good Doian artifacts on the surface; (ii) 2–3 ft. red-brown sandy alluvium with occasional derived Stillbay artifacts at the base; (iii) coarse torrent-bedded gravel with occasional pockets of false-bedded grits 2 ft. to 3 ft. 6 in.; (iv) disconformity: grey sandy clay into which (iii) has deeply eroded; Stillbay artifacts *in situ*; (v) gravel in two places underlying grey sandy clay; base not seen. (Pl. 34, no. 4.)

eroded grey gravels above Eil, the former channel of the Nogal being cut below the present bed, and filled with resorted gravel. Below Eil, where the gorge narrows before reaching the estuary mouth, the grey gravels have all been entirely eroded, except at one point on the left bank $\frac{1}{4}$ mile from the estuary mouth where, at water level, was seen a remnant of the calcareous grey sandy alluvium overlain disconformably by torrent gravels and red sandy alluvium which was in turn covered by red wind-blown sand.[1] In this narrow part of the gorge and apparently at the mouth of the estuary also, the channel in which the redeposited sediments lie must be at some depth below the present river bed, and when the seasonal rainfall became less, this channel was completely filled by the redeposited gravel and red alluvium which formed a bar and partially blocked the mouth of the Nogal. Increasing arid conditions resulted in the accumulation of orange and yellow unconsolidated dune sands across the river mouth, the water from which percolates through the base of these sands to the sea about $\frac{1}{4}$ mile distant.

On the surface of the red alluvium overlying the torrent gravels as well as on the surface of the earlier deposits, where these are exposed, are found numerous working and living floors of the Doian and Wilton Cultures and, at the coast, the strandlooping elements of this latter culture. Where both cultural material and dune sands occurred, the Late Stone Age artifacts were found to underly the unconsolidated dune sands.

SUMMARY OF SUCCESSION

The sequence of events, climatic conditions and cultures in the lower Nogal may be summarized, therefore, as follows:

(i) Cutting of the main and tributary gorges during a major erosion cycle of long duration, through marine sediments (Hafun Series) of local extent.

(ii) Development of the 75–100 ft. bench and the deposition of estuarine beds to this level; presumed to be contemporary with a period of high sea-level.

(iii) Prolonged calcification of the surface of these beds following ferruginization during an arid phase of long duration, contemporary with uplift and earth movement. An industry of Acheulio-Levalloisian facies is thought to be contemporary with this phase.

(iv) Erosion and cutting of a channel, now buried, below the present tug bed during a predominantly wet climatic phase, followed by the deposition

[1] While at this time the Nogal succeeded in removing the greater part of the calcretes and grey gravels it was unable to erode the cemented 100 ft. estuarine beds.

of probably two series of grey gravels, sands and clays to approximately 40 ft. above the tug. Levalloisian artifacts are found throughout these beds but a Somaliland Stillbay industry occurs at the top. Tufa formation by fossil springs begins.

(v) Calcification of the upper part of these beds during an arid period.

(vi) Minor wet cycle and erosion of grey sands and clays, which at the estuary mouth appears to have resulted in the cutting of a deepish channel below the present bed and below sea-level, followed by the deposition of torrent-bedded gravels overlain by red sandy alluvium. Continued tufa formation at springs.

(vii) Unconsolidated orange/yellow dune sands underlain by artifacts of the Somaliland Wilton and Doian cultures.

(viii) Erosion to present contours with some local shifting of dune sands.

CHAPTER 4

THE WEBI SHEBELI AND ITS TRIBUTARIES

This river was examined at a number of points throughout its middle course, and again in its lower reaches at Afgoi and Balad. The Tug Fafan, which at one time joined the Shebeli by the Faf depression, but which now loses itself in sands before reaching the parent river, was studied at Gabredarre, and again in the depression; other tributary tugs of the Webi Shebeli were examined between Gabredarre and Melka Taka on the left bank and again at God'dere, and at various points downstream on the right bank. The Baidoa plateau between the Webi Shebeli and the Juba received careful study, while the Mudugh plateau was examined between Ferfer and Dusa Mareb and again between Bulo Burti and Jesomma. Good sections with artifacts were obtained at a number of widely separated points which have thus given a fairly complete sequence of events and cultures for the southern part of the Horn.

The Webi Shebeli valley can be divided into three main, and for the most part clearly defined, areas.

(i) The upper reaches between the source, in the Ghedeb basin on the high plateau within the Abyssinian boundary, and Malca Dube above Imi.

(ii) The middle reaches, between Malca Dube and Bulo Burti (the southern limit should possibly be carried further south to Jallalaxi) that is to say to its junction with the coastal plain.

(iii) The lower reaches between Bulo Burti (or Jalallaxi) and where the river disappears into the sand at Ballei yer.

I. THE UPPER VALLEY

The upper reaches were studied by the Duc d'Abruzzi's Expedition in 1928–9.[1] The river was found to have its source in the western part of the Ghedeb basin, an undulating depression in the extreme west of the Somali plateau at an average altitude of some 2500 m. and bounded on the north and west by mountain ranges of volcanic traps. The Expedition considered that this basin was once occupied by a large but shallow fresh-water lake, similar to Lake Zwai and the other Galla lakes in the Abyssinian Rift Valley. Some degree of confirmation is given by the discovery of Quaternary diatomaceous clays, the product of a closed

[1] Parona, C. F. (1932).

fresh-water lake basin.[1] Inferred to be of later age, though the two have not been found stratified, are volcanic tuffs which formed at some period subsequent to the disappearance of the lake, presumably during a period of crustal instability. The bed of the Shebeli, at this point near its source, presents a typical youthful profile and meanders north-eastwards over the Ghedeb plain between steep banks usually of volcanic rocks, for a distance of 90 km., after which it descends near Scedala (2330 m.) into a deep gorge by a series of rapids and two waterfalls estimated to be 20 m. and 70 m. high respectively. The gorge was estimated to be some 700 m. wide and over 1000 m. deep with steep sides near its commencement; so that a major erosion cycle or cycles of long duration must be assumed to have accounted for its excavation to this great depth. Parona inferred that the retrocession of the falls and gorge beyond the lip of the Ghedeb basin had caused the draining of the lake. It is most probable that this process was hastened by the eastward tilting of the plateau. The plateau through which this gorge is cut is composed of Jurassic and possibly Cretaceous horizontally bedded sediments overlain by volcanics of the trap series which form the mountain ridge. The plateau itself is gently inclined eastward at approximately the same degree of slope as the river bed so that the strata appear to be running parallel with the river course.[2] At Sheik Hussein the profile of the gorge assumed the form typical of that throughout the greater part of its course to Malca Dube and shows a wide bench or plain said to be some 1000 m. on an average below the plateau. Through this bench level a steeply incised gorge has been cut, which must be between 300 and 400 ft. deep. It is presumed therefore, that this profile represents two distinct erosional phases, an earlier phase during which the plateau remained unaffected by crustal movement for a long period thus enabling the river to widen its valley plain, and a later phase resulting from uplift of the plateau surface and the cutting of the V-shaped gorge. It is also probable that the excavation of the later gorge was contemporaneous with the draining of the Ghedeb Lake. The course of the river between Imi and its junction with the Dakata was examined by the du Bourg de Bozas Expedition in 1901 and collections of artifacts made from a number of sites on the left bank;[3] but little is known of the deposits from which they were derived. The presence of one bifaced tool considered by Breuil to be Micoquian on typological grounds suggests the existence of aggradations of gravels containing an Acheulio-Levalloisian industry.

[1] The following diatoms are listed: *Fragilaria, Rhopalodia, Epithemica, Cymbella, Synedra, Suriella, Cymato pleura.* Stefanini, G. (1936, p. 81), compares this deposit with the Kariandusi diatomite beds which are of Kamasian (Middle Pleistocene) Age.

[2] In some 130 km. the plateau falls from 2420 m. at Sciarida to 1489 m. at Sheik Hussein. The tilting of this western part of the plateau would then be approximately 7·16 m. per km. or 37·59 ft. per mile.

[3] See p. 16.

II. THE MIDDLE VALLEY

The Shebeli in its middle course meanders through a wide, fertile valley plain flanked by escarpments marking the edges of the plateau, the junction of which with the valley is usually effected by a steep scarp. The river now flows over limestones and anhydrite beds of Jurassic and Cretaceous age, which are alternately exposed for considerable stretches or obscured by thick alluvial deposits, as for instance in the Audamboi plains between Melka Taka and Bur Eghei and particularly in the lower reaches below Bulo Burti.

(1) PLATEAU DEPOSITS

The plateau surface on the left, i.e. the north, bank is composed of Jurassic limestones in the western part, and Cretaceous limestones and sandstones in the eastern half. The height of the plateau descends eastwards from an average of 500 m. near Imi to approximately 200 m. at Jesomma some 320 miles to the south-east. On the right, or southern bank, the plateau is composed of Jurassic limestones (overlying in the upper reaches gypsiferous beds assigned to the Trias-Lias) which bend round to form the western limit of the coastal plain past Tijieglo and Isha Baidoa.

The plateau to the north forms the southern limit of the Haud peneplain with average altitudes of 1300 ft. in the south-east and 2300 ft. in the north-west, while the Hoddur/Baidoa plateau, altitude 1400 ft. at Tijieglo to 2000 ft. near Hoddur, represents the continuation of this surface between the Webi Shebeli and Juba rivers. The plateau surface is very gently undulating with occasional low and isolated residuals rising above it, as for example the Gumburu hills and Bur Agar Wein[1] between Galadi and Wardere, also the Durua hills between Gherlogubi and Gorrahei (altitude 3000 ft.), and Bur Corraile near Schillave, in the southern Haud; also Bur Eschunle near Baidoa and Bur Ghederwein, Cal-ie-Corar (1857 ft.), Bur Gheider Idda, and the Ted surface (2290 ft.) and a number of others in Hoddur and Baidoa districts.

On this warped peneplain surface are found deposits—particularly in the vicinity of well areas and pans—of calcareous pebble beds which, in wells, record depths of 40 ft. and more.[2] The top levels of these beds are cemented by a thick deposit of secondary limestone or kunkar of rock-like consistency. While this is not necessarily an indication of great age, these deposits were

[1] Dr Macfadyen informs me that it is possible that these may be original Eocene limestone knolls, and not residual hills.

[2] The components are usually limestone pebbles of fist size set in a highly calcareous matrix. It is considered that such a deposit could not have formed entirely as a result of residual weathering of the underlying rock.

nowhere found to be fossiliferous, and overlying them in several of the well areas were tufas and calcretes of Upper Pleistocene Age. It is presumed, therefore, that these plateau pebble beds should be equated with the pebble beds of western British Somaliland and the Mijertein.

The later deposits of the well areas on the plateau were sometimes found to be fossiliferous and to contain human artifacts. Stratification of individual sites will be dealt with in detail under the section on archaeology, but a few general remarks may be made here. Almost the only areas on the plateau where cultural material was found were in the immediate vicinity of the wells, situated for the most part in the centre of shallow pans varying in size from 100 yd. to $\frac{1}{2}$ mile or more across. After rain these pans may contain water for several weeks and it is certain that during the Upper Pleistocene, if not earlier, they formed small lakes, some of which, such as those at Ted and Burdo, still contain water for most if not all the year even today.[1] Within the basins of these miniature closed lakes were laid down two deposits of tufas and calcretes which, in a few pans, have yielded Levalloisian implements. The surface of these tufas, on the Baidoa plateau and in the western Haud has formed a secondary limestone often difficult to distinguish from the Jurassic and Cretaceous series on which it rests; while in the Mudugh, and eastern and southern Haud where they rest on Eocene and anhydrite beds, these tufas are highly gypsiferous throughout. The surface of the upper calcrete is almost always reddened and near the edges of the pans may be overlain by bright red aeolian sandy soil, which has to some extent been leached and redeposited in the bottoms of the pans. At some places Doian living sites are located in the upper levels of this sand and these are also particularly numerous round the edges of the pans. The permanent flooding of these pans can thus clearly be ascribed to favourable climatic conditions during the Upper Pleistocene (dated by means of cultural material) and their subsequent desicca-

[1] These pans are of two main forms: (i) Shallow pans in the gently warped plateau surface—as for example the western and southern Haud 'ballehs', i.e. the Balleh Sirrau, Goffado, etc., which are situated for the most part on the Cretaceous sandstones. (ii) Originally deep pans, for the most part in limestones and anhydrite beds, the formation of which was due initially to sink-holes, which have become filled by thick tufaceous and gypsiferous secondary deposits. Such pans in their original form must have been very like that of the Mullah's Pool, near Galkayu in the Mudugh, a recent sink-hole with almost vertical sides some 40 ft. deep to water-level and 100 yd. across. The opening in the bottom of these sinks became clogged and resulted in the development of ponds and lakes thus allowing for the formation of tufa deposits. In a number of these pans secondary sinks have formed, which were in turn blocked by unconsolidated deposits. The Somali rarely dig deep wells through the first type of pan, such wells as do exist being shallow holes for the collection of surface water when the 'ballehs' are about to dry up. Within the second type of pan wells are of two kinds, (*a*) re-excavated sink-holes, and (*b*) true deep wells through the secondary tufaceous beds where no secondary sinks had formed. It was noted that while Late Stone Age material is usually to be found in the vicinity of pans of types (i) and (ii) (*a*), it hardly ever occurs in pans of type (ii) (*b*).

tion to progressive aridity and the depression of the water table marking the end of the Pleistocene. Mollusc remains recovered from the tufas of the 'saha' at Ircut included *Limnea* aff., *Laurenti*, *Phyra* aff., *Coulboisi*, *Limicolaria* sp., *Georgia* sp., and *Paludina badiella*,[1] while *Achatina* sp. was found by the writer at El Dubbo and Bugda Qosa. It was suspected that the basal levels of the tufas were composed of gravels which passed down into the calcareous plateau pebble beds for on the Hoddur/Baidoa plateau excavation material from wells showed numerous pebbles of limestone and very occasionally a pebble of quartzite. Examination of such material from wells in the western Haud at Ubitale, Afdub and Gherlogubi appeared to confirm this, as these wells had been excavated through calcretes containing, in the lower levels, numerous pebbles of lava, some of quite large size, which, as the underlying rock is a Cretaceous limestone, can only have been transported from one or other of the rare, intrusive lava outcrops farther to the west. A typical section is that seen in the wells at Afdub, a small pan between Wardere and Gabredarre. On the surface was some 2 ft. of grey redeposited sand, leached in the upper levels. Underlying this was a thin sheet of kunkar approximately 6 in. thick under which was a red semi-stratified, consolidated, sandy wash with small pebbles to 5 ft. This rested disconformably on an uneven surface of white calcrete or tufa approximately 3 ft. or so thick at the surface, which passed gradually down into a less compact calcrete to some 40 ft. (the base of which was not seen however), containing in its lower levels lava pebbles.[2] A variation was to be seen in the pan at Ual further to the west on the Wardere/Gabredarre road where the well had been freshly redug. The well is here within a typical funnel sink which had formed through the secondary calcrete. This sink had been filled with some 20 ft. of grey secondary clay, calcareous in the upper levels, overlain by approximately 6 ft. of red-brown alluvial wash with occasional lenses of silt and grit. This indicates that sufficient time must have elapsed between the deposition of the calcretes, and the grey alluvial clays, to allow for the formation of the sink-hole. At Wardere,[3] where the pan is approximately 1 mile long and some 50 ft. deep, the grey clays have yielded flakes and a unifaced point of Developed Levalloisian to early Stillbay type. On the surface lie Doian artifacts and factory debris. At Saha Gurral near Bugda Qosa, the lower grey calcrete or tufa yielded Levalloisian artifacts,

[1] Stefanini, G. (1936, p. 87).

[2] Dr Macfadyen informs me that at Sirau and other ballehs of the same nature, mostly lying to the west of Sirau, volcanic quartz tuffs overlie or are interdigitated with the uppermost layers of Nubian (Cretaceous) sandstone.

[3] Some of the eastern wells in the Wardere depression are, Dr Macfadyen informs me, dug in solid basalt which is only seen exposed thereabouts in these wells. Up to 40 ft. of fresh basalt was measured and the base was not seen.

as did that at El Dubbo, while the upper red calcrete yielded a very few flakes of thick white patina of Somaliland Stillbay or Magosian type; there was little or no evidence on the surface of occupation during Doian times, though the saha must have contained water at that time. Tufa formation now appears to have ceased in the plateau pans, but tufas are still forming around springs along the south-eastern edge of the Baidoa plateau, known as the Jir. Here at Deileb, Wanei and Isha Baidoa (the springs having built themselves up on these beds) thick tufa deposits have formed and that at Baidoa has yielded numerous plant and molluscan remains notably casts of leaves of *Ficus somaliensis* together with *Achnanthes microcephala*, *Synedra affinis*, *Coscinodiscus radiatus* and *Trachelomonas volvocina*[1] (Pl. 35, fig. 1). These tufas are presumably in part contemporary with those of the plateau pans but, as is the case with the Eil springs, tufa growth has persisted until the present time. These spring tufas have not yielded any artifacts *in situ*, but numerous Doian surface sites were located in the vicinity.

(2) VALLEY DEPOSITS

Within the Shebeli valley and those of the tributaries it was possible to check the sequence of events and cultures at a number of widely separated sites. I shall describe first, the sequence within the main valley and then show how this is paralleled within the tributaries.

In its middle course the Webi Shebeli winds through a flat-floored valley some 6 miles wide, opening out in places to as much as 25 miles or more, and cut to an average depth of 700 ft. (1000 ft. in the lower reaches), usually bounded by short, steep cliffs where the plateau limestones abut abruptly upon the valley (Pl. 35, fig. 2). Elsewhere in areas of gypsiferous beds, not capped by limestone, the junction of plateau and valley is effected by a very gentle gradient. The wide and deep valleys of the Webi Shebeli and Juba are undoubtedly ancient features, graded to the coastal peneplain and must be at least as early as the Late Tertiary or Lower Pleistocene, possibly older. At least two divisions of this major erosional phase can be seen; the earlier marked by certain mature hanging valleys cut to an average depth of 400 ft. below the surface of the plateau, and seen in profile on the valley sides (Pl. 36, fig. 1). Valley sections are seen in Text-fig. 9.

(i) *Older gypsiferous series.* Between Melka Taka and Beletwein and again south of the latter to Bulo Burti and probably beyond, are a series of gypsiferous beds of some considerable thickness.[2] These beds almost certainly extend further up-river than Melka Taka, but they were not examined beyond this point. Downstream as far as Beletwein the gypsiferous beds may, and usually do,

[1] Stefanini, G. (1936, p. 87).

[2] Their distribution is best seen on the Italian military map (1933), by G. Stefanini.

lie on both sides of the river, but south of Beletwein they are for the most part confined to the right or south bank of the river. In the lower reaches bed rock is obscured by deep alluvial sediments so that whether or not these beds extend further to the east could not be determined. On the Italian geological maps no clear division within the beds comprising the older gypsiferous series is made, and these are described for the most part as being of Trias or Lias Age and underlying the Jurassic limestones. Dr F. Dixey informs me, however, that

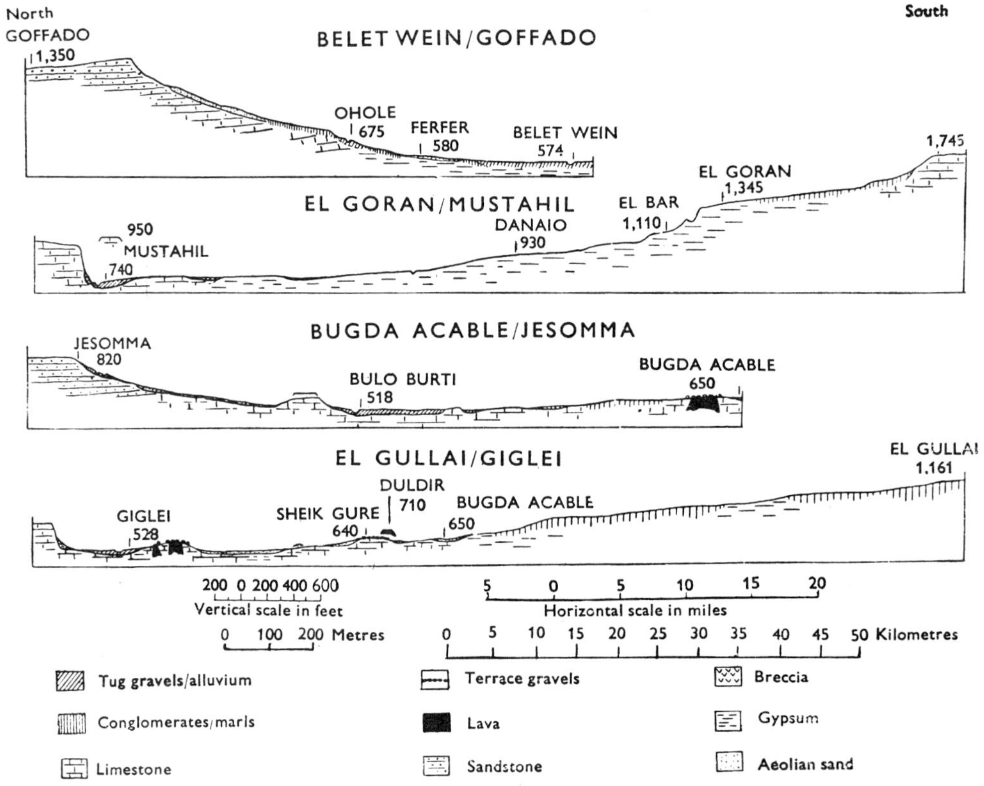

Text-fig. 9. Sections across Webi Shebeli valley.

an examination of the Lugh Series (Triassic-Jurassic) beds at Lugh Ferrandi, within the valley of the Juba, showed that there existed a younger gypsiferous series banked disconformably against the Triassic and Jurassic beds. Whereas the Jurassic gypsiferous beds were always fossiliferous, the post-Jurassic series contained no fossils so that distinction between the two series should usually be clear. While no clearly defined disconformity, such as exists at Lugh Ferrandi, was noted between the gypsiferous series in the valley of the Webi Shebeli, it was nevertheless very evident that these gypsiferous beds were confined essentially to the valley itself or to the large tributaries such as the Fafan (where similar beds were seen in the Faf depression). On the north bank of the Shebeli near Beletwein, and for some distance up the western side of the Faf depression could be seen an earlier gypsiferous series underlying limestones (identified as Cretaceous)[1] which form the flanking hills. The southern scarp, after leaving the Ted surface developed on Jurassic limestones in the neighbourhood of El Goran, is composed of a gypsiferous series of beds (unfossiliferous) consisting of horizontally bedded gypsum marls alternating with beds of massive gypsum and some selenite. This series may underlie the Jurassic beds, but Stefanini and others consider it possible that it may be younger.[2] The present tributary systems are developed within these beds as are, apparently, the middle courses of the valleys of the Shebeli and Fafan, though it must be borne in mind that part of this gypsiferous series may represent valley filling, particularly in the reaches from Mustahil to Melka Taka and beyond. The thickness of the series in the El Goran/Mustahil section must be at least 600–700 ft. Within the valleys of the larger rivers this older gypsiferous series is overlain by later, secondary gypsiferous deposits which can be dated by artifacts to the Upper Pleistocene.

(ii) *Intrusive volcanics.* At several isolated sites throughout the middle Shebeli and Juba drainage basins, notably at Legullei between Hoddur and Yet, at Cura Ba'ad between Melka Taka and Bur Eighei, at Duldir and Sheik Gure near Bugda Acable, and on the left side of the Fafan valley on the edge of the Gorrahei plain, are found isolated instances of intrusive lavas which have been planed and have overlying them locally beds of lava pebbles.[3] The Duldir and Guretca lavas are for the most part olivine basalts intercalated with liparitic tuffs and rest, in the case of the Shebeli group, on Cretaceous sedimentary rocks of marine origin, and, in the case of the Guretca group, on sandstones of the Lugh (Jurassic) Series. There is also an isolated intrusion of tinguaite at Allengo on the plateau near Curetca. Aloisi considers that these alkaline lavas should be referred to the earlier Trap Series of volcanics of Tertiary Age, and compares them with the Bihendula volcanics in British Somaliland (Older Basalts of the Aden Series). There also occurs at Duldir a small outcrop of acid lavas and the intrusions found by the present writer at Cura Ba'ad are also thought to be similar. These lavas are intruded through the older gypsiferous series as well as through the Jurassic and Cretaceous sediments, and they have been peneplained, their flat surfaces apparently corresponding with a stage of erosion of

[1] Stefanini, G. (1936, p. 47).

[2] Stefanini, G. (1936, pp. 41 and 48).

[3] See also the Guretca series of lavas in the Juba valley near Lugh Ferrandi. Aloisi, P. & de Angelis, A. M. (1938, pp. 27 *et seq.*); and a small group at Bio Soro in the Ogaden near the junction of the Webi Shebeli with the Dakata.

the valley which is marked by a usually definite terrace level some 400 ft. at the edges and 200 ft. in the centre of the valley, below the plateau surface and best seen in profile in certain hanging valleys on both sides of the river between Callafo and Bulo Burti. The lavas certainly antedate the formation of a terrace at 50–100 ft., as at Duldir and Cura Ba'ad much of the ingredients of the terrace are lava pebbles from these basalts, while on the flat surface of Bur Wein, an isolated limestone hill in the valley some 20 miles south of Bulo Burti, occur rare lava pebbles presumably derived from Duldir, about 45 miles upstream to the north. The surface of the lavas at Cura Ba'ad and Duldir, which stand approximately 50–75 ft. above the terrace, were probably planed during the long erosional cycle during which the present valley was cut. It is fairly certain, therefore, that they could not have been intruded during the set of earth movements marking the end of the Middle Pleistocene or Last Interpluvial, as they were planed at the time of the formation of the 400 ft. terrace and must, therefore, have been contemporary with an earlier set of earth movements.

(iii) *The Pleistocene sedimentary deposits.* As has been stated above the initial phase of valley cutting which must have extended over a long period of time, lowered the valley some 600–700 ft. below the plateau in two major phases, leaving a terrace at approximately 400 ft. On this terrace there accumulated pebbly calcretes, marls and breccias apparently of considerable depth and very similar to the older calcretes and pebble beds of the well areas on the plateau. For example, the well at El Gullai between Tijieglo and Bugda Acable is sunk some 30 ft. through deposits of this kind without reaching bed rock. The age of these beds is uncertain but they presumably antedate the formation of the 50–100 ft. terrace (to be described), with its accumulation of gravels, and represent a major depositional phase. Subsequent erosion was followed by the aggradation of a gravel terrace lying, in the upper reaches, at heights of approximately 50–75 ft. or less above the Webi Shebeli on the left bank of the river; and this terrace is fairly continuous in the area between Melka Taka and Bur Eghei. It was examined in the region of Cura Ba'ad some 14 miles from the Callafo turn off from the Melka Taka/Gabredarre road, and in the lower reaches of the tributary tug Darchaua, a distance of some 20 miles. At Cura Ba'ad (Pl. 36, fig. 1) it is banked against the volcanic series and the gravels are largely composed of very well-rounded and smoothed pebbles of chert, jasper, and volcanic rocks (tuffs and basalts) with some smaller pebbles of silcrete, agate and limestone. Nearer the Darchaua Tug the terrace has been considerably dissected by the present tug system and especially near its junction with the Shebeli where it is represented by small flat-topped, isolated hillocks covered by the gravels. The gravels are for the most part unconsolidated in their upper levels and they appear at one time to have been subjected to ferruginization with some evidence of desert varnish; the lower levels of the gravel, where seen in shallow erosion gullies, are calcareous. The thickness could not be determined with accuracy but it is probably not more than 15 ft. at the most. No implements were found within the gravel proper, but on the surface were found a number of rolled and unrolled specimens of Levalloisian form, made predominantly from pebbles of chert, which have given to the artifacts a false appearance of primitiveness: a few Late Stone Age artifacts were also found always unrolled and fresh. The opposite, or right bank of the river is here formed by the alluvial Audamboi plains, which may have been due to flooding or ponding of the river upstream of the Bur Eghei barrier, so that no opportunity occurred of tracing the terrace on this bank. On the left bank at Melka Taka several flat-topped limestone residuals rise out of the alluvial plain to a height of a little over 50 ft. Occasional pebbles overlie the surface of these and are presumed to represent the 50–75 ft. terrace level.

A terrace composed predominantly of limestone pebbles of fist and larger size set in a calcareous matrix is seen at heights of approximately 30–120 ft. above the river between Callafo and Mustahil on the left bank, as also near Beletwein. It is probable that this represents the southward continuation of the Cura Ba'ad terrace.

At Sheik Gure (Pl. 36, fig. 2) near Bugda Acable were found the extensive remains of a terrace developed within the Cretaceous series and banked against or sometimes planing the Duldir lavas. These gravels are now considerably denuded, leaving the gravel in shallow mounds but in their original form they must have been some 5 ft. or more thick. Around the base of the main lava outcrops, north of the wells at Duldir, and overlying other outcrops between Giglei and Bugda Acable the gravel is composed of well-rolled and smoothed pebbles of lava with some chert examples exhibiting typical desert varnish. Approximately 1 mile north of Duldir the terrace is seen stretching north-west/south-east and was examined for approximately ½ mile: here the gravel is composed predominantly of well-rounded chert pebbles exhibiting typical desert varnish together with large and small fragments of fossil wood. The whole has been intensely ferruginized and set in a matrix of pisolitic ironstone.[1] Professor Zeuner believes it to be indicative of a weathering horizon of the lateritic or 'terra rossa' type. From an examination of the fossil wood it would seem that this is not in fact contemporary with the aggradation of the gravel, but that it has been derived from an earlier deposit and was rolled and

[1] For the composition of this deposit, known by Italian writers as 'Bohnerze', see Aloisi, P. & de Angelis, A. M. (1938, p. 118).

redeposited at the time of formation of these terrace gravels. Chiarugi recovered from these 'bohnerze' beds the following identifiable fossil woods: *Palmoxylon*, *Dipterocarpoxylon*, *Evodioxylon*, which are said to present Malayan affinities, and the presence of the last species suggests on North African evidence a post-Miocene date for these woods. The original bed from which these remains were derived was not seen, but insufficient time prevented a proper search being made. It is probable also that the majority of the chert pebbles have been derived from some earlier deposit or else they must have been subjected to very intense rolling to reduce them to their present smooth and flattish form. The local nature of the lava pebbles, suggests that the ingredients of the gravels have not travelled far.

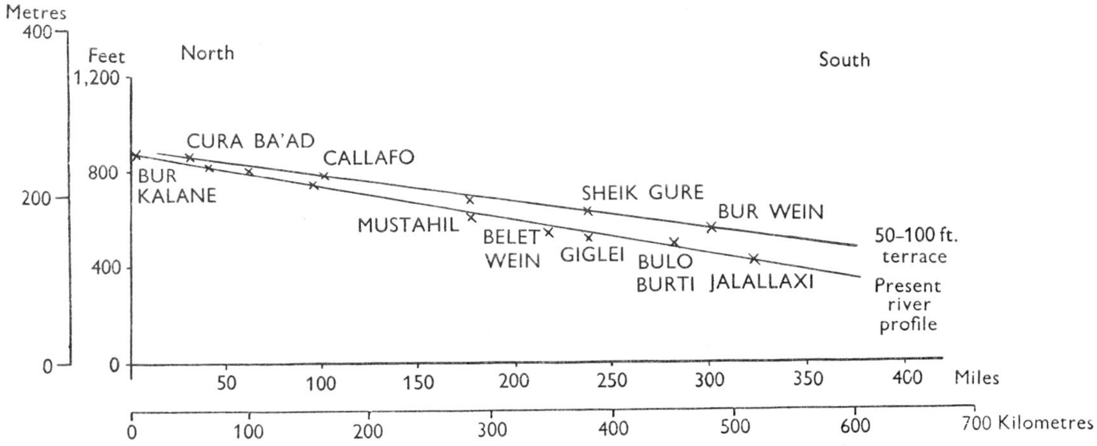

Text-fig. 10. Profile graph, middle Webi Shebeli.

This terrace stands at Sheik Gure some 100–115 ft. above the river, but gravel patches stand considerably below this near Giglei. Overlying the gravel and to some degree incorporated in redeposited patches are artifacts of Levalloisian type again made from pebbles as at Cura Ba'ad—a few of these specimens also are rolled, but there was no definite evidence to suggest that they are contemporary with the aggradation of the 100 ft. terrace. It is considered, therefore, that the Sheik Gure terrace should be correlated with the Cura Ba'ad/Darchaua terrace, and with the flat top of the isolated hill of Bur Wein at about the southern limit of the middle course of the Shebeli. This hill is estimated as being approximately 100 ft. above the river. It is composed of grey, somewhat crystalline, limestone with a few small grains of (?) lava. On its surface are found occasional pebbles of basalt.

In Text-fig. 10 the position of the 50–100 ft. terrace gravels, or gravels believed to represent this terrace or terraces, are plotted on a graph against the profile of the present river. It will be seen that the terrace and profile of the present river tend to converge upstream while downstream the height of the terrace above the river becomes progressively greater. This suggests some rejuvenation at the mouth of the river, such as might result from a lowering of sea-level possibly accompanied by some warping of the surface, ponding and perhaps local deflexion of the river. Subsequent to the formation of the 50–100 ft. terrace or terraces renewed erosion resulted in the excavation of a buried channel. The average depth of the bed of the channel below the present river bed is difficult to estimate but it would seem to be greater in the lower reaches near Jalallaxi, where a depth of 40 ft. of alluvium is recorded. At the confluence of the Darchaua Tug with the Shebeli, bed rock (gypsum) was seen in one section and it is probable that the thickness of the alluvial deposits filling the channel is not so great.

The filling of this channel, to an average thickness above the river bed of some 12–15 ft., comprises two series of usually consolidated sands and silts with some gravel at the base, which are separated by a red bed or sometimes by a slight disconformity. Disconformably overlying the upper series is a reddish brown powdery alluvium sometimes with grit and clay bands and lenses. This deposit is contemporary with the formation of a bench cut in the earlier alluvial series approximately 5–8 ft. above the bed of the present river—this bench is most noticeable on the inner side of meanders and is not merely a flood platform of the river as it has in some places been cut through by subsequent erosion. This bench presumably represents the floor of the channel in which the river was running during an erosion phase immediately anterior to the present cycle.

This sequence can be seen in a number of sections in the Webi Shebeli itself and within lateral gullies or tugs within the valley.

At Melka Taka (Text-fig. 11) the section exposed in the right bank of the river at the pontoon crossing is as follows:

Top: (iv) Brown powdery alluvium cut by gully erosion and varying in thickness from 15 ft. to 3 ft.
(iii) Disconformity. Red gypsiferous alluvium 8 ft. thick in places.
(ii) Gypsiferous kunkar bed, average thickness 6 in.

Base: (i) Grey clay and alluvium, total thickness could not be estimated as base was not seen, but some 4 ft. remained exposed.

On the opposite bank was seen a bench, cut presumably in the series (iii) and (iv) and overlain by a reddish brown alluvium. The present river bed is composed of a sandy alluvium with pebbles.

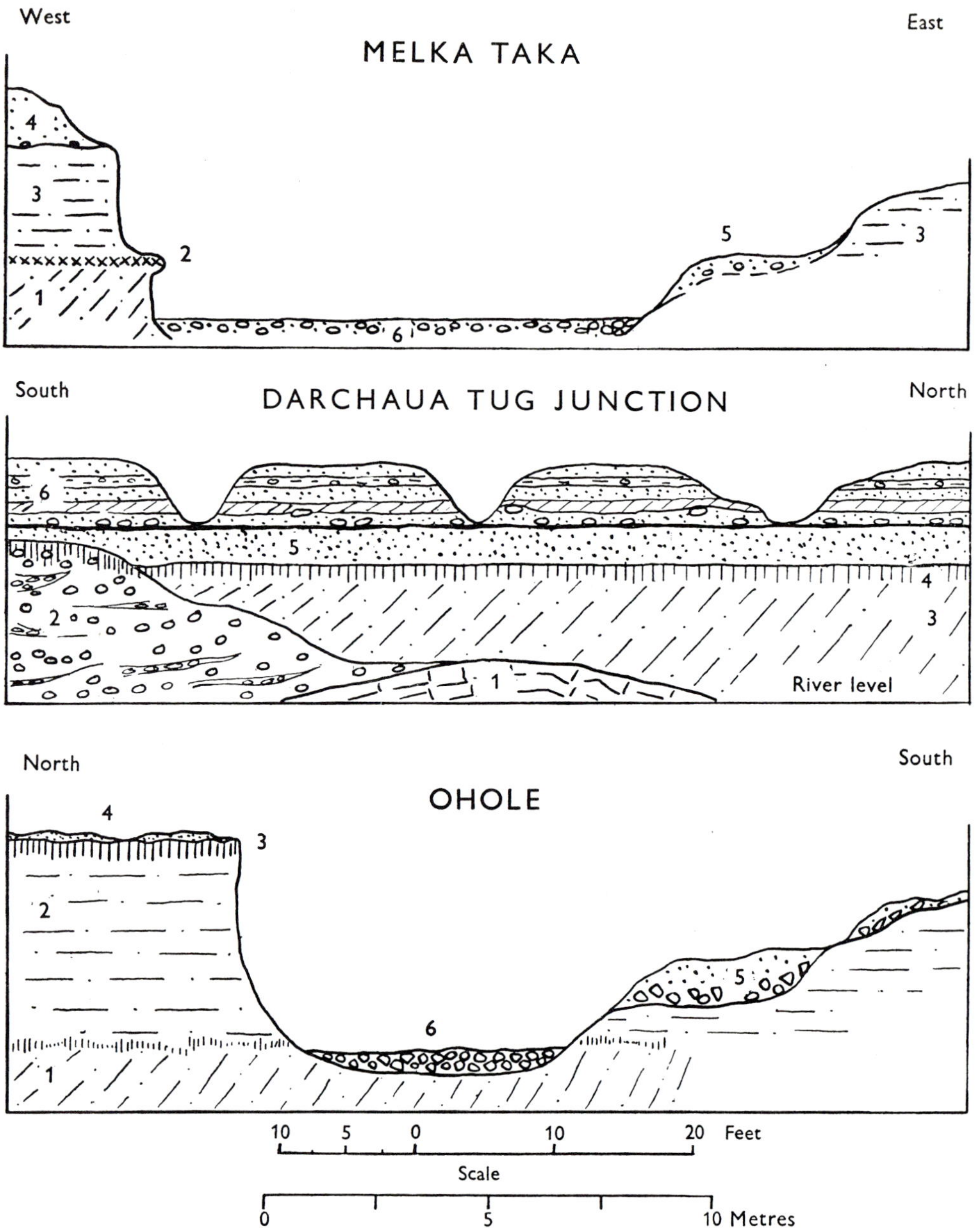

Text-fig. 11. Sections, Webi Shebeli at Melka Taka, Darchaua Tug junction and Ohole.

Key to Sections

Section across the Webi Shebeli at Melka Taka

1. Grey clay and alluvium: Lower Tug Alluvium.
2. Gypsiferous kunkar bed—reddened.
3. Red gypsiferous alluvium: Upper Tug Alluvium.
4. Brown powdery alluvium. }
5. Reddish brown alluvium. }
6. Redeposited gravel in river bed.

Section exposed in right bank of the Webi Shebeli at its junction with the Darchaua Tug

1. Gypsum rock.
2. Current-bedded gravels: Lower Tug Alluvium.
3. Grey alluvium: Upper Tug Alluvium?
4. Red gypsiferous kunkar.
5. Red aeolian/alluvial sand.
6. Banded clays, grits and gravel and aeolian sand.

Section across tributary tug at Ohole: middle Webi Shebeli

1. Calcareous marls: Upper Tug Alluvium.
2. Calcareous marls: Upper Tug Alluvium.
3. Oxidized land surface.
4. Red sands.
5. Torrent gravels and sandy alluvium.
6. Resorted gravel.

These deposits were sterile but approximately $\frac{1}{4}$ mile from the river on the right bank is a low outcrop of limestone, on the surface of which are occasional pebbles of quartz and lava in addition to the usual limestone and chert pebbles and around the edge of this outcrop, sometimes overlying the powdery brown alluvium, is a sparse factory debris of Late Stone Age facies.

At the junction of the Darchaua Tug the following section was exposed on the right bank (Text-fig. 11):

Base: (i) Gypsum rock, warped, possibly as a result of undersurface drainage.
(ii) 8 ft. or so of current-bedded, fine gravels (components derived presumably from the 50 ft. terrace), forming a bank against which are laid the later finer-grained sediments.
(iii) Grey alluvium with some clay, banked against (ii). Thickness 6–10 ft.
(iv) Gypsiferous kunkar band 6 in.
(v) Bright red deposit probably mixture of alluvium and aeolian sand, 3 ft. thick, also banked against (ii).
(vi) Banded silts, clays, grits, some gravel, red sand, and chocolate-brown alluvium 5–6 ft. thick.

Top: (vii) Truncated erosion gullies presumed contemporary with or immediately anterior to the existing cycle.

On the opposite bank at the mouth of the Darchaua, the aeolian deposit (v) was not seen but a bench had been cut at approximately 6–7 ft. above the tug bed, the top of which was formed by deposit (iv), from which was derived a typical Upper Levalloisian flake *in situ*. The chocolate-brown alluvial silts were here some 15 ft. thick and dissected by erosion gullies of the present cycle. From the surface and probably also from the upper levels of this alluvium was derived a somewhat indeterminate Magosio-Doian (?) factory debris.

An exactly similar sequence to that at Darchaua is also seen some 5 miles up river on the left bank at Fara'anou with the difference that the gravels (ii) were not exposed.

On the right bank of the river at Mustahil at the junction of the side-road leading to Graziani's villa, with the main Hoddur/Mustahil road, is a borrow pit which showed the following sequence. From the surface down:

Top: (3) 3 ft. of hillwash and scree deposit with angular fragments of limestone.
(2) A banded brown sandy alluvium. The upper 6 in. to 1 ft. comprises a coarse black-brown, gritty alluvium. Below this is a band some 6 in. thick of fine light-brown deposit, and below this again about 6 in. to 1 ft. thick of chocolate-brown alluvium of medium texture. Maximum thickness not more than 2 ft. 6 in.

Base: (1) A white to grey sandy alluvium, highly calcareous, and evenly bedded, but with pebbles and angular fragments of limestone and chert distributed evenly throughout. A maximum of 6 ft. of this deposit is exposed, the base not being visible. On the right-hand side of the pit a lense of black clay (having a maximum thickness of 1 ft.) intrudes some 3 ft. within this calcareous alluvium.

Approximately 40 ft. above the river bed are compact, coarse gravels consisting of rounded and angular pebbles and boulders of limestone, which are found for the most part flanking the base of the residual hills near the centre of the valley. A red sandy alluvium to a depth of a foot or so sometimes overlies them. Away from the immediate base of the hills the gravels fan out and shallow tugs draining into the Webi Shebeli have resorted them so that, in some places, they overlie the red alluvium: all rest on the gypsiferous series.

On the left bank at Mustahil the gravels are again in evidence, but are rarely more than 3 ft. thick and are sometimes cemented by a calcareous matrix and overlain by a brecciated scree deposit at the base of the scarp. Here the brown alluvium lies several feet below these gravels in the flood plain. Resorted hillwash from these gravels on the left bank sometimes overlies the brown alluvium in tributary gullies between Mustahil and Beletwein.

An important area on the left side of the valley some 5 miles from Ferfer on the road to Dusa Mareb, is situated at Ohole (Fa'arra'ad) where a short tributary tug once flowed to join the Webi Shebeli. The sequence here is as seen in Text-fig. 11. The calcareous marls, which are divided into two parts, an upper and a lower, by a slight disconformity only recognizable in some of the higher sections, are evenly bedded and contain in the upper half occasional artifacts of the Upper Somaliland Levalloisian. This deposit is apparently a marl with some sand, possibly a lake or 'pan' deposit. The present feeble erosion phase has failed to cut through it and the bed of the tug overlies it. The surface of the marl, where not removed by subsequent erosion, is highly kunkarous and piped, and has been stained to an orange colour by iron oxides. It is overlain in places by thin orange to red sands, which appear to be of a residual nature resultant from the breaking-down of this oxidized surface. Elsewhere, predominantly on the left bank of the tug, erosion has resulted in the cutting of a bench in the marls some 8 ft. above the present bed, on which is deposited a coarse, loosely cemented gravel, with numerous angular components and overlain by a red-brown sandy alluvium, which passes into sandy hillwash on the upper slopes. In the tug bed was noted some 2 ft. of resorted gravel.

From below the oxidized land surface at the top of the marls were recovered implements, flakes and cores of Somaliland Stillbay type, and washouts from the gravel yielded similar specimens, presumably derived.

South of Beletwein and above Giglei the Webi Shebeli flows over a barrier of Cretaceous limestones, which are seen in several sections outcropping in the bed of the river. At Giglei itself the following section is exposed, reading from the surface downwards:

(3) Dirty-white sand dunes to 4 ft., with Late Stone Age living sites on their surface.

(2) Chocolate-brown alluvium some 4 ft. thick, passing down into 2 ft. of angular gravel with shells of the freshwater oyster.

(1) 25 ft. of grey, clay-like alluvium with a gypsiferous base. This deposit passed below water-level and bed rock was not seen.

At Jesomma, at the foot of the Cretaceous sandstones forming the northern scarp of the Webi Shebeli valley, near Bulo Burti, the writer found extensive occupation sites. The site at Jesomma lies about $\frac{1}{4}$ mile below the wells and consists of a series of eluvial compact sands with temporary land surfaces and occupation levels (Text-fig. 12). These beds are believed to represent a terrestrial deposit, which accumulated over a long period as a result of surface wash and possibly some wind action: except possibly in the lowest levels there was little to suggest that they were laid down in water and in fact, except at the base (exposed in a small tributary gully), no water-deposited gravels were seen. However, in section in this gully, overlying bed rock, was seen 2 ft. or so of angular gravel, the accumulation presumably of torrential streams. The compact sands and overlying deposits slope gently down towards the tug which here has a south-easterly direction. These sands, composed largely of rounded grains, are yellow in colour in the lower levels, changing to red in the upper in some places and are ferruginous with pisolitic ironstone nodules: this pisolitic material never occurs as a compact mass however, but only as lenses of small, scattered pisolites. Bedding planes are clearly visible in places where land surfaces have existed at different levels and are further accentuated by poorly developed columnar and pipe structure of the upper foot of the underlying sand: these temporary surfaces are rich in Levalloisian artifacts, which are quite fresh in appearance. The thickness of these sands varies from 5 ft. to 25 ft. or more. Occasionally in the resorted material lying on the exposed surfaces is found a chalcedony or quartz pebble or a well-rolled nodule of chert. These have probably been derived from the Cretaceous beds of the scarp, but it is not impossible that they may have been derived from a terrace gravel of the Webi Shebeli not seen by the writer. These occasional pebbles are found in association

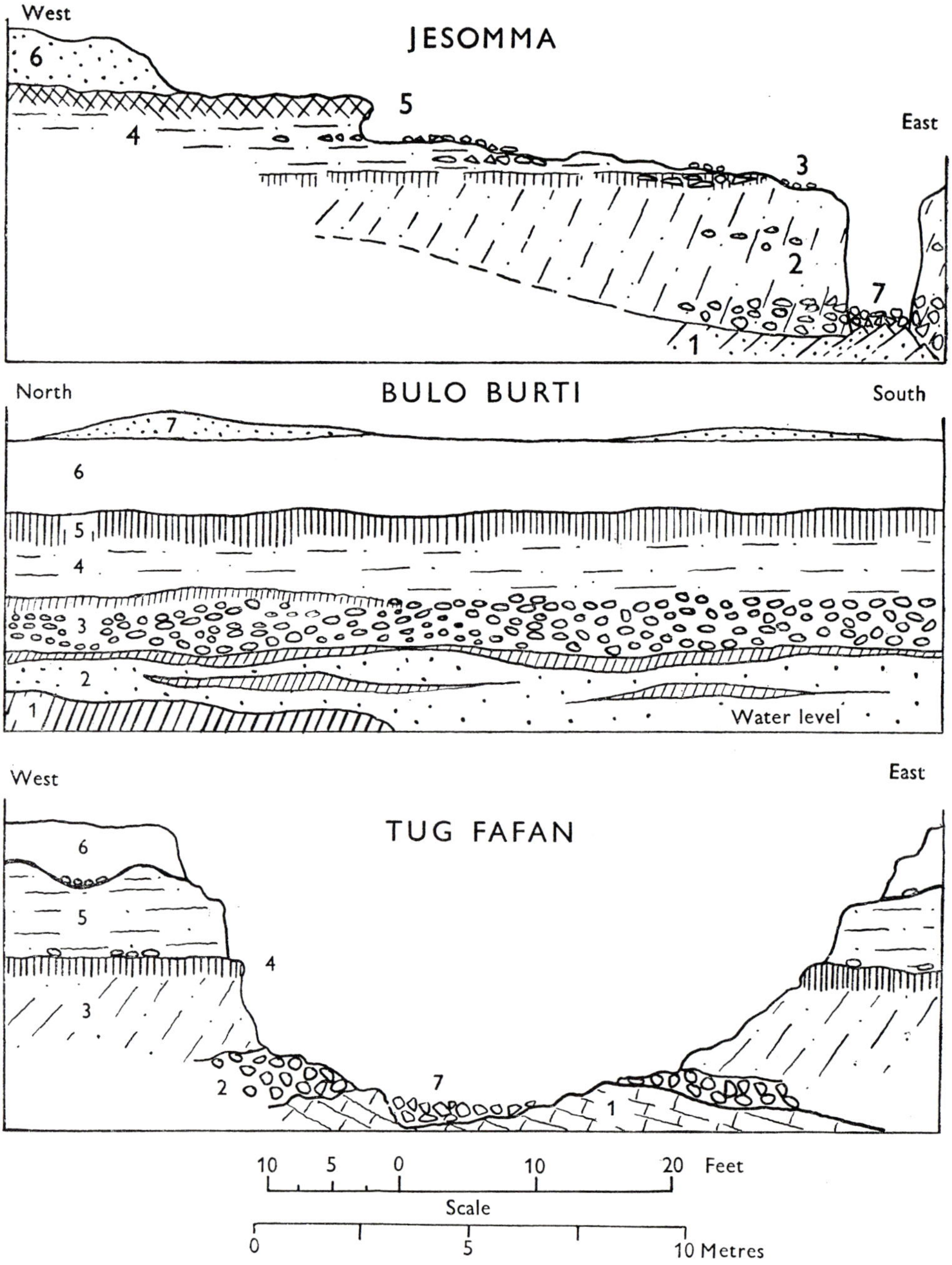

Text-fig. 12. Sections, Webi Shebeli at Jesomma and Bulo Burti and Tug Fafan at Gabredarre.

with fresh and angular fragments of sandstone, sometimes weathered, but showing no evidence of water action. There is evidence in places of an irregular red, or rather pink, layer suggesting a break in the deposition of these sands.

Overlying these yellow and red sands with Levalloisian and a few Stillbay artifacts is a highly ferruginous horizon, possibly an incipient ironcrete with pipe weathering. It was presumably formed under wetter climatic conditions than pertain today in this area. The deposit, having a maximum thickness of 2 ft. is widely developed and contains in its lower levels numerous artifacts of the Somaliland Magosian. This deposit is overlain by 5 ft. and more of red, compact, aeolian sand. The sequence found at Jesomma is seen in the section drawing at Text-fig. 12.

The Jesomma sequence from the edge of the Webi Shebeli valley is reflected in that seen in section in the bed of the river at Bulo Burti, at the back of the District Commissioner's house, and recorded in section drawing, Text-fig. 12. Here calcareous sandy gravels dip below the water-level and lie presumably in the buried channel. Overlying these lower gravels is a second, evenly bedded, gravel in a yellow sandy matrix, passing up into a grey alluvium, the upper part of which has been reddened and presumably equates with the ferruginous deposit at Jesomma. This in turn is overlain by a buff to grey alluvium on the surface of which is a fine, dirty-brown deposit probably of aeolian origin.

Certain, as yet unrelated, deposits in the form of dune sands now consolidated and fossil, were seen on the higher slopes of the valley on the right bank between Mustahil and Bulo Burti. These fossil dunes, which rise to heights of 30 ft. or more above the valley plain, rest on the gypsum rock and are themselves

---

Key to Sections

Section at Jesomma Wells: middle Webi Shebeli valley

1. Cretaceous sandstone.
2. Lower ferruginous sands and land rubble: Lower Levalloisian.
3. Red bed representing a fossil land surface.
4. Upper ferruginous sands and land rubble: Upper Levalloisian and Stillbay.
5. Ferruginous horizon: Magosian and derivative industry.
6. Red aeolian sand.

Section exposed in the left bank of the Webi Shebeli at Bulo Burti

1. Massive kunkar.
2. Gravels, sands and kunkarous lenses.
3. Evenly bedded gravels overlain in places by kunkar.
4. Grey alluvium.
5. Red bed.
6. Buff to grey alluvium.
7. Dirty windblown sands.

Section exposed in the bed of the Tug Fafan at Gabredarre: Ogaden

1. Gently dipping limestones.
2. Lower Tug Gravels.
3. Lower calcareous alluvium, possibly tufaceous.
4. Red bed representing a fossil land surface.
5. Upper calcareous alluvium, possibly tufaceous.
6. Brown river silt and sand ('flood loam') resting on an eroded surface.
7. Resorted gravel in tug bed.

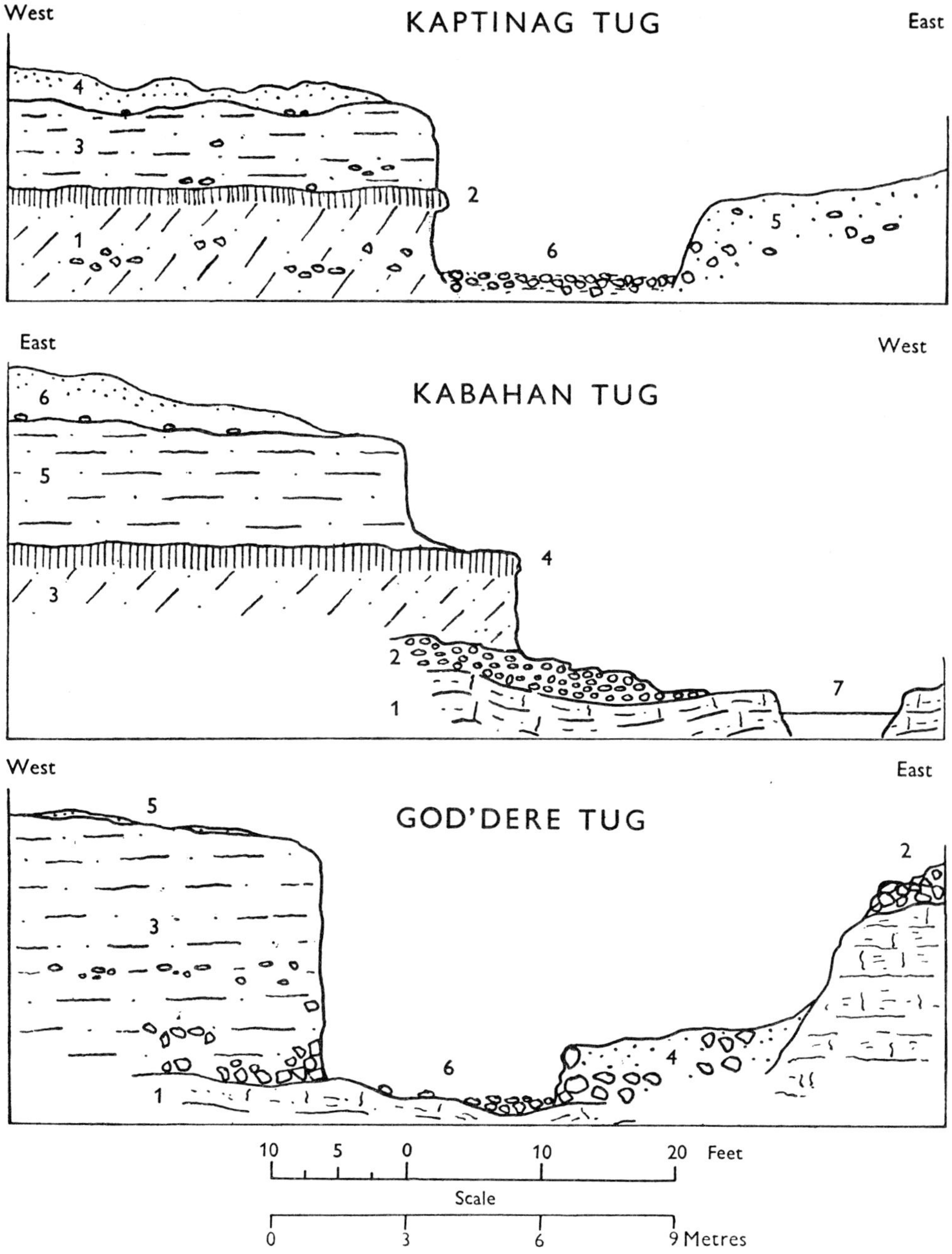

Text-fig. 13. Sections, Kaptinag, Kabahan and God'dere Tugs.

gypsiferous. From their very local occurrence and the fact that they are usually grouped round wells or what were formerly well areas, it would seem that they represent wind-blown sands which accumulated against the vegetation of water-holes. Such dunes are those at Dudum Har, on the edge of the valley plain south-east of El Goran, composed of a core of compact white gypsiferous sands, the surface of which has been broken down into loose white sands: these dunes are heaped round what must have been at one time a water-hole. Nearer the centre of the valley at El Ferrug, near Mustahil, are similar dunes, but here wells still provide saline water; others occur also at Kirkirri and El Abdulla. Again some $7\frac{1}{2}$ miles from Duldir on the left of the Giglei/Bugda Acable road is a long fossil dune composed of gypsiferous sands overlain by a few feet of the later, compact, red dune sands. There is no evidence of a former water-hole here and the dune has the appearance of a fixed 'seif' dune: the Far Abou dunes are also of this kind. While both the Dudum Har and El Ferrug sites were entirely devoid of evidence of any Late Stone Age occupation, almost ubiquitous at the well areas of the plateau and valley, the Duldir dune yielded, from the red sand, a sparse factory debris of this cultural stage. The Duldir dune also overlay gypsiferous beds which represent a surface below, and therefore later than, the 100 ft. terrace at Sheik Gure. It would seem, therefore, that some, if not all, of these fossil dune sites postdate the 100 ft. terrace and antedate the deposition of the red sands: they may be considered as contemporary in part with the calcareous gravels and other alluvial deposits of the buried channel, and it is assumed that in this broad, gently sloping valley strong winds caused some accumulation of dune sands on the valley flanks at the same time as the alluvial deposits of the river were being laid down.

Key to Sections

Section across Kaptinag Tug: Ogaden

1. Grey alluvium and fine gravel: Lower Tug Alluvium.
2. Gypsiferous horizon.
3. Red alluvium: Upper Tug Alluvium.
4. Fine brown alluvium.
5. Gravel and brown alluvium: aggradation terrace.
6. Resorted gravel in tug bed.

Section across Kabahan Tug: Ogaden

1. Warped gypsum rock.
2. Evenly bedded fine red gravels: Lower Tug Gravels.
3. Grey alluvium: Lower Tug Alluvium.
4. Red gypsiferous horizon.
5. Red alluvium: Upper Tug Alluvium.
6. Fine brown alluvium on eroded surface.
7. Pool of standing water.

Section across God'dere Tug near Water Point: middle Webi Shebeli

1. Gypsum rock.
2. Grey torrent-bedded gravel: Lower Tug Gravels (?).
3. Grey alluvium and torrent gravel passing up into red alluvium: Upper Tug Gravels.
4. Low-level aggradation terrace.
5. Aeolian sand.
6. Resorted gravel in tug bed.

(3) THE NORTHERN TRIBUTARIES

Text-figs. 12 and 13 show sections through the more important tributary tugs examined in the Ogaden, from which it will be seen that the same, broad sequence—of two calcareous alluvial deposits (sometimes gypsiferous) overlain disconformably by a loose, coarse gravel and alluvium—exists here also, the only difference being that in the tributaries between the Fafan and the Webi Shebeli subsequent erosion has cut deeply through the calcareous alluvium and exposed the base (Text-figs. 12 and 13).

(*a*) The Tug Fafan (Text-fig. 12). On the eastern edge of the Gorrahei plain and some 50–150 ft. above it are localized banks of lava pebbles. These appear to be derived from intrusive lavas and are of undoubted fluviatile origin. They compare with the 75–100 ft. terrace in the Webi Shebeli and with similar pebble banks in the southern tributaries.

The calcareous gravels and later deposits are well seen in the Tug Fafan, but only in one place was the base exposed and for the most part the kunkarous top of the lower calcareous alluvium forms the floor of the present tug bed (Pl. 37, fig. 1).

Sections near the road crossings in tugs between the Fafan and Melka Taka on the Shebeli show a similar sequence. Two of these sections, those at Kaptinag (66) and Kabahan (69) are shown in Text-fig. 13.

(*b*) At Kaptinag the grey alluvium yielded Levalloisian artifacts, which also occur derived in the resorted gravel of the present tug bed. Some sheet erosion of the brown to red, friable alluvium has taken place, and in the patches exposed, resting on the underlying red gypsiferous alluvium, are scattered Late Stone Age, Magosio-Doian artifacts and factory debris.

(*c*) At Damburweini (67) an almost similar sequence to that exposed at Kaptinag is seen. The base of the grey alluvium was not exposed but at the base of the red gypsiferous alluvium is an intermittent, thin, torrent-bedded gravel having a maximum thickness of 1 ft. The eroded surface of the brown and red alluvia yielded typical Doian artifacts. The low-level bench, or denudation terrace, is again in evidence.

(*d*) At Kabahan the base of the deposits was well seen and rests on gypsiferous beds, presumably of Cretaceous (?) age. The sequence here from top to bottom was as follows:

(7) Present tug erosion to depth of 25–30 ft., cutting a channel through the gypsum bed rock. Permanent water in pools.

(6) Friable and powdery brown alluvium (sterile), overlying a land surface.

(5) Red gypsiferous alluvium (sterile) well consolidated and conformably overlying.

(4) Red gypsiferous horizon (sterile).

(3) Highly gypsiferous grey alluvium passing to light red in upper levels (sterile).

(2) Consolidated gypsum cemented, red and yellow gravels with small, well-rounded pebbles of sandstone and limestone and occasional quartz and tuff pebbles together with some well-smoothed, but angular, larger pebbles and rock fragments. These gravels are evenly bedded and contain rolled and unrolled Levalloisian flakes and cores.

(1) Gypsum rock—warped due to local subsidence.

There is again evidence of a low-level bench cut in the gypsiferous alluvium: it is possible that at this site it represents a flood platform of the present tug, but in view of a similar bench at several other sites, this is considered unlikely.

The small and well-smoothed quartz and tuff pebbles in the basal gravels are presumed to have been derived from an earlier deposit not identified. On the surface of the plain, through which these tugs have eroded their beds, can be found occasional sandstone, quartz and limestone pebbles overlying the gypsum rock, which composes the floor of the plain. From this plain rise occasional isolated, low residual hills of limestone, which rise to a height of 30–40 ft. above the plain. One such is at Durualeh (68) where at the foot was found, in a scree or surface wash, a Levalloisian assemblage of artifacts, weathered but unrolled. Further south, towards the Webi Shebeli at Melka Taka, low hills of trap rock appear, and it is probable that these rocks intrude on a N.N.W.–S.S.E. axis.

(4) THE SOUTHERN TRIBUTARIES

Brief mention must be made of tributary tugs examined on the southern side of the Webi Shebeli valley between God'dere and Beletwein. Here the tributaries have developed steep-sided channels through beds of massive gypsum, sterile of fossils. This gypsiferous series has been extensively planed during a long erosion cycle antedating the tug erosion, and occasional pebbles of quartz, limestone and lava can be recovered from the surface. The only lava intrusions in this area that were seen were further to the south-west in the headwaters of tugs draining into the Juba; two lava outcrops, both in wide, shallow, mature valleys, similar to those described in the central Haud (see p. 82) and with pebbles of this rock covering the surface in their vicinity, were examined—one at Legullei (intruded through gypsum) and one at Madeida on the plateau south-west of Hoddur (intruded through limestone).

The tug at Awal Tirre has eroded a channel 30 ft. deep with vertical sides: no deposition of sediments was seen. The Bohol Gilgil, near God'dere, has cut a similar channel to a depth of 50 ft. within the gypsum series.

At God'dere itself, in the tug bed below the fort, which is built on the eroded gypsum surface, the sequence is as shown in Text-fig. 13. From the gypsiferous alluvium were recovered Levalloisian implements, while on the surface of this deposit was found a home and factory site of the Doian Culture. The bohols between God'dere and Danaio show the same sequence as that at God'dere, but, where examined, were sterile of artifacts. At El Goran, the same sequence of two gypsiferous alluvia cut through by a gorge some 30 ft. deep, which debouched upon the Shebeli valley plain, was noted; but to the south-east near El Abdulla, in the Bohol Washaga, artifacts were eroding from the upper levels of the gypsiferous alluvium, here almost white in colour.

III. THE LOWER WEBI SHEBELI AND THE 'BUR COUNTRY'

The lower reaches of the Webi Shebeli are characterized by deep alluvial sediments which, to date, have yielded little or no cultural material. Sections through these deposits in the main river are almost non-existent, but a few exist in tributaries that were examined, notably in the Bur Hakaba area.

Characteristic of the lower reaches is a greyish brown alluvium, the upper levels of which can be seen in section in the river at Jalallaxi, Villagio, Balad and Afgoi. At Afgoi (282 ft.), below the new road bridge, bed rock (limestone) appears to outcrop in the bed of the river, but it is more probable that this is the deceptively hard secondary limestone surface of a Pleistocene deposit. In a well at Giuhar, near Villagio, de Angelis records a depth of 40 ft. of grey-brown alluvium, and deduces from a chemical analysis that the components were derived from the Basement Complex rocks, which form the floor of the southern coastal plain.[1] At Balad (351 ft.), some 15 ft. of the brown alluvium is exposed in the river bed and overlying this, in a pit near the Residente's house, on the right bank, is some 2–3 ft. of lenticular, loose, white sand, within which occurred pottery and some unfossilized bone. This deposit must, therefore, be a comparatively recent one. Distant some 300 yd. from the river, the ground gently rises and there is exposed in road ditches a compact, unbedded, white to yellow sand, which is probably of aeolian origin and is redeposited sand from the coastal dunes, here distant some $\frac{1}{2}$ mile from the left bank. It is thought that

[1] Aloisi, P. & de Angelis, A. M. (1938, p. 110).

this sand overlies an earlier terrace of the Webi Shebeli. While the upper levels of the grey-brown alluvium are composed of very fine material, which may well be wind-blown,[1] the lower levels are undoubtedly laid down by water and it is necessary to suppose extensive flooding of the lower reaches of the Webi Shebeli at the time that this deposit was laid down. The reason for the flooding was the formation of the coastal dune barrier, through which, unlike the Juba, the river was unable to force its way. This flooding must at one time have been extensive in the elbow or bend comprising the alluvial plain of Dafet, caused by the diversion of the river to the south-west. The wells at Wanle Wein (308 ft.) some 40 miles to the west are over 40 ft. deep and are excavated through the same grey alluvium which gives place, apparently gradually, to current-bedded, alluvial grits and fine gravels, which are often calcareous; the total thickness of the alluvium could not be recorded, but that this is considerable is certain. For the most part, therefore, the later geology is obscured in the lower reaches of the Webi Shebeli by a thick mantle of alluvium.

The 'Bur Country' is, broadly speaking, that part of the coastal peneplain between the alluvial plain and the foot of the plateau proper, at an average elevation of between 500 and 1500 ft. (A section can be seen in Text-fig. 14.) Here the granites and schists of the Basement Complex rocks are exposed by the removal of the overlying Jurassic surface,[2] which has been denuded and planed at the time of formation of the coastal peneplain. Only where the more resistant elements have withstood erosion is the monotony of the plain broken by the isolated 'bur': between Bur Hakaba and Dinsor over 140 such burs are known to exist (Pl. 37, fig. 2). Over wide areas this plain is covered by a white sandy calcareous deposit, apparently sterile, which is sealed by a deposit of kunkar with irregular pipe structure. Sometimes within the sand and kunkar are found cemented pebbles of limestone and quartz. This deposit varies considerably, but in the stretch between Isha Baidoa and Bur Hakaba it is over 20 ft. deep and is probably much more. Stefanini states that the depth is considerable but gives no details. Overlying this kunkarous deposit is a sandy ferruginous formation characterized by a bright red to orange sand usually with pisolitic nodules in its lower levels: this sand varies greatly in thickness from more than 25 ft. in some wells near Nunei, to a mantle of a few feet only

[1] Professor Zeuner has examined a sample of this alluvium which contains a large amount of finely divided calcium carbonate. In grain size the matrix resembles loess which supports the view that it was transplanted by wind.

[2] In a letter to the writer Dr F. Dixey informed him that the sub-Jurassic surface was eroded mainly in the later Trias, since the overlying beds are of the lowermost Jurassic. Inland, however, the base levelling continued through a large part of the Jurassic, and the main peneplain was not terminated by uplift until some time later in the Jurassic.

covering the kunkar: it is thought to be of mixed aeolian and eluvial origin.[1] This orange sand is known as the 'Doi' by the Rahanwein and Digil Somalis of southern Somalia and covers quite two-thirds of the southern coastal plain: it carries a thickish acacia growth, constitutes the stock-grazing area common to these tribes, and is the home of the Bon and Ribi, the free hunting groups of the south. Manasse, Artini and de Angelis have examined and analysed these sands and the pisolitic and ferruginous lower levels, which the latter describes as being analogous to the formation of 'terra rossa'.[2] Stefanini states 'from this eluvial covering, in parts perhaps aeolian and eluvial, emerge as isolated points the inselbergs (burs) witness of an intense and prolonged period of erosion, of decomposition and breaking down, that must have been concluded by the Plio-Pleistocene, since the modest incised valleys of the region, with their Holocene and Recent alluvium are excavated in it, and are therefore more recent'.[3]

Before passing to a consideration of the sequence in the tributary rivers, mention must be made of two further formations. First, the pediment of conglomerates and breccia, cemented to rock hardness at the surface by calcium carbonate, which flanks the base of the Jir or scarp, joining the southern plateau with the coastal plain. The thickness of this deposit (at an average altitude of 1000 ft.) is considerable, and wells at El Ille are said to be over 200 ft. deep. Apart from wells, no deep sections exist through the conglomerates and no cultural material was found within them: they spread out at the foot of the scarp and merge imperceptibly with the calcretes of the plain proper. Secondly, scatterings of iron-stained quartz pebbles—normally an inch or less in diameter, well rolled and smoothed, but occasionally mixed with larger specimens up to 3 in. diameter. These scatterings of pebbles overlie the granites and schists where these are exposed at the surface, or are sometimes covered by a thin layer of orange sand. They were found at the rifle range, Isha Baidoa, on the very edge of the southern plateau; widely spread in the area of Bur Dinsor; in the Garas Got area of Bur Hakaba district, and at several other places on the coastal plain. The writer would identify them with the erratics and pebbles incorporated in the sandy kunkars and surface limestone of the Doi; they probably represent the remains of one of the earliest deposits on the coastal peneplain, having been incorporated later in the sandy kunkars, which, where they overlie thinly the Basement Complex rocks, have been leached by the acid nature of these latter.

Round the foot of the burs, notably those of Bur Hakaba, Bur Eibe, and Bur Degis, bed rock is very uneven and some well shafts show a considerable depth

[1] Stefanini, G. (1916, p. 255).
[2] de Angelis, A. M. (1938, pp. 115 *et seq.*).
[3] Stefanini, G. (1936, p. 84).

of alluvial deposits. Some 30 ft. or more of coarse pebbles and boulders of quartz and quartzite set in a grey calcareous matrix pass up into fine-grained deposits of grits and fine gravels, set in a matrix of grey sand with some clay, and finally into wind-deposited yellow sands at the surface. A few Levalloisian artifacts and some fauna have been recovered from the upper levels of this deposit, and quantities of Late Stone Age tools overlie them.

The sequence in the tributary tugs of the Bur region is best seen in that of the Bohol Bandoi on the road between Bur Hakaba and Bur Eibe and again at the road bridge on the Audegle road. These 'bohols' have eroded wide shallow valleys through the crystalline rocks. Some 50–75 ft. above the river bed and from ½ mile to 1 mile on either side, almost at the junction with the surface of the plain, is a coarse gravel composed of well-rounded pebbles of quartz mainly of larger size than those found on the plain itself and not so iron-stained. This gravel is some 2 ft. deep at the most and, where examined, proved to be quite sterile of cultural material. Within the bed of the river is exposed 10 ft. of yellow, kunkarous alluvium. The upper levels are sandy, the lower of more clay-like consistency. The base of this deposit was not exposed and lay buried below the present river bed. The top of this alluvium yielded some fresh flakes and one core of Levalloisian type, in quartz and limestone. Overlying the yellow alluvium, but with no very pronounced stratigraphical break, is 4–5 ft. of red-brown alluvium of sandy texture. The reddish colour is most marked in the upper half of the deposit, the lower part being lighter in colour. Below these deposits a denudation terrace has been cut in the yellow alluvium some 6 ft. or less above the bed, on which rests 3–4 ft. of gritty torrent gravel and grey sandy alluvium containing some derived material. The sequence is completed by erosion to present contours. On the eroded surface overlying the red alluvium was found a scattering of Doian factory debris. A section exposed in a new well in the grounds of the Fiturari's Rest House at Bur Hakaba is indicative of swampy conditions surrounding the bohols in this area. Here, from the surface down, was found 3 ft. of a chocolate-brown alluvium, passing down into laminated sand and fine gravel lenses. This rested disconformably on the uneven surface of the underlying bed, which comprised 6–7 ft. of black clay, gritty in places and with practically no pebbles. This clay rested on an undulating bed, 2 ft. thick, of fine then coarse gravel which in turn rested on an uneven surface of partially decomposed granite, with veins of calcium carbonate running diagonally downwards. This gravel deposit shows no evidence of bedding except that the largest pebbles are usually at the base, and it was characterized by well-smoothed quartz pebbles similar to those described as occurring in places on the surface of the plain: it was not markedly calcareous and was sterile.

Surrounding the Burs, and in rock-shelters, were found numerous Late Stone Age occupation sites, the stratigraphy of which will be described when considering the archaeology of the region.

SUMMARY

We may now summarize the stratigraphical succession in the three sections of the Webi Shebeli, and on the southern and central parts of the plateau:[1]

(i) UPPER WEBI SHEBELI

(*a*) The river originally probably had its source in a wide, but shallow lake which filled the depression on the high plateau known as the Ghedeb Basin. The outlet from this lake was through a shallow gorge, then through a wide valley cut into the Cretaceous sediments. Diatomaceous deposits of Middle (?) Pleistocene Age were formed in the basin of the lake presumably under wet climatic conditions.

(*b*) Earth movement, volcanic activity and tilting of the plateau to the south-east followed during a dry period equated probably with the Last Interpluvial, or beginning of the Last Pluvial, the main Abyssinian Rift Valley acting as an axis of elevation.[2] Volcanic tuffs accumulate within the basin of the former lake, which was drained at this time.

(*c*) The elevation of the surface accompanying (*b*) rejuvenated the erosive force of the river which commenced to cut a 300–400 ft. deep gorge with V-shaped profile; some aggradations of sediments thought to contain artifacts of Upper Pleistocene age are known from the lower parts of the upper valley.

(ii) MIDDLE WEBI SHEBELI

(*a*) Major erosional phase of long duration in which a wide flat valley was cut some 600–700 ft. deep in at least two stages with a halt at 400 ft. This initial and long period of valley cutting took place at least as early as the end-Tertiary or Lower Pleistocene.

(*b*) Major depositional phase, again of long duration, in which thick conglomerates and some breccias, capped by thick deposits of surface limestone spread down the upper slopes of the valley. Some of the older gypsiferous beds may date to this time.

(*c*) A further erosion phase during which a terrace, or terraces, was formed at from 50 to 100 ft., on which was deposited gravel subsequently calcified.

[1] No opportunity occurred of examining, other than very superficially the deposits of the Juba valley, but the sequence at Bardera showed the following succession and is closely similar to that found in the Webi Shebeli:

(*a*) Gently sloping valley eroded in Trias limestones and marls on the denuded surface of which occur, on the upper slopes, occasional quartz pebbles and other erratics.

(*b*) On the left bank a 30–40 ft. calcareous aggradation terrace of fine current-bedded gravels, grits and sands on which Bardera fort is built: the lower levels are obscured by the later deposits.

(*c*) A deposit of reddish brown alluvium aggraded in a channel cut within the calcareous gravels and grits, and some 20 ft. below their surface. This alluvium has been analysed by Artini & de Angelis (de Angelis, 1938).

[2] Nilsson, E. (1940, p. 52).

(*d*) Ferruginization of the 50/100 ft. terrace gravels, during an arid or semi-arid period; accompanied possibly by some gentle warping resulting in ponding of certain areas during the subsequent phase.

(*e*) Rejuvenation and renewed erosion resulting in the cutting of a deep channel, the floor of which is now, for the most part, buried beneath the present tug and river beds.

(*f*) Aggradation of bedded gravels in the main river (torrent gravels in the tributaries) overlain by a sandy alluvium, the whole having been affected by calcification. Levalloisian Culture.

(*g*) A dry period of short duration temporarily halts aggradation. During this time the surface of the alluvium was subject to more pronounced calcification. Levalloisian Culture.

(*h*) Renewal of semi-arid conditions and of aggradation at which time a further gravel and/or alluvium was deposited. The markedly redder colour of this alluvium indicates the onset of a dry phase. Levalloisian and Stillbay Cultures. Some fossil dune formation on the higher slopes.

(*i*) Formation of incipient ironcrete and pisolitic ironstone, probably contemporary with (*j*), followed by red wind-blown sands denoting arid conditions. Somaliland Magosian and Doian Cultures.

(*j*) Minor erosional phase and a return to semi-arid conditions during which a denudation terrace was eroded in the earlier deposits some 8 ft. above present river level, on which accumulated torrent gravels and fine sandy alluvium. Some flooding of ponded (?) areas, and the formation of swamps. Somaliland Magosian and Doian Cultures.

(*k*) Some minor erosion, deposition of orange and dirty sands, and erosion to present contours. Doian Culture.

(iii) LOWER WEBI SHEBELI AND COASTAL PLAIN

(*a*) Formation of coastal peneplain to which the valley of the Webi Shebeli is graded, during the later Tertiary.

(*b*) Aggradation of conglomerates at the foot of the plateau scarp during the Lower (?) Pleistocene.

(*c*) Feeble erosion on the coastal plain by tributaries, followed by the aggradation of a high level gravel terrace, which shows evidence of weak ferruginization.

(*d*) Accumulation by wind and eluvial causes of a thick deposit of sand (incorporating earlier erratics) on the coastal plain. Renewed erosion within the tributary valleys, followed by the aggradation of two deposits of alluvium separated by a minor disconformity and with evidence of some swampy areas. Increasingly dry conditions resulted in the calcification of these deposits.

(*e*) Deposition of ironcrete and pisolitic ironstone probably contemporary with (*f*) followed by red and orange sands during an arid phase of some intensity.

(*f*) Minor erosional phase accompanied by extreme flooding in the lower reaches of the Webi Shebeli and the formation of swamps. Some erosion by tributary tugs and formation of a low-level bench followed by the accumulation of torrent gravels and grey-brown alluvium.

(*g*) Dirty wind-blown sand and erosion to present contours.

(iv) THE SOUTHERN AND CENTRAL PLATEAU

(*a*) Formation of deep sinks and solution valleys over a long wet period followed by the accumulation of thick conglomerates filling these sinks and spreading extensively over the plateau peneplain surface. Prolonged calcification follows.

(*b*) Further development of sinks, later blocked and resulting in formation of small lakes in which accumulated two tufa deposits, sometimes containing fauna and a Levalloisian industry. Tufa formation by springs round the edge of the plateau.

(*c*) Extensive deposition of aeolian red sand during an arid phase.

(*d*) Renewed tufa formation by springs: pan areas form small temporary lakes and ponds. Doian Culture.

CHAPTER 5

THE DANAKIL RIFT

Under this name is included the wedge-shaped graben of semi- and complete desert lying between the walls of the Abyssinian and Somaliland fault scarps and the Danakil Horst. The evidence from this area is meagre and therefore inconclusive so that considerably more detailed work will be necessary before the later geological and cultural sequence can be finally established. However, certain cultural evidence recovered by the writer from the southern Danakil region has prompted the following review of the succession in this inhospitable part of the Horn of Africa.

I. THE EARLIER VOLCANIC SERIES

Any study of the sequence within the Danakil Rift is dependent upon the history of the rift movements. The core of the rift has been shown to be composed of rocks of the Basement Complex edged with Mesozoic sedimentary deposits, both series being extremely faulted. Overlying these rocks, over wide areas on the plateau and in the rift, are volcanics of different ages, which do not seem yet to have been conclusively dated. Krenkel[1] has summarized the Late Tertiary volcanic movements, but does not give any definite evidence as to their age. Blandford[2] divides the volcanics into two phases (1 *a*, *b* and 2).

(1 *a*) An earlier Trap Series (Ashangi Series) comprising dolerites and amygdaloidal basalts, similar to the Deccan Traps, then thought to be of Upper Cretaceous, but now of Tertiary (Miocene) age.

(1 *b*) A later Trap Series (Magdala Series) comprising felspathic trachytes, resting unconformably on (1 *a*) and intercalated with sedimentary strata containing the fossil *Nicolia aegyptiaca* which is considered by Stefanini[3] to indicate a Cretaceous rather than a Tertiary age.

(2) The Aden Series (of quite distinct lithological character) comprising basalts, in part intercalated with sedimentary rocks of Recent age.

[1] Krenkel, E. (1925).

[2] Blandford, W. T. (1869, p. 190).

[3] Stefanini, G. (1933, p. 104).

Dainelli & Marinelli[1] similarly differentiate between the two Trap formations, the first comprising doleritic and olivine basalts, the second trachytes (Senafe Traps), and show them to be quite distinct from the volcanics of the Aden Series, comprising doleritic basalts with some porphyry (distinguished from the volcanics of the plateaux by their external appearance), also liparitic-spherulites, pumice or breccia, and dacite: these are intercalated with sedimentary beds attributable to a Plio-Pleistocene Age.

Gregory,[2] in his work on the Rift Valley has a section on the Abyssinian Rift Valley, and draws up a correlation between the volcanics of this area and those of the East African Rift. The Ashangi Group (Trap Series) he equates with his Kapitian (Upper Cretaceous) Series. The main Magdala Series, he equates with the Laikipian (of Miocene age),[3] while the upper part of the Magdala Series and the volcanics in the Awash (or Hawash) valley, he correlates with the Naivashan rocks, which he considered to be of Lower Pliocene Age, but which have since been proved to date from the end-Middle Pleistocene, as they overlie Kamasian sediments containing the Chelles-Acheul Culture and allied fossils. Although he excludes the Awash eruptions from the Aden Series, he identifies certain basalts, and some alkaline and acid lavas (trachytes and rhyolites) which he considered to be of Upper Pliocene and Pleistocene Age, as belonging to this Series.

Arsandaux[4] records a sequence of three volcanic phases; first, basalts and porphyries, secondly acid rocks of alkaline nature, and thirdly doleritic rocks. Lamare[5] identifies two main series, a stratoid series (identified with Blandford's Trap Series) and a Recent series (identified with the Aden Series). His sequence for the Awash Gorge, shows the Stratoid Series, which antedates the period of major rifting and is comparable to the Deccan Traps, resting on faulted sedimentary rocks (Mesozoic?) and overlain by the Recent series composed of lavas derived from simple volcanoes, their state of preservation giving them a more or less Recent date. Stefanini[6] considers that the Stratoid Series in the Awash Gorge represents only the upper part of the earlier (Trap) series.

[1] Dainelli, G. & Marinelli, O. (1912, pp. 73 *et seq.*). [2] Gregory, J. W. (1921, pp. 329–44).

[3] In this connexion see Fuchs, V. E. (1939, p. 229 and footnote).

Fuchs and McInnes found that the Laikipian plateau eruptions overlay the Kamasian sediments in the Lake Baringo area, but that in these sediments were fossils which showed them to be not earlier than Pleistocene in age and most probably Middle Pleistocene. It must be stressed, however, that the age of the lower Kamasian beds is not really known and may be found to be of pre-Pleistocene Age. Arambourg's discovery in the Lake Rudolf area of a Burdigalian fauna associated with lavas claimed to be of similar type and occurrence as Gregory's Laikipian should be noted (Arambourg, 1933).

[4] Arsandaux, H. & Neuville, H. (1905).

[5] (*a*) Teilhard de Chardin, P. *et al.* (1930). (*b*) Lamare, P. (1935).

[6] Stefanini, G. (1936, p. 102).

There seems to be little doubt, therefore, that the main Trap Series (Ashangi Traps, etc.) is comparable to the Deccan Traps, both in composition and age, but that the upper part (Magdala Series) is very probably later.

II. THE EARLIER SEDIMENTARY SERIES

Blandford's[1] sections at Arogye and Magdala show, from the base up:

(i) Basalt.

(ii) Trachytes with crystals of glassy felspar.

(iii) Coarse brown sandstone 15–18 m. (with black shale and silicified tree trunks in the sandstone. These were later identified by Unger as *Nicolia aegyptiaca*, vide Stefanini (1936), footnote on pp. 103–4).

(iv) Volcanic ash and basalt.

(v) White, argillaceous rock of sedimentary origin, 6–7½ m.

(vi) Compact basalt. (Top of section.)

This sequence is remarkably similar to that recorded by Nilsson[2] for his Lake Yaya series on the Abyssinian high plateau, which showed basalt overlain by sands and clays, overlain by volcanic ash and pumice, overlain by thick sandstones (with, at the top, an abundance of fossil tree trunks (not identified)), overlain to the surface by basalt lavas. The Magdala sedimentary beds might also be compared with the Upper Daban Series, described by Macfadyen[3] also with fossil wood (*Capparidoxylon geinitzi* Schenk). Undoubtedly connected with the Lake Yaya sediments are the Chilga Beds, south-west of Gondar, composed of sandstones and lignite with fossil trees.[4]

Stefanini,[5] on Gregory's correlation, places the Magdala Series as Miocene to Lower Pliocene in age. A Middle Pleistocene date must now be considered, however, if the upper part of the series correlates with Gregory's Naivashan lavas. Macfadyen states that the age of the Upper Daban beds (conglomerates) is not known. Owing to the similarity between the sediments of the Lake Yaya series and the Kamasian sediments, Nilsson at first referred the former to the Kamasian Pluvial, but later considered it more probable that they date to the preceding Pluvial, the Kageran, or even earlier. The Chilga Series is tentatively placed by Stefanini in the Pliocene.

[1] Blandford, W. T. (1870, p. 186).
[2] Nilsson, E. (1940, p. 50).
[3] Macfadyen, W. A. (1933).
[4] Stefanini, G. (1936, p. 74); Dainelli, G. (1939, p. 22).
[5] Stefanini, G. (1936, p. 104, footnote).

It would seem, therefore, that the latest lavas of the Trap Series overlie sedimentary beds, which may date to the end-Tertiary or Lower Pleistocene, with a possibility of a Middle Pleistocene date for the end of the series.

The deposition of the latest volcanics of the Trap Series was followed by the major rift movements in the Danakil and Abyssinian Rift Valleys, and later by the main Aden Series volcanics which are concentrated for the most part within the Rift, and are more localized in their distribution than the earlier lavas.

III. THE LATER VOLCANIC SERIES AND RIFT MOVEMENTS

The Aden Series volcanic rocks have been studied in some detail by Vinassa,[1] Arsandaux,[2] and Teilhard, Dreyfuss & Lamare[3] and the sequence appears to be as follows:

(i) Lower Basalts—basalt—porphyries (at Duanle Horst, Awash Gorge, etc.).

(ii) Rhyolites—tilted and overlying (i) unconformably.

(iii) Upper Basalts (felspathic), very fresh and horizontal.

It is of interest to note that Teilhard, Dreyfuss & Lamare state the recent volcanics (iii) were deposited only after a phase of quiescence and of intense erosion of the rhyolites. The same authors state that the Upper Basalts at Jibuti are always earlier than the raised beaches, which they never overlie, and they conclude that the age of these lavas is relatively early and almost certainly antedates the development of the Gulf of Tajura. However, Teilhard de Chardin, Breuil & Wernert[4] consider it possible that the earliest (upper) sheet gravels of the Mount Mabla/Tajura area may be intercalated with the latest of the recent basalts which are undoubtedly of Pleistocene age. If this should prove to be the case, then the earliest artifacts in this terrace (Acheulio-Levalloisian), which are rolled, must be contemporary with the later part of this period of volcanicity. Lamare[5] places the rhyolites in an intermediate series between the Traps (to which the Lower Basalts are presumed to belong), and the Recent series. Arsandaux[6] places the rhyolites, including the pantelerite and basalt-porphyries, in the earlier, Trap Series. The Afar plateau, the southern scarp of which forms the northern limit of the Dire Dawa 'corridor' is described by Arsandaux as being composed of Recent doleritic basalts, which have been derived from

[1] Vinassa de Regny, P. (1924). [2] Arsandaux, H. (1906).

[3] Teilhard de Chardin, P. *et al.* (1930).

[4] Teilhard de Chardin, P., Breuil, H. & Wernert, P. (1940, pp. 497–522).

[5] Teilhard de Chardin, P. *et al.* (1930).

[6] Arsandaux, H. (1906).

the volcanoes of Afdem, Assabot and Fantale, and join with the Recent Basalts of French Somaliland.

The faulting in French Somaliland is divided by Dreyfuss[1] into three phases:

(i) Faulting developed prior to the deposition of the Recent Basalts affecting the Duanle Horst (direction N./S.; N.W./S.E.; W.S.W./E.N.E.)

(ii) A series affecting the Recent Basalts, but not those formations attributed by him to the Quaternary, with a direction E./W.

(iii) Faulting with a N.W./S.E. direction with some warping affecting the gypsiferous deposits of Lake Assal and the coast of the Gulf of Tajura: of Pleistocene age.

IV. THE LATER SEDIMENTARY SERIES

(1) NORTHERN AND CENTRAL DANAKIL

Blandford[2] in his definition of the Aden Series of volcanics, states that they are interstratified with sedimentary rocks, particularly sandstones and conglomerates. Baldacci[3] distinguished two series on the Eritrean coastal plain. An earlier series comprising (*a*) calcareous and gypsiferous marls which have been faulted and which he attributes doubtfully to the Miocene; (*b*) clays, with veins of gypsum and yellow sands, containing *Pecten jacobaeus*, *Corbula gibba* and other Mediterranean fossils, referred tentatively to the Pliocene. This series was overlain unconformably by a later series comprising conglomerates and calcareous coralline deposits. Into these deposits (it is not clear whether the conglomerates are included) are intruded volcanic tuffs and acid lavas.

Dainelli & Marinelli[4] consider the age of these deposits as being more probably Late Pliocene to Early Pleistocene and correlate them with other beds containing *Melanoides tuberculatus* (*Melania tuberculata*) and *Corbicula fluminalis*. The marls and sandy clays are markedly faulted, notably in the Juba and Dongale areas, while at Naba Ramoda and on the coastal plain generally, the conglomerates lie unconformably upon the faulted sands, with local intrusions of tuffs and lavas which become more frequent from north to south.

In the Danakil Rift, Nesbitt[5] describes, between Mount Currub and Mount Ascoma (particularly near Tandatro), sandstone beds which have been step-faulted. Basalts intrude through these beds especially near Ascoma. At Assassi Babi Foro and in other areas Nesbitt found well-consolidated, light-coloured beds of gypsum or calcareous tufa, intercalated with and overlain by magnetic lavas.

[1] Dreyfuss, M. (1932, pp. 285–91).
[2] Blandford, W. T. (1870, p. 190).
[3] Baldacci, L. (1891).
[4] Dainelli, G. & Marinelli, O. (1912).
[5] Nesbitt, L. M. (1935).

Other faulted sedimentary beds are recorded in the south-west. Kamasian-like sediment in the basin of the Galla lakes and in the Awash valley is recorded by Nilsson[1] while Weber[2] records from Ghiorgis, on the north side of the Awash, diatomaceous beds with *Orthosira*, and yellow calcretes with numerous quartz grains.

The conglomerates overlying Baldacci's faulted sands of the Eritrean coastal plain, and in the northern part of the Danakil Rift are correlated by Stefanini[3] with a series of pebble and boulder beds, conglomerates and sands which are found to form a ring round the more central parts of Danakil at the base of the fault-scarp. These beds, with a large percentage of pebbles and boulders of crystalline rocks, extend from the foot of the Abyssinian scarp in the latitude of Massawa as far as latitude 14° N., and are also found to occur in the escarpment hills. The deposits are terraced and are considered by Dainelli & Marinelli[4] to represent beaches of an early Danakil lake. Three terraces are recorded at 180, 120 and 30 m. above sea-level in the Danakil depression. The highest terrace is considered to be of doubtful marine origin, but the remainder represent shore-lines of an early freshwater lake which filled the depression and in which torrent rivers have deposited their load and have formed deltas. On the eastern side of the depression Vinassa[5] found similar fluvio-lacustrine formations which were traced as far as latitude 13° N., where the terraced pebble beds spread out for 10–12 km. from the foot of the scarp towards the depression and continue northwards as far as Saada and Beleita, where they form the divide between the Danakil and Red Sea depressions: their height is approximately equal to that of the terraces on the western side. The terraces here are composed of predominantly quartzitic and limestone pebbles, blackened by desert varnish, and a little volcanic material: these terraces are intercalated with recently formed basalt flows. Vinassa finds no indication that the terraces are of marine origin, as they are lacking in any faunal remains. Krenkel[6] records in the Lake Alel Bad (Assale) depression, four terraces of a former freshwater lake which filled the depression, and which he says show no indication of marine conditions; but Stefanini[7] points out that special conditions of salinity might be cited to account for this absence of fauna, though such conditions would probably require a closed lake basin. If Vinassa is correct the freshwater lake so formed must have had very considerable proportions. Dainelli & Marinelli and Vinassa all

[1] Nilsson, E. (1940, p. 38).
[2] Weber, M. (1906).
[3] Stefanini, G. (1936, p. 79).
[4] Dainelli, G. & Marinelli, O. (1912, pp. 154–7).
[5] Vinassa de Regny, P. (1924, pp. 97–100).
[6] Krenkel, E. (1925, p. 82).
[7] Stefanini, G. (1936, p. 78).

agree, however, that these deposits are characteristic of a greatly increased precipitation such as would persist during a 'Pluvial period'.

Later in age than the freshwater (?) lake deposits there has been found by Vinassa[1] on the west side of the Danakil Horst, a rocky, limestone raised beach which he describes as a very recent shore-line full of shells of echinoids and very fresh coral, identical with those of the present-day shore of the Red Sea. Other authors cite *Laganum depressum* as coming from this beach. On the opposite side (west) of the Danakil depression there is no evidence of a similar deposit. On account of its constant level of 0 m., Stefanini considers that the coastal barrier of the Danakil depression is probably more recent than the raised beach seen along the coast and which is found at a height of from 10 to 20 m. or more, though the contemporaneity of the two beaches could be accounted for by the fact that uplift occurred along the east coast while the surface remained relatively stable within the Danakil depression.

This period of marine invasion was followed, in the depression, by extensive formations of gypsum and some basalt lavas. These gypsiferous beds form a more or less uninterrupted ring round the depression, and are found as terraces, the lower ones being present only in the basins of the small, saline lakes of the depression. Near Deroma at about 0 m., Stefanini records three terraces at −20 to −30 m., a second at −41 to −50 m. and a third at −50 to −54 m. Below this terrace the floor, which varies between −95 and −114 m., is seasonally flooded and forms a saline lake. At Lake Badda Dainelli & Marinelli[2] record gypsum terraces from 5 to 6 m. to nearly −100 m., and Stefanini[3] quotes Bibolini as finding, in these same series of terraces an older level of gypseous conglomerate with at the top a horizon of gypseous marl with pockets of *Diplodonta*, and a more recent level of gypsum and gypseous clays with abundant corals and echinoids and, locally, (?) volcanic intrusions. Similar terraced gypsiferous and saline deposits occur from the latitude of Lake Afreda to 13° S. ending in the region of the Magorros plains.[4]

In some places this gypsiferous series is overlain by fans of pebbles and mud and recent dune sands.

Stefanini[5] has summarized the later evidence from the Danakil depression:

(i) The deposition of conglomerates and terraced pebble beds of a fluvio-lacustrine origin resulting from pluvial conditions during the Pleistocene.

[1] Vinassa de Regny, P. (1924, p. 101).
[2] Dainelli, G. & Marinelli, O. (1921, pp. 157–60).
[3] Stefanini, G. (1936, p. 93).
[4] Vinassa de Regny, P. (1924). Nesbitt, L. M. (1935).
[5] Stefanini, G. (1936, p. 95).

(ii) Marine invasion and the formation of the 0 m. raised beach and deposits of normal salinity during the end-Pleistocene or early Holocene. Communication with the sea was cut off by gradual uplift of the coastline.

(iii) The formation of gypsum-saline deposits during the Holocene.

(2) LAKE ASSAL

The later geological history of Lake Assal in French Somaliland, which has been studied by Degoutin[1] and Dreyfuss[2] presents a close correlation with that of the Danakil lakes. The earliest formations of Lake Assal were freshwater deposits, about 15 m. thick, comprising tufas, marls and calcretes, with *Melanoides tuberculatus* (*Melania tuberculata*), *Corbicula consobrina* and *Coelatura teretiuscula*: these lie at altitudes of approximately 100 m. and rest on basalts. Overlying these freshwater beds at a height below present sea-level rests a stratum of marls with *Ervilia purpura* and *Cerithium* sp. which denote a gradual change to saline water, probably to lagoon or marine conditions. Overlying this saline stratum there follows a bed of diatomite with *Navicula* sp., kunkar and nodules of gypsum passing into gypsiferous beds followed by rock salt down to present lake level at −170 m. Degoutin considers that the sea invaded the lake, via the Goubett Charab, at the time that the marls with *Ervilia* were laid down: but Dreyfuss thinks that this was not the case, but that a rise in sea-level caused subterranean feeding of the lake. Stefanini[3] suggests that there may have been some marine invasion during the initial stages, but that increased aridity (accompanied presumably by a lowering of sea-level) isolated the lake resulting in the super-salinity of its water. Dreyfuss rather doubtfully refers the freshwater series to the Upper Pliocene and the gypsiferous series to the Pleistocene and Recent.

On the French Somaliland border, the freshwater series has also been found similarly terraced, and containing the same molluscan fauna in the Gobad plain (200–300 m.) as far as Lake Abbe, in the Gagade plain, and in the Henle and Hasbare plains, all of which are said to represent old infilled lakes, analogous to those of Lafofle and Amarti in Danakil. Stefanini suggests a possible correlation between the freshwater lacustrine beds of Lake Assal and the pebble beds of the Danakil depression.

(3) THE AWASH VALLEY

Nilsson[4] has examined the Awash valley which opens into the south-west corner of the Danakil desert and the railway strip between French Somaliland and Addis Ababa. He records an 'ancient coast zone' between Ourso and Errer

[1] Degoutin (1922, pp. 5–54).
[2] Dreyfuss, M. (1932, pp. 36 *et seq.*).
[3] Stefanini, G. (1936, p. 95).
[4] Nilsson, E. (1940, p. 40).

with associated delta plain, and traced this along the foot of the Cercer scarp, at altitudes of between 1220 m. and 1460 m.; this he considers denotes an uplift of the surface which increases towards the Awash valley and Lake Zwai. He also records raised beaches, apparently marine, at Moggio (Modjo), Bole and Mount Bora. These are cut into volcanic rocks and Kamasian-type sediments at altitudes of 1802, 1835 and 1832 m. respectively, and, if they represent the same beach, show tilting of approximately 2·3 m. per km. towards Dire Dawa. Nilsson suggests that the continuation of this beach may be found in the 200 m. raised beach at Dubar, near Berbera. His conclusions are of considerable interest and may be quoted here:

> The beaches show that there must have occurred an upheaval of the Rift Valley zone and also a corresponding tilting of the Somaliland plateau from this zone towards the Indian Ocean. It is highly probable that these earth movements must have been brought to an end before the oldest freshwater lake of the Zwai-Shala basin cut out its terrace, which is horizontal, as is also the case with all the lower beaches of the basin.
>
> If the Kamasia-type sediment encountered in the Abyssinian Rift valley is of the same age as the Kamasia sediment of the B.E.A. (*sic*), the highest raised beach at the former district (for the sake of brevity referred to below as 'the Modjo Beach') is younger than the Great Pluvial. It may, therefore, belong to the Last Inter-Pluvial Epoch (or perhaps be contemporaneous with the first lake of the Last Pluvial). I have encountered the Abyssinian Kamasia sediment at a considerably higher elevation than the Modjo Beach. Thus I have seen it along the railway as far up as an altitude of about 2000 m. near the Akaki Station. Somewhat to the south of the terrace at Bole I have traced it up to nearly 2100 m., and east of Lake Langana to about 2200 m. above the sea level.[1]

The present writer also noted Kamasian-type sediment on the road between Adama and Moggio (1640–1780 m.) along the main road from Dire Dawa to Addis Ababa, similar to the yellowish Kamasian deposits on the Mau escarpment in Kenya. These Kamasian-type beds are of considerable thickness and have been intensely faulted, they are overlain and intercalated with lavas in the region of the Galla lakes. It is tempting to suggest a correlation between the Kamasian-type sediments of the Abyssinian Rift and the Awash valley with the faulted sandstones, marls and clays of the Danakil Rift which are intercalated with, and overlain by basalt lavas.

(4) SOUTH-EASTERN DANAKIL

The writer was able to make a brief examination of the deposits of the Aisha Plain, and of sections on the road between Aisha and Dire Dawa from which some cultural material was obtained.

[1] Nilsson, E. (1940, p. 42).

An important section is exposed in the valley of the Garasleh, or Gildessa, Tug on the Jibuti road 30 miles north-east of Dire Dawa, above and below the road crossing (Text-fig. 7; Pl. 39, fig. 1). The road from Aisha reaches the tug bed by a short but steep scarp cut in the deposits of the right bank, which are here some 50 ft. or more thick. About $\frac{1}{4}$ mile below the road-crossing the following sequence was exposed in the right bank:

(i) Two gravel deposits each overlain by some 4–5 ft. of alluvial clay, red to greyish white in colour. The gravel consisted of well-rounded pebbles of sedimentary and crystalline rocks and some lava (3–6 in. average across the major axis) and were set in a sandy, clayey matrix, grey to reddish brown in colour, which sometimes turns to purple, due possibly either to the subsequent vulcanicity or to the decomposition of pebbles of tuff. These gravels are on an average 10–15 ft. thick with no very marked bedding, but with some banding of pebbles and alluvium in the uppermost part, where also the gravels have a more calcareous matrix with some crystals of gypsum. The base of the lower gravel is usually not exposed, but in a section about $\frac{1}{4}$ mile upstream from the road-crossing on the right bank this gravel is seen resting on a poorly consolidated tuff.

(ii) Overlying the upper red, clayey alluvium is a 'pebble bed' showing no sorting of the constituents. This bed is some 2–5 ft. thick, often of a compact nature and with a calcareous matrix and undoubtedly represents a land surface accumulation.

(iii) Above this 'pebble bed' is a layer some 6 in. thick of red 'sandstone' and above this a lava 4–10 ft. thick: the red 'sandstone' is believed to be the result of baking of the underlying deposit by the lava. This basalt (?) lava covers these gravels for several miles towards Aisha and is presumed to be derived from vents in the volcanic hills some 15–20 miles to the north-east.

These sedimentary beds were, as far as could be ascertained, horizontal and had not been affected by faulting: on the right bank they proved to be sterile.

On the left bank the river has considerably eroded the gravels, which appear to surround and to be banked against intrusions of purplish tuffs. The whole of the upper part of the deposits has been removed by erosion, but the lower levels remain exposed in good sections up to $\frac{1}{2}$ mile or more downstream. Here *in situ* in the lower gravels were recovered one large flake of lava with primitive faceting of the butt, an unprepared core or chopper, and primary flakes with inclined, unfaceted platforms. Overlying these denuded gravels on the left bank was a deposit some 2–3 ft. thick of redeposited gravel which, by its ill-sorted nature, is suggestive of a hillwash. Within and on the surface of this bed were found Levalloisian artifacts of two distinct ages and made from lava. Below this hillwash and stretching down into the tug bed is a low-level terrace which is well exposed in section in the road cutting. Here was found some 4 ft. of unsorted, compact gravels set in a calcareous matrix, overlain by evenly banded, thin, fine gravels, grits and sandy alluvium, buff to brown in colour. From this were recovered three small, faceted flakes, of advanced form and one possible disc core, made from chert. Overlying this terrace and the earlier deposits were artifacts and factory debris of a microlithic industry of Somaliland Wilton facies: the majority of these microlithic tools were made from obsidian and chert.

Other sections in this area, nearer Dire Dawa, show the same torrent gravels of resorted pebbles up to 15 ft. thick, resting on a grey, sandy clay or on volcanic tuffs. The gravels and clayey alluvium underlying the lava have the appearance of deltaic material deposited in fresh water, and there can be little doubt that they are comparable to the terraced pebble beds and conglomerates round the edge of the Danakil depression: they are complementary to the sediments found in the Aisha plains.

The Aisha plain is a flat but gently tilted plain of sand and lava, bounded on the south by the fault-scarp (Pl. 40, fig. 1), on the south-west by lava plateaux, and on the north-east by the Duanle Horst. On the west it stretches to the Danakil depression proper. The floor of the plain is, in the main, composed of sedimentary deposits, through which are intruded lavas, which in places also overlie the sedimentary beds. Occasionally, on the plain proper, wide and shallow erosion channels have been cut through the fine-grained sediments and drain towards the north-west. These channels are filled in places by coarse torrent gravels and brown alluvium overlain by wind-blown sands. The tilt of this plain can best be seen to the right of the road at Bio Kaboba near its south-eastern limit, where the direction appears to be north-west. The deposits were examined at Adele 7 miles north of Aisha, and again at points 1 mile north and 8 miles south-east. At Adele is an extensive, recently eroded area and a good section is exposed. (Pl. 40, fig. 2.) In a cliff on the right of the road at Adele is a 10–15 ft. thick lava, with an appreciable amount of rubble at the base. This rests on a very uneven surface of bright red to grey sands and clays which contain small pebbles of sandstones, quartzite, (?) porphyry and lava scattered sparsely through it. The total exposed thickness of the sediments is about 60 ft., but the base of these sands and clays was not seen. In a second section where only the uppermost part of these beds is exposed, the sequence is from the surface down:

(i) Lava to 8 ft.

(ii) Red friable sandstone 3–6 ft. thick.

(iii) Sands and clays with some gravel bands to 5 ft., with pebbles of sandstone, lava, quartzite and quartz. Veins of kunkar pass downwards through the sands. No artifacts were found.

These beds appear to be gently inclined to the north-west. They have been subjected to fairly intensive erosion in the Adele area and the tugs have cut steep-sided channels, which in some cases, notably at Adele itself, are filled with a deposit of gravels of lava pebbles, overlain sometimes by a brown alluvium. Nearer Duanle the tugs appear to be still degrading their beds, and little or no sediments were found in their valleys, other than resorted boulder gravel in the tug bed itself. On this denuded surface are found occasional obsidian artifacts of late form.

One mile north of Aisha there is exposed, in a railway cutting, 1–3 ft. of coarse boulder gravel of lavas and tuffs, with a very small percentage of quartz and chert pebbles. The surface of this deposit is irregular and rests on a correspondingly irregular surface. The lava pebbles are angular, but rounded, and show a gloss due to desert patina. This bed is interpreted as an eluvial deposit. It overlies disconformably, in the cutting, the fine-grained sedimentary series and elsewhere in the vicinity similarly overlies intrusions of the volcanic rocks. In the cutting the underlying sands and clays are grey to dark brown in colour, with false-bedded grit and fine gravel lenses. The gravels are composed of quartz, chert, silcrete pebbles and some badly weathered pebbles of lava. Some 4–7 ft. of sands and clay are exposed containing fist-size nodules of kunkar, and thin lenses of calcrete. The gravels and grits become more frequent towards the base of the section and on one side some 6 in. to 1 ft. 6 in. of gravel is exposed. Bed rock was not seen. The deposits were sterile but on the surface in the vicinity were found living sites of the Somaliland Wilton Culture.

Eight miles south-west of Aisha on the left of the road to Dire Dawa is a second extensive erosion area, which was briefly examined: the plain falls away to the east and there can be little doubt that this is due to later earth movement (Pl. 39, fig. 2). Volcanics can be clearly seen here intruded through the gently tilted sedimentary series, and in one place they were seen to overlie them. Sections show, at the top, a ferruginous pebble bed with a gritty, sandy matrix, some 3 ft. thick, composed of angular but rounded pebbles of lava, chert, silcrete, limestone, sandstone and some crystalline rocks: there is no bedding of this deposit, which is considered to correlate with the eluvial pebble bed in the previous section. Underlying this was seen some 20 ft. of grey to buff grits with occasional bands and lenses of fine gravels. This passes down gradually into a consolidated white to buff sand, with small pebbles, some 15 ft. thick:

in places this bed appears to be absent. The lowest levels show 8 ft. of bedded gravels and grits, brown in colour: the base of the deposit was not seen, but in the broken country between Aisha and Adjin at the south-east corner of French Somaliland, it would seem that bed rock is Cretaceous sandstone overlain by lavas.

Unfortunately, these sediments, where examined, proved to be sterile, but it is certain that a proper survey would record cultural material here. On the surface of the eluvial oxidized pebble bed were found rare Middle Stone Age prepared cores and a very few flakes. One much-weathered chopping tool was also recovered. These deep, fine-grained sediments have not the appearance of lake deposits but rather of alluvial fan and delta material deposited by regenerated streams and rivers coming down from the fault-scarp, and depositing their load in the plains of the Rift in shallow freshwater lakes. Further, it must be supposed that this could not have taken place under arid or semi-arid conditions, but during a period of considerably increased precipitation. Where examined by the writer, these deposits are nowhere really consolidated and they bear no resemblance to the Kamasian-type deposits. The archaeological evidence, meagre as it is, is yet of value, as it gives for the first time some indication of the age of these alluvial, freshwater deposits, and points to an Upper Pleistocene date.

The only other cultural evidence from this area (apart from that of the French Somaliland coast to be discussed later) comes from the neck of land separating Lake Assal from the Goubbet Charab. Here Wernert found a 2 m. thick deposit of gravels on the eroded surface of which were a large number of cores, discs and flakes usually very weathered and even rolled, and which he found difficult to date. In view of Dreyfuss' observations it might be possible to equate these gravels with the 'freshwater' series further inland. The molluscan fauna favours a fairly recent, post-Kamasian, age for these beds. This freshwater series may, therefore, tentatively be placed at the very end of the Middle or more probably at the beginning of the Upper Pleistocene, if the Kamasian-type sediment of the Abyssinian and Danakil Rifts is of the same age as the Kamasian in Kenya.[1]

It would be of interest to know whether the lavas that intrude and overlie these freshwater beds in some places can be distinguished from the younger (upper) basalts of the Aden Series. They are presumably more recent, in view of Dreyfuss' observations in the Lake Assal basin, and indeed this unstable area

[1] It is of interest to note that Nilsson (1940, p. 39) found, in the highest horizontal beach, cut out of Kamasian-type sediment, in the Zwai-Shala basin, numbers of obsidian artifacts, the earliest of which have been rolled and are therefore earlier than the beach. It is understood from Dr L. S. B. Leakey, that these artifacts are of fairly developed Levalloisian facies.

must have been affected until very recent times by earth movement and volcanic activity. Dainelli[1] has described the active volcanic areas of the Danakil Rift, and in the country between the Awash and Addis Ababa, the road passes through an area of very recent cinder cones and incipient craters, on which there is little or no vegetation.

SUMMARY AND TENTATIVE CORRELATION OF LATER GEOLOGICAL EVENTS IN THE DANAKIL RIFT

(1) The main Trap Series of volcanic rocks is deposited on the plateau and in the Rift, which was already in existence.

(2) Early freshwater lake beds are laid down on the Abyssinian plateau, interstratified and overlain by lavas of Laikipian and Naivashan type, followed by faulting and the major period of rifting. Presumed contemporary, either with these early lake beds on the plateau, or with the succeeding Kamasian-type deposits (or perhaps with both), is the old coast zone noted by Nilsson at the foot of the fault-scarp, the earliest conglomerates which line the foot of the fault-scarp, and the faulted marine marls, sandstones, and clays of the Eritrean plain and northern Danakil. End-Tertiary to Lower Pleistocene (?).

(3) Deposition of Kamasian-type sediments in the Abyssinian Rift and south-western part of the Danakil Rift, interstratified with basalt lavas and tuffs (main Aden Series volcanics). Earth movement, rifting and faulting on a large scale followed, affecting the Kamasian-type beds. Middle Pleistocene (?).

(4) The formation of freshwater lakes within the Danakil depression fed by streams and rivers from the plateau, which deposited pebble and boulder beds round the shores of the lakes and deltaic material within their basins. The Garasleh Tug shows two phases of deposition of coarse and fine-grained sediments, containing artifacts of early Levalloisian form. Cultural and faunal evidence suggests that these events should be dated to the Gamblian Pluvial and perhaps also to the close of the preceding Last Interpluvial: they are therefore of early Upper Pleistocene age.

(5) Marine invasion of the northern part of the Danakil depression and the formation of the 0 m. beach, along the west side of the Danakil Horst, together with deposits of normal salinity. These were followed by earth movement and volcanic activity on a reduced scale, which has affected the freshwater series and resulted in further volcanic intrusions. It is suggested that this 0 m. beach of northern Danakil may be contemporary with the 8 m. raised beach on the Berbera coast and the 'lagoon' coral phase of the Obok coast and is therefore

[1] Dainelli, G. (1936, p. 121 and plate).

of (?) early Gamblian age. The difference in height is accounted for by the gradual uplift of the coastal region, which thus cut off the 'Danakil gulf' from the sea, and a possible gentle down-warping of the floor of the depression. In this connexion it will be noted that Dreyfuss records folding of the Quaternary strata of Lake Assal, as does Stefanini of the coastal formation of Gowein on the east coast.

(6) The formation of gypsum saline deposits in the basins of the slowly receding lakes of the Danakil depression and of Lake Assal. Renewed degradation round the periphery by tributary tugs draining into the depression followed by the accumulation in some cases of a coarse boulder gravel and sandy alluvium. Post-Pleistocene.

It must be pointed out, however, that if the 0 m. beach of the depression is to be equated with the 0 m. beach of the south-east coast, then it is probably of pre-Gamblian not early Gamblian age. This would presuppose that the freshwater series belongs to the previous Pluvial, the Kamasian. While it is not impossible that this may later be found to be the case, it must be emphasized that this freshwater series of deposits bears little or no resemblance to any of the Kamasian beds that I have seen, but is always closely analogous to Gamblian deposits. Moreover, the cultural evidence, meagre though it is, is even more difficult to reconcile with an earlier dating and it would be necessary to suppose that a pure Levalloisian element was present in this area at a time earlier than its appearance elsewhere on the continent.

A further possibility is that this 0 m. beach of northern Danakil may correspond to the ± 2 m. beach of the north and east coasts. Such a date would be in keeping with the recent appearance of the beach while the uplift that cut off the depression from the Red Sea is only in the nature of 100 m. and in such an unstable region this gentle up-warping might well be of quite recent date.

Until therefore cultural material is found in association with these deposits in the depression it is not possible to determine other than in very general terms the periods to which they belong.

CHAPTER 6

THE NORTH AND EAST COASTS

THE NORTH COAST

I. MARINE AND TERRESTRIAL DEPOSITS

The coastal areas are prehistorically little known but such stratigraphical and cultural evidence as exists is given here in order that the position may be clarified for future workers.

Within the Gulf of Aden Rift Macfadyen[1] has dated the Dubar marine limestones, which are confined to the fault, as of Miocene age, and Cox has identified fossils of Miocene age collected by Barrington Brown[2] from Benda Ziada. The former has shown that Gregory[3] was incorrect in considering that the Gulf of Aden faulting did not commence until the Late Pliocene. He concludes that there were two main periods of faulting in each of three main trends—the East African, the Red Sea, and the Gulf of Aden trends—in that order, and places the initial faulting in Late Eocene to Oligocene times, and the later in the Late Pliocene.

The deposits of the Zeila plain have been studied by Parkinson[4] and more recently by Hunt[5] who notes the following sequence which, he says, divides the plain, stratigraphically, into four concentric zones truncated by the sea coast:

(i) Conglomerates with boulders of crystalline rocks, Jurassic limestone and Nubian sandstone, and soft buff false-bedded calcareous sandstones. At Geriado these beds are cut by a basalt dyke feeding a flow which bakes and overlies them. The average height of these conglomerates and boulder beds is 311 m. a.s.l. flanking the foot of the fault-scarp.

(ii) Coarse conglomerates, sandy clays, gravels and sands—all mainly calcareous and set in a matrix of sandy red calcareous clay. In addition to pebbles from (i) occur volcanic pebbles—rhyolites and acid volcanics. The acid lavas are contemporary with this period as they penetrate a belt of the sediments in a series of vents from Jibuti to Geriado. Parkinson noted a spherulitic lava flow which overlay 292 ft. of the sedimentary beds, here

[1] Macfadyen, W. A. (1933).
[2] Barrington Brown, C. (1931).
[3] Gregory, J. W. (1921).
[4] Parkinson, J. (1932).
[5] Hunt, J. A. (1946).

comprising corals, shelly sands and clays. An acid volcanic vent, comprising rhyolites and ash near Geriado, has uptilted the phase (i) sediments and basalt dyke as well as some overlying sediments of phase (ii). Average height above sea-level 200 m.

(iii) 'Laterite'—a subdivision of the phase (ii) sediments comprising red sandy clays, volcanic ash and clayey sands with lenses of conglomerate and sandstone usually with impermeable calcareous cement. At Ameruta (152 m. a.s.l.) a bore-hole shows 128 m. of sandy clays and sands with pebbles and some bands of gravel, much of the strata coloured rusty red. A similar sequence is seen at Silsil (74 m. a.s.l.) with, at 68 m., freshwater gravels and sands. A *Helicoid* snail is the only fossil recorded from these beds. The 'laterite' zone extends from the phase (ii) sediments at approximately 200 m. a.s.l. to about 15 m. a.s.l. where there is a marked flattening of the gradient.

(iv) Coastal marine beds, presumed to begin at the 15 m. contour. They are overlain by salt silts and loess.

(v) Red and white dune sands and coral and shell-rag reef.

Hunt notes that these Pleistocene and Recent deposits have been affected by warping on a south-west to north-east axis, which has caused deflexion and ponding of the north-east flowing rivers.

On the Berbera coastal plain Macfadyen[1] distinguished in the field one well-developed highly fossiliferous raised beach at ±8 m. At Berbera, this is composed of more or less consolidated coral limestone, while half-way to Bulhar it was seen to consist of a well-cemented pebble bed set in sandy cement with sparse fossils. Thought to be the same beach, is an occurrence recorded at Ras Khanzir by Gregory and a further occurrence is reported by Macfadyen at Raguda; Barrington Brown[2] records a similar beach at Benda Ziada $4\frac{1}{2}$–6 m. above sea-level.

Macfadyen also records two rather problematical raised beaches as existing at 85 and 200 m. respectively. These have been determined from fossils collected at these heights and subsequently identified as being of Pleistocene to Recent age. The existence of these beaches still requires verification in the field however. Dr Macfadyen has informed me by letter that he has recently proved the existence of a low-level raised beach at an average height of ±2 m. above sea-level between Berbera and Zeila. This beach consists of quite unconsolidated marine sands with shells and some shell banks.

I am indebted also to Dr Macfadyen for the sequence recorded by Parkinson from two deep bores at Zeila, which show conclusive evidence that the coastline

[1] Macfadyen, W. A. (1932, 1933). [2] Barrington Brown, C. (1931).

at Zeila was at one time at least 250 ft. lower relative to the sea-level than it is at present. It is tempting to correlate this phase of marine regression with the later regression noted on the east coast of Somalia. Both wells were sunk on the south-west outskirts of Zeila, about 100 yd. from each other and at an altitude of between 5 and 10 ft. above sea-level. The sequence was found to be as follows:

Well 1

(ft.)	
0– 66	Coral sand, quartz, shell fragments, finer towards base
67– 70	Clay, much fine sand
71– 80	Rather coarse coral sand
81– 90	Clay, much fine sand
91–116	Coral sand, pebbles up to 5 mm.
117–120	Sticky yellow and white clay
121–145	Coarse coral sand, little quartz, shell fragments
146–154	Sticky, pale terra-cotta clay
155–185	Pale sandy clay
186–207	Dark brown clay
208–215	Rather coarse coral sand: fine quartz
216–225	Sticky grey clay
226–235	Greyish brown sandy clay
236–242	Very fine sand, minute mica flakes

Well 2

(ft.)	
0– 90	Coral sand, shell fragments, becoming finer at depth
91–110	Pale yellow clay: rare shell fragments
111–125	Light brown clay: shell fragments
126–170	Pale yellow clay: rare shell fragments
171–201	Brownish clay, much silt: quartz grains, small pebbles, 5 mm.
202–210	Pale yellow clay but whiter than 90–100 ft.; shell fragments
211–226	Brown clay, much silt, shell fragments
227–235	Pale yellow clay as 201–210 ft.; probably rather more silt
236–260	Brownish clay as 170–201 ft.; slightly darker mica showing: small quartz pebbles at 260 ft.
261–287	Sandy clay, quartz grains, small pebbles, shell fragments
288–292½	Coarse quartz sand
at 292½	Rhyolite (i.e. bed rock)

If the ± 2 m. raised beach is represented by the top levels of the uppermost coral sand, it would seem to indicate that this beach represents a comparatively recent fluctuation in sea-level and cannot be representative of a still-stand in the major regression from the ± 8 m. raised beach. The former alternative is incidentally more in keeping with the unconsolidated nature of its sediments.

On the coastal plain of eastern British Somaliland, Barrington Brown found a thick series of conglomerates, backed against the fault-scarp, which pass nearer the coast into partially consolidated micaceous sands and clays with gypsum lenses. They are shown doubtfully as Pliocene on his map and section and are well developed in the region of Benda Ziada and again at the Karin Gap to the west.[1]

There can be little doubt that they should be correlated with the phase (i) conglomerates and sands noted by Hunt in the Zeila plain.

The later coastal deposits, which he dates as Pleistocene, Brown divides into two distinct stages:

(i) An older stage with a slope of 1° to the north, based on the foot of the slopes of the conglomerates, and covering a wide area between the principal streams.

(ii) A younger stage forming tongues and terraces along the flanks of the streams 40–50 ft. below stage (i).

This sudden change of slope between the conglomerates and stage (i), Brown considers represents the height of the sea-level at that time and is now 450 ft. (143 m.) above present sea-level. He also shows anticlinal folding affecting the conglomerates and sands, but does not record whether this has also affected his Quaternary sediments. These alluvio-marine deposits he describes as a composite flood-cone of detritus laid down upon marine sands (containing some recent shells) and consisting of sands, gravels and boulders commonly consolidated by calcareous cement. These merge with the $4\frac{1}{2}$–6 m. raised coral beach already mentioned.

Stefanini[2] lists a number of recorded raised beaches predominantly of coral formations from the Gulf of Aden and Eritrean coasts. In the Dahlak Islands coral formations are found up to 40 m. above sea-level. South of Massawa there is a similar formation along the shore, notably at the Buri peninsula, where fossils from two separate formations have been identified. Vinassa[3] records two raised beaches along the east (Red Sea) side of the Danakil Horst—a coral formation overlain by cemented dunes, and a (presumed earlier) deposit of gypsiferous sands and coral, found at a height of 10–20 m. at Assab: here the coastal plain, composed of gypsiferous dune sands and concretionary deposits, rises gently from south to north.

[1] Similar deposits are also to be seen in the Dudub Gorge draining the Asseh Gap northwards to Ankor. (Hunt, J. A., 1944.)

[2] Stefanini, G. (1936, p. 87).

[3] Vinassa de Regny, P. (1924).

In French Somaliland Teilhard de Chardin[1] recognized at Obok three coral formations:

(i) An Old Red Coral formed apparently in relatively deep water and possibly earlier than the post-rhyolite basalts of Mount Mabla.

(ii) A Younger Coral alternating with gravels and resting disconformably upon (i). With this coral are equated the raised corals of the Tajura coast, always of post-Younger Basalt age.

(iii) A 'Lagoon' Coral characterized by *Cerithium palustre* (*Tenebralia palustris*).

Teilhard equates the 60 m. gravels of Mount Mabla with the Younger Coral and the 30 m. gravels with the Lagoon Coral.

At Jibuti, Dreyfuss[2] records a coral with *Cerithium palustre* as well as a more recent and lower coral. At Khor Ambado the same author established the existence of three beach levels at 10–15 m., at 20–25 m. and at 65–70 m., but he is doubtful whether these represent genuine raised beaches.

Revoil[3] noted a raised beach of coral and shells at Las Gore some 15 m. thick, which continued along the northern Mijertein coast. Raised beaches are also recorded at Kandala (coral) at 5–6 m. at Bosaso (nullipore), at 12–15 m. at Alula (limestone and coral). At Ras Filuc, Bosaso and Ras Hantara a second, lower, raised beach of recent date is also recorded.[4]

The writer was able to investigate briefly the sequence on the coastal plain between Dubar and Berbera, but no opportunity occurred of properly examining the deposits here.

At Dubar, banked against the foot of the faulted Cretaceous and Eocene rocks which form the coastal hills, and at a lower level than the Dubar limestones, was found a conglomerate of pebbles and boulders of limestone, quartz, quartzite and derived fossil wood cemented by a hard calcareous matrix. This bed appeared to be of considerable depth but, with the exception of the uppermost levels which were exposed in small tugs draining north, the total thickness could not be estimated. The deposit appeared to follow the foot of the coastal hills to the east and west and stands at an altitude of just under 200 m. At the gap in the coastal hills through which the main road to Hargeisa passes, these conglomerates rest against the Basement Complex rocks. At Dubar the conglomerates pass beneath a thick series of sands, clays and boulder and pebble beds which have the appearance of a vast alluvial fan spreading out from the

[1] Teilhard de Chardin, P. (1932, *et al.*, 1940).
[2] Dreyfuss, M. (1932, p. 31).
[3] Revoil, G. (1882, pp. 99, 237).
[4] Stefanini, G. (1936, p. 90).

coastal hills (Pl. 41, fig. 1). These alluvial deposits slope at first more steeply towards the sea for the first 2 miles from the hills, and then flatten out to a very gentle slope and appear to merge with the 8 m. raised beach at the coast itself: they begin at heights of approximately 175 m. and pass down to approximately 10 m. above sea-level.

The 8 m. raised coral beach is easy to identify and forms a pronounced but low cliff which in places, noticeably near the lighthouse, abuts on to the sea. Elsewhere, at the rifle range and again east of Berbera town, the 8 m. beach is some distance inland, at the latter place some ½ mile or so, and between it and the sea is a low coastal flat some few feet only above sea-level composed of fairly compact marine sands and some surface deposits of salt or gypsum (Pl. 41, fig. 2). Overlying it at the coast are unconsolidated white dune sands. East of the town, the underlying rock is a tilted coral sandstone of yellow colour which outcrops at the shore itself. The shore here is protected by a coral reef, the formation of which must be contemporary in part with the low-level beach or coastal flats. Near the rifle range, west of the town the underlying rock is a red to pink sandstone. Here only a few yards from the sea, shallow wells yield brackish water which the Somali use for watering their stock. These wells are about 3–5 ft. deep, often less, dug through unconsolidated sands.

The cultural evidence is of interest. At Dubar was found an important site situated about ½ mile west from the water catchment area with its large potable hot springs issuing from the Nubian (Cretaceous) sandstone. The site is approximately 100 yd. square, where there occurs an outcrop of ferruginous quartzite and silicified sandstone, within the Cretaceous beds. Implements are here found scattered in considerable numbers over the surface at the base of the hills and overlying the 200 m. conglomerates. They show varying degrees of weathering and some are rolled, presumably by torrent streams and scree action, and may be divided into two main series, an earlier, much weathered and iron stained of Acheulio-Levalloisian type, and a later comprising typical Levalloisian artifacts. Unfortunately, it was not possible to prove whether similar artifacts occurred in the alluvial fan deposits, but it can be stated that, wherever examined, none were ever found to overlie them. While, therefore, the artifacts of this site are all later than the formation of the conglomerates, it is not improbable that they will be found to be contemporary, at least in part, with the alluvial series.

Overlying the alluvial series at this site, at the pass through the coastal hills on the Hargeisa road, and again at Mile 6 on this road from Berbera, were found scattered artifacts of Somaliland Stillbay and Late Stone Age type resting on the eroded surface of the sandy alluvium. At the latter site they had been eroded from beneath unconsolidated dune sands: all show wind and sand weathering to varying degree. At the lighthouse, occasional small flakes with faceted platforms overlie the 8 m. raised beach; they show some signs of abrasion.

Two rock-shelters were found at the pass through the coastal hills on the Burao road. They are both developed in Eocene limestone and the deposits they contain reflect well the later climatic sequence. They will be described when considering the cultural succession in this area. (See p. 269.)

SUMMARY AND CORRELATION OF COASTAL DEPOSITS OF THE GULF OF ADEN COAST

The following correlation is only very tentative but may serve as a guide for future work:

(i) Subsequent to the deposition of the Miocene Dubar Series, the coastal peneplain was developed probably in Pliocene times: it must be supposed that some uplift of the surface took place between this period and the subsequent depositional phase.

(ii) On this plain were laid down conglomerates, and boulder beds at 200–300 m., flanking the coastal hills, and passing into marine sands nearer the coast; these have been affected by the older volcanics and faulting of the Aden Series. (Zeila plain, Berbera plain, Karin and Asseh gaps, and Benda Ziada plain.)[1] Lower Pleistocene (?).

(iii) Conglomerates, sandy clays, gravels, sands and shelly corals of an alluvial-marine nature, interstratified with acid lavas and affected by faulting. (Zeila, Berbera, Benda Ziada plains.) Middle Pleistocene (?).

(iv) Continued aggradation of sediments of an alluvial-marine nature comprising conglomerates, gravels, sands and red sandy clay. The Acheulio-Levalloisian and Levalloisian material from Dubar may be contemporary with this phase in which case these deposits are probably of early Upper Pleistocene age. With this phase is correlated the 60 m. and 30 m. gravel terraces of Mount Mabla in French Somaliland.

(v) Contemporary in part with phase (iv) is considered to be the formation of the ±5–25 m. raised beach (or beaches) of the north coast (8 m. at Berbera), which is overlain by 'epi-Levalloisian' implements of late form. The 'Lagoon' Coral of the Obok coast may be correlated with this beach,

[1] It is suggested that these conglomerates and boulder beds may perhaps be identified with the 200 m. raised beach tentatively identified by Macfadyen from fossils collected at this altitude at Dubar, but further field-work is necessary before this can be established with certainty.

and the 'Younger Coral' of the same area with the earlier part of phase (iv). It is probable that future work will show that more than one beach level, probably two or more levels, are present here. On a gently sloping shore a steadily receding or encroaching sea will presumably leave strand lines at a number of different levels. All would belong to the same epoch but the significant level would be that at which the maximum halt was made. This level could only be obtained after an exhaustive survey of a given area. The lower part of the ±5–25 m. raised beach or beaches is probably contemporary with the beginning of the Upper Pleistocene, and the fixed sand dunes of the central plain perhaps represent accumulations by strong monsoon winds during this period.

(vi) Marine regression to at least −76 m. as evidenced by the deep bores at Zeila. Early part of the Gamblian Pluvial (?).

(vii) Marine transgression and formation of the ±2 m. raised beach of Jibuti, Berbera-Zeila, Ras Filuc, Bosaso and Ras Hantara. This beach possibly equates with the latest Stone Age industries of the coast.

(viii) Formation of the coral reef, and accumulation of coastal dune sands.

We may interpret this sequence in terms of changes in sea-level as indicating perhaps:

(*a*) A period of high sea-level during stage (iii).

(*b*) A fall (?) followed by a gradual rise in sea-level which reaches its peak during stage (v).

(*c*) A fall in sea-level to well below that existing today, followed by a gradual rise and then a slight fall.

II. THE BAB-EL-MANDEB LAND-BRIDGE

Of importance for the interpretation of the Stone Age cultural connexions of the Horn is the question of the existence of a land-bridge at the Bab-el-Mandeb Strait in Pleistocene times. Macfadyen[1] quotes L. R. Cox, who, on the palaeontological evidence concludes that the breakdown of the barrier at Bab-el-Mandeb is of Middle Pliocene age. Gregory,[2] writing, however, ten years earlier than Cox, dates the subsidences which admitted the waters of the Indian Ocean into the Red Sea as having taken place in Late Pliocene or Pleistocene times. These subsidences caused the eruptions which built up the volcanic islands in the southern part of the Red Sea and the extinct volcanoes of

[1] Macfadyen, W. A. (1933, p. 15).

[2] Gregory, J. W. (1921, p. 352).

Aden, Eritrea and Arabia, and resulted in the extinction of the Mediterranean fauna. Prior to these earth movements, the southern part of the Red Sea was a land area. If Gregory is correct then these movements would have to be correlated with those at the end of the Lower Pleistocene, which brought about the extinction of the Kaiso fauna in the basins of the western Rift[1] and in the Lake Rudolf basin and lowered the lake floor there.[2]

Nilsson's evidence seems to suggest that, since the formation of the Kamasian-type sediments, an uplift of the surface of about 1800 m. must have taken place in the Moggio area, and of 200 m. at the coast (if his contention is correct that the 200 m. beach at Dubar is the continuation of the Moggio beach): this uplift must have been completed at latest before the main part of the Gamblian Pluvial. A drop in sea-level of 180 m. is needed, according to Ball,[3] to establish a land-bridge at the Strait of Bab-el-Mandeb. It is fairly certain, on evidence from the east cost of Somalia, that a fall in sea-level occurred at the beginning of the Gamblian so that it is not impossible that a land connexion between Danakil and southern Arabia may have existed for a time at this period. The evidence at present is too scanty, however, to allow more than a hint at such a possibility and detailed research will be necessary before the existence of such a connexion could be conclusively proved. At Zeila there is record of the sea-level having fallen to 89 m. below the present level, while, as we have seen, Nilsson suggests that there has been isostatic movement of several hundred metres in the southern part of the Danakil Rift. It is, however, considered most unlikely that any land connexion could have existed at any time later than the Lower Gamblian.

That there was probably contact by sea between the African and southern Arabian coasts of the Red Sea during Late Stone Age times is evidenced by the discovery of an obsidian microlithic industry of Somaliland Wilton type in the Dahlak Islands which are situated off the Eritrean coast at the southern end of the Red Sea.[4] This corroborates Miss Caton-Thompson's suggestion that the microlithic obsidian industries of southern Arabia owe their appearance there to African influence. In the Hadhramaut geometric microliths were found by her in association with the pre-Islamic temple at Hureidha in the Wadi 'Amd which is dated as being no earlier than the sixth century B.C.[5] Such a date lends weight to my contention that the Late Stone Age in the Horn probably lingered on into historic times there.[6]

[1] Wayland, E. J. (1934).
[2] Fuchs, V. E. (1939, p. 250).
[3] Ball, J. (1939).
[4] Blanc, A. C. (1952).
[5] Caton-Thompson, G. & Gardner, E. W. (1939, pp. 31–32).
[6] See pp. 293, 289, 250.

THE EAST COAST

From Cape Guardafui to Garad, in the northern Mudugh, the coastal plain does not exist as a continuous feature, but it can be traced uninterruptedly, from a short distance above Garad southwards to the Somalia/Kenya boundary and beyond, gradually increasing in extent and width from north to south. For most of the distance along the Mijertein coast the plateau either abuts directly upon the sea or upon a much restricted and isolated section of coastal plain, the rock here being predominantly Eocene limestone. On the Mudugh coast the plateau—here composed of Eocene anhydrite beds—joins the plain by a gradual and gentle slope, as in the Obbia area, while in the south the junction with the plateau (on Jurassic limestones), now several miles inland, is effected usually by a short but steepish scarp, as for example near Isha Baidoa. There is no evidence of rifting or of intrusive volcanic rocks anywhere down this coastline.

The recorded evidence for high-level coastal deposits similar to the conglomerates and marine sands of the Gulf of Aden coast is slight. Certain beds of conglomerates and limestones with silicified bands are found on the edge of the plateau in the neighbourhood of Warandi, Dolobscio and Ilbehla at altitudes ranging from 140 m. to just over 100 m. Similar deposits were examined by the writer in the Mirsale and Afgudut areas further west, at an altitude of approximately 170 m. and were noted again, but not examined, up to approximately 200 m. near the road between Galkayu and Obbia. These rocks lie below the level of the Eocene limestones of the plateau proper. From the neighbourhood of Eman Saharer (220 m. approximately) eastward to the coast the Eocene limestones have been eroded, bed rock here being composed of the underlying Eocene anhydrite series. At Mirsale the conglomerates form low hills due south of the well area and rise some 20–30 ft. above the surrounding gypsum plain. That they were once of wide extent is evident from their present extended distribution, but they have been subjected to fairly intense erosion which has removed them from much of the area. The sequence noted shows some 2 ft. of banded chert overlying 15–20 ft. of grey to creamy limestone with thin chert bands. This bed rests on sandy conglomerates in some places 15 ft. or more thick, composed of small quartz and limestone pebbles set in a limestone matrix. The chert beds were extensively mined by late Middle, and Late Stone Age man and were the source of his raw material in the eastern Mudugh. These deposits have been described by Stefanini[1] who considers them to be formed of derived material from broken-down Eocene formations and to assume the character of

[1] Stefanini, G. (1936, p. 84).

a superficial deposit (partially terraced at Warandi), which has been redeposited by freshwater and torrent action: he concludes that these beds are Pleistocene in age. Their structure has been examined by de Angelis[1] and there seems to be little doubt that they are of freshwater origin.

The only other high-level deposits that I am able to record from the coastal plain are the conglomerates at the foot of the Isha Baidoa scarp which have not been studied in any detail; also certain boulder and pebble remnants round the base of some of the 'burs' in the vicinity of Bur Hakaba. The Baidoa conglomerates stand at approximately 300 m. and those in the 'Bur Country' at between 170–200 m.

The most prominent feature of the coastline proper is, as has already been stated, the long and well-developed chain of dunes which lines the coast from latitude 6° N. to the southern boundary of Somalia at latitude 2° S. The only break in this formation is at the mouth of the Juba. Between the mouth of the Juba and Mogadishu the dunes are close to the shore and rarely more than 8–10 km. wide but north of Mogadishu they spread inland, until, in the hinterland of Itala, they attain a maximum width of 100 km. Here they are some 15 km. from the sea, which distance gradually increases northwards. The dunes attain to heights of 152 m. at Merca, 137 m. near Mogadishu and 100 m. between Warsheik and Mareg. North of Mareg the height decreases further, but near Obbia rises again to the hog's-back of Gowein which reaches a maximum height of 378 m. North of Gowein, the height decreases more rapidly. With the exception of the Gowein hog's-back, it will be seen that altitude gradually decreases from south to north. Stefanini[2] considers that the Gowein formation is probably the result of folding similar to that noted by Dreyfuss in the Lake Assal basin, and may be of similar age. If this is the case, it would account for the great height of the sands in this area and would indicate that the accumulation of, at any rate, the earlier part of these sands antedates this final period of earth movement. As will be shown the sands were not all accumulated at one time; in some places two series of parallel dunes were found and at least three phases of deposition were noted by the writer.

The composition of these sands has been studied by Artini[3] and de Angelis[4] but it would seem that only specimens from the latest (red) sands on the fossil dunes, and from the mobile sands were examined. Artini notes the similarity of their composition to sands deposited by the Juba, and this is confirmed by de Angelis. According to Artini, the sands were in the first instance off-loaded

[1] Aloisi, P. & de Angelis, A. M. (1938, p. 92).

[2] Stefanini, G. (1936, p. 141).

[3] Artini, E. (1915, pp. 137–68).

[4] Aloisi, P. & de Angelis, A. M. (1938, pp. 105–9).

at the mouth of the Juba and other rivers in the south, from there were distributed by very strong currents along the Benadir (southern Somalia) coast, and subsequently accumulated as dunes along this coastline by violent monsoon winds.

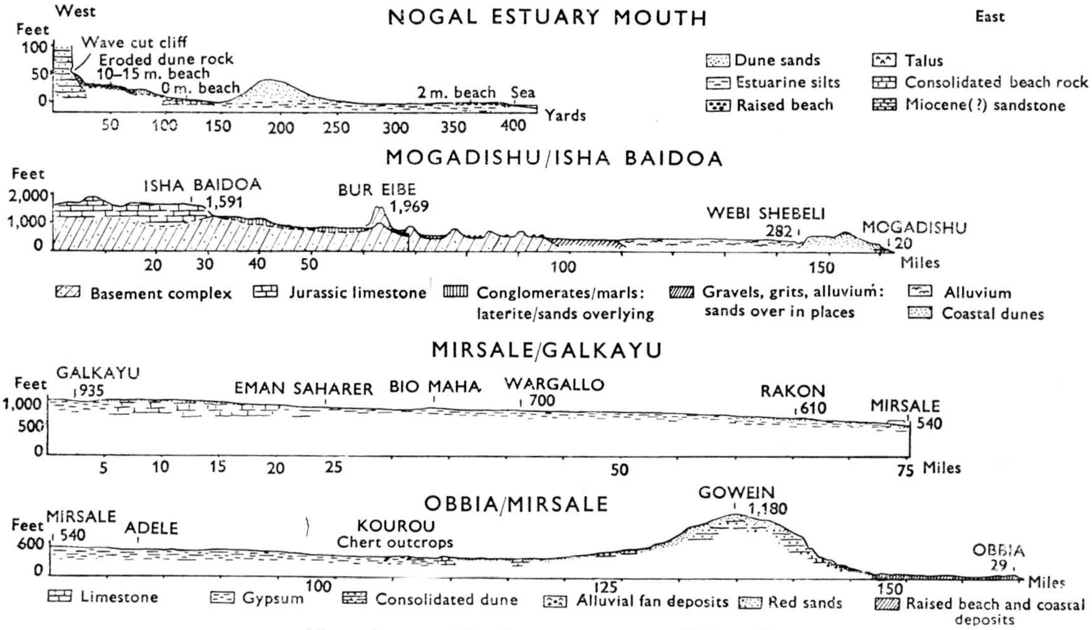

Text-fig. 14. Sections, east coast of Somalia.

The coastline itself was examined by the writer at Biede at the mouth of the Nogal, at Obbia in the central Mudugh, and between Mogadishu and Danane in the south. At Biede, the following sequence was found (see Text-fig. 14):

(i) Cut into the estuarine beds of the Nogal Gorge is a consolidated beach at an average height of 12 m. above present sea-level: this beach is well seen in section at its highest point at the back of the Customs House (14 m.), at the Customs House itself (12 m.) and along the coast to the north (11 m.) and south (11 m.) of the Nogal estuary (Pl. 42, figs. 1, 2 and 3). It comprises beach cobbles interstratified with layers of current-bedded silts and sands in a calcareous cement. The highest point at which the beach was found in the estuary approximates to a bench mark or notch at the base of the coastal cliffs at 17 m. This is presumed to be a wave-cut line of the 12 m. beach, but it is possible that it may represent an earlier sea-level: a talus deposit falls steeply from this bench to the raised beach proper. The bench mark has been accentuated by subaerial weathering in the soft Miocene (?) strata, and in places narrow shelters have been formed open to the sea. These were occupied by Late Stone Age strandloopers and now by the present day Rer Magno fishers of this coast. The 12 m. raised beach was found to be sterile where examined.

(ii) At a height of 0 m. was found a small outcrop of secondary limestone with beach cobbles, which was found to slope gently down towards the sea: this deposit is thought to be earlier than the 12 m. beach which appears to overlie it near the Biede rock shelter. It is, therefore, considered probable that it represents an earlier low-level beach (Pl. 42, fig. 3), though it might of course be a still-stand in the recession of the sea from the $\pm$12 m. beach.

(iii) Overlying this low-level beach at a height of 2 m. and upwards is a much eroded, consolidated, fossil dune of white to yellow quartz sand and shell fragments. The surface of this deposit is a reddened kunkar in fine sand, and shows irregular pipe structure. Large masses have been tilted by the undermining of the lower levels by wave action. Its relationship to the $\pm$12 m. beach was too obscure to be really certain whether the dune overlay also this beach or not.

(iv) Overlying the 12 m. beach is a red, sandy, detrital deposit for the most part unconsolidated, which has yielded, in the lower and middle levels, a few weathered Levalloisian flakes and some fresh Somaliland Stillbay specimens: these are important as they suggest an upper limit for the age of the 12 m. beach. The uppermost levels and surface yielded strandlooper artifacts.

(v) An unconsolidated deposit of marine sand, shells, and pebbles about $\frac{1}{4}$ mile wide between the sea and the 12 m. beach, is elevated some 3–4 ft. and possibly represents a third and later beach. The estuary mouth is filled by deepish, red, sandy, gypsiferous or saline clays and silts which apparently pass below present sea-level, as could be seen in borrow pits at Biede. These estuarine, sandy clays certainly postdate the 12 m. beach and are thought to be contemporary with the red, detrital deposit which overlies this beach.

(vi) Overlying the whole sequence are white to yellow dune sands which block the estuary mouth. If these sands are contemporary with the sterile, yellow sand in the Biede rock-shelter (see p. 282) their accumulation marks the end of the Late Stone Age strandlooper occupation.

The above sequence can perhaps be interpreted as showing firstly a low-level beach at 0 m., then a high-level beach or perhaps two—one at 17 m. and the other at 12 m.—followed by marine regression and a second low beach just above sea-level.

In the central Mudugh at Obbia the outstanding features are (*a*) the consolidated fossil dune of Gowein, which can be well seen on the section (Text-fig. 14), and (*b*) the very gentle slope of the plateau which merges imperceptibly with the coastal plain. The plain itself is composed of gypsiferous beds and surface-limestone and on these have been accumulated the Gowein sands. Deep gullies have been eroded in these sands both on the landward and seaward sides, those on the latter side being particularly deep. Sections show 10 ft. and more of red and buff sands, overlying very thick fossil dunes of white to yellow sands, the surface of which has been cemented into a hard calcrete or kunkar, oxidized in places, and pierced by irregular tubes and pipes, and has yielded to Stefanini and to the writer land molluscs similar to those found in the Mogadishu fossil dunes. Below the hard surface-limestone the sands are friable and appear to be several hundred feet thick, with occasional harder and more calcareous levels. The great depth of this formation accounts largely for the height of Gowein. The overlying red to buff sands are compact but not consolidated, the disconformity with the cemented fossil dune being most marked. They are on an average about 10 ft. or more thick, except on the inland flank of Gowein, but they are sometimes found filling old erosion gullies in the cemented dune, and then attain to considerable depth. On the top of Gowein they sometimes form long dunes now fixed by grass and scrub. On the inland slope of Gowein, however, these red sands attain their greatest depth of probably 50 ft. and more and appear to have been banked against the foot of the earlier formation by winds blowing from the land. In this area between the gypsiferous beds of the plateau and Gowein are still found mobile dunes, which in some places cover surface limestones. The surface of these red dunes on Gowein has been subjected to some minor degree of erosion which has resulted in an uneven, wavy horizon yielding sparse Late Stone Age artifacts and factory debris and which is overlain by a few feet only of loose orange to buff sand.

The descent from Gowein on the seaward side is steep, and at the foot of the slope are found stony, alluvial, fan deposits which spread out from the base of the scarp for a distance of $1\frac{1}{4}$ miles. These occur as mounds of pebbles and boulder fans in a matrix of grit and orange to buff sand, and presumably represent detrital material washed down from the slopes of Gowein by seasonal torrents. They appear to be related to the torrent-bedded gravels seen in the gullies where these debouch upon the coastal plain. Both deposits are sterile but rare Late Stone Age factory debris occurred on the surface.

At the foot of Gowein on this side was noted an outcrop of conglomeratic limestone. This seemed to pass beneath the dune formation and was at first thought to be Eocene limestone which had formed the 'core' of Gowein and so accounted for the great height of this formation. Italian geologists make no mention of any deposits earlier than the 'dune formation' here, and indeed, the grey colour of the limestone does not compare with the creamy colour of the Eocene beds further inland. Although no fossils were found in this bed during the very short examination possible to the writer, it is likely that this conglomeratic limestone at the foot of Gowein represents an early marine terrace, as it occurs in open position to the sea. The coastal plain proper, between Gowein and Obbia, is composed of gypsiferous and calcareous beds, which are thought to be of considerable depth and of marine origin, but this could not be determined with accuracy as sections were absent. At mile 6 from Obbia two samples of surface limestone yielded marine fossils, identified by Professor Zeuner, while samples of limestone taken at mile 10 from Obbia have been identified by him as representing almost certainly a beach deposit. These beds are overlain locally by compact dune sands, buff to yellow in colour (which are identified with the red and buff, compact sands of Gowein), and also by some mobile dunes of small size which increase in numbers towards the coast. These calcareous and gypsiferous deposits grade into a raised beach, standing some 15–20 m. above sea-level. This beach consists of a white concretionary kunkar or calcareous sandy coral formation passing down into calcareous conglomerates of rolled shell and coral fragments and derived blocks and large fragments of beach sandstone, cemented, presumably by calcium carbonate, but with a distinctly siliceous appearance.[1] This beach has yielded the land mollusc *Otopoma guillianopsis* and the marine molluscs *Spondylus* sp., *Arca lacerata*, *Lucina* sp. and *Pholas* sp. It is distant just over ¼ mile from the shore at Obbia and can be traced for several miles up and down the coast. Captain H. J. Courtney informs me that inland from Garad, between Obbia and Eil, there are two well-marked steps or benches; an upper bench composed of gypsiferous rock and selenite, overlain by consolidated dunes, descending by a short, steep scarp to the lower bench which is a white sand plain with some sparse grass and scrub abutting directly upon the sea.

On the coast itself just south of Obbia town where a small promontory juts out into the sea the following sequence was seen (Pl. 42, fig. 4):

(1) At the base, yellow, marine sands of fine grain with comminuted shell fragments. No complete shells were seen, but the bed showed irregular

[1] This sandstone would appear to be similar to the beach sandstone described by R. B. S. Sewell from islands in the Indian Ocean. Sewell, R. B. S. (1935, p. 500).

tube and pipe structure (which may, however, be due to living organisms as this bed is semi-submerged). Some 2 ft. were exposed, the remainder of the deposit being submerged.

(2) Coarser, horizontally bedded and laminated, marine sandstone, yellow in colour, with larger sand grains and shell fragments and occasional complete shells embedded in the upper levels. The thickest section seen was 5 ft.

(3) Coarse, calcareous, marine sands of a similar composition to (2) but with curious, vertical and oblique jointing, containing large shell fragments and whole shells, which show no clear evidence of bedding. This bed is brownish in colour and rests conformably on the underlying sandstone, though the junction between the two is fairly well marked. The uppermost levels have been calcified and weathering has eroded them into a series of sharp needle-like points. This upper part of the bed also contains derived blocks of the cemented beach sandstone previously described. The maximum thickness seen was 3 ft., the base being approximately at high tide level.

(4) A little to the south, on the shoreward end of the promontory, there rests disconformably a partially consolidated and false-bedded deposit of orange to buff, wind-blown sands, some 4 ft. thick, and this is in turn overlain by a midden of oyster and mussel shells set in a black earth matrix, covered in some places by several feet of loose wind-blown sand. Bed (3) is thought to represent a low-level beach and to pass under and perhaps into the 15–20 m. beach. Whether this was the case, however, could not be definitely proved as sections were absent, and as at Biede it is possible that it represents a still-stand in the recession of the sea from the 15–20 m. beach or beaches. The derived blocks of beach sandstone indicate the presence of a yet earlier beach from which they were derived.

Approximately ¼ mile south again of the above site on the edge of a narrow but longish valley, which runs parallel with the coast a few feet above sea-level and which must at one time have formed a lagoon, the sequence was found to be:

(1) At the base, yellow sand with a coating of gypsiferous kunkar into which had been dug shallow wells yielding brackish water only a few feet from the surface. These sands were thought to have been accumulated by wind action, but this is by no means certain and they definitely contain some pebbles and shell fragments: some 3 ft. of this deposit was exposed.

(2) On the surface of (1) were found artifacts of Late Stone Age strandlooper type.

(3) Unconsolidated whitish aeolian sand, 1–3 ft. thick: sterile.

(4) Oyster and mussel shell midden, which elsewhere yielded some worked iron fragments.

(5) Unconsolidated dune sands 2–3 ft. thick.

Later in date than the 0 m. and 15–20 m. beaches, is the raised, but low, sometimes partially submerged, cemented coral and sand deposit which runs parallel to the coast at Obbia. Stefanini describes this deposit as a coral reef and considers that the corals of this, sometimes fringing, sometimes barrier, reef were probably killed as a result of a general lowering of sea-level. Stefanini[1] concludes that the breaching of this reef at Obbia and the fact that it is here partly submerged may be taken as indicating some subsequent rise in sea-level. To the writer, however, this deposit appeared to resemble the consolidated kunkarous fossil dunes of the Benadir coast which in places are seen to pass down to and even below sea-level.

The sequence at Obbia may be interpreted as indicating:

(i) A conjectured high-level beach of which the derived calcified sandstone blocks and fragments in the 15–20 m. and 0 m. beaches are evidence.

(ii) A fall in sea-level marked by the low-level beach at 0 m., if this is not merely a still-stand in the 15–20 m. regression.

(iii) A rise in sea-level to the 15–20 m. raised beach.

(iv) A second fall in sea-level to below that existing today, and a possible still-stand at just above present sea-level when small tidal lagoons and the dead offshore (?) coral reef were formed.

The sequence of deposits seen north and south of Mogadishu, at Gesira, and again at Danane is exactly similar; the best sections being 1 mile north of the Lido, at the military Details Camp and at Forte Cecchi (all at Mogadishu), at Gesira between the island and the top of the second line of dunes, and in erosion gullies north of the prison and sanatorium at Danane.

A general section is that shown in Text-fig. 15. It will be noted that there are three main phases of dune formation:

(i) A basic, consolidated, deep, fossil dune composed of sands of marine origin (overlying a low-level beach). The maximum depth of the dune is always on the inland side. These dunes pass downwards into the sea where they are being eroded and form small islands and low reefs—notably between the harbour and Gesira.

[1] Stefanini, G. (1936, p. 90).

(ii) A second compact dune of similar composition overlying (i) nearer the coast.

(iii) Red and orange dune sands overlying (i) and sometimes (ii).

The Lido sections, Mogadishu, show the following sequence from the base upwards:

(i) Three feet of yellow calcareous beach conglomerate full of shell and coral fragments, passing below present sea-level. Also containing some humic matter apparently derived from contemporaneous soil. The latter, Professor Zeuner suggests, indicates perhaps that it was a storm beach.

(ii) Low-level beach 0–2 m. (0–6 ft.) above sea-level, with pebbles, fragments of coral and shell and whole shells (see Appendix G), set in a brownish red loamy matrix; maximum thickness 2 ft., average 1 ft. (Pl. 43, fig. 1).

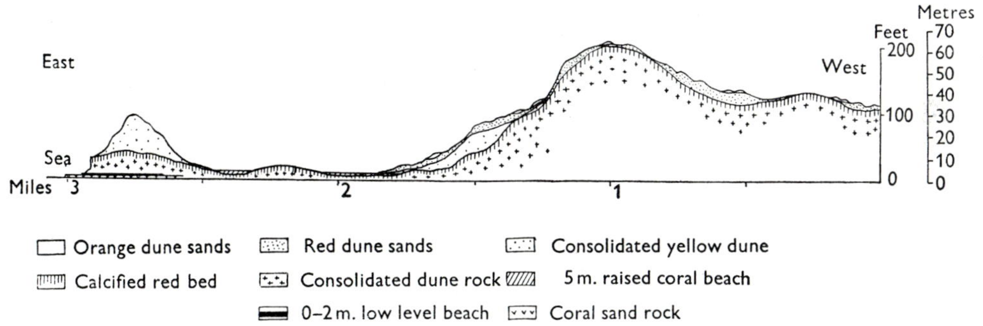

Text-fig. 15. Section across coastal dunes at Mogadishu.

(iii) Consolidated, white, calcified sands: thickness 15–20 ft. at coast—considerably more inland, probably 100 ft. (i.e. Forte Cecchi). The surface of this deposit is markedly undulating and sections show typical, aeolian false bedding. This bed is considered to represent a coastal dune formation. The surface for a depth of 6 in. to 1 ft. is a kunkar-like concretionary deposit of even-grained limestone-sand showing irregular pipe structure and is markedly reddened. It contains shells and casts of land molluscs, notably *Achatina* sp.: and is considered to represent a former land surface (Pl. 43, fig. 2).

(iv) This land surface is sometimes exposed today by wind action, but at the coast itself it is usually overlain by a second dune formation of friable but compact, marine sands, grey to yellowish white in colour. It shows typical false bedding but is sterile except for scarce, broken shell fragments. Thickness 10–70 ft.

(v) Compact red to buff dune sands 3–10 ft. thick overlie (iii) inland. They contain some derived shells from the land surface at the top of (iii) but are otherwise sterile.

(vi) The surface of these red dunes is undulating and is marked by a thin, detrital deposit in which are found scattered Late Stone Age implements and factory debris. This deposit is overlain by loose, orange to buff dune sands.

The same sequence is seen at Gesira, but at Danane the second accumulation of dune sands is seen stratified between the first dune sand and the latest red and orange sands. The section here, in an erosion gully at the back of the sanatorium, shows (Pl. 43, fig. 3):

(i) At the base a deposit of kunkar or calcrete, piped and fissured, and with casts of *Achatina*. 4 ft. thick, base not seen. (This bed is considered to be the same as that seen in section at the coast itself where 50 ft. or more of false-bedded dune sands rest upon the coral sand platform at sea-level.)

(ii) 15 ft. of friable, grey to buff, calcareous, clayey sands, of coarse granitic nature and fine gravel cemented by lime with a partially formed kunkar capping and 'veins' of kunkar running vertically downwards. This is overlain conformably by 15–20 ft. of white to grey, calcareous sands with, at the top, more concentrated kunkar formation.

(iii) Disconformably overlying (ii) is a compact, red sand 15–20 ft. thick in depressions (apparently old erosion gullies) and 10 ft. when on the top of ridges. The surface of this red sand is again markedly irregular, forming an old land surface with Late Stone Age factory debris and detrital material, and overlain by a varying thickness of loose, orange sands. These dunes have been markedly affected by differential weathering particularly in the area between Danane and Brava (Pl. 43, fig. 4).

The above sections indicate the presence of a low-level beach, overlain by a fossil dune composed of sands of marine origin perhaps in part water-deposited; a climatic break evidenced by the red bed; further deposition of dune sands of similar origin, followed by red, and later, orange sands.

A section exposed on the shore at the Mogadishu military Details Camp, however, adds further evidence. Here on the shore, at present sea-level, the following sequence was exposed:

(i) Marine, coral sandstone passing below sea-level. The surface of this deposit contains broken and some whole shells and fragments of coral, and is characterized by some irregular piping.

(ii) A cemented, reddish-brown, sandy loam, on an average 1 ft. thick, with land and some marine molluscs.

(iii) A cemented, calcareous deposit of marine sand, broken and whole shells and fragments of coral. Approximately 1 ft. of this deposit overlay (ii) at the section seen, but the thickness is probably greater farther inland.

(iv) Overlying this section are unconsolidated white, beach and dune sands with some shells at the base.

From the nature of the deposit there is absolutely no doubt that (i) and (ii) at the above section can be correlated with the low-level beach (ii) at the Lido section and with the similar 0 m. beach at Obbia, in which case deposit (iii) at the Details Camp clearly indicates a rise in sea-level subsequent to the formation of this beach. This marine transgression must have converted into lagoons or tidal marshes the valleys protected from the sea by the grey to buff dunes (iv) of the Lido sections. These dunes and protected valleys can be seen at intervals along the coast between Mogadishu airport and Gesira. The flooding of these valleys would necessitate a rise in sea-level of approximately 5 m. only: in some cases, notably at Gesira, appreciably less. In support of this evidence for a phase of marine transgression on this coast later than the 0 m. beach it may be noted that Artini[1] describes a marine sandstone from a raised terrace at Brava. Stefanini[2] records two raised beaches near Jumbo at the mouth of the Juba. An upper beach composed of limestones and clays, with *Ostrea cucullata*, *Murex*, *Lucina* and *Cerithium palustre*; and a lower beach, almost at sea-level, composed of beach sands with *Solenotellina ruppelli* and *Thiara vouamica*. On a level with the lower beach at Gobwein banks of oyster-shells occur. Further beaches are recorded at Kismayu and Ras Kiambone.

Dr Macfadyen also informs me by letter that in a road-metal pit, about 200 yd. south of the airfield at Kismayu, he found a section showing 2 ft. of red, earthy sand overlying 1 ft. of secondary limestone (kunkar). Near the base of the red earthy sand and in the kunkar he recovered the following molluscs—identified by Dr L. R. Cox as follows:

LAND SHELLS:
- *Achatina fulica* Férussac
- *A. daroliensis* Kobelt
- *Otopoma* sp. (? aff. *obtusum* (Pfeiffer))

FRESHWATER:
- *Theodoxus natalensis* (Reeve)
- *Theodoxus* sp.

BRACKISH WATER:
- *Terebralia palustris* (Linné) (i.e. *Cerithium palustre*)

MARINE:
- *Anadara* cf. *antiquata* (Linné)
- *Ostrea* sp. indet.

Terebralia palustris also occurs in the 2 m. raised beach on the north coast between Berbera and Zeila.

[1] Quoted by de Angelis, A. M. (1938, p. 96).

[2] Stefanini, G. (1936, p. 91).

These recent and low-level occurrences of marine deposits and molluscs at Jumbo, Gobwein and Kismayu suggest the presence of a beach level similar to the 2 m. raised beach on the north coast.

Of the dune formations, the sand in the earlier, cemented, dunes is undoubtedly of marine origin; and the writer is of the opinion that these coastal dunes were accumulated by strong monsoon winds, of which, on Artini's evidence, the prevailing wind was probably from the south-west. It is considered, moreover, that these earlier dunes were accumulated during a wet phase, for, they may be correlated, most probably, with alluvial deposits in the interior which were known to have been formed under wet conditions. Only the latest, red and orange sands of the coastal dunes, which may be identified with similar sands in the interior, appear to have been formed under more arid conditions. It is suggested, therefore, that the earlier, consolidated, coastal dunes were formed during a period or periods, of increased wet conditions as a result of strong monsoon winds, while the later, red and orange sands were formed as a result of increasing desiccation.

Most regrettably, none of these earlier coastal deposits, with the exception of the land surface above the 12 m. beach at Biede (and the rock-shelter at this site containing a Late Stone Age strandlooping industry to be described in Part IV) contained any cultural material. The dating of these deposits must, therefore, remain conjectural until such time as artifacts are found within the earlier beds. By correlation, however, we may arrive at an approximate date for these deposits. The orange and red sands can be dated as they extend over considerable areas of the interior in association with cultural material. The red-brown alluvial deposits in the Webi Shebeli are clearly later than the first and second phases of dune formation at the coast, as in the Afgoi area the former are banked against the landward side (see Text-fig. 14) of the dunes.

I would correlate tentatively, therefore, these two accumulations of calcareous, leached dunes with the two phases of calcareous gravels and alluvium in the Webi Shebeli and other valleys, and in fact the kunkarous sands of the Bur country are almost certainly the inland counterpart of these coastal deposits.

The following sequence may be suggested, therefore, from a correlation of all the sections recorded:

(i) Possible high-level conglomerates at altitudes of 100–200 m.

(ii) A hiatus of long duration.[1]

(iii) A low level, 0–2 m. beach, representing a period of regression or low sea-level followed and in part contemporary with wide-scale dune formation

[1] Stefanini, G. (1936, p. 90), records high-level terraces in the Hafun peninsula, which may, after further work, be found to fill this gap in the sequence.

due to strong and wet monsoon winds. Possibly some anticlinal movement at this time produced the hog's-back of Gowein and may have also affected, by uplift, the marine formations.

(iv) Marine transgression to a height of 11–24 m.

(v) A fall in sea-level (probably to below present sea-level) followed by the formation of a low-level beach at ±2 m., and of shallow lagoons contemporaneous with the accumulation of a second line of coastal dunes, similarly of monsoon origin.

(vi) Marine retrogression to present sea-level and the accumulation of red and orange dune sands on the earlier formations.

Since the above tentative succession of events was suggested the Cambridge Archaeological Expedition to Cyrenaica has published its first report on work in the coastal region of the Gebel Akdar between Tocra and Derna.[1] The topography there appears to be closely parallel to that on the east Somaliland coast. The same association occurs of alluvial fan deposits and consolidated coastal dunes superimposed on a basic system of marine terraces. The authors warn against the dangers of estimating sea-levels on isolated exposures of deposits which is in fact what we have been attempting in Somaliland. From the number of different recorded heights for raised beaches in the Horn it is obvious that the majority of the actual heights recorded are of very little value as in not a few instances they must surely represent temporary resting stages in a period or periods of marine regression or transgression, and few, if any, of the references cited give much indication of the beach element to which the given height refers. In Cyrenaica the consolidated coastal dune deposits were found to descend below sea-level and to overlie the 5 m. terrace which contains within it a Levalloiso-Mousterian industry.

In the light of this Cyrenaican evidence, therefore, an alternative and preferable interpretation may be suggested for the later sequence of events on the Somaliland east coast:

(iii) A high sea-level at ±11–24 m. (though here again more than one beach may be present). Inferred contemporaneity with a Levalloisian phase of culture.

(iv) and (v) A gradual regression of sea-level, with still-stands at various levels (i.e. at 0–2 m.) down to a height certainly below present sea-level. Contemporary with this regression would be the accumulation of the kunkarous fossil dunes (which pass below sea-level) followed by a gradual transgression of the sea to at first ±2 m. then to present sea-level.

[1] McBurney, C. B. M., Hey, R. W. & Watson, W. (1948, pp. 33–45).

CHAPTER 7

CORRELATION OF GEOLOGICAL DEPOSITS, EVENTS AND CULTURES WITHIN THE HORN

Owing to the lack of adequate palaeontological evidence any correlation and dating of the Pleistocene sequence in the Horn must be based first, on parallel sequences of similar geological phenomena and secondly, on the archaeological evidence contained within the sequence of deposits. Stone implements occur in quantity throughout the Somalilands and where a sufficiently representative series is found they have been used as zone fossils as a means of establishing a relative chronology.

On the Somali plateau which has been affected to a considerable degree by earth movement on a large scale it is a matter of difficulty to estimate the amount of climatic fluctuation that may have taken place there during Pleistocene times. Much of the denudation that has taken place, and much of the depositional phenomena could have been the result of earth movement. While earth movement must have intensified the erosional and depositional activities of the river systems, there is, however, evidence which more definitely points to climatic changes during this time and which the writer finds difficult to interpret in any other way. The coarse nature of the terrace and sheet gravels in the Webi Shebeli and Nogal valleys indicates a precipitation greater than that pertaining today, and the evenly bedded nature of much of the Lower Tug Gravels is evidence for a further period of increased precipitation. Some of the most conclusive evidence, however, comes from the closed basins of the pans and small lake areas on the plateau. The deposits within these basins might be expected to reflect any climatic change that may have taken place. In two of these miniature lakes, Ted and Burdo on the southern plateau, two (and at Ted three) terraces are seen, the upper of which is marked by thick tufa deposits. In other pans, Saha Gurral, Morogavi and El Dubbo, for example, two tufa deposits are seen which have yielded Levalloisian artifacts. Tufa formation has now ceased in Somaliland except at certain springs and the conditions necessary for the formation of tufa would require to be wetter than those at the present day. Again on the east coast the barrier dunes must have accumulated as the result of strong monsoon winds which again indicates a greater precipitation.

The writer is of the opinion, therefore, that evidence exists which indicates that definite changes of climate took place during Pleistocene times in the Horn of Africa: whether this change was of the intensity suggested by the term Pluvial is doubtful, but it is certain that at particular times the climate was wetter and drier than that which pertains there today. Zeuner[1] has pointed out that a northward displacement of the caloric equator would simultaneously bring about a northward shift of the northern limit of the monsoons. As the fluctuation of the caloric equator in the tropical zone is said to have been as much as 5–8° N. and S. of the present geographical equator it can be seen that a northward swing would provide exactly the wetter climatic conditions and strong monsoon winds found in the Horn, while a similar swing to the southward would result in more arid conditions and the desert sands.

The inland areas show a close parallel sequence of events and deposits. The sequence in each area commences with a major erosional phase of long duration when broad deepish valleys were eroded through the more or less perfect peneplain which forms the plateau.

It is presumed that the earliest plateau conglomerates antedate this erosion cycle, though it is not improbable that they may in part be contemporary with it, as they appear as high-level deposits which have been cut through by the present drainage system.

During this first and major erosional phase, valleys of the main river systems of the Webi Shebeli and Juba have been graded to the coastal plain which on comparative evidence[2] probably dates to the end-Tertiary or Lower Pleistocene. This erosion phase is therefore equated with the first, or Kageran, Pluvial of East Africa.

At the coasts at this time it would seem that a period of movement resulted in the gradual emergence of the coastal plain, and similar movement is projected for the emergence of the plains of Danakil.

This long period of valley cutting is followed by a major phase of deposition when thick aggradations of conglomerates, boulder beds and breccias, set in a marly matrix, were laid down within these valleys. The evidence from western British Somaliland indicates that these beds form the lowest infilling in the valleys and the deep nature of this first erosion phase is also seen in the gorge of the lower Nogal, though the fluctuations of sea-level presumably prevailed over the climatic rhythm in this area. Elsewhere, however, it would appear

[1] Zeuner, F. E. (1946, p. 266).

[2] Paterson, T. T. (1940, p. 397).

The coastal sequence along the Gulf of Aden Coast is closely parallel to that of the Madras coastal plain of southern India. The latter is marine in origin and emerged in Upper Siwalik (Lower Pleistocene) times.

that these conglomerates and boulder beds were deposited only on the upper slopes of the valleys which were subsequently deepened by later erosional activity. If, however, the basal gypsiferous and calcareous beds (which have so far proved unfossiliferous) in the Webi Shebeli, Juba and upper Nogal should in fact prove to be of a secondary nature as has been suggested, then the sequence: (i) major and initial period of valley cutting, (ii) major depositional period (the sediments of which provide the lowest infilling of the valleys), (iii) subsequent erosional and depositional phases of decreasing intensity—will prove to be identical throughout the Horn. It must be remembered, however, that the Gulf of Aden Rift was already in existence in Miocene times and the effect of posthumous earth movement along this trend might be expected to have affected the hydrographic systems of the northern plateau and might therefore account for this physiographical difference between the north and south-east—the climatic cycle being able to prevail over isostatic movement in the river systems of the more stable parts of the latter area.

The earliest deposits of the northern and eastern coastal plains (apart that is from the Miocene marine beds) and of the Danakil Rift are similar conglomerates and boulder deposits; there can be little doubt that they are contemporary with their counterparts in the major river valleys, and the conglomerates at the foot of the scarp delimiting the south-eastern coastal plain pass into those on the south side of the Webi Shebeli valley.

The upper levels of these conglomerates and boulder beds have everywhere been affected by calcification and the formation of an extremely hard surface limestone, several feet thick, which effectively seals the deposits, and which was presumably formed over a long period of time under semi-arid to arid climatic conditions by the evaporation of ground water bringing to the surface carbonates from the underlying rocks. Faulting and basalt lavas affecting the Gulf of Aden conglomerate beds indicate a period of earth movement at this time.

This climatic break is followed in the major valleys of the Webi Shebeli and Nogal by renewed erosion and the aggradation of coarse terrace and sheet gravels. In the lower Nogal they take the form of estuarine deposits which fill the deep channel in the gorge. These beds have been ferruginized and the gravels often show a red to black glaze which is closely similar to a 'desert varnish'. In the Nogal and in some parts of the Webi Shebeli (Callafo: Mustahil) these terrace gravels have been calcified and are covered by a hard limestone of a secondary nature, which is considered to represent an arid phase with sharply defined seasonal change similar to that which is reflected by the surface limestone sealing the marly conglomerates of the initial phase of deposition.

This dry period appears to have been coincident with one of renewed earth movement and it is to this time that I would date the renewed tilting of the Somali plateau to the south-east away from the Abyssinian Rift and towards the Indian Ocean. This tilting caused the draining of the Ghedeb freshwater lake on the high plateau in which the Webi Shebeli and Juba had their source and increased the intensity of the subsequent denudation phase.

On the coastal plains of the Gulf of Aden (and perhaps on the east coast also) is found a second aggradation of coarse conglomerates with gravels, sands and red clays interstratified with acid lavas, which are provisionally correlated with the terrace and sheet gravels of the second aggradation phase in the main valleys. These deposits have also been affected by earth movement which has faulted them on the Zeila plain. The events in the Danakil Rift at this time are difficult to determine as earth movement particularly at the south-west corner has obscured the sequence. It is possible that the Kamasian-type deposits of the Awash Gorge, and the marine sandstones of the Aden Series in northern Danakil which have both been affected by earth movement on an immense scale represent the Middle Pleistocene beds in this unstable area but further work is necessary in order to confirm this and to determine to what extent the conglomerates and boulder beds which ring the depression have been affected by this movement.

At this stage and without archaeological and palacontological evidence it is not possible to separate the Lower and Middle Pleistocene deposits, but it is considered that the long arid period which affected the terrace and sheet gravels of the second aggradation phase probably represents the long and dry inter-Pluvial before the return of wetter conditions during the Upper Pleistocene. None of these deposits so far described has yielded stone implements, but this is probably because no really detailed search for them has yet been made.

This Interpluvial break is a most pronounced one, and is followed by renewed denudation which marks the initiation of the present tributary tug systems and enabled the rivers to erode a deep channel, now buried, in which were accumulated two alluvial deposits of gravels, sands and silt, well documented by stone artifacts. This sequence of (i) buried channel erosion, followed by (ii) the alluvial aggradation deposits, is found throughout the river valleys that have been examined and only in north-western Somaliland is there any variation of this sequence. These aggradation deposits have been termed the Tug Gravels and the Tug Alluvium respectively. They are divided into Older and Younger phases (the latter subdivided into substages I and II) in the case of the north-western Somaliland sequence, and into Lower and Upper phases in the case of the remaining areas. The term 'Tug Alluvium' was first used by Parkinson: it aptly describes these deposits and in modified form has been retained as the climatic conditions which contributed to its formation must have been of a semi-arid to arid nature.

In north-western Somaliland, antedating the Younger Tug Gravels and Alluvium which contain the Levalloisian culture sequence, are found one, sometimes two, accumulations of coarse torrent gravels and kunkarous alluvium—the Older Tug Gravels and Alluvium. These beds, which have in some instances been calcified, contain the earliest cultural material yet found in the Horn of Africa—artifacts of the Acheulio-Levalloisian Culture. This culture is found stratified only within the Older Tug Gravels of the north-western Somaliland tugs and its stratigraphical position and date is therefore dependent upon the date of these gravels. From the position in which they occur, on the flanks of the valleys and passing down into the buried channel overlying the 'Malas', it would seem that they should be assigned to the beginning of the Last, or Gamblian, Pluvial when wet conditions were gradually becoming more intense, and not to the end of the preceding Kamasian Pluvial. Confirmation for such a date is provided by the alluvial torrent nature of the deposits. The implementiferous parts of these gravels are found mainly on the upper slopes of the valleys but there is no reason to dissociate these higher level occurrences from those in the buried channel: they have accordingly been shown on the Correlation Table (p. 152) as representing the first phase of the Gamblian Pluvial. The Acheulio-Levalloisian Culture is, therefore, dated to this time—a position which correlates well, as will be seen later, with its occurrence in other parts of the Continent.

An end-Kamasian, or rather Last Interpluvial, date must, however, be considered as a possibility, and some corroboration for this is given by the sites in the gorge of the lower Nogal and at Dubar. The Nogal estuarine beds have been heavily calcified but are overlain at one point near the side of the gorge by an Acheulio-Levalloisian site in which the artifacts show heavy calcareous incrustation and are much weathered. These estuarine beds are cut through by the Tug Gravels and Alluvium which yield Levalloisian assemblages. At the Berbera site the few Acheulio-Levalloisian artifacts found are all considerably more weathered than the Levalloisian tools, and in some cases show signs of ferruginization. Should, therefore, the Older Tug Gravels and Alluvium prove to be contemporary with the Nogal estuarine beds, which clearly belong to a cycle anterior to that which resulted in the aggradation of the Tug Gravels and Alluvium, then they must be assigned to the end of the Kamasian Pluvial, and the Acheulio-Levalloisian Culture must accordingly be put back in date to a time immediately postdating these beds. In any case further work may prove the existence of similar deposits to the Older Tug Gravels in our other main areas and will thus provide the corroborative evidence, which at present is lacking.

The main Tug Gravels and Tug Alluvium need little explanation. They clearly denote a single aggradation phase of long duration which is divisible into two subphases separated by a short arid phase. In each subphase gravels are overlain by sandy, silty alluvium which has been calcified and sometimes reddened. They yield Levalloisian artifacts in some quantity and indicate that cultural development was very slow. The nature of the alluvium is such that much of the sandy content is thought to be of aeolian origin and the similarity with the 'aeolian-silt' deposits in the southern Arabian valleys is probably a close one.[1] At the same time as this alluvium was accumulating in the river valleys there was some dune formation (i.e. middle Webi Shebeli) on the upper slopes of the valleys, with similar sands of wider extent on the plateau and coastal plain. Down the east coastline strong winds built up an accumulation of barrier dunes of sufficient magnitude to cause the deviation of the Webi Shebeli.

In the tugs of the Danakil Rift we find the aggradation of two alluvial fan deposits of Tug Gravels and Alluvium in a deep channel which discharged their load into shallow freshwater lakes, some of which gradually became filled and silted up with this fine-grained fan material as the gradient was built up: the Danakil freshwater lakes are on admittedly rather scanty archaeological evidence, correlated with the Tug Gravels and Alluvium of the river valleys.

The dry phase which marks the end of the period of aggradation of the Tug Gravels and Alluvium (with which the Somaliland Stillbay Culture is contemporary), was also accompanied by some earth movement albeit on a local and much reduced scale. This reached its maximum activity in the north-west, and shows a gradual decline in strength in the south-east. In Danakil further faulting and local lava spreads are recorded, but this diminishes to folding on the north and east coasts of Somaliland.[2] This dry period is considered to mark the end of the Gamblian Pluvial in the Horn of Africa.

There is evidence for two further periods when the climate was wetter than that existing today. The first, of greater magnitude than the second, resulted in the accumulation of coarse torrent gravels and fine-grained, partially aeolian, alluvium. It was in part contemporary with, but in the main followed by, the accumulation of red wind-blown sands[3] both in the interior (where they cover wide areas in the south and centre) and on the coast.

[1] Caton-Thompson, G. & Gardner, E. W. (1939, p. 25).

[2] Should the existence of the anticlinal fold of Gowein be confirmed it is possible that this may have had something to do with the deflexion of the Webi Shebeli.

[3] It is just possible that some of the red sands may antedate the torrent gravels, in which case they would be contemporary with the dry period marking the end of the Gamblian Pluvial. At present, however, the stratigraphical evidence is in the main against such a date and this is largely confirmed by the cultural material.

More developed, predominantly microlithic, cultures are contemporary with this First Post-Pluvial Wet Phase as it has been termed. In the south and centre we find the Somaliland Magosian, which develops into the more local Doian, and, in the north, a culture with a preponderance of true blades and angle-burins, which appear to represent intrusive elements—termed the Hargeisan Culture. In the Ogaden there are indications of an early true Wilton.

A second minor wet phase follows (Second Post-Pluvial Wet Phase) with which are contemporary several Late Stone Age variants. The Doian in a more developed form is present in the south with a strandlooping variant on the coast.

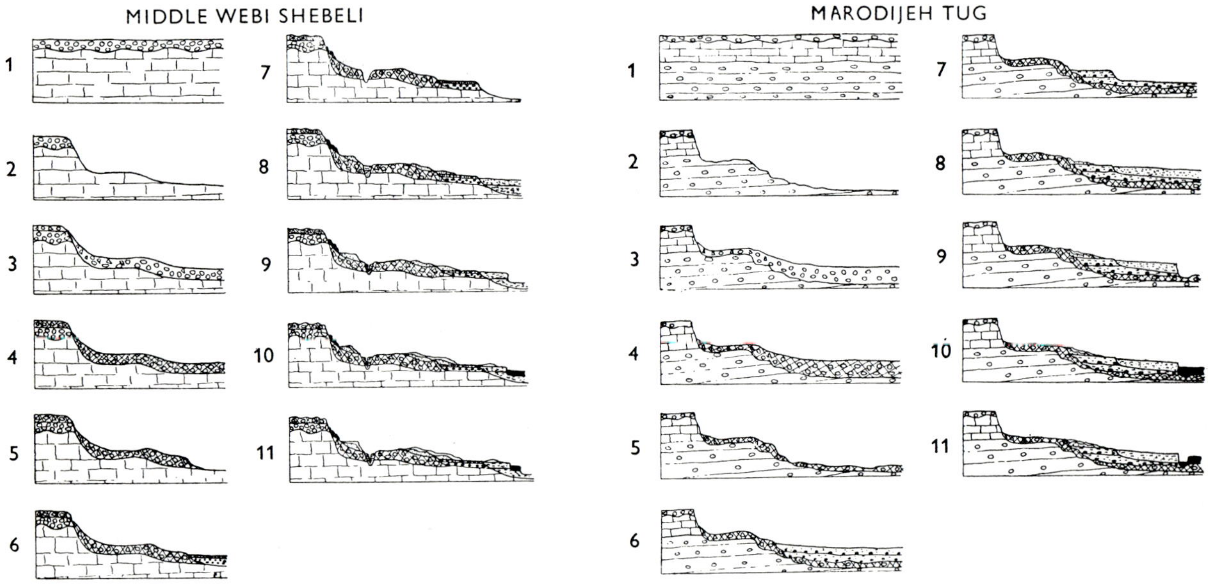

Text-fig. 16. Diagrammatic sections illustrating stages of the Pleistocene and Recent succession in the Webi Shebeli and Marodijeh Tug.

In the north the Hargeisan develops into the Somaliland Wilton and into a variant with hollow-based arrow-heads which give it a 'neolithic' appearance. This form of the Hargeisan spreads southwards into the Mudugh where it produces hybrids with the Doian. In the Ogaden, however, are found Wilton forms, probably intrusive from the Galla high plateau and varying somewhat from the true Somaliland Wilton of the north. Strandlooping industries on the northern part of the east coast, include elements which suggest a Middle Stone Age ancestry.

The correlation of the Upper Pleistocene and Recent coastal sequence with that in the interior is made possible by two areas, those at Obok on the French Somaliland coast, and at Eil on the eastern coast of Somalia. On the north

Text-fig. 16. Description

Middle Webi Shebeli

1. Deposition of pebble beds and conglomerates on the plateau during the Late Tertiary.
2. Major erosional phase of long duration during which a wide flat valley was cut some 600–700 ft. deep in at least two stages. This initial period of valley cutting probably took place during the end-Tertiary or Lower Pleistocene or earlier.
3. Major depositional phase during which thick conglomerates, marls and breccias were deposited in the valley, possibly during the Lower and Middle Pleistocene.
4. Consolidation and calcification of these deposits during a long arid period.
5. Renewed denudation and the erosion of the valley conglomerates and marls. Formation of the 50–100 ft. terrace, or terraces.
6. Aggradation of gravels on the 50–100 ft. terrace, or terraces. Ferruginization and sealing of these gravels by a deposit of pisolitic ironstone. Contemporary earth movement and gentle warping. end-Middle or early Upper Pleistocene.
7. Renewed erosion resulting, in the main river, in the excavation of a channel the floor of which is now for the most part buried, and in the formation of steep V-sided valleys by tributary streams. Some formation of tufa at springs on the scarp. Upper Pleistocene.
8. Aggradation of two deposits of gravels, sands and silts separated by a red bed denoting an arid, or semi-arid, phase of short duration. These deposits contain the Lower and Upper Levalloisian Culture. Continued tufa formation on the scarp, and the accumulation of dune sands on the upper slopes in the main valley. End of the Upper Pleistocene.
9. Minor erosional phase during which the river failed to re-excavate its bed to the former level. Formation of a ferruginous horizon on the upper slopes of the valley. Somaliland Stillbay and Magosian Cultures.
10. Aggradation of coarse gravels and alluvium under conditions of intermittent rainfall. Accumulation of red wind-blown sands on the upper slopes. Somaliland Magosian and Doian Cultures.
11. Further minor erosion, some deposition of dune sands, and erosion to present contours.

Marodijeh Tug: Hargeisa

1. Deposition of plateau pebble beds during the end-Tertiary. Previous to this period faulting had already taken place in the Gulf of Aden Rift trend.
2. Major erosional phase during which a deep valley was cut in the Eocene limestones and Cretaceous sandstones. End-Tertiary or Lower Pleistocene.
3. Major aggradation phase when marly pebble beds and breccias were deposited on the slopes and in the bottom of the valley. Lower and/or Middle Pleistocene.
4. Consolidation of these deposits during an arid period of long duration.
5. Renewed erosion and excavation of a deep channel in the earlier deposits. Beginning of the Upper Pleistocene?
6. Aggradation of one, sometimes two, torrent-bedded gravels and alluvium (Older Tug Gravels and Alluvium), containing the Acheulio-Levalloisian Culture.
7. Further erosion, again to below present tug level.
8. Aggradation of gravels and red alluvium in the channel so formed (Younger Tug Gravels and Alluvium), containing the Levalloisian Culture, and at the top Somaliland Stillbay tools. Some accumulation of wind-blown sands on the upper slopes of the valley. End of the Pleistocene.
9. A minor erosional phase accompanied by some further accumulation of wind-blown sand. Somaliland Stillbay and Hargeisan Cultures.
10. Aggradation of coarse gravels, grits and sands. Hargeisan and Somaliland Wilton Cultures.
11. Further minor erosional phase and excavation to present contours.

coast the two sheet gravels from Mount Mabla are correlated by Teilhard de Chardin, Breuil & Wernert (1940) with a phase of gradual marine transgression: these gravels contain the Levalloisian complex. Two high-level beach levels may therefore be inferred.

The subsequent sudden drop in sea-level in early Gamblian (Upper Pleistocene) times along this coast, is evidenced by the deep bores at Zeila and the gradual return to just above present sea-level by the deposits in the Zeila bores and the ±2 m. raised beach.

On the east coast the deposits in the Nogal gorge which were presumably formed under the influence of changing sea-level can be correlated with the coastal beds. The earliest sequence of events is obscure but during the erosion of the first buried channel (during a wet phase or pluvial) the sea must have stood at a low level. The aggradation of the estuarine beds during the decline of this pluvial indicates a gradual rise in sea-level to ±30 m. (the height at which these beds stand at the estuary mouth). The sea remained at this high level during the calcification of these beds. The Acheulio-Levalloisian Culture dates to a period postdating this event and apparently antedating that which followed. The succeeding denudation phase marking the beginning of the Last Pluvial and the cutting of the second buried channel indicates marine regression to a low sea-level represented perhaps by the ±0 m. beach. The subsequent aggradation deposits with their Levalloisian assemblages may be taken as indicating a gradual rise in sea-level—represented by the ±12 m. raised beach. Later degradation—to below sea-level—is indicated by the red sands and silts at the estuary mouth which fill a channel cut deeply below the present floor, and correlate with the torrent gravels and alluvium further upstream from the mouth. The subsequent transgression to just above the present level is indicated by the ±2 m. beach.

The provisional correlation of deposits, events and cultures is set out in the Correlation Table.

For comparative purposes the sequence in the middle Webi Shebeli and in the Marodijeh Tug is correlated in diagrammatic form at Text-fig. 16.

PART III

THE PREHISTORY OF THE SPECIAL AREAS

CHAPTER 8

TERMINOLOGY

It will be appreciated that there exists at present a fundamental difference of outlook as regards the terminology of Palaeolithic cultures in the Horn of Africa when we take into consideration the direction from which prehistoric problems there are approached. Prehistorians from South, Central and East Africa would employ a terminology which, particularly in its later stages, is fundamentally of southern African derivation. On the other hand workers in the North African field would use a terminology based on Europe and the Near East. The Horn of Africa is one of the meeting places of those two schools, the one working up from the south, and the other down from the north. It has therefore been found necessary to define the cultural and technical terms used in order to eliminate any confusion that might result from the too strict adherence to one or other of these schools.

The application of cultural names to describe technical processes of manufacture has been purposely avoided, and the terms proposed by the Pan-African Congress on Prehistory, 1947, have been adopted here. For example, the expression 'faceted-platform technique' is employed in place of the term 'Levallois technique' to describe this process. Similarly, the expression 'block-on-block technique' has been substituted for the term 'Clacton technique'. Consistently with the above description of processes of manufacture on a technological basis, the terms 'wood technique', 'blade technique' and 'bi-polar technique'[1] are used to describe those processes rather than chronological or cultural terms which are only too liable to be misunderstood. Only where the employment of cultural terms to describe a specialized technical process has proved unavoidable has it been necessary to deviate from the above practice. Where, therefore, terms such as 'Tabelbalat technique' or 'Heluan retouch' are used these are defined in footnotes.

For the major subdivisions of the cultural succession the purely general terminology of Early, Middle and Late Stone Age is used in preference to Lower, Middle and Upper Palaeolithic; the latter terms, which Professor Garrod[2] has advocated should be used exclusively in a chronological sense, still retain

[1] Percussion technique whereby, if a core is held vertically on an anvil and struck at the opposite end with a hammer, flakes are simultaneously removed from both ends of the core.

[2] Garrod, D. A. E. (1938, p. 2).

for many prehistorians as Dr Paterson has pointed out,[1] a purely technical or cultural significance. The Early Stone Age in Africa may be taken as being co-terminate with the Middle Pleistocene and with the final development of the Chelles-Acheul Culture complex *sensu stricto*. Certain complexes such as the Acheulio-Levalloisian are in every sense transitional cultures between the Early and Middle Stone Age, but, as has been shown, it is more probable that chronologically such complexes should be assigned to the Upper Pleistocene. The Middle Stone Age in the Horn includes the Levalloisian Culture complex together with its ultimate development into the Somaliland Stillbay. The Middle Stone Age is co-terminate with the Upper Pleistocene. Transitional between the Middle and the Late Stone Age is the Somaliland variant of the Magosian and the intrusive blade and burin industries of the Hargeisan. The Late Stone Age includes the development of those two cultural phases, the former into the Doian and the latter into the Somaliland Wilton.

The cultural terms employed are as follows:

THE ACHEULIO-LEVALLOISIAN

This is the earliest culture of which we have definite evidence in the Horn, although there is strong evidence to presuppose the prior existence of an advanced form of Upper Acheulian in which the prepared-core technique may be lacking. Wood technique is found in indivisible association with the faceted-platform technique with the Acheulio-Levalloisian. Tool types include hand-axes and cleavers with unprepared cores and the flakes struck from these which combine to provide the Acheulian element. The Levalloisian element is provided by large, radially flaked, prepared cores and faceted flakes, together with smaller disc cores and the flakes struck from these. The term 'Fauresmith', which is employed in South and East Africa to describe a culture complex of predominantly Chelles-Acheul tradition in association with a fairly developed faceted-platform technique and an increased use of flake tools, cannot as yet, except in a general sense, be applied in the Horn; there are typological and technical differences which require more complete evidence before they can be explained.

THE LEVALLOISIAN CULTURE

This culture complex is based on the small, radially flaked core and faceted flake, the products of a faceted-platform technique. The complex was of prolonged duration and of slow development. In all except its latest phase it displays the typological and technical attributes of a succession of pure Levalloisian industries and having regard to its northern associations, the writer is of the

[1] Paterson, T. T. (1945, p. 1).

Table showing corresponding divisions of the Stone Age in Europe and the Horn of Africa

Europe	The Somalilands		
Typological equivalents	Main divisions	Subdivisions	Main tool types
Neolithic	Late Stone Age	{Somaliland Wilton, Doian}	Microliths, pressure-flaked arrow-heads, burins, etc.
Mesolithic		{Hargeisan	Straight-backed bladelets, burins, end-scrapers
		Somaliland Magosian}	Microliths, small pressure-flaked points, etc.
Palaeolithic	Middle Stone Age	Somaliland Stillbay	Pressure-flaked points, scrapers, burins, backed flakes, etc.
		Levalloisian	Simple points, faceted flakes and prepared cores
		Acheulio-Levalloisian	Hand-axes, cleavers, flake tools, large circular prepared cores
	Early Stone Age	(Upper Acheulian) ?	Hand-axes and cleavers
Eolithic	—	—	—

opinion that the cultural use of the adjective 'Levalloisian' to describe the culture complex is fully justified. Its counterpart in North Africa is the Levalloiso-Mousterian.

THE SOMALILAND STILLBAY CULTURE

In the latest stage, more advanced technical processes resulted in the development of bifaced retouch with semi- and fully-bifaced points, backed flakes and other tool types which characterize the widespread Stillbay Culture of the subcontinent and East Africa. A Stillbay industry has already been recorded from northern Abyssinia, probably near the northern limit of this culture complex, and as the industries of this type within the Horn differ in no major and fundamental degree from the East and South African expression it has been necessary only to add the adjectival qualification and describe this ultimate expression of the Levalloisian tradition, whether the outcome or not of culture contact, under the term 'Somaliland Stillbay'.[1]

[1] No exact comparison with any of the South African or East African cultures is implied. Certain technical affinities with each of the regional variants of the widespread Stillbay culture-complex (all believed to be derived from a common parentage) are considered to justify, however, the inclusion of those Somaliland industries within this culture-complex.

THE SOMALILAND MAGOSIAN AND WILTON

The later industries, from the end of the Upper Pleistocene onwards, show a decidedly more local divergence, though their cultural associations are in most cases sufficiently clear. Thus we find a Somaliland Magosian and a Somaliland Wilton. In two instances, however, local divergence and specialization has necessitated the introduction of new cultural terms, but this has been done only as a temporary measure and because it is considered unavoidable. When the gaps in our knowledge of the sequence in adjoining areas are filled, it will probably prove possible to discard such localized terms.

Certain industries which first make their appearance in the north-west during the dry period which marks the end of the Upper Pleistocene, exhibit clear blade and burin affinities, but include also a small element of tool types that may have been acquired as a result of culture-contact, but may equally well be a basic part of such industries. These elements—occasional small disc cores and Magosian-like points—distinguish these industries from other blade and burin cultures of Africa—from the Kenya Capsian, and the true North African Capsian. While it is clearly recognized that these industries obviously represent a local variant of the blade and burin culture-complex, it has, nevertheless, by reason of this regional difference and specialization, been thought necessary to refer to them under the term 'Hargeisan Culture' after the site where characteristic flaking floors were first found *in situ*.

The term 'Neolithic', where applied, is used in a technical and typological sense only and does not imply any cultural or economic significance.

The word 'variant' has been employed to describe regional differences within a culture group.

In the south and centre of our area the Somaliland Magosian Culture gradually develops into a specialized series of industries showing interesting local variants. Local specialization is here more apparent than in any other culture group so that the term 'Doian' has been applied to these regionally restricted, Late Stone Age industries. The name is derived from the Somali (or rather Rahanwein) word 'Doi' and is used to describe the flat, scrub-covered, red sand country of southern Somaliland, in which many of these factory and living sites have been found. A general term such as this has been used rather than a site name, as the range of cultural material varies, sometimes not inconsiderably, from locality to locality.

While the typological differences which distinguish certain of the Doian and Somaliland Wilton industries, the one from the other, have warranted their classification as 'regional variants' it is stressed that these 'variants' almost certainly represent nothing more than local divergence within a single culture.

Where contemporaneity can be proved or inferred we may and do find the normal industries in one area, 'neolithic' variants in others and strandlooping variants in yet others. It is not suggested that the men who made these belonged to three different physical types. Almost certainly the essential features of the physical make-up of the men who made all these Late Stone Age industries, and possibly the cultures also, was the same throughout the Horn. Naturally we cannot be entirely certain of this until sufficient human skeletal material has been found, but it seems most probable that these Late Stone Age typological variations should be interpreted rather as indicating tribal differences in material culture, the most important of the contributary causes being environment and suitability and availability of raw materials. A brief examination of, for example, the material culture of the present-day Somali inhabitants will show sometimes considerable variation—one group will use the bow and poisoned arrow, another only the spear, and so on. While there may also be some divergence in minor physical characteristics from tribe to tribe and from confederacy to confederacy, first and foremost, however, all are Somalis; similarly it is thought that there was probably one basic racial stock in Late Stone Age times in the Horn. Whether this physical type was Hamitic, negroid or Bush remains still to be determined.

CHAPTER 9

THE ACHEULIO-LEVALLOISIAN CULTURE. THE LEVALLOISIAN CULTURE

THE ACHEULIO-LEVALLOISIAN CULTURE

This is the earliest culture of which definite evidence was recovered in the Somalilands, though it is possible that some of the material collected by Seton-Karr from the Tug Issutugan (143) may be earlier and thus be representative of a pure Upper Acheulian phase. It is certain, nevertheless, that very strong Acheulian elements are present in the Horn, best seen in the collections from the Tug Issutugan and Sheik, so that although the writer was not fortunate enough to recover conclusive evidence to this effect, it may not be too rash to postulate the presence in north-western Somaliland of a pure Upper Acheulian cultural stage antedating the Acheulio-Levalloisian. It must be remembered that raw material played a very great part in determining the technical development of a culture in a given region. Van Riet Lowe[1] has shown that there is little doubt that the evolution of the prepared-core technique in the Vaal River basin was directly connected with the fact that Chelles-Acheul man used almost exclusively large boulders of diabase for his tools, which necessitated the invention of a suitable method of detaching large flakes for manufacture into hand-axes and cleavers. The same writer[2] has also shown how a prepared-core technique was similarly developed in Uganda where large boulders of quartzite were the usual medium for implements during Chelles-Acheul times. At Kharga Oasis and certain of the North African sites (e.g. El Ma El Abiod) flint or chert cobbles were the usual material, the tools being made from the 'core' itself thus obviating the need for a prepared-core technique. It is therefore not only the material itself but the form in which the material occurs which is such an important factor in influencing the technique found with the Chelles-Acheul culture in the African continent. At Sheik where chert cobbles provided the larger part of the raw material the large circular prepared-cores have not yet been found with the hand-axe and cleaver industry, but at Hargeisa, where the makers of these industries used quartzite in the form of large nodules or lumps struck from outcrops for most of their tools, the prepared-core is well represented. Generalizing, therefore, one may suggest that where cobbles were used the

[1] Lowe, C. van Riet (1945). [2] Lowe, C. van Riet (1952, pp. 52–54).

result was a 'core tool' and where flakes struck from prepared-cores were the basis the hand-axes and cleavers were 'flake tools'. It is by no means improbable, therefore, that the Upper Acheulian in the Horn will be found (where the raw material so dictates) to exhibit a knowledge of the prepared-core technique,

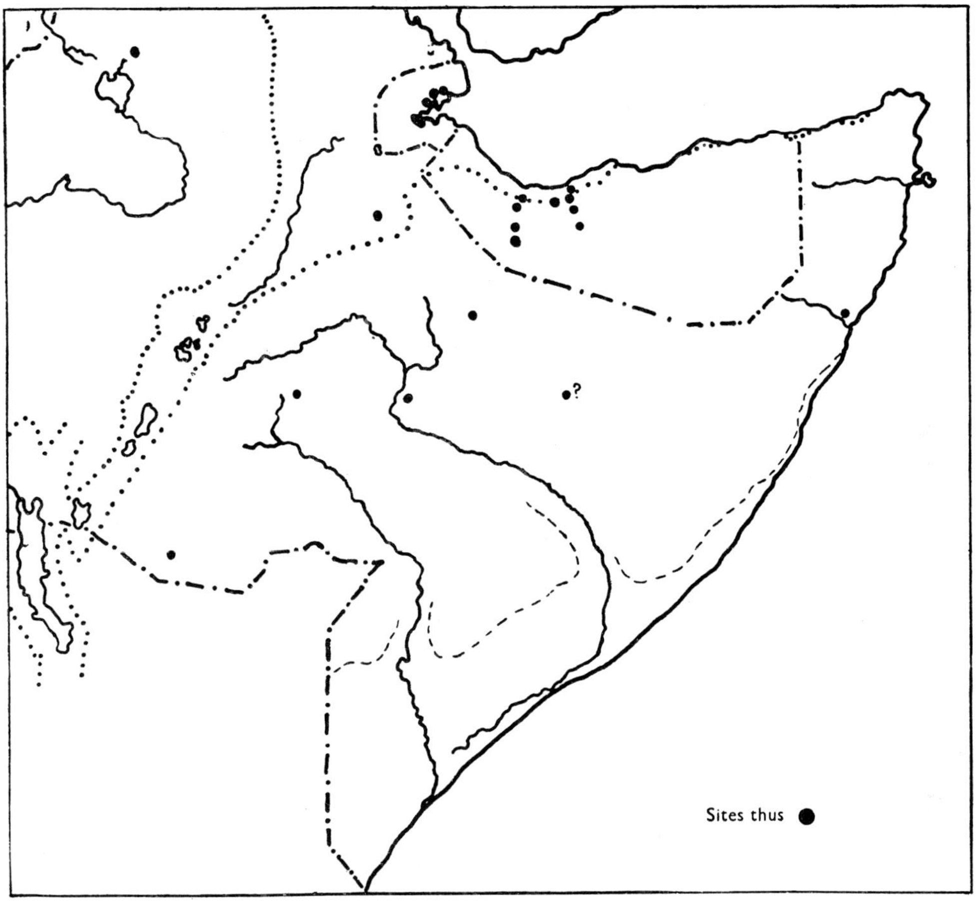

Text-fig. 17. Distribution of Acheulio-Levalloisian industries.

and in fact it may even be that some of the material that has been described below as Acheulio-Levalloisian may in fact be of Upper Acheulian age.

The centre of diffusion for the Acheulio-Levalloisian in the Horn appears to have been the north-western part of the Somali plateau, as it is here that sites are commonest and the highest development is seen. From this area the culture appears to spread into the Danakil Rift, the Guban, or Gulf of Aden coastal plain, to Dagahbur, and perhaps to the Webi Shebeli, in the south, and to the lower Nogal in the east. It is regrettable that all too few sites of this culture exist, and it is at first glance surprising that not a single site has yet been found in the south and centre of our area. The writer considers, however, that its apparent absence from these areas can be explained by the fact that insufficient search has been made in the deposits likely to yield evidence of this culture, rather than that physiographical or climatic causes excluded it from those parts of the peninsula. For example, a proper examination of the 15–30 m. terrace gravels in the middle Webi Shebeli will probably produce evidence of its presence. Text-fig. 17 shows the cultural distribution of the Acheulio-Levalloisian.

Stratigraphy. The type site for the Acheulio-Levalloisian Culture is at Hargeisa (3) where tributary gullies have exposed two deposits of the Older Tug Gravels and Alluvium. On stratigraphical grounds it is here divisible into a Lower and an Upper phase respectively.

THE LOWER ACHEULIO-LEVALLOISIAN. (Pls. 1, 2)

Tools are found in the basal deposit of the Older Tug Gravels in two tributary tugs in the Marodijeh valley at Hargeisa: these sites are known as H. 8 and H. 12 respectively (see Text-fig. 4) and the tools were found to occur in an unrolled or slightly rolled state at all levels within the gravels. The grey, kunkarous, Older Tug Alluvium which overlies them is sterile. There can be little doubt that the tools are contemporary with the gravels, and several specimens show incrustations of lime, identical with that cementing the gravel. The Lower Acheulio-Levalloisian belongs, therefore, to the earlier half of the decline of the first peak of the Gamblian Pluvial: the climate was of a semi-arid nature. The range of implements recovered is not great (site H. 12, twenty-seven tools; site H. 8, six tools) as only specimens actually found *in situ* have been used as evidence, but from those available there does not appear to be any great technological distinction between the industry contained in these gravels and that from the Older Tug Gravels II.

UPPER ACHEULIO-LEVALLOISIAN. (Pls. 5–7)

The Older Tug Gravels and grits at Hargeisa site H. 12 have yielded a greater number of artifacts *in situ*—these are contained within the Older Tug Gravels II and in the overlying sandy grits in which flaking floors were located. One such working floor is seen in Pl. 31, fig. 1, and from it a number of artifacts and flakes were recovered. The Old Tug Alluvium II which overlies these gravels at some points is always sterile of tools. Tools are predominantly unrolled, but there are also a fair proportion of rolled specimens present, some of which may be derived. Incrustations of lime or kunkar are not uncommon. The Upper Acheulio-Levalloisian is assigned to the later half of the decline of the first peak of the Gamblian Pluvial.

UNDIFFERENTIATED SITES

Owing to the lack of stratigraphical evidence it is not possible to subdivide the material from the following sites and it has been thought unwise to rely on a purely typological division.

SHEIK. (Pls. 3, 4.) Regrettably we know nothing of the stratigraphy either at Upper or Lower Sheik, but it is presumed to be of a similar nature to that existing at Hargeisa. With few exceptions tools are fresh and unrolled and show a matt patina.

SKUTAR. Specimens are derived from a coarse gravel exposed by recent erosion in the bed of the Tug Der, some 20 miles north of Burao. They are for the most part unrolled. The gravel showed a kunkarous cement in which several of the implements were embedded.

EIL. (Pls. 8, 9.) Specimens come from a factory site in the immediate vicinity of a small isolated hill near the northern edge of the Nogal Gorge, approximately 1 mile north-west of Eil. This hill is of a marly limestone, capped by a deposit, some 5 ft. thick, of ferruginous, silicified sandstone. It is this ferruginous sandstone that has been used for the manufacture of the tools. Artifacts are found in the scree and at the foot of the hill, eroding from a deposit of thick surface limestone or kunkar: this deposit is overlain nearer the centre of the gorge by orange, wind-deposited sands. When incorporated in the tufa, tools are fresh but those that have been long exposed are weathered. Satisfactory stratigraphical evidence is absent as, while these tools overlay the estuarine beds, their relationship to the Tug Gravels could not be established.

BERBERA (DUBAR). (Pl. 10.) Artifacts overlie the Boulder conglomerates at the foot of the coastal hills at Dubar, approximately ½ mile north-west of the water catchment area, near an outcrop of Cretaceous quartzite. Some fragments of fossil wood occur mixed with the implements, and derived presumably from the Boulder Conglomerates. The tools lie on the surface and there is evidence that the site was occupied over a long period of time as tools divisible on physical condition, into at least two series, occur. The majority of specimens have been abraded by eluvial action. Those of the earlier and more heavily abraded series sometimes show signs of ferruginization.

Geological horizon. The first peak of the Last or Gamblian Pluvial.

Climate. Semi-arid.

Implements. (Pls. 1–10). Tool types include well-made hand-axes of pear-shaped, La Micoque, and ovate forms (Pls. 1, 3, 4, 5, 8, 10). The dominant technique used in their production is wood technique, but resolved flaking is also found. U-shaped cleavers made on large end- and side-flakes are also a characteristic feature (Pls. 1, 4, 5). A large number of flake tools, struck by block-on-block and faceted-platform techniques also occur (Pls. 2, 6–8, 10) together with the cores from which they were struck. Large, circular, prepared cores (Pls. 2, 7, 9, 10) and disc cores (Pl. 7) occur in association with various forms of unprepared core (Pl. 6). Where possible chert was used by the makers of this culture, but quartzite provided a good substitute at Hargeisa where apparently 95 per cent of the artifacts are made from this. Quartzite and lava were occasionally used at Sheik in addition to chert. At Eil tools were made from a rather poor ferruginous sandstone, while quartzite was used at Berbera (Dubar).

An inventory of tool-types from the main sites is given opposite.

Hand-axes—132. These comprise 28 per cent of the cultural material. They vary between 9 and 3 in. in average length, and may be divided into two main groups—finished and unfinished specimens. The greater number of the latter come from the factory site H. 12 at Hargeisa, but it is probable that several of those from Eil and Berbera are also unfinished as they are considerably cruder in appearance, even allowing for the difference in raw material. The main site at Hargeisa H. 12 is adjacent to an outcrop of quartzite and the immediate area is littered with factory debris. The most characteristic form is the Micoquian type (Pl. 4, no. 2; Pl. 5, no. 8) which grades into the small, subtriangular, pointed form (Pl. 1, no. 5; Pl. 5, no. 1): while not comprising a large series they are, however, sufficiently distinctive. Pointed pear-shaped examples are typical (Pl. 1, no. 1; Pl. 3, no. 1; Pl. 5, nos. 2, 4; Pl. 10, no. 3) and grade from the sharp, pointed form to the wider angled variety. A few elongated specimens also occur (Pl. 8, no. 6). Ovates are common but are predominantly elongated, approximating to the limande form (Pl. 1, no. 8; Pl. 5, no. 6; Pl. 8, no. 5), true, well-made ovates are rare, but a few fine examples do exist (Pl. 3, no. 2): the usual forms are cruder in execution (Pl. 1, no. 3; Pl. 5, no. 6).

The technique employed on the finished examples from Hargeisa, Sheik and Skutar is essentially wood technique, though resolved flaking is not infrequently found, particularly on some of the smaller, subtriangular examples. Hand-axes with unworked butts are much more common than those where the butt has been worked: but in the latter examples the butt is always carefully shaped and rounded. The quality of workmanship is undoubtedly high and can best be seen on the fine and thin points and edges of the specimens in chert. At Eil and Berbera, however, there is little evidence of wood technique and hand-axes are cruder and executed usually by a stone-hammer technique. The greater number of hand-axes are core tools, but Sheik in particular has a number of examples made on large flakes: these latter were struck from both prepared and unprepared cores, although the number is insufficient to indicate which technique was most universally employed.

Inventory of tool types of the Acheulio-Levalloisian Culture

Tool types	Lower Ach.-Lev.	Upper Ach.-Lev.	Undifferentiated Ach.-Lev.*				Totals
	Hargeisa H. 12 and H. 8	Hargeisa H. 12	Sheik	Skutar	Eil	Berbera (Dubar)	
Hand-axes							
Pear-shaped pointed	2?	14	5	—	5	4	30
Pear-shaped elongated	—	—	1	—	2	—	3
La Micoque type	—	2	6	1	—	—	9
Subtriangular, pointed	3	12	16	—	2	—	33
Subtriangular, round-ended	1	—	3	—	—	—	4
Ovate	2	3	22	—	1	—	28
Unfinished, pear-shaped	—	7	—	1	—	—	8
Unfinished, ovate	—	9	3	1	—	—	13
Unfinished, subtriangular	—	2	2	—	—	—	4
Cleavers							
U-shaped on side-flakes	2	2	1	—	—	—	5
U-shaped on end-flakes	2	3	4	1	—	—	10
Choppers							
Discoidal	1	—	5	—	—	—	6
Scrapers							
Oblique-bulb type, single edge	—	—	5	—	—	—	5
Oblique-bulb type, all-round retouch	—	—	4	—	3	2	9
End-bulb type, single edge	—	—	5	—	—	—	5
End-bulb type, double edge	—	4	2	1	—	—	7
End-bulb type, end	—	2	—	—	—	—	2
Points	—	—	5	—	—	—	5
Cores							
Prepared oval and circular	3	17	4	—	9	3	36
Prepared disc	—	9	6	6	—	1	22
Prepared subtriangular	—	6	—	—	4	—	10
Prepared flake-blade	—	1	—	—	—	—	1
Unprepared flake-blade	—	2	1	—	—	—	3
Unprepared spindle	—	—	5	—	—	—	5
Unprepared pebble	—	3	—	2	—	—	5
Flakes							
Faceted, some with retouch	5	36	4	6	7	4	62
Unfaceted, some with retouch	4	23	2	8	2	9	48
Wood-technique type	—	5	—	—	—	—	5
Flake-blades							
Faceted, some with retouch	3	19	—	1	3	2	28
Unfaceted, some with retouch	4	21	7	4	4	3	43
Bolas (?)	—	2	3	—	—	—	5
Polyhedral stones	—	—	—	—	—	1	1
Totals	32	204	121	32	42	29	460

* Includes only the more important sites.

Cleavers—15. These tools are present in both the Lower and Upper Acheulio-Levalloisian at Hargeisa and also at Sheik and Skutar, but there is none in the collections from Eil and Berbera. Specimens are all of the U-cleaver type and are made either on side-flakes or, more commonly, on end-flakes. The butts usually have been carefully trimmed and the bulb and platform removed by retouch on the main flake surface. The edges are either parallel (Pl. 1, no. 7; Pl. 4, no. 1) or converge slightly at the working edge (Pl. 1, no. 6; Pl. 5, nos. 7, 9), flaking being directed from either one or both sides. In section, tools are usually bi-convex or plano-convex but two specimens from Sheik, which may be unfinished, exhibit parallelogram sections. The cleaver edge may be set either at right-angles or obliquely to the long axis. Examples showing parallel edges are usually longer in proportion than those with converging edges. However, in the former type, the butt is invariably carefully rounded, and although the latter type does occur in the Lower Acheulio-Levalloisian, the U-cleaver with parallel edges is thought to be the earlier form. In the Upper Acheulio-Levalloisian are found several end-flakes, some quite small and usually struck from unprepared cores, which have been trimmed down both edges and sometimes also round the butt to form a kind of cleaver-flake. The edges converge slightly at the working end which is thin and cleaver-like, sometimes with evidence of utilization. An example is illustrated at Pl. 5, no. 9. This form appears to represent the ultimate development of the cleaver technique.

Choppers—6. A few artifacts are found which have been described as discoidal choppers. They are approximately discoidal to ovate in shape. While the edges suggest use as choppers it is not improbable that they are unstruck, prepared cores, though the latter are usually high-backed. The form is, however, typical of this culture and occurs both in the Lower (Pl. 1, no. 4) and Upper (Pl. 7, no. 5) Acheulio-Levalloisian.

Scrapers—28. It is not always easy to distinguish between true scrapers and flakes with utilized edges; so that only where there can be no doubt that a carefully trimmed edge was intended, have specimens been described as scrapers. These may show either prepared or unprepared striking-platforms; in the former the butt is usually at right-angles, and, in the latter, oblique. The commonest scraper form is where secondary retouch has been carried down one edge (Pl. 6, nos. 6, 8), but examples with double edges are also found (Pl. 6, no. 4) as well as end-scrapers (Pl. 6, nos. 1, 2) and occasional notched tools. Flakes worked into large, crude forms of end-scraper are found at Eil (Pl. 8, no. 7) and also at Berbera (Pl. 10, no. 2) and while this form is not typical of the Hargeisan or Sheik sites there is some evidence that it occurs here also (Pl. 2, no. 3). Scrapers with all-round retouch made on large, broad flakes are also found (Pl. 8, nos. 1, 2, and Pl. 10, no. 1).

Points—5. Under this term are included a small series of unifaced points made on oblique-bulb flakes, and showing varying degrees of resolved, secondary retouch.

Prepared cores—69. The most characteristic form is the large circular prepared core with high back, radially flaked on the upper side. Some eight good examples occur at Hargeisa, both struck and unstruck specimens being present (Pl. 2, no. 1; Pl. 7, no. 6). There is none in the collections from Sheik, but it is certain from a study of the flakes that they are present here also. The forms from Eil and Berbera are more angular in shape (Pl. 9; Pl. 10, no. 4) and while the Hargeisa specimens show usually the removal of one large flake only, those from Eil and Berbera show that sometimes several small flakes, instead of one large flake, were obtained from the cores. These prepared, circular cores are entirely typical of the radially flaked, Levallois core and closely resemble the South African Victoria West II, high-backed, or horse-hoof cores from the Vaal river. In addition to the above are found smaller, oval and circular prepared cores, usually with high back (Pl. 1, no. 2; Pl. 7, no. 2), which yielded short and broad flakes with faceted platforms. The small disc core is also present in the Upper Acheulio-Levalloisian (Pl. 7, no. 3). The true subtriangular, prepared core is not found at the Hargeisa sites, but several examples approximate to this type (Pl. 7, no. 2). The small hand-axe at Pl. 1, no. 5, shows signs of possible use as a subtriangular, prepared core. The specimen illustrated at Pl. 6, no. 3, is the upper half of a well-made hand-axe, which had become snapped across the short axis and subsequently re-used as a core. There is no indication of preparation of the butt, however. Eil yielded four true, subtriangular, prepared cores, again with high back (Pl. 8, no. 4). One prepared core for the removal of flake-blades was recovered from the Upper Acheulio-Levalloisian at Hargeisa (Pl. 6, no. 7).

Unprepared cores—13. From an examination of debitage it is apparent that unprepared cores are quite as common if they do not in fact predominate over the prepared cores. The commonest form was probably the pebble, or proto-biconical core, but the longish, spindle-shaped cores from Sheik appear to be characteristic, as is also the flake-blade form at the other sites.

Flakes. These can be divided for convenience into large and thick specimens, usually over 5 in. long, and short and thin specimens. Noticeable at Hargeisa is a kind of 'duck-head'[1] flake which may be a roughout for a hand-axe; the

[1] Term employed by T. P. O'Brien to describe a plano-convex sectioned tool with thick butt and thin sectioned point. O'Brien, T. P. (1939, p. 214).

striking-platform is inclined and unprepared: some half-dozen of these were recovered. The smaller flakes are either oval or pointed, with either radial, convergent or parallel flaking of the upper face.

Flakes struck from prepared cores—62. (Pl. 2, nos. 2, 4; Pl. 6, nos. 1, 2, 4, 6, 8; Pl. 7, no. 4; Pl. 8, nos. 1, 3; Pl. 10, no. 1.) These usually show faceted striking-platforms but some with plain platforms set at right-angles to the main flake surface also occur (Pl. 2, no. 2; Pl. 7, no. 4). Faceting on the large flakes is usually fairly crude, but in the smaller specimens it is more carefully produced. Butts are usually straight with the bulb centrally placed, but specimens where the platform is convex are not uncommon.

Flakes struck from unprepared cores—48. (Pl. 8, no. 2; Pl. 10, no. 2.) Both large and small forms occur but the latter are predominantly debitage from primary flaking and rarely show signs of utilization. Platforms are usually set obliquely and at a steep angle to the main flake surface.

Flake-blades—71. (Pl. 2, no. 3; Pl. 6, nos. 4, 5; Pl. 8, no. 7.) Under this term have been included a few large and long flakes, in addition to true flake-blades, in order to distinguish them from the true, broad flake which is the more typical form with this culture. Some twenty-eight examples occur with faceted striking-platforms and forty-three with plain inclined platforms.

Wood-technique flakes—5. (Pl. 5, no. 3.) A small number of typical flakes showing controlled flaking by wood technique were kept, out of a larger number found with the Upper Acheulio-Levalloisian at Hargeisa. These flakes are thin, with flat, negligible bulb and platform, and thin, wavy scars on the upper surface. They never show evidence of secondary retouch.

ADDITIONAL SITES

Several additional sites may be briefly mentioned, since, although the number of tools recovered is usually very small, yet they extend the known distribution of this culture.

MANJASSEH (233). At the foot of the Gulf of Aden fault-scarp, has yielded one well-made, ovate hand-axe.

GED DOBO (202). Thirteen miles from Manjasseh, yielded one pear-shaped, pointed hand-axe, one ovate, one unfinished ovate, and one U-cleaver, made on an end-flake.

HALEYA (203). Five miles east of Hargeisa. Finds include one subtriangular hand-axe, one oval, prepared core, and one faceted flake.

DAGAHBUR/WEBI SHEBELI CONFLUENCE (163). One sub-triangular pointed hand-axe.

TUG ISSUTUGAN AND GOBAN (143). See pp. 22–3, 160.

THE FRENCH SOMALI COAST. Series I from Mount Mabla (149), Obok (150), Kor Ali (148), see pp. 29–30.

DAGAHBUR (116). One elongated, ovate hand-axe.

GHERLOGUBI (64). Seventeen miles south-east of Wardere. One crude, side-scraper or hand-axe in lava from dumps of material excavated from wells.

GARASLEH TUG (8). One large flake with crudely prepared striking-platform, one primary, unfaceted flake, one small, unfaceted flake-blade and one chopping tool. These specimens are fresh or only slightly rolled and are derived from the eroded Lower Alluvial Fan Gravels at this site. More probably they should be assigned to the Lower Levalloisian rather than the Acheulio-Levalloisian, but the material is too scanty and atypical to determine this with certainty.

The distribution list of sites yielding a typologically Acheulio-Levalloisian industry is completed by the addition of three sites in Abyssinia: Gondar,[1] Mega (see p. 39) and Robabouta ((173), see p. 16).

There is little to distinguish the Upper from the Lower Acheulio-Levalloisian at Hargeisa. The Upper phase has perhaps a greater number of smaller flake tools, and the addition of the disc core. Also a higher proportion of hand-axes have the butts untrimmed, and it is probable that the parallel-sided U-cleaver does not occur; but there is no typological or technological change of any importance.

It will be seen, therefore, that the Acheulio-Levalloisian is essentially the result of the adoption of the faceted-platform technique by a developed form of the African Acheulian. These Acheulian elements are purest at Sheik and Hargeisa and weakest at Eil and Berbera, where greater stress is laid on flake production and the bifacial technique is relatively poorer. Whereas Acheul wood technique, and fine resolved retouch, characterize the more finished tools at the former sites, the underlying technique at the latter, is predominantly stone-hammer technique, though resolved flaking is also present. It may be postulated, therefore, that the plateau sites (Sheik, Hargeisa, Skutar, Gondar, etc.) are earlier and more typically Acheulio-Levalloisian than the industries from Eil, Berbera and the French Somaliland Coast, which, by their poorly made hand-axes, absence of cleavers, and extended development of the faceted-platform technique, appear to represent the ultimate phase of this culture and to be transitional between the Acheulio-Levalloisian and the Lower Levalloisian.

[1] Clark, J. D. (1945*a*).

THE LEVALLOISIAN CULTURE

This culture is both widespread and long-lived throughout the Horn, as can be seen from the Distribution Map (Text-fig. 18)[1] and its possession of our area

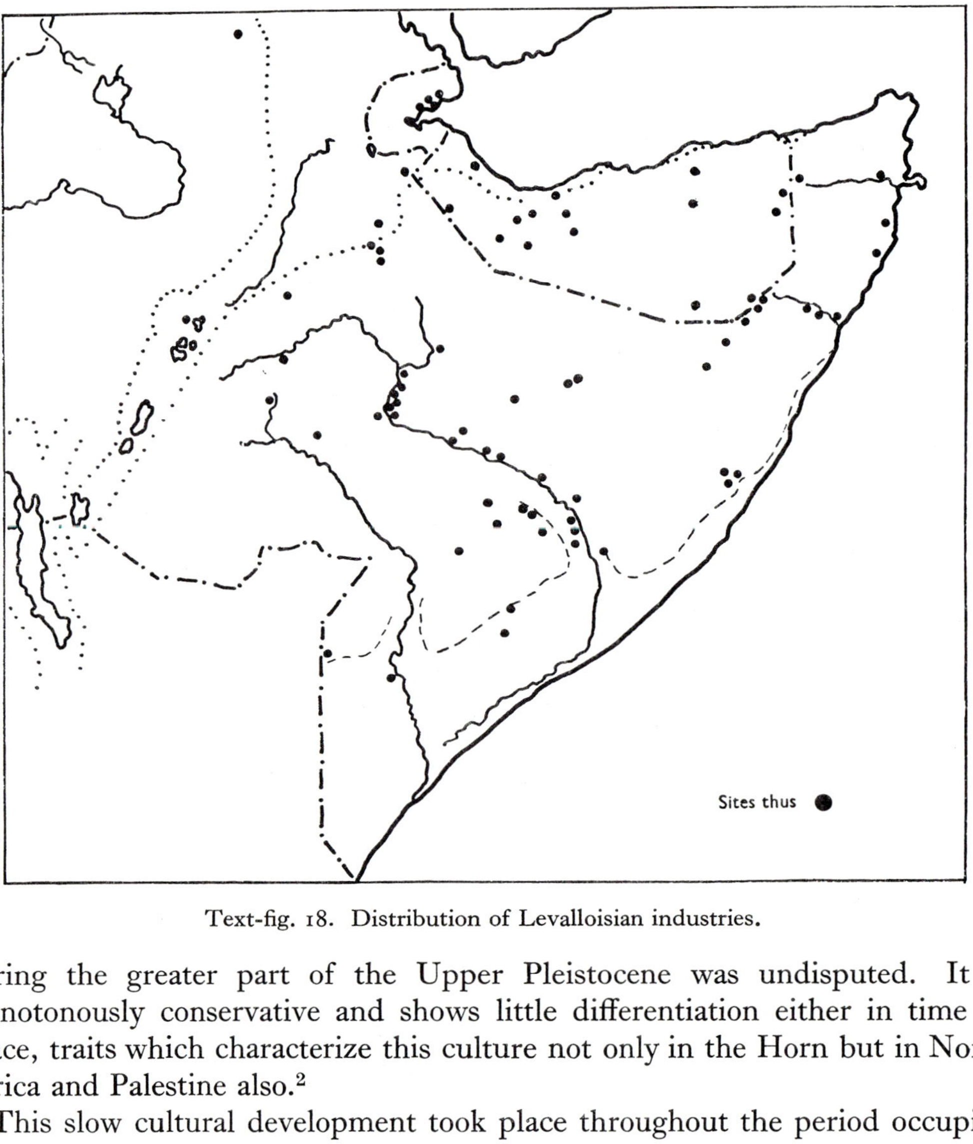

Text-fig. 18. Distribution of Levalloisian industries.

during the greater part of the Upper Pleistocene was undisputed. It is monotonously conservative and shows little differentiation either in time or space, traits which characterize this culture not only in the Horn but in North Africa and Palestine also.[2]

This slow cultural development took place throughout the period occupied

[1] All sites known to have yielded material of Levalloisian affinities have been included, except those the location of which is uncertain. The Abyssinian sites of Lake Adili, Araue (Errer valley), Amba Alagi, and Dagha Dima are also shown for the sake of completeness.

[2] Garrod, D. A. E. (1938) pp. 113–24.

by the aggradation of the Tug Gravels and Alluvium (or the Younger Tug Gravels and Alluvium in the case of the north-western Somaliland sequence): that is to say during the major part of the Upper Pleistocene.

On stratigraphical evidence the culture may be divided into a Lower and an Upper Levalloisian, being contained within the Lower and Upper Tug Gravels respectively.

For the most part the Levalloisian is remarkably unspecialized, but a certain refinement in technique and a general diminution in the size of artifacts in the Upper division is noticeable, as well as a tendency towards more deliberate, secondary retouch and specialization of tool types in the later stages. The predominant forms are prepared cores, either discoidal, subtriangular, or sub-rectangular, and faceted flakes and flake-blades, some of which have been worked into points and scrapers of various kinds.

The ultimate development of the Levalloisian in our area is to be found in the Somaliland Stillbay in which specialization and technical refinement show a considerable advance.

THE LOWER LEVALLOISIAN

Stratigraphy. The culture has been found *in situ* within the Lower Tug Gravels and Alluvium at sites at Garoe (54), Eil (56), Donkukoq (52), Hargeisa (3), Gabredarre (65), Kabahan (69), God'dere (35), and Jesomma (85). It was stratified immediately below the Upper Tug Gravels, which, at the majority of these sites, yielded Upper Levalloisian assemblages. Tools occur in both abraded and fresh condition, thus confirming their contemporaneity with the gravels. There is some definite evidence of bedding in these fine gravels which indicates that they were laid down under comparatively humid conditions.

The raw material employed was chert, wherever this could be locally obtained: at Garoe, Eil, Donkukoq and Gabredarre all the artifacts collected are of this stone; but at Hargeisa and Kabahan the medium is predominantly quartzite and silicified sandstone, and at Jesomma only sandstone was used.

The chert specimens almost without exception show a thick white patina which has penetrated the stone to such an extent that it has left only a very small central core of the unaltered chert. This form of patination, I believe to be indicative of sharply alternating, seasonal, climatic conditions.

Geological horizon. The decline of the second, and major, peak of the Gamblian Pluvial.

Climate. Wet to semi-arid.

Implements. (Pls. 11 and 12, nos. 5–14.) An inventory of tool types recovered from the sites detailed above is given on the following page.

Inventory of tool types of the Lower Levalloisian Culture

Tool types	Garoe and Don-kukoq	Har-geisa	Gabre-darre	Kaba-han	God'-dere	Je-somma	Totals
Points: unifaced	2	2	1	—	1	1	7
Scrapers, various	1	—	—	—	—	—	1
Notched flakes	3	1	—	—	1	—	5
Retouched fragments	1	3	3	—	—	—	7
Flakes, faceted, some retouched	8	2	4	6	2	7	29
Flakes, unfaceted, some retouched	4	1	6	3	4	5	23
Flakes, indeterminate, some retouched	3	—	—	—	7	1	11
Flake-blades, faceted, some retouched	2	2	8	7	—	—	19
Flake-blades, unfaceted, some retouched	3	1	2	4	1	—	11
Trimming stones	1	1	2	1	—	1	6
Hand-axes	3	—	—	1	—	1	5
Flake cores, prepared	14	2	3	7	8	4	38
Flake cores, unprepared	1	—	—	—	—	—	1
Flake-blade cores, prepared	—	—	1	—	1	—	2
Flake-blade cores, unprepared	1	1	—	1	—	—	3
Totals	47	16	30	30	25	20	168

The most important items in the Lower Levalloisian cultural make-up are the prepared cores and faceted flakes and flake-blades: but trimming stones, or steeply trimmed core-scrapers, are a type tool, as are also a few crude hand-axes or, more accurately, choppers. Owing to its unspecialized nature the Lower Levalloisian contains very few scrapers and points, but a few largish examples of the latter do occur, as well as certain notched and retouched fragments of flakes and flake-blades.

'Stone technique' is the form of flaking employed both for primary and secondary flaking. A little squamous flaking is present but the predominant form of secondary retouch is of a 'nibbled' and resolved or stepped nature.

The proportion of flakes and flake-blades with faceted and unfaceted striking-platforms is about even—faceted examples form 29 per cent and unfaceted examples 21 per cent of the culture assemblages. The proportion is greater, however, when a comparison is made between prepared and unprepared cores: the former comprise 24 per cent of the total and the latter only 2 per cent. This difference cannot be explained entirely on discrimination in collecting specimens, and must be taken as indicating that prepared-core technique was that most commonly employed. Prepared striking-platforms are usually centrally placed but specimens are also found with the platform inclined obliquely to the major axis: the angle between the striking-platform and the main flake surface is approximately 90°. Unfaceted flakes show a restricted platform nearly always set obliquely to the major axis and inclined at an average angle of 129°.

Points—6. (Pl. 11, no. 9; Pl. 12, no. 12.) A few unifaced points are found on large flakes, which have been struck from the unprepared cores. Secondary retouch is almost always confined to the edges of the tool and is of a crudely resolved nature. The specimen illustrated at Pl. 11, no. 9, however, shows a higher degree of secondary retouch.

Scrapers—1. (Pl. 11, no. 8.) Although flakes with secondary retouch are not uncommon only one specimen was found with a sufficient degree of secondary retouch to warrant designation as a scraper. This example is made on a tabular piece of Eocene chert, which has been carefully retouched round most of the circumference.

Notched flakes—5. (Pl. 11, no. 5.) Notches are a characteristic feature, and they are formed usually by steep, secondary retouch. They were formed not only as the principal purpose of the tool but also incidentally on the edges of other tools. The specimen illustrated is an example of the former type, but notches are also found on other specimens such as those at Pl. 11, nos. 1, 2, 6.

Retouched fragments—7. (Pl. 11, no. 1.) Fragments of faceted and unfaceted flakes which have been snapped across or which have had the butt purposely removed are found with somewhat amorphous retouch round the edges; they are characteristic.

Faceted flakes—29. (Pl. 11, no. 2; Pl. 12, nos. 7, 8.) Flakes show either radial or, more rarely, parallel flaking on the upper surface. Striking-platforms are carefully prepared and usually straight or convex. The edges show a minimum of secondary retouch and 'nibbling' due to utilization: a little squamous flaking is in evidence (Pl. 12, no. 8).

Unfaceted flakes—24. (Pl. 11, nos. 3, 5; Pl. 12, no. 13.) These call for little mention. They have apparently been struck from either flat, single-platformed cores, or from proto-biconical cores, but some specimens are derived from prepared cores, on the evidence of the preparation of the upper surface of the flake. The edges show the same 'nibbled' retouch.

Indeterminate flakes—11. A few examples were found in which the bulb and platform have been removed.

Faceted flake-blades—19. (Pl. 11, no. 4; Pl. 12, no. 6.) These have been struck from either single-ended or double-ended cores and the preparation on the upper surface shows either parallel or opposed (or alternate) flaking, depending on the form of core from which the flake was derived. The secondary retouch is typical and confined to the edges. Striking-platforms are usually centrally placed and straight.

Unfaceted flake-blades—11. Struck from single-platformed cores of the long and narrow variety, these require no particular mention. Parallel flaking is that most usually found on the upper surface.

Trimming stones or core-scrapers—6. (Pl. 12, no. 11.) This characteristic tool is plano-convex in section and the flat, ventral side may often be a natural surface. The circumference shows steep, almost vertical, retouch round most of the edge: this retouch is of a primary rather than a secondary type, the finished artifact being in the nature of a 'push plane' or crude, core-scraper. The term 'trimming stone' is preferred.

Hand-axes—5. (Pl. 11, no. 10; Pl. 12, no. 14.) These tools are in the nature of crude, bifacial choppers and bear little or no resemblance to the hand-axes associated with the typical Acheulio-Levalloisian. The edges are formed by alternate primary flaking and show no secondary retouch other than that due to utilization.

Flake cores: prepared—38. (Pl. 11, nos. 6, 7, 11; Pl. 12, nos. 5, 9.) These may be subdivided into the following forms:

Oval and disc	21
Subrectangular—single-ended	2
Subrectangular—double-ended	7
Subtriangular	8
Total	38

The oval to disc variety (Pl. 12, no. 5) which is that most commonly found, varies in size from 4–5 in. to 2 or 2½ in. across the major axis. The larger specimens are usually biconvex or high-backed in section but the smaller are almost always flattish: more than one flake was usually removed from the upper surface of the larger specimens, but rarely more than one from the smaller examples.

The double-ended, subrectangular core is a characteristic form. One large specimen (Pl. 11, no. 11) shows the removal of three flakes from the upper surface, but examples as large as this are rare, the usual specimens being smaller and flat, and made, where possible, from chert (Pl. 11, no. 6).

The subtriangular core (Pl. 11, no. 7; Pl. 12, no. 9) employed for the removal of a single flake of this form is either flat or high-backed and nearly always of smallish size.

Flake cores: unprepared—1. These call for no special mention. The one example collected is of the proto-biconical variety.

Flake-blade cores: prepared—2. These may be either single- or double-ended, judging from the flake-blades. Both the cores collected are of the double-ended variety, and are really an elongated form of the double-ended sub-rectangular core.

Flake-blade cores: unprepared—3. (Pl. 12, no. 10.) Single- and double-ended forms of this core must also have been employed, the example figured being of the double-ended variety.

ADDITIONAL SITES. Mention must be made of the few flakes and prepared core derived from the basal Tug Gravels and Alluvium in the Nogal Gorge at Eil. Five prepared flakes, and one discoidal core of chert were recovered. Owing to the few artifacts recovered from this site it has not been included in the inventory, but the specimens are all typical of the descriptions given above.

THE UPPER LEVALLOISIAN

Stratigraphy. The upper division of the Levalloisian has been found stratified within the Upper Tug Gravels and Alluvium at a number of sites, the most important of which are Garoe (54), and Donkukoq (52) in the upper Nogal; Hargeisa (3) in the Marodijeh valley; Gabredarre (65) on the Tug Fafan; and Jesomma (85) on the northern flank of the Webi Shebeli near Bulo Burti. The above sites, where representative cultural material is found stratigraphically above the Lower division, have been used as the basis for the description of the Upper Levalloisian. A number of other stratified sites also occur, however, which, although they have not yielded the quantity of tools recovered from the main sites, are nevertheless of importance, and provide valuable corroborative evidence. Such sites are found at Madeida (30), El Dubbo (40), Bohol Washaga (34), Saha Gurral (37), Ohole (42), and Bohol Bandoi (78) all in southern Somalia; Durualeh (68) and Darchaua (73) in the Ogaden; and Eil (56), and Tug Babani (59) in the lower Nogal.[1]

The Upper Tug Gravels usually show little evidence of bedding and are more in the nature of torrent gravels accumulated during short rainy seasons, the climate remaining predominantly dry for the greater part of the year; the Upper Tug Alluvium is of a fine nature, with no recognizable bedding planes, and is nearly always stained to a reddish colour by oxides. The nature and colour of this deposit suggests a comparison with the 'aeolian silt' deposits in the Wadi Hadhramaut in southern Arabia.

Artifacts are contained within the gravels, but rarely, if ever, in the overlying alluvium: they are usually unrolled, but a few abraded specimens occur. Where

[1] See also the sites at Ted (38), p. 228; Bur Eibe (18), p. 197; Laba Higlo (53), p. 292; and Hararile (55), p. 26.

they are made from chert, as at Garoe, Donkukoq, and Gabredarre, tools show usually a thick, white patina, similar to that found with the Lower Levalloisian and due presumably to similar climatic conditions. There is no change in the raw material employed at any of the main sites.

Geological horizon. The decline of the third and last peak of the Gamblian Pluvial.

Climate. Semi-arid to arid.

Implements. (Pls. 13, 14.) The following is an inventory of the tool types found with the Upper Levalloisian.

Inventory of tool types of the Upper Levalloisian Culture

Tool types	Garoe and Donkukoq	Hargeisa	Gabredarre	Jesomma	Totals
Points, unifaced	2	2	1	7	12
Points, bifaced	—	—	—	3	3
End-scrapers	1	—	—	8	9
Scrapers, various	1	3	1	2	7
Notched flakes	—	3	—	2	5
Retouched fragments	19	2	—	—	21
Flakes, faceted, some retouched	25	5	20	18	68
Flakes, unfaceted, some retouched	12	4	5	17	38
Flakes, indeterminate	3	3	—	—	6
Flake-blades, faceted, some retouched	10	3	8	9	30
Flake-blades, unfaceted, some retouched	7	1	13	10	31
Trimming stones	—	—	4	5	9
Flake cores, prepared	17	3	18	29	67
Flake cores, unprepared	5	1	—	2	8
Flake-blade cores, prepared	1	2	—	—	3
Flake-blade cores, unprepared	1	—	2	1	4
Totals	104	32	72	113	321

A wider range of more specialized tools is now found, and occasional, well-made implements occur. The majority of the artifacts still retain, however, the unspecialized nature characteristic of the Lower Levalloisian. The same forms of prepared core occur, though the subrectangular and subtriangular forms now take precedence over the oval and disc form. Prepared flakes show, on the whole, a general diminution in size and pointed forms with convergent flaking on the upper face make their appearance. Of interest, also, are a few flakes showing constriction of the edges towards the butt, and intentional removal of the bulb and striking-platform by a form of Tabelbalat retouch.[1]

[1] The removal of the striking-platform and reduction of the bulb of percussion of a flake, or flake-blade, by retouch directed from the upper face of the tool, at the butt-end of the main flake surface; cf. Emireh retouch.

Primary flaking shows little or no change from that found with the Lower Levalloisian, but there is a development in secondary retouch and a tendency can be seen towards flat percussion flaking and bifacial flake technique which result in thinner sectioned tools. This can best be seen on the few points or foliates which are found.

Points—15. (Pl. 13, nos. 6, 14; Pl. 14, nos. 1, 2, 10, 13.) The commonest form is a unifaced or parti-bifaced point which may be either subtriangular (Pl. 13, nos. 6, 14; Pl. 14, no. 10) or foliate (Pl. 14, nos. 1, 2).

The former may be either flat in section with retouch on either face (Pl. 13, no. 6; Pl. 14, no. 10) or else subtriangular in section, with steepish retouch, always on the upper face (Pl. 13, no. 14). Butts may be left untouched or else reduced by incipient Tabelbalat retouch.

The foliate forms are parti-bifacial, but rare, and represent, with the fully bifacial forms, the highest attainment in pure Levalloisian technique. Such examples are usually thin sectioned, comparatively speaking, and have had the bulb and platform reduced.

The fully bifacial forms are confined to three broken fragments from Jesomma, of which Pl. 14, no. 13, is an example. These specimens are probably similarly foliate in form.

End-scrapers—9. (Pl. 13, no. 13; Pl. 14, nos. 4, 5, 6.) A few rather crude, but undoubted end-scrapers are found, nearly all from Jesomma. They are made from short flakes and flake-blades, with either plain or faceted striking-platforms. Two double-ended examples were also found (Pl. 14, no. 5).

Scrapers, various—7. (Pl. 13, no. 7; Pl. 14, no. 11.) The true side-scraper is very rare, but two undoubted examples (illustrated) were found. The remaining forms are atypical and consist of flakes, partially trimmed to straight, convex or sometimes concave, scraping edges.

Notched flakes—5. These are similar to the notched flakes found with the Lower Levalloisian.

Retouched fragments—21. The majority show nibbling and notching by a rather steep retouch.

Faceted flakes—68. (Pl. 13, nos. 1, 2, 4, 6, 7, 10, 11, 15; Pl. 14, nos. 3, 7.) These vary in length between 1½ and 3½ in., but some of the specimens from Gabredarre are as small as 1 in. Striking-platforms are usually carefully prepared but a few specimens are found with one or two facets forming the platform, which is set at right-angles to the main flake surface. Striking-platforms may be placed either centrally or obliquely to the major axis and may be either straight, concave or of *chapeau-de-gendarme* form: bulbs are clearly seen but are not particularly prominent. Flaking on the upper surface may be either radial (most common), convergent, alternate, or parallel, depending upon the types of core from which the flake was removed. Hinge or plunging fracture is a characteristic feature. Flakes from disc cores show radial flaking, those from subrectangular, single-ended cores show either radial or alternate flaking, and those from subtriangular cores show convergent flaking.

Secondary retouch is confined to the edges of the tools and much of this appears to be the result of utilization: squamous flaking is present on a few examples.

A typical, but rare, form is the flake in which the edges have been made to converge at the butt end, sometimes accompanied also by Tabelbalat reduction of the bulb and platform. Such forms (Pl. 13, nos. 1, 9, 10) suggest a tendency towards pedunculate development, but there is absolutely no evidence that tanged, or stemmed, flakes were ever produced in the Somalilands, and it must be stressed that these forms with constricted butts are abnormalities of the Upper Levalloisian in the Horn. At Garoe four specimens were found with constricted edges at the butt, and three with Tabelbalat retouch; at Gabredarre, none of the first type and four of the second; at Jesomma, three of the second type. At Hargeisa no such specimens were found. The only other sites known to the writer[1] which have yielded flakes with convergent edges at the butt are Bihen (60) and the surface site found by Gilliland at Las Anod (181) which have also yielded specimens showing crude Tabelbalat retouch.

Unfaceted flakes—38. (Pl. 13, nos. 3, 8.) These were struck, for the most part, from pebble, proto-biconical, and asymmetric, biconical cores, the latter either single- or double-ended. Some specimens, however (Pl. 13, no. 8), were undoubtedly derived from prepared cores, though the platform may be unfaceted, and were probably flakes removed in the course of repreparation of the upper surface of such a core.

Flakes, indeterminate—6. A few specimens occur in which the bulb and platform have been removed, and the origin of these cannot be determined.

Faceted flake-blades—30. (Pl. 13, nos. 4, 9; Pl. 14, no. 6.) These show, usually, either parallel or alternate flaking on the upper surface, depending upon whether they have been removed from single- or double-ended flake-blade cores, and are typical.

Unfaceted flake-blades—31. (Pl. 13, no. 5.) These are derived from asymmetric biconical cores with either a single, or two opposed platforms. Except in the absence of preparation, they show little difference from the faceted flake-blades.

[1] With the exception of the atypical, tanged scraper of late date found by Teilhard de Chardin, Breuil & Wernert, associated with hut circles on Djebel Djinn, in French Somaliland. Teilhard de Chardin *et al.* (1940, fig. 3).

Trimming stones or core-scrapers—9. (Pl. 14, no. 15.) These continue as a typical tool of the Levalloisian, and show little variation from the forms found with the Lower Levalloisian.

Flake cores: prepared—67. (Pl. 13, nos. 16, 20, 22; Pl. 14, nos. 8, 9, 12, 14, 16, 17.) These may be subdivided as follows:

Oval and disc	15
Subrectangular, single	4
Subrectangular, double	34
Subtriangular	14
Total	67

It will be seen that the double-ended, subrectangular core is now the commonest form (Pl. 13, no. 19; Pl. 14, nos. 8, 14). It should be emphasized, however, that, except at Gabredarre, these cores are not as markedly rectangular as those from Egypt and the Nile Valley, and that they grade gradually into the discoidal form of core. Of note is the diminutive, prepared core illustrated at Pl. 14, no. 8. The subtriangular core is also well represented (Pl. 13, nos. 16, 18, 20) but is usually of small size: the backs are usually rounded while those of the subrectangular forms are flat. The oval and disc cores are well made, with either flat or rounded backs. They are now made predominantly for the removal of not more than one faceted flake, before repreparation. (Pl. 13, nos. 17, 22; Pl. 14, nos. 9, 12, 14, 17.) Some examples (Pl. 14, nos. 9, 12) have been carefully trimmed to form discs, possibly for some secondary purpose.

Flake cores: unprepared—8. (Pl. 13, nos. 12, 21.) The usual form appears to have been the pebble or proto-biconical core, though asymmetric forms with one, or opposed, platforms are also found.

Flake-blade cores: prepared—3. No very typical examples were found, but of those recovered two show a single prepared platform at right-angles to the long axis down which flake-blades were removed; the third specimen is a double-ended core with one platform prepared and the other plain.

Flake-blade cores: unprepared—4. Similar to the prepared flake-blade core with the difference that the platform is not prepared.

ADDITIONAL SITES

SOUTHERN SOMALIA

MADEIDA (30). This is situated on the Hoddur/Baidoa plateau in the wide and flat Madeida plain, which is some 15 miles across and lowered some 100 ft. below the plateau surface. The superficial deposit on this plain consists of a red sandy loam with numerous pisolites of kunkar and an occasional pebble of quartz and trap rock. This deposit was more than 5 ft. deep and appeared to pass into calcrete or kunkar. From the surface were recovered a few Doian artifacts. Weathering out from the lower levels of the red sand were found faceted flakes and discoidal cores with a mottled or a white to ochreous patina. Artifacts included one large, plain-platformed flake, with scraping edge; fourteen faceted flakes; nine unfaceted flakes; one faceted flake-blade; ten unfaceted flake-blades; four indeterminate flake-blades; three redirecting flakes; two plunging flakes from disc cores; eight discoidal, prepared cores; five single-platformed, unprepared, flake-blade cores; three double-platformed, unprepared, flake-blade cores. The assemblage approximates to an Upper Levalloisian stage of culture. Material: chert.

EL DUBBO (40). Situate near the edge of the Hoddur plateau, east of Tijieglo, this well area, approximately $\frac{1}{4}$ mile wide, is fed by springs, seeping through the Jurassic limestones. During the dry season it is dry except for three small water-holes, but during the rains the pan is filled by a small, shallow lake. Round the edges of this pan are found tufas and surface limestones. The lower levels are bluish grey in colour, and are laminated, while the upper levels and surface show tubular structure, the pipes averaging $\frac{1}{2}$ in. in diameter and passing vertically down into the lower levels. Subaerial weathering has caused 'needle-point' weathering of the surface. The thickness of this tufa, which rests on the Jurassic limestone, was difficult to determine, but it was certainly 3 ft. thick and sloped downwards from the edges towards the centre of the pan. In the immediate vicinity of one of the springs the tufa was less hard and compact and may represent a somewhat later formation. Occasional pebbles of chert and lava are found within the tufa.

A sparse Doian factory debris is found upon the surface. From within the tufa were recovered the following chert artifacts, for the most part fresh and unpatinated and of small size: one faceted flake, one flake with unfaceted platform derived from a prepared core, three short, thick flakes with steeply inclined, plain platforms—one worked to a crude unifaced point, and the other two with scraping edges, one unfaceted flake-blade, five retouched fragments.

In addition was recovered a sparse fauna, the only identifiable material being the molar teeth of a large Equid, probably *Equus grevyi* and fragments of *Achatina* sp.

The cultural material should be compared with that overlying the 50–100 ft. terrace of the Webi Shebeli and from Saha Gurral: it is considered to represent a local form of Upper Levalloisian.

BOHOL WASHAGA (34). Near El Abdulla, on the road between El Goran and Burdo. Here two white gypsiferous marl or tufa deposits were found exposed in section in the tug bed. A brief examination showed small prepared cores and flakes, made for the most part on small chert pebbles, eroding from the top of the upper tufa: these artifacts showed a brick-red patina.

SAHA GURRAL (37). At Bugda Qosa on the northern edge of the Hoddur plateau. Here is found a shallow water-hole, formed probably round a funnel sink through gypsum, some 200 to 150 yd. across and 15 ft. deep. The sequence of beds, exposed in the sides of the 'saha' shows, at the top 3–5 ft. of red, gypsiferous sand with cellular kunkar, near the base. Near the base of this bed was recovered a small faceted flake, with thick white patina, of late, probably Stillbay, form. This was weathered but unrolled. Underlying this cemented red gypsiferous bed was some 10 ft. of white to grey calcareous marl containing sparse lenses of fine gravel and grit. This bed contained the following artifacts: four butt-ends of faceted flakes; two broad, unfaceted primary flakes; one flake-blade with good faceted butt; one well made, flat disc core; one poor, double-ended, sub-rectangular prepared core; five retouched flake and flake-blade fragments. All these specimens are fresh and are either unpatinated or show a thin ochreous patina. Material: chert.

The specimens recovered are typical of the Upper Levalloisian.

OHOLE (42). Near Ferfer, on the north side of the Webi Shebeli valley. From the upper calcareous marls were recovered five flakes in chert, with faceted striking-platforms and thick white patina, and one unifaced point similarly patinated.

BOHOL BANDOI (78), near Bur Hakaba. The following artifacts in quartz were recovered from the top levels, and from the surface of the lower, yellow and kunkarous Tug Alluvium, upstream of the Audegle road bridge: one pebble hammer-stone, one bifaced chopper, one disc core, four faceted flakes, one core scraper, thirteen flake-blades, or long flakes with parallel or converging flaking on the upper surface, and some marginal retouch, one polyhedral core. Tools were either fresh or slightly abraded.

OGADEN

DURUALEH (68). A small isolated bur of chert and limestone on the west of the road between Danan and Melka Taka. From the scree at the foot of this bur were recovered the following artifacts: four disc cores, one side-chopper, one flake with Tabelbalat retouch at the butt, one end-bulb type of side-scraper. Artifacts are in chert, and are heavily patinated and weathered.

DARCHAUA (73). The site is situated at the junction of this tributary tug with the Webi Shebeli. Within the top of the upper calcareous alluvium were found the following artifacts: one core chopper, two small subrectangular cores, two small subtriangular cores, five small faceted flakes, seven flake-blades and two flakes with unprepared striking-platforms. Material: chert.

THE LOWER NOGAL

EIL (56). From the grey Upper Tug Gravels and alluvium were recovered the following chert artifacts both in a fresh, and slightly rolled state, with thick white patina: one faceted flake, one unfaceted flake from a prepared core, one primary, unfaceted flake, one retouched fragment, one flake-blade with faceted butt, one single-platformed, prepared, flake-blade core, two small prepared disc cores.

TUG BABANI (59). From the red Tug Gravels and Alluvium was recovered the following Upper Levalloisian assemblage: four faceted flakes with radial flaking and straight butt, one worked into a crude, unifaced point and one with a marked notch; also two retouched flake-blades. These artifacts are fresh or slightly abraded and exhibit a buff patina, some lime incrustation is present.

UNDIFFERENTIATED LEVALLOISIAN SITES

Although it has not proved possible to determine, except typologically, whether the material found at these sites should be assigned to the Lower or Upper Levalloisian, by reason of the lack of adequate stratigraphical evidence; yet several of them are of considerable importance and have yielded a fully representative series of artifacts. It is probable that in most cases where we are dealing with a surface assemblage, both the Lower and Upper stages of this culture are represented.

NORTH-WESTERN SOMALILAND (Pl. 15, nos. 1–11)

BERBERA (DUBAR) (14). This surface site has been described on pp. 126–7. The earliest series, on typological grounds, has been assigned to the Acheulio-Levalloisian. The main series of artifacts, on the evidence of typology and *état physique*, belongs to a pure Levalloisian stage of culture. With very few exceptions, the tools, which are made from Cretaceous quartzite, show abrasion by surface weathering. The site is extensive and contains a large number of artifacts.

The inventory of tool types is given here, and from this it will be seen that almost the full range of artifacts is present, as at the stratified sites described above.

Faceted flakes—30. (Pl. 15, nos. 3, 8, 9, 10.) These vary in length from 3 to 1¼ in., and are for the most part short and broad and thick for their size. The subtriangular flake is a typical form. Striking-platforms usually show careful preparation, are wide and centrally placed. In one instance the platform has been removed by incipient Tabelbalat retouch (Pl. 15, no. 3). These flakes have been trimmed by marginal retouch to form scrapers and unifaced points.

Unfaceted flakes—9. These are usually large and broad and of a primary nature, but four pointed flakes with convergent flaking and plain striking-platform, set obliquely, are also included.

Flake-blades—10. These fall more into the class of long flakes, rather than flake-blades. One only shows a prepared striking-platform. Four larger specimens also occur on three of which the butt has been removed.

Retouched fragments—57. (Pl. 15, nos. 5, 6, 11.) Several of these may be due to the nature of the site and represent the work of natural abrasion, but this cannot be the case with the majority. Flakes with the butt removed have been worked by steep resolved retouch round the edges of the fragment to form notches, and straight, convex, and concave working edges. Pl. 15, no. 6, shows an example with two opposed notches forming a beak, or blunt point. Length ranges between 3 and 1 in. or even less.

Trimming stones—2. (Pl. 15, no. 1.) Two typical examples of these plano-convex sectioned tools occur.

Hand-axes—4. (Pl. 15, no. 2.) These all exhibit crude bifacial working and are small and subtriangular in form. Flaking is usually confined to the point, while the butt is quite unworked.

Choppers—5. Crudely made core tools, the retouch being confined to the whole or part of one edge to form a side-chopper.

Flake cores: prepared—13. (Pl. 15, nos. 4, 7.) These range in length from 4½ to 1½ in. The oval to discoidal, core is the commonest form but the double-ended, subrectangular type is also represented; the discoidal forms having usually rounded or high backs, and the subrectangular forms being flattish.

Flake cores: unprepared—3. Three multiple-platformed, or formless, cores, from which short and broad, or long flakes were removed.

THE DANAKIL RIFT

GARASLEH TUG, NEAR DIRE DAWA—(8). (Pl. 15, nos. 12–19.) The artifacts recovered from the alluvial fan gravels have been described on p. 169. From the eluvial gravel which rests unconformably on the denuded surface of the fan gravels were recovered two series of artifacts which can be divided on physical condition. The earlier series is abraded, and is found within the eluvial gravel, while the later series is weathered but unrolled and comes from the top and surface of the gravel. The material is predominantly lava but a few specimens of sandstone also occur.

First series. Comprises two short and thick flakes with unfaceted striking-platforms and a minimum of secondary retouch suggesting use as side-scrapers, two flake-blades with unfaceted platforms, one broad flake with faceted striking-platform, two crudely bifacial core tools or hand-axes with unworked butts.

Second series. Contains the following assemblage—*Flakes derived from prepared cores*—6. (Pl. 15, nos. 12, 13, 17.) Two are from disc cores, and the remaining four are pointed flakes, with convergent flaking, struck from subtriangular cores: the butts of these latter show inclined and plain striking-platforms. Secondary retouch is usually confined to the point and edges.

Indeterminate flakes—9. The butt ends of these flakes have been intentionally removed usually by a crude form of Tabelbalat bulbar retouch.

Retouched nodules—3. One example (Pl. 15, no. 14) is a good example of a side-scraper, the retouch being of a steep, resolved nature.

Flake-blades—5. These are struck mainly from unprepared cores and vary in length from 5½ to 2½ in., but one example with crudely prepared striking-platform occurs. (Pl. 15, no. 16.)

Flake cores: prepared. Three discoidal to subtriangular specimens. (Pl. 15, no. 18.)

Flake-blade cores: unprepared. Two typical examples were recovered with single platforms. (Pl. 15, no. 15.)

Hand-axes—2. (Pl. 15, no. 19.) These are crude, bifaced choppers or hand-axes similar to those found rolled in the eluvial gravel.

The second series is typical of the Upper Levalloisian and it is tempting to refer the first and earlier series from the eluvial gravel to the Lower division of this culture.

SOUTHERN SOMALIA

WEBI SHEBELI: SURFACE OF THE 50/100 FT. TERRACE AT CURA BA'AD (72). (Pl. 12, nos. 1–4.) A number of chert artifacts were collected from the surface of the 50/100 ft. gravel terrace at Cura Ba'ad. Apart from a small number of fresh tools of Doian affinities, the artifacts all belong to a single series of Levalloisian type. Six of these specimens are rolled but the remainder are weathered or only very slightly abraded: this abrasion is considered to be due to long exposure on the surface. The material consists of the following tools made from small chert pebbles.

Faceted flakes—13. These are all of small size and exhibit radial flaking on the upper surface, one example has had the platform removed by Tabelbalat technique. Some show steepish marginal retouch to form scraping edges and notches.

Unfaceted flakes—1. A single small example was found.

Flake-blades—14. Striking-platforms may be either faceted, plain or intentionally removed: all are small and sometimes show marginal retouch.

Flake cores: prepared—9. These are all small and vary from the discoidal to the subrectangular in shape (Pl. 12, no. 4), and are limited in size by the material used: the platforms appear normally to have been plain but one pebble core with prepared platform also occurs.

Flake-blade cores—1. Unprepared, single-platformed, flake-blade core.

Pebble artifacts—4. These consist of small pebbles, bifacially worked down one edge. They may represent unfinished cores.

WEBI SHEBELI 50/100 FT. TERRACE AT SHEIK GURE (83). (Pl. 12, nos. 1–3.) The eroded surface of the ferruginized 100 ft. terrace gravels at Sheik Gure yielded cultural material of two distinct ages. A Late Stone Age, fresh and unweathered, series of Doian tools and factory debris with white patina, and an earlier, weathered and sometimes abraded series, patinated to a chestnut or reddish brown colour and of Levalloisian affinities.

The Levalloisian series of artifacts are made from small pebbles of siliceous sandstone, quartzite, and chert, and in some rare cases, of fossil wood. Of the total number of specimens six only show evidence of rolling, but the remainder are considerably weathered, due presumably to long surface exposure. The collection comprises the following artifacts:

Flakes—10. (Pl. 12, nos. 1, 3.) Two of these are long flakes approximating to flake-blades, the remainder being broad and thick. They are nearly all primary flakes struck from pebble cores, and show varying amounts of cortex on the upper face. Butts are plain and unprepared, but two larger specimens show crude preparation of the platform. A little secondary retouch is present but is confined to the edges of the flake. Retouch may be either flat or steep, the latter form usually associated with notches.

Retouched nodule—1. One example of a plano-convex sectioned high-backed scraper or 'push-plane' was found, with steep, marginal retouch.

Pebble and nodular artifacts—5. These are all small and show bilateral, primary flaking, sometimes steep, down one edge or across the end of a pebble or nodule.

Pebble cores—13. (Pl. 12, no. 2.) These are all small of the discoidal to subrectangular, single—or sometimes double—platformed variety. The majority show some indication of preparation of the striking-platform. From these cores were removed single, short, broad flakes with primitively faceted, or plain, striking-platform.

These artifacts resemble the series from the Cura Ba'ad 50 ft. terrace gravel, from El Dubbo and Saha Gurral and on typological grounds may be considered to be of Upper Levalloisian age, though it must be borne in mind that the often small size of the specimens is primarily determined by the raw material available. It is also probable that this was a factory rather than a home site.

MILE 75, BARDERA/WAJIR ROAD (93). The main road here passes through a series of shallow depressions on the peneplained plateau surface. It is possible that these depressions at one time formed shallow ponds or lakes, as round their edges were found a few heavily weathered and rolled artifacts of ferruginous sandstone and quartzite, patinated a deep chocolate colour. Artifacts comprise the following: one pebble core for the removal of flake-blades, one crudely faceted, broad flake with steep scraping edge, one prepared flake with crudely reduced butt, one pointed flake with crude reduction of butt, two retouched fragments with notches, three thick flakes with plain striking-platforms inclined at a steep angle to the main flake surface.

This series has the appearance of a Lower rather than an Upper Levalloisian stage of culture.

MANDERA NORTH SITE (11). On the surface of denuded gravel in the camp area were recovered two faceted flakes of chert with white patina and one prepared, oval and high-backed core of quartzite, which typologically should be referred to the Levalloisian.

AISHA PLAIN (6). From the ferruginized desert pavement overlying the freshwater alluvial deposits on the Aisha plain 8 miles north-west of the town, was recovered a weathered, crudely prepared disc core, and a bifacial chopper in quartzite with ferruginous patina.

MELKA TAKA (71). From a small outcrop of limestone rising above the deposit of Tug Gravels and Alluvium were recovered, besides Late Stone Age Doian tools, two weathered and white-patinated Levalloisian flakes of late form. Material: chert.

MILE 18, CALLAFO/MUSTAHIL ROAD (DURDERE) (76). From the surface of the 50/100 ft. terrace gravel was recovered one weathered and white-patinated, subtriangular prepared core.

BIHEN (60). From Bihen to some 8 miles south towards the foot of the Nogal scarp are found numbers of low, isolated hills of anhydrite beds, capped by Eocene limestone and containing, in the lower levels, bands of black to grey chert. These chert bands have been exploited as a source of raw material from Levalloisian times onwards, and a thickish scree surrounds the foot of these hills, and contains an enormous quantity of factory debris and some artifacts. The Levalloisian debitage is patinated chestnut brown to buff in colour, while the Late Stone Age material is either unpatinated or a mottled white colour.

Faceted flakes—5. These consist of three radially flaked specimens, and two with convergent flaking. In addition are seven flakes with unprepared butts which have been removed from prepared cores. Four of the latter show constriction of the edges at the butt.

Unfaceted flakes. Three only were retained but a large number existed. These for the most part were thick, primary flakes and show a steeply inclined platform.

Flake-blades. Nine large and eleven small flake-blades were collected. The majority show plain striking-platforms but these are usually at right-angles to the main flake surface. One specimen shows Tabelbalat retouch.

Prepared flake and flake-blade cores—4. These are typical and include two subtriangular specimens, one oval and high-backed specimen, and a double-ended, flake-blade core.

Several of the flakes and flake-blades show signs of marginal retouch. While the series undoubtedly represents in the main debitage from a factory site, it is, however, typical of the Upper Levalloisian.

THE NOGAL VALLEY

GOVERNOR'S CAMP AREA, LAS ANOD (181). This site was discovered by Captain H. B. Gilliland, in 1944. Las Anod is situated on the edge of the southern scarp of the upper Nogal valley, on gypsum/anhydrite beds of Eocene age. The material from this site is derived from scree at the base of two hill slopes, one at the side of the District Commissioner's house, and the other about a mile to the north-east. That at the former site appears to be eroding from secondary gypsum beds, which are being cut into by many small erosion gullies broken irregularly by ledges on which the implements have weathered out. From Gilliland's description and photograph it would seem that the deposit is a scree, cemented in places by gypsum, from which are derived factory debris and a few implements showing pink to grey patina: these tools are for the most part fresh and unweathered. No thickness of deposit was noted, and there would appear to be no connexion with the alluvial beds of the Las Anod Tug. On the evidence of typology and physical condition the material can be divided into two series, only the earlier of which concerns us here.

Material. Brown chert from the upper levels of the Eocene, anhydrite beds.

Condition. Mainly fresh and unabraded. Buff to pink patina, varying with the degree of exposure and weathering.

Implements. Points—3. (Pl. 16, nos. 2, 4, 5.) Two of these specimens are unifaced points which have been made on flakes with faceted striking-platforms, one of which shows constricted-edge technique at the butt. The third specimen is a completely bifacial, foliate point, worked carefully by flat percussion technique over both faces: it has been broken and reworked at one end.

Faceted flakes—3. These are typical examples showing radial flaking on the upper surface, and right-angled, faceted striking-platforms.

Flakes with reduction of the butt—5. (Pl. 16, no. 7.) Three of the specimens show deliberate basal retouch as in the specimen illustrated.

Unfaceted flakes—12. These are broad and thickish showing a steeply inclined striking-platform. The majority show marginal retouch, notching and signs of utilization at the edges. One example is a plunging flake.

Flake-blades—10. (Pl. 16, no. 6.) These, for the most part show parallel and alternate flaking. The largest is 5 in. long, the smallest 2 in. Platforms are inclined and plain. The majority show signs of utilization.

Hand-axes—2. (Pl. 16, no. 3.) Both specimens are probably made from flakes, one being only partially bifacial. The second specimen (illustrated), shows more careful working of a flat, percussion nature: this specimen is undoubtedly an unfinished tool, however, and it is believed, as the site from which it is derived is probably a factory site, that both these tools represent roughouts for the type of bifacial, foliate point described above.

Choppers—2. One specimen is made on a piece of tabular chert, the other on a nodular fragment and shows one, bilaterally flaked, working edge.

Prepared core—1. (Pl. 16, no. 1.) One good example of a double-ended prepared core for the removal of flake-blades.

It can be clearly seen that this assemblage falls within the Levalloisian culture complex, and the main series of tools compares well with the Upper Levalloisian from Garoe and Jesomma. However, the bifacial foliate point is of a more developed form than is usually found with this culture stage. The writer would not necessarily exclude it from the assemblage for this reason alone, however, and it is possible that we have here an example of the survival and development of the bifacial tradition of flaking from the Acheulio-Levalloisian. If our interpretation of the 'hand-axes', as being roughouts for such foliate points, is correct, this would provide substantive evidence. The bifacial point associated with the Somaliland Stillbay is nearly always subtriangular in form with a broad butt end: moreover, the fact that portions of parti- and entirely bifacial points of the form of this Las Anod specimen do occasionally occur with the Upper Levalloisian in the Horn, strongly favours the inclusion of the Las Anod example as an integral part of this culture complex.

SUMMARY

In spite of the pronounced conservative nature of the Levalloisian Culture a gradual but steady techno-typological development can be discerned. In the Lower Levalloisian, tools are on the whole larger (average length for flakes is $2\frac{1}{2}$–3 in. and for flake-blades 3 in.) than those found with the Upper stage of this culture[1] (average length for flakes 2–2½ in. and for flake-blades 2½–3 in.), this is due to the diminishing size of the cores. There is little specialization in the Lower stage, with the exception of rare forms of unifaced point, while in the upper stage more controlled, secondary retouch and flat percussion flaking are coincident with various scraper forms, and the unifaced point with marginal retouch. The highest technical attainment of the Upper Levalloisian is to be seen in the parti- and completely bifacial foliate points, the beginnings of Tabelbalat technique and the flakes with narrowed or constricted butts. The flakes and flake-blades from the later Levalloisian sites are usually thinner; the striking platforms are more carefully faceted and the *chapeau-de-gendarme* form occurs. Of interest is the persistence of crude, bifacial technique in the form of the core biface or chopper, presumably derived from an Acheulio-Levalloisian ancestry. A comparison of the numbers of the different forms of prepared core from the main sites is of interest:

Core form	Lower Levalloisian	Upper Levalloisian
Oval and disc	21 (52·5%)	15 (21·4%)
Subrectangular—single	2 (5%)	4 (5·7%)
Subrectangular—double	7 (17·5%)	34 (48·6%)
Subtriangular	8 (20%)	14 (20%)
Flake-blade cores—single	—	2 (2·9%)
Flake-blade cores—double	2 (5%)	1 (1·4%)
Totals	40	70

In the Lower Levalloisian the perhaps less specialized oval to disc core comprises over 50 per cent of the core forms, while in the Upper Levalloisian it comprises only just over 20 per cent. A considerable increase, on the other hand, is noticeable in the double-ended, subrectangular form. The percentages of subtriangular and flake-blade cores remain approximately the same.

The writer believes that it is possible to trace the development of the Levalloisian Culture in the Horn from an Acheulio-Levalloisian ancestry with large, high-backed, circular and disc cores, the bifaced core hand-axe and cleaver (as at Hargeisa and Sheik); through the transitional stages, as found at Berbera and Eil to the Lower Levalloisian with a preponderance of oval and disc cores and a very few bifacial core tools; and so to the Upper Levalloisian with a higher percentage of specialized tools and cores, and the inception of a more developed, bifacial technique which is perfected in the succeeding culture.

[1] The small size of the majority of the cores and flake-blades from the Upper calcretes at Gabredarre suggests that we are dealing here with a later stage of the Levalloisian complex than is present at Garoe, Hargeisa and Jesomma.

CHAPTER 10

THE SOMALILAND STILLBAY CULTURE. THE MAGOSIAN CULTURE AND THE MAGOSIO/DOIAN TRANSITIONAL INDUSTRIES. THE HARGEISAN CULTURE AND DERIVATIVE INDUSTRIES

THE SOMALILAND STILLBAY CULTURE

This culture is best represented in the plateau country of north-western Somaliland. The most important sites, and those by which the culture is dated in the north, are situated in the area of Hargeisa. Industries of Stillbay form are widespread, however, within the Horn and show a distribution closely related to that of the Levalloisian. In Somalia the culture is dated by the three stratified sites at Bur Eibe (18), Ohole (42) and Jesomma (85). Very important, but regrettably scanty evidence, comes from the East coast, from Biede (57), at the mouth of the Nogal, where the Stillbay can be tentatively dated in relation to the ± 12 m. raised beach. Additional sites of importance were found at Mandera North (11), Borama (5), Donkukoq (52), Garoe (54) and by other workers at Dagahbur (116) and Dubato (230).

On superficial examination of the cultural material it would seem that the Somaliland Stillbay is a direct derivative from the Upper Levalloisian. While the writer is of the belief that this is basically the case, a more detailed examination of tool types and technique suggests that this view must be modified to the extent that new forms of tool (notably the subtriangular unifaced and bifaced points), and the introduction of pressure-flaking, in conjunction with the almost complete absence of certain Levalloisian forms (notably the double-ended subrectangular core, and the flakes with partial thinning at the butt), are strongly suggestive of the introduction of new elements within the Horn. At Hargeisa the stratigraphical evidence at one site suggests a division into a Lower and Upper phase though it is improbable that any great period of time separates the two. At the remaining sites such subdivision is not possible. The culture is dated to the dry period which marks the end of the Gamblian Pluvial. The inventory of the cultural material recovered from the main stratified sites is given opposite.

Inventory of tool types of the Somaliland Stillbay Culture

Tool types	Lower Stillbay Hargeisa	Upper Stillbay Hargeisa	Eil (undiff.)	Bur Eibe (undiff.)	Ohole (undiff.)	Totals
Points, unifaced	55	69	3	30	9	166
Points, parti- or totally bifaced	58	37	—	32	1	128
Scrapers, side: on side-flakes	—	1	1	9	—	11
Scrapers, side: on end-flakes	1	20	2	14	—	37
Scrapers, round	6	—	—	14	—	20
Scrapers, end	3	7	2	—	5	17
Scrapers, various	1	17	—	12	—	30
Burins, single-blow	1	4	—	1	—	6
Backed-blades	—	—	4	3	—	7
Backed-flakes	18	22	1	5	—	46
Serrated flakes	12	8	—	—	—	20
Diminutive core bifaces	12	8	—	1	—	21
Flakes, faceted	23	43	8	44	50	168
Flakes, unfaceted	6	4	28	6	30	74
Flake-blades	4	10	18	—	15	47
Blades	—	—	20	—	—	20
Flake cores, prepared	12	32	8	49	11	112
Flake-blade cores	—	—	—	5	—	5
Blade cores	—	2	11	1	—	14
Miscellaneous	3	3	10	13	3	32
Totals	215	287	116	239	124	981

THE LOWER SOMALILAND STILLBAY (HARGEISA)

Stratigraphy. At site H.7, Hargeisa, overlying the Younger Tug Gravel, and the intermittent land rubble which covers it in some places, there is found a red sandy alluvium—somewhat gritty in its lower, but fine-textured in its upper levels. The lower part of this alluvium (which is shown as deposit (6) on Text-fig. 4) is considered to be largely of eluvial origin: the upper part is more definitely of aeolian origin.

From the lower levels are eroding large numbers of artifacts and of debitage which are for the most part quite fresh and unabraded. Over three-quarters of the artifacts are of chert, but quartzite was also used and the difference of material appears to have made little or no difference to the degree of finish and technical skill exhibited. The chert specimens show either a white or grey patina, which is not particularly thick: small incrustations of lime on the specimens are not infrequently met with. These specimens provide the material for the description of the Lower division of the Stillbay. Upper Stillbay tools are eroding from the upper levels of the alluvium.

Geological horizon. The first half of the dry period marking the end of the Gamblian Pluvial.

Climate. Semi-arid to arid.

Implements. The list of tool types will be found in the inventory. The main forms are points and scrapers of various kinds, but it must be emphasized that

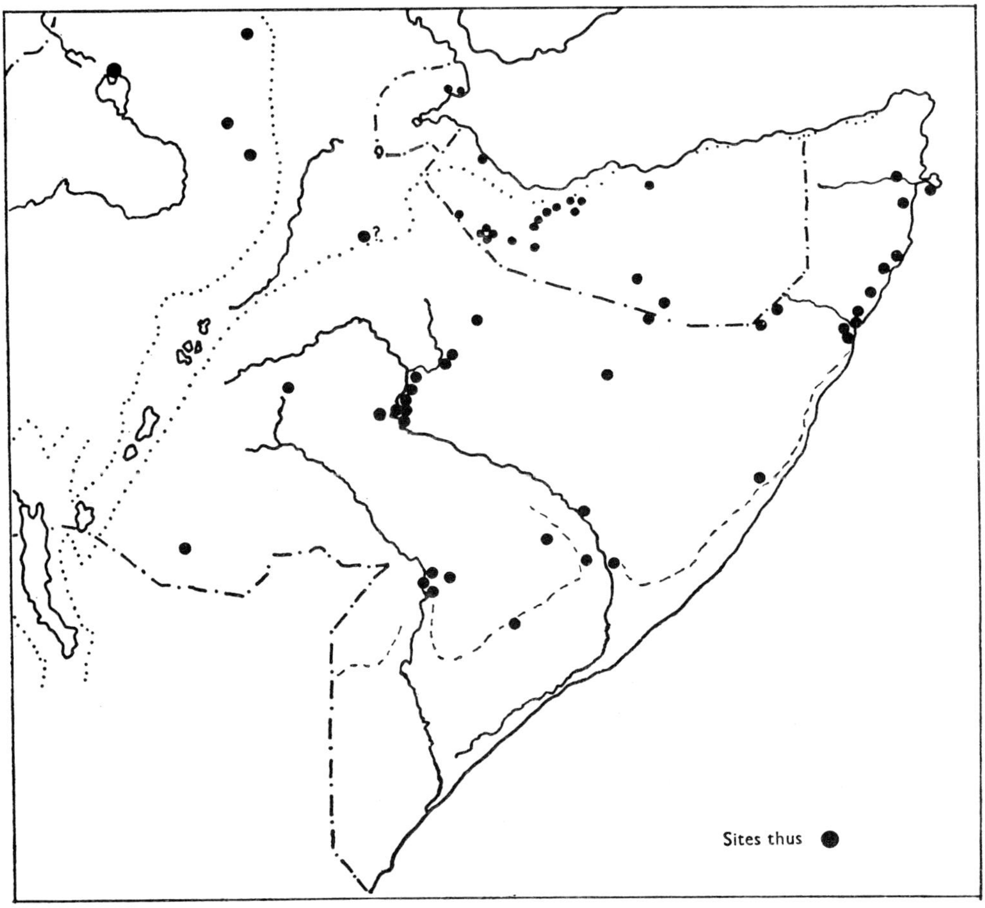

Text-fig. 19. Distribution of Somaliland Stillbay industries

associated with the finished implements at all the Hargeisan sites there was a quantity of factory debris. As it was necessary to limit strictly the amount of material collected, only such pieces of debitage as adequately demonstrated technical processes could be retained.

The predominant technique found on the points is flat percussion flaking, but tools exhibiting undoubted evidence of pressure flaking are found. The scrapers, however, exhibit a rather fine, resolved retouch. The basic technique is that of the prepared core in association with percussion technique.

Points, unifaced—55. (Pl. 17, nos. 5, 6, 7.) Of these, thirty-one specimens are subtriangular in form, sixteen are foliate, nine are asymmetric, one has a rounded butt, and one is ellipsoid. The majority have been made on pointed flakes exhibiting converging flaking and, where this has not been removed, a well-prepared, faceted striking-platform. The foliate forms (Pl. 17, no. 7) continue the Upper Levalloisian tradition. The majority of the specimens recovered are broken, but the average length of the tool must have been about 5 in. The subtriangular forms (Pl. 17, no. 6) are entirely typical of the Stillbay complex. The butt is usually formed by the striking-platform. Retouch is usually restricted to the edges but in rare cases may be carried over the entire upper face. There is also evidence of some trimming of the underside at the point. The asymmetric points (Pl. 17, no. 5) are a small but distinctive group. They are somewhat thicker than the other forms and not infrequently show serrated edges.

Points, bifaced—58. (Pl. 17, nos. 1, 2, 3, 4, 8.) Twenty are subtriangular in form, eighteen foliate and seventeen have rounded butts.

The foliate forms, exhibiting flat percussion flaking (Pl. 17, nos. 1, 3) have been carefully worked, either entirely or partly, over both faces and, even more than the unifaced forms, are the logical outcome of the bifaced tradition of the Upper Levalloisian, best seen at Jesomma. The subtriangular specimens (Pl. 17, nos. 2, 4) approximate to the classic Bambata form and are usually finely worked, sometimes by pressure. Striking-platforms may be either reduced as in no. 2, or retained as in no. 4. Sections are usually lenticular. Intermediate between the subtriangular and the foliate, is the rounded butt form (Pl. 17, no. 8). The main characteristic of these examples is the careful (Tabelbalat) reduction of the platform and bulb.

Scrapers—11. (Pl. 17, nos. 9, 17, 18, 19.) The true *racloir* or side-scraper on a side flake is not found, the commonest forms being the side-scraper on an end-flake, having either one or both edges trimmed by straight or convex marginal retouch (Pl. 17, nos. 9, 19): the round scraper on a broad, prepared flake with retouch round the greater part of the circumference (Pl. 17, no. 18); and the end-scraper on an end-flake or flake-blade (Pl. 17, no. 17). They are all, especially the first, typical Stillbay forms.

Burins—1. (Pl. 17, no. 16.) One example of a single-blow burin on a subtriangular prepared core, and three other more doubtful specimens exist. In the opinion of the writer it is not unlikely that these were not intentional burins but the result of incidental flaking.

Backed flakes—18. (Pl. 17, nos. 9, 10, 12.) These are an important group of tools which are made on faceted flakes of the pointed variety. The backing is not usually vertical but curves over to meet the upper face: working is of a resolved nature and is always directed from the main flake surface. Pl. 17, no. 10, is a typical example, no. 9 has been retouched down the opposite edge to form a straight, but serrated scraping edge; while in no. 12 the backing is confined to the upper end and the point has been worked into an awl or *perçoir*: in this instance the backing is almost vertical. Of note also is the Kasouga retouch[1] seen on the underside of the cutting edge on some of these specimens.

These tools recall the epi-Levalloisian backed flakes from north-east Africa and are probably the prototype for the Stillbay backed blades in the sub-continent.

Serrated flakes—12. These are for the most part faceted flakes which show serrations, worked from the undersurface, down the whole or part of one edge. Such serration is seen in Pl. 17, no. 9.

Diminutive core bifaces—13. (Pl. 17, no. 11.) An interesting group of small hand-axes $3\frac{1}{2}$–$1\frac{1}{2}$ in. long made from cores and thick flakes. They exhibit block-on-block percussion flaking and are limande (6) and subtriangular or pear-shaped (6) in form. Sections are usually biconvex.

Faceted flakes—23. These are predominantly long—14, rather than broad—9. The former show parallel or converging flaking directed from the butt, while flaking is more radial in the broad types. All are either thin or subtriangular in section. Striking-platforms are carefully prepared and usually symmetrical, whether centrally or obliquely placed. Platforms may be either straight, convex, or of *chapeau-de-gendarme* form.

Flakes, unfaceted—6. A large number of such flakes existed. They show parallel flaking on the upper face, the platform is small, usually set obliquely and inclined at an angle of 100–110° to the main flake surface.

Flake-blades—4. These are typical and faceted. It is now much more difficult to distinguish between long flakes and flake-blades, the typical form being that of the specimen illustrated at Pl. 17, no. 10.

Prepared flake cores—12. (Pl. 17, nos. 13–16.) Eight are subtriangular and four are discoidal. Those collected are nearly all small and vary in maximum length between 3 and $1\frac{1}{2}$ in. The subrectangular double-ended core of the Upper Levalloisian is conspicuous by its complete absence. The diminutive specimens (nos. 13, 15) are exceptional, the usual form being flattish and of the dimensions

[1] So named from certain small flakes from an industry at the mouth of the Kasouga river, near Grahamstown, which exhibit marginal, squamous flaking down the length of the underside of one edge.

of nos. 14 and 16. Larger, discoidal cores were noted and it is obvious that the majority of the artifacts above were derived from cores of a diameter of 3 in. or more.

Various—3. Included under this term are three crudish and pointed fabricators, or tap-borers, with plano-convex section.

UPPER SOMALILAND STILLBAY (HARGEISA)

Stratigraphy. Eroding from the red sandy alluvium ((6) on Text-fig. 4), at sites H. 1, 2, 3, 4, 5, 6, 8, 12 and 13 and from the middle to upper levels at site H. 7, there are found artifacts and factory debris in large numbers. These are made from chert but a few specimens in quartzite were also found. They are all fresh in appearance but exhibit a fairly uniformly thick, white patina. Some lime incrustation is present.

Geological horizon. The second half of the dry period between the end of the Gamblian Pluvial and the onset of the first post-Pluvial Wet Phase.

Climate. Arid to semi-arid.

Implements. Similar forms to those found with the lower division occur, with the addition of blades and blade cores in small numbers. A greater degree of pressure flaking is now seen, and in fact this is now the predominant technique for secondary retouch.

Points, unifaced—69. (Pl. 18, nos. 18, 24.) These are now the commonest form of tool. Fifty-one are subtriangular, only twelve are foliate, three are asymmetric, and four are of the rounded butt form: these latter are more lanceolate than before (Pl. 18, no. 18). The subtriangular form still predominantly retains the wide and straight, faceted butt (Pl. 18, no. 24), while the foliate points approximate more to the 'slug' form (Pl. 19, no. 1).

Points, bifaced—37. (Pl. 18, nos. 19, 20; Pl. 19, nos. 2–4.) About half the number of bifaced to unifaced points exist, but the former all show a high degree of skill in manufacture and a high proportion of pressure flaking has gone into the secondary retouch of these tools. Thirteen are subtriangular (Pl. 18, no. 19), thirteen are more pointed leaf-shaped forms with broad rounded butts (Pl. 18, no. 20; Pl. 19, nos. 2–4), and only five are of the true elongated foliate type. Equally fine specimens are made in quartzite as in chert.

Scrapers—45. (Pl. 18, no. 21; Pl. 19, no. 5.) The typical form is a single- or double-edged scraper on an end-flake, but end-scrapers are also found, worked on the ends of long flakes or flake-blades, as well as various hollow and notched forms and crude core-scrapers.

Burins—4. These are somewhat doubtful single-blow burins.

Backed flakes—22. (Pl. 18, nos. 22, 23.) These are nearly all on faceted flakes and the backing is more markedly oblique and truncates the flake. Kasouga retouch on the cutting edge is also present. The specimen illustrated at Pl. 18, no. 21, is intermediate between a side-scraper and a backed flake and anticipates the 'backed blade' form of these Stillbay complex tools.

Serrated flakes—8. (Pl. 19, no. 8.) These are typical.

Diminutive core bifaces—8. Similar to those found with the Lower division of the Somaliland Stillbay.

Faceted flakes—43. These show little variation from those previously described. There are, however, very few of the broad, radially flaked specimens. Platforms are always carefully faceted and the flaking on the upper face is either parallel or converging. Tabelbalat technique is that usually employed for the reduction of the bulb and platform.

Unfaceted flakes—4. Of the long variety with parallel flaking, these have been struck from single-platformed, unprepared cores.

Flake-blades—10. Approximately half show faceted striking-platforms and are derived from single-platformed, flake-blade cores.

Prepared flake cores—32. (Pl. 18, nos. 25, 26.) These consist of five large discoidal cores, usually with high backs (Pl. 18, no. 26), approximate diameter $3\frac{1}{2}$–4 in.; sixteen flattish disc cores, diameter $2\frac{1}{2}$–$1\frac{1}{2}$ in.; ten subtriangular cores of an average length of 2 in. (Pl. 18, no. 25); also one double-ended core, discoidal to subrectangular in form recovered from site H. 5. The smaller cores show careful preparation and in every instance one flake only was removed before repreparation.

Blade cores—2. These are small ($1\frac{1}{2}$ in. long) and have been used for the removal of diminutive blades. One shows a single unprepared platform, the other is of the bipolar variety. Both come from site H. 1.

Various—3. Includes two redirecting flakes of the long variety and one indeterminate flake.

It is abundantly clear from a comparison of the Upper and Lower Stillbay from Hargeisa that they differ in no major essentials. The divergence is a matter rather of degree, the Upper leading imperceptibly from the Lower and developing a gradual perfection and modification of certain types of tool—notably the bifaced points and the backed flakes. Significant is the greater number of bifaced and unifaced foliate points in the Lower division, the predominant use of flat percussion flaking in the manufacture of these, and the presence of a few typical broad faceted flakes. In the Upper Stillbay the characteristics are points with broad and rounded butt, the increase in pressure flaking, and the suggestion of the introduction of the use of true blades.

The other main sites have yielded a material culture closely similar to that described from Hargeisa, but show that local preference for one or other form of tool may lead to a certain amount of localized specialization.

EIL. (Pl. 19, nos. 6, 7.) Stillbay artifacts are derived from the top of the grey Upper Tug Alluvium and from the calcrete or tufa which cements the surface of this deposit. Tools are in chert and exhibit the characteristic thick white patina: they are fresh and unabraded. Tools are also found embedded in the tufa surrounding three springs at Eil: these are quite fresh and unpatinated.

The points and scrapers (Pl. 19, nos. 6, 7) are typical and require little mention. Of interest, however, is the presence of four small, straight-backed blades, of microlithic proportions, twenty blades of similar size, and eleven blade cores. Nine of these latter are bipolar cores and the other two show single platforms. The faceted flakes and prepared cores are typical: Tabelbalat technique is that usually employed for the reduction of the butt as at Hargeisa.

BUR EIBE. (Pl. 18, nos. 1, 3–5, 7–17.) The implements are eroding from a ferruginous horizon apparently an incipient ironcrete some 1–3 ft. thick which covers the lower slopes of the 'burs' at this site. The main site is situated in the immediate vicinity of Bur Yassin, a small outlier at the south-east corner of Bur Eibe (see Sketch-map, Text-fig. 20). This deposit suggests semi-arid rather than arid conditions and it is not improbable that this area of southern Somalia enjoyed at this time a heavier rainfall than the north and centre as is in fact the case today.[1]

Tools are made from chert nodules and pebbles derived from the Jurassic limestones of the plateau, here some 43 miles distant. Quartz, preferably crystalline was also used and could be obtained locally. With occasional exceptions, tools are quite fresh and exhibit varying degrees of buff, yellow and red patina. The need for care in the use of raw material has resulted in a generally smaller size for the artifacts from this site.

Four specimens are undoubtedly derived as they are weathered or rolled and are more typically Levalloisian. One is a large end-flake of limestone and the other three are faceted flakes: one of the latter has been broken across the middle line and reworked as a hollow-scraper during Stillbay times.

Points, unifaced—30. (Pl. 18, nos. 5, 8, 9, 10.) These are well made on flakes with either faceted or plain platforms. Resolved retouch is that most usually employed. Twenty-seven specimens are subtriangular in form (no. 8), two are foliate (no. 5), though it is possible that this specimen may be derived. One is oval (no. 10) and four asymmetric (no. 9).

[1] The average rainfall for the Isha Baidoa/Bur Hakaba area is, on Italian records, from 30 to 40 in. a year. The average for Hargeisa would be between 10 and 15 in.

Points, bifaced—32. (Pl. 18, nos. 1, 3.) Usually only partly bifacial, these are almost evenly divided in numbers between the subtriangular form with straight butt and that with rounded butt. Pl. 18, no. 1 is a particularly fine specimen made on a thin flake of chert and is only $4\frac{1}{2}$ mm. thick at the centre.

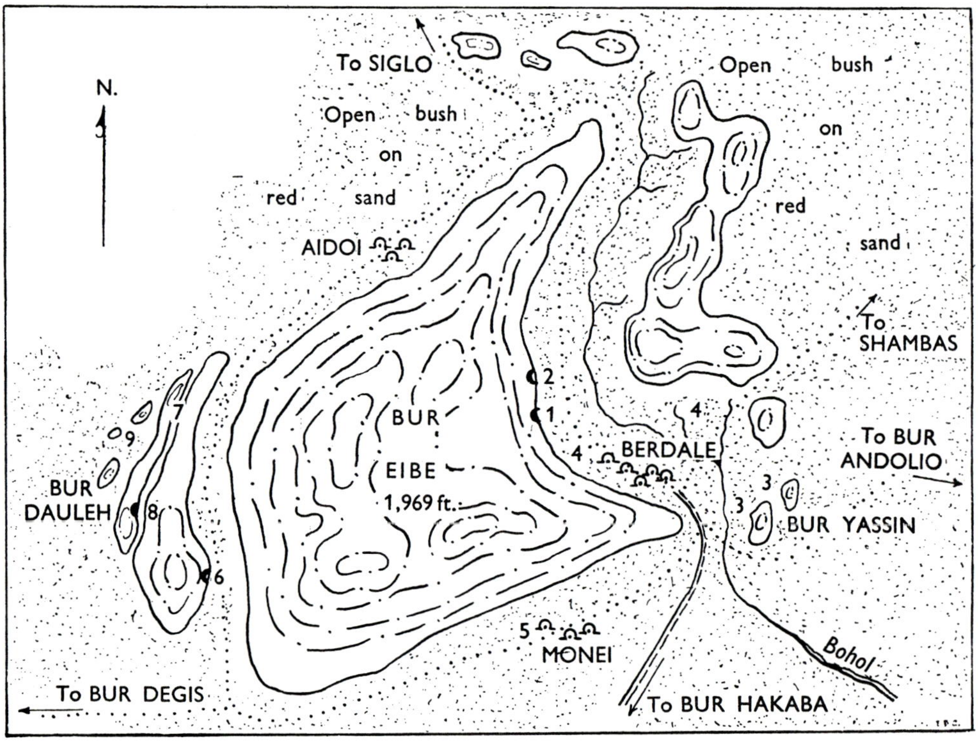

1 Gure Makeke
2 Gure Warbei
3 Magosian and Stillbay open station site
4 Doian and Magosian open station sites
5 Doian open station and shelter site
6 Magosio/Doian shelter site
7 Doian open station site
8 Painted rock shelter
9 'War' diggings Magosian and Stillbay

Scale 1 inch = 880 yards = 804 metres

Text-fig. 20. Sketch-map of sites at Bur Eibe.

Scrapers—49. (Pl. 18, nos. 13, 14, 16, 17.) These include typical single- and double-edge scrapers on end-flakes (no. 16), as well as some genuine *racloirs* on side-flakes (no. 13). Included under the term round scrapers are several forms on short, broad, faceted flakes, showing trimming round most of the circumference. Some of these tend to pass into the end-scraper class (nos. 14, 17). Hollow and notched scrapers also occur.

Burins. One single-blow burin.

Backed blades—3. (Pl. 18, nos. 11, 12.) Two of these are genuine, blunted, backed blades, but the third is a redirecting flake (no. 12) that has been used as a backed blade: the backing is straight.

Backed flakes—4. (Pl. 18, no. 7.) These are on broader faceted flakes than are those at Hargeisa.

Diminutive core bifaces—1. (Pl. 18, no. 4.) An entirely typical specimen but somewhat smaller than the Hargeisa examples.

The factory material is entirely typical. Of the prepared cores one is a large disc, thirty-eight are small discoidal cores and nine are subtriangular. One single-platformed, unprepared blade core was found.

Among the miscellaneous tools are included four crude single-edge choppers, one *perçoir*, six *lames écaillées* and two redirecting flakes.

OHOLE. (Pl. 18, no. 6.) The tools are derived from just below and in the ferruginous surface of the upper calcareous marls (=Younger Tug Alluvium) on the right bank of the tug, and from the eroded surface of these marls on the left bank. Specimens are typical of the Stillbay and include unifaced points (Pl. 18, no. 6), a partially bifaced example, and the usual flake and scraper forms. Of the prepared cores five are discoidal with rounded backs, three are flat discs, and three are subtriangular. Among the miscellaneous artifacts are one trimming stone or plano-convex scraper (possibly derived), and one unilateral chopper.

The material used was silicified sandstone, limestone and poor quality chert.

ADDITIONAL SITES

NORTH-WESTERN SOMALILAND

MANDERA NORTH (11). From the junction of the grey alluvium with the red alluvium and above the land rubble (see Text-fig. 5) were derived a small, but typical, Stillbay series, which comprised the following artifacts in chert: two unifaced points, four disc cores, seven faceted flakes, four unfaceted long flakes, two backed blades (one in obsidian), seven diminutive blades, seven small unfaceted flakes, one *lame écaillée*, four single-platformed, diminutive, blade cores.

BORAMA (5). From the red sandy alluvium (8) (shown at Text-fig. 5) in the tug at Borama, was recovered the following Stillbay assemblage. Tools are mainly made of chert, but quartz and an occasional piece of obsidian also occurs: all are fresh and unweathered: five faceted flakes, five flake-blades, eight unfaceted primary flakes, six single- and double-edged side-scrapers on end-flakes, one redirecting flake, three small disc cores, one *lame écaillée*, one unifaced point on a thick flake and one fine, thin, pressure-flaked example, and two backed flake-blades.

DUBATO (230). Very complete surface collections were made from this area, near the edge of the north-western plateau, by both Trevor and McInnes. On a very brief visit the writer was able to establish that Stillbay artifacts were eroding from red, sandy alluvium on the upper slopes of a small tug; this alluvium was similar to that at the Hargeisa sites and it is presumed that Trevor's and McInnes' collections were derived from this deposit. Tools are made from chert and show the characteristic white patina.

A fairly full inventory of tools types is represented:

Points. Forty-five unifaced points and thirty-two bifaced specimens. These grade gradually from the triangular point with straight butt, through the more pointed, triangular form with rounded butt, to the pointed or even lanceolate forms with rounded or straight butts. Of the bifaced points seven are triangular in shape with straight butts, nineteen are subtriangular with rounded butts, five are foliate, and one is lanceolate. Of particular mention is one very finely made lanceolate, bifaced point, most carefully pressure flaked over both faces. Three small 'tear-drop' unifaced points $1\frac{1}{2}$–2 in. long with rounded butts and triangular section are of the form sometimes found with the Magosian and Hargeisan industries and are presumed to be later. The majority of the points have straight, untrimmed butts, and lenticular or plano-convex sections. Only a comparatively few of the unifaced specimens exhibit pressure flaking and the majority show a fine resolved retouch.

Scrapers—11. These comprise one true *racloir*, three side-scrapers on end-flakes, six round scrapers, and one end-scraper.

Burins. One polyhedral specimen.

Backed blades. Two small, straight-backed blades, one being a redirecting flake retouched at the point. Also one micro-blade.

Small core bifaces. Three specimens.

Faceted flakes—16. Five of these are triangular and pointed, three are large, radially flaked, broad flakes, and eight are small radially flaked and broad specimens.

Flake-blades, faceted. Seven typical specimens.

Cores—24. Thirteen are subtriangular and eleven are oval and disc cores.

Miscellaneous—2. One awl and one core chopper.

NOGAL VALLEY

DONKUKOQ (52). From the tug which yielded the Levalloisian artifacts were recovered a few flakes and other debitage attributable to the Stillbay. The specimens were *in situ* at the junction of the Lower and Upper Alluvia. They are in chert and have a somewhat glossy, light brown patina: they are quite fresh. Specimens include one retrimmed disc core, one hollow scraper on a derived flake, two side-scrapers on end-flakes, one with a straight and the other with a concave edge, seven unfaceted primary flakes, one large flake with bulb and platform purposely removed.

GAROE TUG (54). From the torrent gravels and gritty brown alluvium which overlie the Upper Tug Gravels and Alluvium was recovered a sparse collection of tools, some of which show signs of abrasion. This is in fact what might be expected as the torrent gravels and alluvium are ascribed to the First post-Pluvial wet phase, the accumulation of this deposit would thus be later than the Stillbay Culture.

Tools are in chert and show a mottled to buff patina; four specimens, besides certain derived flakes, are abraded, the remainder are fresh. Specimens include one disc core, one broken, subtriangular or subrectangular core, four unfaceted primary flakes, one broad, radially flaked, faceted flake with all-round retouch, one crude side-scraper, two flake-blades with unfaceted platforms and marginal retouch and one unifaced triangular point on a faceted flake.

THE EAST COAST. (Pl. 20, nos. 39–42)

BIEDE (57). The few tools that were found *in situ* in the detritus deposit overlying the ±12 m. raised beach at Biede are of considerable importance as they are of value in helping to date the Somaliland Stillbay in relation to changes in sea-level.

Overlying the deposits of the ±12 m. beach on either side of the Nogal estuary are certain subaerial, land-rubble and blown-sand deposits, which are on an average some 4–6 ft. thick but are deeper at the foot of the cliffs where they merge into a cemented scree. From a depth of 3 ft. within this deposit there were recovered, at a point ½ mile north of the village, one small flat disc core (no. 40), and one end-scraper (no. 41). From the cemented scree near the Biede rock-shelter were recovered one well-made, diminutive core biface (no. 39) and one broad, faceted flake, worked to a notch on one edge. This latter specimen is somewhat abraded, and may be derived but the other tools are fresh. All are in chert and exhibit a white patina.

Although we could wish for further specimens, there is, in the opinion of the writer, little doubt that on the north-east Somali coast the Stillbay Culture postdates the ±12 m. raised beach.

SOUTHERN SOMALIA AND THE OGADEN. (Pl. 18, no. 2)

DAGAHBUR (116). A small series was collected by Lieut. B. Tremain. These were eroding from a sandy loam overlying the gravels in the main tug. The specimens are fresh and of slightly glossy patina. They comprise one unifaced, subtriangular point, three unprepared flakes, one unprepared flake-blade, and one worked nodule.

JESOMMA (85). (Pl. 18, no. 2.) From the top of the upper ferruginous sands and immediately underlying an incipient ironcrete horizon with vertical pipe structure, there was found a small series of tools which are typologically more advanced than the underlying Upper Levalloisian, and which are of typical Stillbay form. From the base of the ironcrete is derived a quantity of Magosian artifacts.

The Stillbay specimens are made from local silcrete and chert: the majority are fresh, but a few show signs of slight abrasion. The series comprises ten unifaced, subtriangular points (average 2–2½ in. long), three broken fragments of foliate specimens (average length probably about 2–3 in.), six bifaced, subtriangular points (Pl. 18, no. 2) average length 2–3 in., three bifaced, foliate points (2¼–3¾ in. long). Also one backed flake, one small, flat disc core, five broad faceted flakes, two pointed flakes with plain platforms, and one flake-blade.

This industry is obviously very close to the Upper Levalloisian which underlies it.

SUMMARY

The Somaliland Stillbay represents the blending of new technical processes with the ancestral Levalloisian tradition. The primary flaking is essentially Levalloisian and type fossils are the disc and subtriangular prepared core and faceted flake. In fact if the end-products were removed from a Stillbay factory site, the assemblage might well be referred to the Upper or Developed Levalloisian.

The Levalloisian ancestry of the Stillbay can also be seen in the broad, faceted flakes with radial flaking, the small core bifaces, the flat percussion flaking and the foliate points of the Lower Stillbay from Hargeisa. A gradual development towards the bifaced subtriangular point can be traced. In the Lower Stillbay 45 per cent of the points are subtriangular, 30 per cent are foliate, 15 per cent are forms with wide, though rounded, butts, the remaining 10 per cent are of various atypical forms. In the Upper Stillbay from Hargeisa 63 per cent are subtriangular, only 16 per cent are foliate, 16 per cent have wide and rounded butts and 5 per cent are atypical. In the Bur Eibe Stillbay development has progressed still further and we find 74 per cent are subtriangular, 23 per cent have wide and rounded butts, and only 3 per cent are foliate.

The Stillbay differs from the Levalloisian in that the predominant form of flake is pointed and not oval; the subrectangular double-ended core is almost completely absent, and certain new technical forms and types of tool make their appearance. Pressure flaking supercedes flat percussion on the most finished end-products. End-scrapers, backed flakes, and a few rare burins make their appearance and the manufacture of small blades is introduced. Apart from some local specialization there appears to be little regional difference, if a general view is taken; though it is possible that the industries from Eil, Borama, and Bur Eibe may be somewhat later than those from the other sites. The main Stillbay areas are situated on the north-western plateau, and in the middle Webi Shebeli. Outside these areas it is not impossible that the Levalloisian may have persisted and is represented by certain local and degenerate industries of which those from Hararile (55), in the lower Nogal valley, and at Aror may be examples. There is reason also to believe that the double-ended, blade core of the Ogaden form of the Wilton is derived directly from the developed Upper Levalloisian which underlies it in the Tug Fafan at Gabredarre. (See p. 270.)

THE SOMALILAND MAGOSIAN CULTURE AND THE MAGOSIO/DOIAN TRANSITIONAL INDUSTRIES

The Somaliland Magosian is found in three main areas, namely, in southern Somalia (with outlying sites stretching into the central Mudugh), in the Ogaden, and in the Danakil Rift (see Distribution Map, Text-fig. 21).

The culture is well represented in southern Somalia by the sites at Bur Eibe (18) and Bur Hakaba (20), and in the Danakil Rift area at the Porc Epic Cave (152),[1] near Dire Dawa, and at Obok (150)[2] on the French Somaliland coast. Its presence in the south-western Ogaden and western Abyssinia is inferred from the collections made there by the du Bourg de Bozas Expedition in 1901–2.[3]

At the southern Somalia sites and at Porc Epic the Somaliland Magosian is stratified and can be accurately dated, but at the majority of sites yielding typologically similar industries, the artifacts usually lie upon the surface and are often mixed with later cultural material. In the Somalia sites this later material is almost invariably of Doian affinities, while in the Ogaden and in the Rift it can be ascribed to the Wilton.

The stratified sites in the south-east provide proof that the Doian Culture is directly derived from the Magosian and several transitional industries have been found which indicate that there was in this area an uninterrupted develop-

[1] Breuil, H. (1934). Teilhard de Chardin, P. (1930). Breuil, H. *et al.* (1951).
[2] Teilhard de Chardin, P. *et al.* (1940). [3] Breuil, H. & Kelley, H. (1936).

ment from the Magosian. Where these industries are found *in situ*, there can be no doubt of the transitional position that they occupy. At Porc Epic on the other hand the local Magosian would appear to evolve into an industry of Somaliland Wilton affinities, again by a process of gradual development.

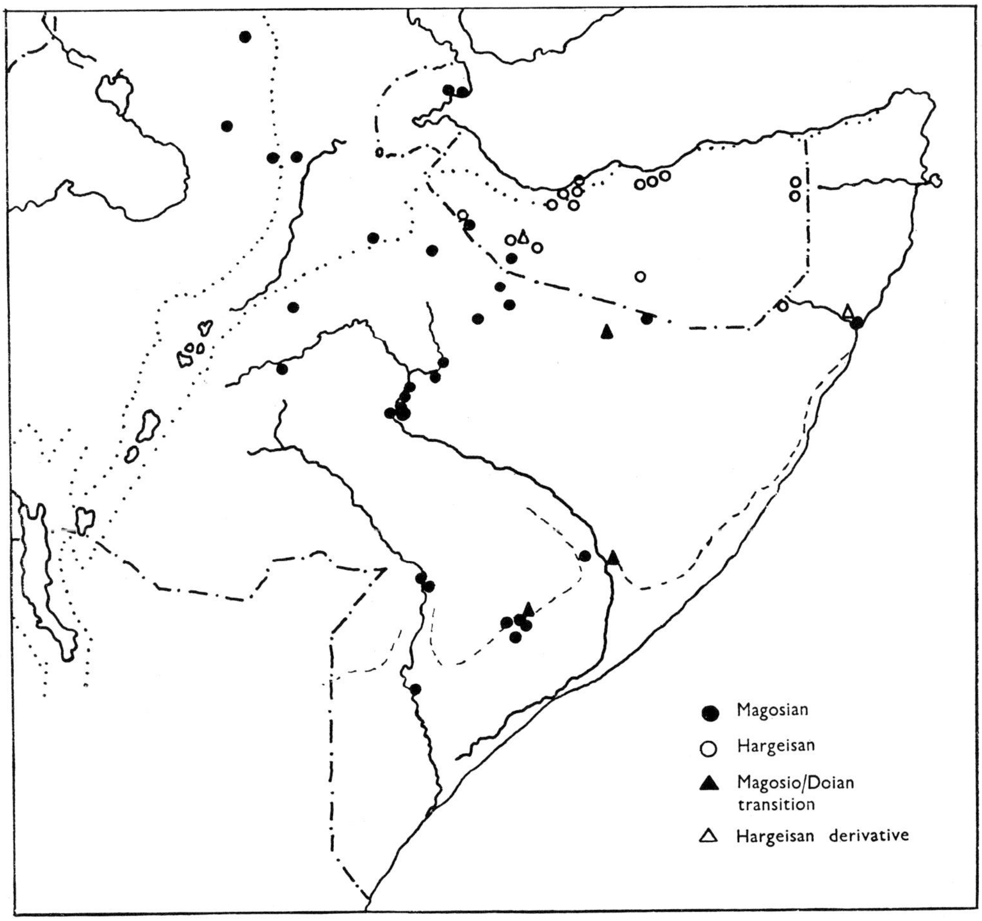

Text-fig. 21. Distribution of Magosian and Hargeisan industries.

In origin the Magosian may be considered to be directly derived from the Somaliland Stillbay which immediately underlies it at some sites (Bur Eibe, Jesomma, Bur Degis, etc.) but it is probable that some of the new cultural elements that are found with the Magosian are the results of culture contact with the intrusive Hargeisan industries (see p. 218) in the north.

Stratigraphy. The type sites for the Magosian in southern Somalia are at Bur Eibe (Bur Yassin and Gure Warbei) (18) and at Bur Hakaba (20). Other stratified sites in this area are at Bur Degis (17), and Jesomma (85) and unstratified surface stations have been identified at Sheik Gure (83), Bun Mess (95) and Lugh Ferrandi (100). Surface sites are also known from stations in the central and northern Haud (200; 10; 183) and from Eil (56) where the artifacts are found *in situ* in tufas.

The Ogaden Magosian is best represented by the sites on the Webi Shebeli (153; 157; 161; 162; 163; 164) and at Roukatcha Wells (219) and Sheik Hussein (178) on the Abyssinian High plateau. The type site for the Rift is Porc Epic (152) but some additional dating evidence for this culture at the coast is furnished by the Mount Mabla (149) and Obok sites (150). We may include with these Rift sites, two sites on the western plateau, namely at Jigjiga (4) and Jifa Ure (205).

SOUTHERN AND CENTRAL SOMALIA

Excavations were carried out at two important sites at Bur Eibe (see Sketch-map, Text-fig. 20) namely at Bur Yassin, an open station site, and at a rock-shelter, the Gure Warbei, and also at the Rifle Range site at Bur Hakaba. The stratigraphy at these sites is given below:

BUR EIBE (BUR YASSIN). On the eastern side of Bur Eibe there are a number of outlying smaller 'burs' lying on a linc running approximately north and south and separated from the main bur by a narrow but shallow valley. At the south-eastern corner opposite Berdale village are several small burs the most important of which is known as Bur Yassin. It is here that the main sites are found but the stratigraphical and cultural sequence here is confirmed for the whole area by a number of subsidiary sites situated at points up and down this valley and on the west and south sides of Bur Eibe. The sequence at these latter sites is exposed in the banks of *wars* (*uars*), or rain-water catchment dams or cisterns, which are excavated by the Eile through the deposits at the foot of the burs. At Bur Yassin the sites are grouped round the western sides of the hill and also at the south-western end of the hill immediately to the north of Bur Yassin: they cover approximately $\frac{1}{4}$ mile in area. On the eastern and northern sides not a single site was found to exist and it would seem as if camping sites were chosen with a view to protection from the north-east monsoon, a point of interest in view of the climatic conditions which persisted at this time.

The valley separating Berdale village from Bur Yassin shows a shallow profile, the bed rock being granite which rises on either flank to form the burs. On the slopes on either side of this valley, particularly on the Bur Yassin side, bed rock is overlain by surface deposits of variable thickness but rarely attaining any great depth. The centre of the valley is filled, however, by a grey, sandy, quartz alluvium and calcareous grit which in wells, and one abandoned prospecting pit showed a depth of more than 15–20 ft. without reaching bed rock.

The sequence on the west side, on the slope at the foot of Bur Yassin, was found to be as follows from the base upwards:

(i) Decomposed and exfoliated granite altered by iron oxides, overlying fresh granite rock.

(ii) Consolidated deposit of incipient ironcrete from 1 to 3 ft. thick containing at the base, numerous artifacts of the Somaliland Stillbay Culture, and in the main and upper part of the deposit quantities of artifacts and debitage of the Somaliland Magosian. This deposit is 'laminated' rather than being of a pisolitic nature and is considered to have formed as a result of seepage water and run-off from the burs during a time when the rainfall was somewhat greater than it is today; but with a climate of contrasted wet and dry seasons. The ironcrete conforms to the slope of the valley. Where this deposit has been exposed by surface erosion the artifacts are often difficult to extract owing to the hard nature of the deposit, but where not previously exposed the deposit is softer and easier to dig. The Stillbay artifacts are usually more heavily patinated, while the Magosian tools are in 'mint' condition and often show a glossy patina, which is a characteristic of artifacts of the Magosian and later cultures throughout the open station sites of Somalia.

(iii) The ironcrete passes gradually up into an immature yellow to red quartz sand, with rounded but subangular grains, stained by iron oxides and containing numerous pisolitic ironstone nodules. This ferruginous sand is compact but nowhere consolidated and there is very little evidence of bedding. On an average it is some 5 ft. thick. It is considered to represent a hill-wash, laid down in uninterrupted succession upon the underlying deposit at a time when the prevailing climatic conditions were becoming drier and the formation of ironcrete could not be sustained; though not sufficiently dry to prevent the formation of ferruginous pisolites. The industry contained within this deposit is for the most part transitional between the Magosian and the Doian, though the upper levels have also yielded true Doian artifacts. Tools are always in quite fresh condition. At one place, at the foot of Bur Yassin, the upper levels of this deposit were found to be calcareous and it would appear that some replacement of the iron oxides by calcium carbonate has occurred here.

(iv) At some points a thin fine quartz sand overlies these deposits—it is quite unconsolidated and is considered to be mainly of aeolian origin.

> This deposit is thickest on the Bur Eibe side of the valley where it passes into a fine grey dust, containing in the upper levels evidence of Eile occupation.[1]

This stratigraphical sequence is confirmed by additional sites at Bur Eibe, at sites 5, 7 and 9 (see Sketch-map, Text-fig. 20). At site 7 the sequence showed 1 ft. of rubble over 2 ft. of yellow, gritty sand with Doian artifacts and fossil bone, over 1 ft. of ironcrete with Magosian debitage, over exfoliated granite. At a section exposed in *war* diggings at the foot of Bur Eibe ½ mile north of the Gure Makeke was seen 2 ft. or more of fine grey quartz sand, over 1 ft. 6 in. of grey sandy clay, over a lenticular bed of compact grey quartz sand, over 2 ft. and more of ferruginous quartz sand with occasional transitional Doian (?) and Magosian debitage, over exfoliated granite.

In a *war* digging at the south-west side of Bur Degis, the sequence exposed showed 1 ft. to 1 ft. 6 in. of fine, grey sand with angular grains (pottery occurred to a depth of 6 in. in this deposit, and below the pottery Doian artifacts), over 1 ft. 6 in. of consolidated, grey, sandy calcareous deposit with ferruginous lenses (this deposit contained some good Magosio/Doian transitional tools, and also true Magosian artifacts), over a sparse gravel some 6 in. or more thick of well-rounded but angular quartz pebbles, filling pockets and passing gradually down into the underlying exfoliated granite rock. (From the surface of the gravel and at some places within it were recovered typical Stillbay tools.)

GURE WARBEI (BUR EIBE). For a full description of this cave and the excavations undertaken there see pp. 230–6. The Magosian industry was contained within the lowest occupation deposit occurring in this shelter (a gritty, brown earth), immediately overlying a sterile, yellow sand formed largely as a result of decomposition and disintegration of the bed rock. The thickness of the Magosian level was found to vary between approximately 1–3 ft. and contained occasional small ferruginous pisolites. It was overlain by deposits containing well-developed Magosio/Doian, and true Doian industries.

BUR HAKABA (RIFLE RANGE SITE). A full description of this site will be found on pp. 236–40. The Magosian industry occurred within the lower occupation

[1] The Eile (lit. 'dog-owners'), though they are today mixed to a large degree with Digil, Rahanwein and Somali blood, are probably the descendants of the shorter-statured negroid hunters who are thought to have been the original inhabitants of the Horn. The main group of Eile live round Bur Eibe, though there is good traditional evidence that they shared with the Galla Wardai, prior to the coming of the Elai and other Rahanwein tribes, a large part of the Benadir coastal plain. Small nuclei of Eile still exist in Dafet and on the lower and middle Webi Shebeli among the Shidle. The Eile are considered as 'bon' (free men) by the Rahanwein among whom they live, and are in several ways similar to the independent groups of Bon and Ribi who live by hunting in the country between the Webi Shebeli and Juba rivers and also in Trans-Jubaland.

level—a ferruginous black earth resting on and in part mixed with an eluvial (?) gravel of well-rounded quartz pebbles. This in turn rested upon granite rock. The Magosian artifacts occurred in an entirely fresh condition and were associated with numbers of comminuted fragments of appreciably fossilized animal bone and occasional teeth. This deposit was overlain and sealed by a compact calcareous sand, white in colour and sterile. On the surface of this white sand was found the Doian occupation level.

MERKANWEIN BALLEH: BOHOTLEH (183). The artifacts from this site were recovered by Capt. H. B. Gilliland from the spoil heaps of red sand thrown out from well diggings. This sand represents the most superficial deposit in this area. The specimens, the majority of which are quite fresh, sometimes show a glossy patina and are made of chert. The greater number represent debitage from a factory site. The faceted flakes are on the whole larger than those usually found with the Magosian, but the presence of *outils écaillés*, and flat disc cores suggests that the assemblage should be tentatively classed as Magosian rather than Stillbay. However, if the red sand at Bohotleh can be identified with that at Alablah and the other ballehs of the central Haud, as the writer believes it can, this would indicate an approximate contemporaneity of all these industries (other than derived specimens) contained within, but not merely on, this deposit. It must be remembered, however, that the accumulation of this sand must have been spread over a considerable period of time.[1]

EIL: LOWER NOGAL (56). The tufas formed by springs $\frac{1}{4}$–$\frac{1}{2}$ mile upstream on the left bank of the Nogal, from the Government Station at Eil, were found to contain numerous examples of debitage and a few implements of Magosian type. These were embedded within the tufas and were often extremely difficult to remove; they were found to occur in the upper levels, which spread down the slope of the valley to a height of 15–20 ft. below the topmost tufa beds.

THE DANAKIL RIFT

PORC EPIC CAVE. The full report on the excavations in this cave has only recently been published[2] but a preliminary note was given previously by Teilhard de Chardin[3] and the paintings have been described by the Abbé Breuil.[4]

This rock-shelter had become filled with a deposit some 1 m. 70 cm. thick consisting of ash and a fine loose earth or dust which showed no evidence of stratification throughout its thickness. Only at the base and close to the rock

[1] See remarks under Alablah Balleh, p. 217 and n. 1 and under Rakon and Mirsale, pp. 254 and 277.

[2] Breuil, H. *et al.* (1951).

[3] Teilhard de Chardin, P. (1930*a*).

[4] Breuil, H. (1934).

walls was a change noted and here a thin stalagmitic and phosphatic crust was found to occur.

The shelter contained abundant cultural material in obsidian, and a reddish chert; basalt, rhyolite and quartz were also more rarely used. My thanks are due to Professor H. V. Vallois and Mr Harper Kelley for permission to examine this material. The industry from the lowest levels in the 'ash deposit' might be termed final Somaliland Stillbay and comprises unifaced and a few bifaced points, scrapers and numerous flakes and flake-blades with faceted striking-platforms. By its, on an average, larger size it can be distinguished from the industry in the middle and upper levels. This, the main industry, is typologically an undoubted Somaliland Magosian and consists of numerous small unifaced and some bifaced points, a few short end-scrapers or large thumbnail-scrapers, scrapers on flakes worked on one or more edges, rare single-blow burins (one angle-burin is recorded), utilized blades with marginal retouch, a few denticulated blades and *perçoirs*, backed-bladelets and microliths of Magosian and Wilton form, and some diminutive bifaces which recall similar tools noted with the Somaliland Stillbay. The authors in the final report illustrate typical specimens but regrettably give no statistics or tool lists. The backed-bladelets and microliths are rare but are present throughout the deposit with the exception of the lowest layers. Both straight-backed and curved-backed forms occur but the more geometric forms are most abundant between 45 and 75 cm. from the surface. The unifaced points are the most abundant implements found and are present at all levels. They may be subtriangular in shape with the faceted striking platform either left in place or trimmed away, or more rarely pointed (willow-leaf) forms are found. Many are parti-bifacial with secondary trimming on the underface at the point or butt end. The bifacial laurel-leaf forms are rarer but again are found at all levels though the greatest number were found between the 45 and 60 cm. layers. The predominant core type is discoidal or ellipsoid yielding flakes with faceted and plain striking-platforms and often with radial flaking on the upper face. A characteristic form is the diminutive disc or discoidal core. Flake-blade cores also occur, however, and are worked often from opposing ends. The stages in the development of this type of core from the discoidal core can be clearly seen. The flake-blades and blades removed from these cores may show plain or faceted platforms. The greater number of these types especially in the upper levels would appear to indicate that they were the main forms produced. Micro-blade cores do occur but are rare. One good example of a haematite crayon or pencil with well-pronounced striae was noted from the lowest layer. A crude pottery occurred in the upper levels but the excavators were uncertain whether this should also be associated with the main stone industry. Although sherds occurred down to the middle to lower levels the excavators appear to think (a view with which the writer is also in agreement) that, owing to the very friable and loose nature of the deposit, they had been carried down by burrowing animals of which there was plenty of evidence in the shelter. No worked bone was found and what animal bone there was consisted of porcupine, pig, hyrax and antelope.

Of interest is the fragment of fossilized human jaw which was recovered by Breuil in a *zone phosphatée* forming a crust on an intrusion of the rock floor which obtruded through the lower part of the deposit at the right of the entrance. The fragment is very heavily mineralized and encased in a concretionary lime deposit which is difficult to remove. Vallois[1] considers this mandibular fragment to show characteristics closer to the Neanderthal form than to modern man. It is thus presumably the earliest human fossil yet recovered from the Horn and must be associated with an earlier deposit and industry antedating that which occurs in the powdery grey earth.

OBOK. Here Teilhard, Breuil and Wernert found that the Stillbay/Magosian artifacts overlay the 'lagoon coral', tentatively correlated by the writer with the 5–25 m. raised beach of early Gamblian age.

THE OGADEN

Little is known concerning the stratigraphy of the du Bourg de Bozas sites but from the evidence available it would appear that the Magosian was contained within deposits similar to those noted in Somalia. For example the Stillbay and Magosian artifacts from the Tug Ensiral, are derived from an *argile rouge* and from an overlying *limon plus claire*. At site 13 (161) on the left bank of the Webi Shebeli one implement shows ferruginous concretions. The tools from station 14 (162) are from an *argile rouge*, as also are those from Sheik Hussein (178). Many of the Magosian implements from Tug Ensiral, and from sites 13, 14, 15 and 16, show the characteristic glossy patina noted on the Somalia specimens.

Geological horizon. The first post-Pluvial Wet Phase.

Climate. Semi-arid.

Implements. An inventory of the tools occurring at the type sites of the Somaliland Magosian and the Magosio/Doian transitional industries is given opposite.

The commonest tools are diminutive points pressure flaked over one or both sides, and microliths of different forms. A quantity of factory debris was associated with the artifacts so that a fairly complete study of the technical

[1] Vallois, H. V. (1951).

Inventory of tool types of the Somaliland Magosian Culture and Magosio/Doian transitional industries

Tool types	Magosian			Magosio/Doian transition			Totals
	Bur Eibe (Bur Yassin)	Bur Eibe (Gure Warbei)	Bur Hakaba (Rifle Range)	Je-somma	Alablah Balleh (Root excavations)	Bur Eibe* (Gure Warbei)	
Points:							
Bifaced, subtriangular	89	2	5	7	1		104
Bifaced, foliate	12	—	1	10	2		25
Bifaced, lanceolate	(3)†	—	—	1	—		4
Unifaced, subtriangular	84	8	1	76	—		169
Unifaced, foliate	—	1	1	19	—		21
Unifaced, lanceolate	1 + (2)	—	—	—	1		4
Unifaced, asymmetric	7	—	—	23	—		30
Hollow-based	(4)	—	—	3	—		7
Scrapers:							
Side on end-flakes	6	1	2	80	—		89
End and side on end-flakes	—	1	—	2	2		5
End	1	1	—	8	4		14
Diminutive round	14	1	1	29	1		46
Thumbnail	9	—	—	—	—		9
Various	5	—	1	7	3		16
Burins:							
Single blow	7	—	—	8	—		15
Bec-de-flute	3	1	2	1	—		7
Gouge	3	—	—	—	2		5
Backed flakes	6	—	—	—	—		6
Backed blades, large	10 + (8)	—	1	22	—		41
Microliths:							
Curved backed	29	3	9	91	11		143
Obliquely backed	6	—	—	—	—		6
Straight-backed	13	—	20	79	13		125
Double-backed	7	—	1	9	—		17
Crescents	4	—	—	19	—		23
Trapezes	3	—	—	5	—		8
Heluan type	1 + (5)	—	—	—	—		6
Outils écaillés	27	4	6	17	3		57
Blades:							
Utilized	40	—	8	Not differentiated	11		59
Unutilized	48	77	109	151	Numerous		385
Flakes:							
Utilized	31	6	3	13	19		72
Unutilized	106	349	192	166	283		1096
Prepared cores:							
Disc	35	14	9	79	6		143
Subtriangular	4	—	2	6	1		13
Blade cores:							
Single and multi-platformed	19	—	12	7	11		49
Bipolar	5	—	13	9	12		39
Various	25	11	11	42	19		108
Totals	682	480	410	989	405		2966

* For list and description of tools from Layer C, see p. 241.

† Figures in brackets indicate surface specimens not found *in situ*.

processes employed was possible. The primary flaking is in many ways a continuation of that found with the Somaliland Stillbay, except that cores and flakes are now smaller, on an average, and more refined. The flake-blade core has progressed almost imperceptibly into a micro-blade core, though examples of the larger forms are still found. The bipolar blade core now makes its appearance, and together with the dimple scarred rubber and anvil stone (Pl. 22, no. 64) indicates the use of bipolar anvil technique. Together with these blade-producing cores there are still found diminutive disc cores, which yielded small triangular flakes with carefully faceted platforms. Secondary work reaches a high degree of efficiency, and is either executed by controlled percussion flaking or by pressure. The delicately made bifaced points in crystalline quartz (e.g. Pl. 22, no. 37) could only have been produced as a result of pressure flaking.

Points, bifaced. (Pl. 22, nos. 34–43, 45; Pl. 23, no. 85; Pl. 24, nos. 33–35.) The majority are small and well-made, subtriangular in form (85 per cent), and worked over the whole of both sides of the tool; they may be either long and pointed with straight butts (Pl. 22, nos. 36, 42; Pl. 24, no. 34), or short and broad (Pl. 22, nos. 35, 37–41; Pl. 23, no. 85; Pl. 24, nos. 33, 35), a few approximate to an equilateral triangle in form, and anticipate the triangular and hollow-based arrowheads of the Doian (Pl. 22, nos. 43, 45). In addition there occur a few foliate points (Pl. 22, no. 34) but these forms are rare. At Bur Eibe (Bur Yassin) were found three lanceolate points (Pl. 22, no. 56) though as these were not found *in situ* it is probable that they were derived from the overlying Doian deposit.

Points, unifaced. (Pl. 22, no. 44; Pl. 23, nos. 80–84.) These are similar in size and form to the bifaced specimens. 85 per cent are subtriangular in form and many retain the faceted platforms at the butt, showing that they were made from small triangular flakes struck from prepared cores. Secondary work may extend over the whole of the upper face (Pl. 22, no. 44) or may be confined to the edges of the tool (Pl. 23, nos. 81–84); in the latter instance often giving it a subtriangular cross-section. Rare foliate examples also occur (Pl. 23, no. 80) and represent, presumably, the ancestral form of the Doian limace point. A particular variant is the asymmetric point smaller but similar to the Stillbay asymmetric form of this tool.

Points, hollow-based. No examples of this point were found *in situ* but four specimens from the surface were collected at the Bur Yassin site. They can confidently be assigned to the Doian Culture.

Scrapers. (Pl. 22, nos. 51, 57–60; Pl. 23, nos. 86, 89, 90.) The typical Stillbay form of scraper, worked down one or both edges, or down both edges and round the end, of an end-flake, is present, as well as occasional true end-scrapers on short blades or flakes (Pl. 22, nos. 59, 60). The most characteristic type is, however, a small round or thumbnail-scraper (Pl. 22, nos. 51, 57, 58; Pl. 23, no. 89). The butt may either be left unworked (Pl. 22, no. 58; Pl. 23, no. 89), or else the working may be carried right round the edge of the tool and the butt trimmed away. These specimens are usually thin and fan-shaped, and present a sharp edge round the circumference of the tool. A few atypical hollow and core scrapers are also found.

Burins. These are rather poor specimens of the single-blow and *bec-de-flute* burin, and are typical of the burins found with the Stillbay/Magosian complex in other parts of the African Continent. Three gouge-burins must be mentioned from Bur Yassin. These are similar to the specimen illustrated at Pl. 20, no. 14; while there can be little doubt that they are primarily bipolar cores it is most probable that they were put to secondary use as burins. It is a point of considerable interest that no single Magosian site has yielded an angle-burin similar to those found with the Hargeisan Culture. The micro-burin does not occur.

Backed flakes. (Pl. 22, no. 48.) These are presumably a development from the Stillbay backed flakes which may perhaps be the prototype for the backed blades which now make their appearance.

Large backed blades. (Pl. 22, nos. 46, 47, 49.) These are confined for the most part to the Bur Eibe sites and are either large crescents with one tapered (no. 46), or sometimes eared (no. 47), end, or else large trapezes (no. 49)—the latter being the rarer form. The backing may be directed from either the under, or from both, faces of the tool.

Microliths. (Pl. 22, nos. 50, 52–54; Pl. 23, no. 87; Pl. 24, nos. 36–42.) These form one of the main groups of tool and are almost equally divided between two main types—diminutive, true blades with curved backing (Pl. 22, nos. 50; Pl. 23, no. 87) or similar blades with straight, blunted backs (Pl. 24, nos. 36–41). The backing is usually directed from the under face, and curves gently over on the upper face, but specimens also occur where the backing has been directed from both faces. Especially is this so in regard to the straight-backed microliths, when the tool is usually wedge-shaped in section. A further characteristic of the straight-backed microliths is the retouch that is sometimes found at the base of the cutting edge of the tool. This is presumably in order to facilitate hafting. A high proportion of microliths show signs of utilization in the presence of small scars on the cutting edge, and in particular by Kasouga flaking along the whole or part of the underside of this edge.

Variations on the above forms are: a few examples obliquely blunted down the upper half of one edge; double-backed microliths which form a small but distinctive type (Pl. 22, no. 52; Pl. 24, no. 42); crescents (Pl. 22, nos. 53, 54);

trapezes (not the *petit tranchet*) similar to the large specimen illustrated; and one specimen from Bur Yassin (Pl. 22, no. 50) with incipient Heluan retouch.

Outils écaillés. These tools are probably the fabricators for producing the backing on the microliths. They may be made on broken flakes, blades, or even small bipolar cores. The typical *écrasé* scars, on the latter specimens, however, preclude them from being classed merely as worked-out bipolar cores, and it is considered that their use was probably as suggested above.

Blades. (Pl. 22, no. 55.) These are thin and vary in length between $2\frac{1}{2}$ to under 1 in., and were struck from small, unprepared blade cores. A number of utilized specimens occur with working down one or both edges (Pl. 22, no. 55). Blades form 29 per cent of the total number of flakes and blades collected.

Flakes. These form 71 per cent of the flakes and blades collected showing that flake production was still the predominant technique, although specialization in the production of blades was beginning to assume importance. Flakes are derived either from flake-blade cores with one or more platforms—in which case the working on the upper face is usually parallel—or else from disc cores yielding, usually, subtriangular flakes, and, more rarely, oval flakes, with finely faceted platforms. Not more than 40 per cent, if as many, of the flakes show previous preparation in this way.

Prepared cores. (Pl. 22, nos. 62, 63; Pl. 23, no. 88; Pl. 24, no. 45.) These are typical disc cores and vary between 1–2 in. in diameter. They are either flat or high-backed and it is not improbable that they were sometimes used for some secondary purpose. In addition to the disc form there occur a few rare examples of subtriangular prepared cores similar to the specimen from Alablah Balleh figured at Pl. 21, no. 14.

Blade and flake-blade cores. (Pl. 22, no. 61; Pl. 24, no. 44.) These may be divided into two main types—single, or multi-platformed examples in which the platform may be straight, convex or circular, with as many as three platforms, usually at right-angles, present on the same core (Pl. 22, no. 61), or else with two platforms at opposite ends of the core (bipolar), either on the same face, on alternate faces, or on both faces of the core (Pl. 24, no. 44). These latter cores are always small, usually never over $1\frac{1}{2}$ in. in length and were used for the production of micro-blades.

Various. Under this group are included two rubbers, one with dimple scarring in the centre showing its use as an anvil in the production of blades by bipolar technique (Pl. 22, no. 64), one piece of polished haematite, one hammer stone, two crude lava (?) flakes, ten redirecting flakes, and one high-backed 'slug' tool from Bur Yassin. A flake from a rubber, one redirecting flake, three pieces of haematite, one piece of red schist (?) and five fragments of ostrich egg-shell from the Gure Warbei. Two hammer stones, numerous fragments of comminuted bone, two of which show signs of utilization (Pl. 24, no. 43), one fragment of specularite, five pieces of haematite, and one piece of ostrich egg-shell from Bur Hakaba.

An inventory of the tools found at additional Magosian sites will be found in Appendix B (p. 356).

From a study of these Magosian assemblages it can be seen that basically this culture is a development from the Somaliland Stillbay and the bulk of the factory debris, the points, the burins and the side- and end-scrapers are indications of this ancestry. On the other hand the impetus that appears to have resulted in an initiation of micro-blade production on a bigger scale than hitherto, and the forms of small, round and thumbnail-scraper may have been the result of contact with the true blade and burin, Hargeisan, culture on the northern part of the plateau.

When comparing the cultural inventory from each of the three main areas it will be seen that the typical tools occur throughout. The subtriangular, diminutive point, backed blades and microliths, diminutive disc cores, *outils écaillés*, faceted flakes worked into side- and end-scrapers, *bec-de-flute* burins, and small round and thumbnail-scrapers are found in each area. The only difference would appear to be that in the Ogaden, where there is a greater number of round and thumbnail-scrapers present, than in southern Somalia. We might also expect to find a few rare angle-burins in these areas, though none has as yet been recovered.

MAGOSIO/DOIAN TRANSITIONAL INDUSTRIES

The three most important of these transitional industries are found at Jesomma, at the Gure Warbei, and at Alablah Balleh.

JESOMMA

At the Jesomma site the cultural assemblage contains many forms which typologically are referable to the pure Magosian but the stratigraphical evidence did not permit of a division being made. The main site is situated approximately 1 mile down the road from the wells at Jesomma and some 300 yd. to the left (see Text-fig. 12). Here there are exposed on the slopes of a small tug, some 5 ft. or more of compact but sterile red sand the surface of which has been subjected to subsequent erosion. This red, wind-blown sand, overlies a ferruginous horizon which shows tubular, or irregular pipe structure. The surface is fairly regular but the base is uneven and passes gradually into underlying compact yellow sands. This incipient ironcrete varies in thickness but is on an average some 2 ft. thick; it would appear to represent a deposit similar to that occurring at Bur Eibe, and to have formed under similar conditions, though it is probable that ironcrete formation continued later in the neighbourhood of Jesomma Springs than it did at Bur Eibe. The Magosian derivative industry is found within and on the under surface of this ferruginous deposit and is in an entirely fresh condition. The raw material is predominantly chert derived from the Cretaceous sandstones, but tools in silcrete also occur. Immediately underlying this industry is found the Somaliland Stillbay and the Levalloisian. From the inventory of tools given above it will be seen that while the typical Magosian points, scrapers, microliths, etc., occur in quantity there are also present certain new forms which indicate a somewhat later date for this assemblage, and point to its transitional nature at the end of the Magosian occupation (Pl. 22, nos. 1–33).

The commonest forms of tool are diminutive, broad subtriangular points predominantly unifaced (Pl. 22, nos. 16–21), but bifaced points also occur (Pl. 22, nos. 15, 22). The markedly triangular form (no. 22) can be seen to have given rise to the hollow-based, bifaced or parti-bifaced point, or arrowhead (Pl. 22, nos. 23, 24). These delicately made tools are of considerable interest and are the first to be found in the Somalilands. Other new forms are the leaf-shaped and lanceolate points (Pl. 22, nos. 25, 26, 28), some with serrated edges: these tools are finely pressure flaked over one or both faces and show either a thin, lenticular or a triangular section. The same forms of microliths and backed blades occur as did with the pure Magosian (Pl. 22, nos. 1–14) but rare pressure-flaked 'rods' (Pl. 22, no. 27) are an innovation. Scrapers are of two main kinds the commonest forms being the side, or side and end, scrapers (Pl. 22, no. 32) and diminutive round scrapers similar to those from Bur Eibe. The factory debris is identical with that found with the pure Magosian and flakes and flake-blades (Pl. 22, nos. 31, 32), blades, and cores (Pl. 22, nos. 29, 30) of the usual form are all represented. Approximately one-third of the flakes recovered show carefully faceted platforms. In addition, under the term 'Various' may be listed a high-backed slug-scraper or fabricator (Pl. 22, no. 33), a rubber with dimple scar, a piece of polished haematite and the usual redirecting flakes (29) as well as nine, crude unifaced 'points' triangular in form and section, with battered edges, which may have been used as some kind of fabricator.

GURE WARBEI (BUR EIBE)

The transitional industry at the Gure Warbei shelter—which here might almost be termed Lower Doian is found stratigraphically above the Magosian and below the Doian proper. It occurs in the calcareous grey earth (altered to a sandy yellow earth under the drip line) (Text-fig. 24), immediately overlying the gritty brown earth containing the Magosian. A series of typical artifacts is illustrated at Pl. 23, nos. 47–79. For a full description and inventory of this industry see p. 242. While the pure Magosian forms persist the new types noted at Jesomma are also found, in particular the lanceolate, or limace, points and the hollow-based arrowhead.

ALABLAH BALLEH

The site at Alablah Balleh (206) in the central Haud, was discovered by Major P. E. Glover. The assemblage was recovered from pits dug for 'root excavations' in connexion with pasture control. These pits were dug about 200 yd. from the western edge of the 'balleh', through compact red sand, which is typical of the red aeolian sand noted on the east coast and over much of the interior parts of the Horn. Specimens were collected at varying depths from the surface to 59 cm. The pits were sunk to a depth of 1 m., apparently to the base of the sand which probably overlay a secondary deposit of gypsum, or calcrete, stained pink in its upper levels. With the exception of a few derived and weathered artifacts of Levalloisian type from near the base of the deposit, all the specimens collected from these excavations are entirely fresh, but occasionally show a glossy, or lustrous, patina. This patina is thought to be due to sand polishing, such as would result from the gradual accumulation of sands blown into the edge of the balleh by strong winds. It is thought that the climate was drier than it was during Magosian times and that strong winds must be considered to have accounted for the accumulation of sands.[1] No stratification within the red sand was noted by Glover. Implements occurred singly or in groups, as might be expected if they were incorporated in sands accumulated in this way. That the artifacts are contemporary with these red sands there can be no doubt. Subsequent denudation has eroded the sands from the more central parts of the 'balleh' and has rolled and weathered the artifacts contained within them.

A series of artifacts from this site has been illustrated at Pl. 21, nos. 1–14 and the complete inventory is given above. The material is chert derived predominantly from the Cretaceous sandstones. Tools include a finely made sub-triangular, bifaced point of Stillbay or Magosian type, a bifaced, foliate point,

[1] As we have seen, there is some indication at Bur Eibe for considering that the ironcrete was formed at a time of strong north-east monsoon winds, which were accompanied by a wetter, or semi-arid climate. It is not impossible therefore that some at least of these widespread red sands may have accumulated during this climatic phase but that local physiographic conditions successfully precluded the formation of ironcrete. In the main, however, these sands, with their well-rounded grains, must be considered to have accumulated under more or less arid conditions.

a unifaced, lanceolate point, straight and curved backed microliths some with basal retouch on the opposite edge, small round and end-scrapers, miniature disc cores and one well-made subtriangular specimen, gouge-burins, *outils écaillés*, flakes with faceted (13 specimens) and plain platforms (270 specimens), microblades and blade cores (circular, multi-platformed and bipolar).

With the exception of the lanceolate point the assemblage might be considered to be typical of the Somaliland Magosian, but its stratigraphical position—in the red aeolian sand which, at Jesomma and in the Bur Eibe/Bur Hakaba area, overlies the deposit containing the pure Magosian Culture—indicates that we are more probably dealing with an industry which has retained its ancestral character to a greater degree than the other transitional industries with which it is considered to be approximately contemporary.

THE HARGEISAN CULTURE AND DERIVATIVE INDUSTRIES

Stratigraphically overlying the Lower, but perhaps in part contemporary with the Upper Somaliland Stillbay at Hargeisa, Mandera, and Borama, occur industries which exhibit clear 'neanthropic'—blade and burin—affinities, but which also include certain elements which are presumably derived from the Magosio/Stillbay complex. These latter elements—occasional disc cores and rare Magosian-type points—distinguish these industries from the other blade and burin cultures of Africa—from the Kenya Aurignacian, or Capsian, and from the true Tunisian Capsian. These Somaliland industries clearly represent a local and probably hybrid form and it has been thought necessary for this reason to raise them to cultural status under the term Hargeisan Culture, after the site where characteristic flaking floors were first found *in situ*.

Sites of this culture have not yet been identified for certain outside the boundaries of British Somaliland and the Mijertein, that is to say they are confined to the northern part of the plateau and to the Gulf of Aden Rift (see Distribution Map, Text-fig. 21). It is likely, however, that they will be found to occur in the western Ogaden and in the Galla provinces of the high plateau as these industries are certainly not of autochthonous origin and are presumed to have entered the northern plateau country from the Harar region, and indeed analogous industries are known to exist on the northern parts of the Abyssinian high plateau.

At first glance it would seem that these northern Somaliland industries are but a local form of the Magosian, but a detailed study shows that in the angle-burins and end-scrapers (which are as good as any found with the true North African Capsian) we have forms which are entirely foreign to the Magosian, and clearly demonstrate that we are dealing with a distinct cultural complex.

Unfortunately the number of sites where these new industries are represented is few and in only two, perhaps three, areas are these sites stratified. The following description of this culture must therefore be considered as tentative only until further work can be carried out and additional stratified deposits are forthcoming, preferably from a cave site.

Stratigraphy. Four fairly well stratified sites exist at Hargeisa (sites H. 12 R, H. 1, H. 4 and H. 7) the most important of which is site H. 12 R. The specimens are found *in situ* and eroding from a red sandy alluvium (Deposit (6) in Text-fig. 4) some 2–3 ft. thick which overlies the land rubble containing Somaliland Stillbay artifacts. This deposit is fine and grey in colour in the upper levels. At H. 12 R artifacts are found from the base of the sandy alluvium to within a few inches of the top. At the other sites the Hargeisan industry is confined to the upper or top levels of this alluvium, and the Somaliland Stillbay is found stratigraphically below it. The tools at the Hargeisa sites H. 1, H. 3, H. 4, H. 6 and H. 7 are for the most part scattered or in isolated groups, but at the site at H. 12 R, which is situated about 50 yd. from the edge of the tributary tug, the artifacts are all contained within a small area some 20 ft. square. At site H. 7 a small flaking floor was found. At Mandera North, which with the Hargeisa sites has yielded the best series of artifacts of this culture collected by the writer, specimens have come from a similar horizon in the top levels of the red alluvial sand: at one place a small living floor was found. Important additional sites are those found at Hargeisa by Lambert, at Buran by Barrington Brown, and at Raguda Tug and Kirit by Gilliland.

Geological horizon. Decline of the dry phase marking the end of the Gamblian, and First post-Pluvial Wet Phase.

Climate. Arid to semi-arid.

Implements. The material culture consists of typical blade cores, both single- and double-ended; large blades with straight or curved, blunted backs (these are, however, rather rare); microliths, predominantly of the straight-backed variety with and without basal retouch; well-made angle, *bec-de-flute* and other burins; end-scrapers on blades, both single- and double-ended; short-end or round scrapers on blades or flakes. There also apparently occur small pointed and subtriangular unifaced points, although it is possible that some are artifacts derived from the underlying Somaliland Stillbay deposits. The inventory of tool types from the main sites is given on the following page.

Inventory of tool types of the Hargeisan Culture and derivative industries

Tool types	Hargeisan							Hargeisan derivatives		Totals
	Hargeisa H.12R	Hargeisa H.3	Hargeisa H.4	Hargeisa H.1	Hargeisa H.6	Hargeisa H.7	Mandera North	Hargeisa H.2	Eil	
Points:										
Bifaced	—	—	—	1	—	(2)	—	1	—	4
Unifaced	1+(2)	1	3	2	2	—	7+(2)	20	—	40
Bifaced hollow-based	—	—	—	—	—	—	—	4	—	4
Microliths:										
Curved-backed	3	—	—	2	1	—	1	2	—	9
Straight-backed	17	7	—	2	3	—	7	12	1	49
Crescents	—	—	1	1	1	—	1	1	—	5
Scrapers:										
End, on blades	17	5	—	5	—	2	5	2	1	37
Short end or round	9	—	—	4	1	—	2	5	4	25
Various	2	—	1	2	—	1	3	9	—	18
Large backed blades	4	1	—	—	—	—	1	—	1	7
Backed flakes	—	—	—	—	—	—	2	5	—	7
Burins:										
Angle	1	—	—	2	—	—	—	—	—	3
Bec-de-flute	—	1	—	—	—	—	—	4	2	7
Single-blow	1	—	—	—	—	—	2	—	—	3
Outils écaillés	—	—	—	1	—	1	1	—	—	3
Blades, some utilized	67	3	1	4	1	—	45	14	4	139
Flakes:										
Prepared	—	1	—	4	—	—	29	4	2	40
Unprepared	6	2	1	1	—	—	53	16	1	80
Cores:										
(Flake), prepared: discs and triangles	—	—	—	3	—	—	6	7	—	16
(Blade), unprepared single and multi-platformed	3	—	—	4	—	—	12	3	—	22
Bipolar	6	—	—	10	—	—	4	5	1	26
Ostrich egg-shell	—	20	—	—	—	—	96	—	—	116+
Potsherds	—	(1)	—	—	—	—	5	1	—	7
Various	13	2	2	7	—	1	34	10	1	70
Totals	152	44	9	55	9	7	318	125	18	737

Points, bifaced—3. (Pl. 20, no. 11.) Only one specimen was found that can with any certainty be ascribed to the Hargeisan. This is a broken pyriform or lozenge-shaped specimen recovered from the top levels of the red-grey alluvium at site H.1. Two broken subtriangular specimens from site H.7 are probably derived.

Points, unifaced—20. (Pl. 20, nos. 16, 17, 18.) These are more typically Magosian in form and are small, pointed or subtriangular in shape and in section, often with secondary trimming confined to the immediate edges of the tool. This secondary work is usually of a fine resolved nature.

Microliths—47. (Pl. 20, nos. 2–9.) The characteristic form has a straight blunted back, the working being directed from either one, or both, faces of the tool. Variants are curved and double-backed forms and occasional crescents. Microliths vary in length between 1 and 2 in. with an average length of $1\frac{1}{2}$ in.

Scrapers—59. (Pl. 20, nos. 22–25.) The most important group consists of end-scrapers on blades, these vary considerably in length and may be single (no. 25) or double (no. 23) ended. Another characteristic group comprises end-scrapers on short blades or flakes which may have been worked either at the end only or else round the end and both edges of the tool to form a kind of round scraper (nos. 22, 24). Other forms are single and double side-scrapers on end-flakes similar to those found with the Stillbay and Magosian and occasional hollow scrapers.

Large backed blades—6. (Pl. 20, no. 1.) The backing may be either straight (or even concave) or curved, and is usually worked from both sides of the tool.

Backed flakes—2. A few flakes with one blunted edge occur similar to those noted with the Somaliland Magosian: they are atypical however.

Burins—7. (Pl. 20, nos. 12–14.) These form a small but important group and comprise genuine angle-burins (concave and straight forms of oblique and transverse angle-burins) as well as good *bec-de-flute*, single-blow, and gouge-burins. The latter are made on small bipolar cores. It is of great significance that not a single micro-burin or microlith made by the micro-burin technique, has yet been found anywhere in the Somalilands. Special care was taken by the writer when examining factory debris for evidence of any by-products of the micro-burin technique, but such specimens were completely absent.

Outils écaillés—3. This tool is the characteristic fabricator.

Blades—121. (Pl. 20, no. 15.) These are in the main microlithic but some larger specimens exist. The majority show no secondary work but a few have been utilized or show secondary trimming down one or both edges (no. 15). All blades are derived from true blade cores.

Flakes: prepared—34. The majority are small, either triangular or oval in shape, with small well-prepared striking-platforms. They appear to have been struck from small disc or oval prepared cores.

Flakes: unprepared—63. Mainly primary flakes derived from blade or flake-blade cores.

Prepared flake cores—9. (Pl. 20, nos. 19, 21.) These are of the small, flat Magosian type and are predominantly discoidal in form, but one triangular example was recovered from Hargeisa (no. 19). A crude pseudo-disc core also exists (no. 21) which may, however, represent a worked-out multi-platformed unprepared core: this latter type is rare.

Unprepared blade cores—39. (Pl. 20, no. 20.) These are usually small, and are either single, or multi-platformed (the former type often being a conical core with circular or semicircular platform), together with bipolar cores.

Ostrich egg-shell—116+. For the first time ostrich egg-shell appears, perhaps as water-containers: at one site (site H. 12R) a cache of several hundred pieces was found. No fragment shows any evidence of engraving.

Pottery—6. The discovery of five sherds of poorly fired red to buff pottery under conditions which suggest their association with the Hargeisan industry at Mandera North is of considerable interest. Two of the sherds, which are undecorated, are in a red to buff ware. One of these sherds shows thick lime incrustation similar to that found on many of the Hargeisan artifacts. These sherds come from site U.25 where the living floor was discovered. The remaining three sherds are in a coarse thick ware, poorly fired, red to buff in colour on the exterior and black to grey inside. These sherds were found together near the top of the red-grey alluvium and are lime incrusted. One sherd shows part of a crude decoration round the shoulder (?) consisting of unevenly spaced stylus impressions made before firing.

Various—59. (Pl. 20, no. 10.) Under this group are included twelve redirecting flakes and one fragment of polished haematite from site H. 12 R. Twenty-three redirecting flakes, four oval flakes with Tabelbalat reduction of the bulb and platform (no. 10), and one triangular piece of rubbed graphite schist with indications of a hole pierced at one corner presumably for suspension (doubtfully associated), from site H. 1. One redirecting flake from site H. 7. Two redirecting flakes from site H. 3. Two 'fabricators' from site H. 4, and seven redirecting flakes, two fragments of marine shells (doubtfully associated), one 'pick', one semicircular trimming stone, two awls, one piece of polished haematite, and twenty unclassified flakes and other debitage from Mandera North.

ADDITIONAL SITES

A list of tool types from other sites on the northern plateau which have yielded Hargeisan assemblages will be found in Appendix C (p. 357). Mention may be made here of a home and factory site found by Lieutenant Lambert on the banks of a small tributary tug approximately ¾ mile east of Hargeisa township, on the left bank of the Marodijeh Tug. The bulk of the material collected by Lambert is factory debris consisting of long blades, and some prepared flakes. Implements include typical *bec-de-flute* and angle-burins, single- and double-ended end-scrapers on blades, a few microliths and subtriangular unifaced points of rare Magosian type. The collection was made from the surface and it is probable that some earlier Stillbay material is included. Other sites are found at Kirit, on the Raguda Tug, and in the Manjasseh area.

Barrington Brown's site at Hormo[1] (see Appendix A) has yielded a good example of what is apparently a pure Hargeisan industry, consisting of straight-backed blades and bladelets, single and double end-scrapers on blades and long utilized blades. This collection is housed in the University Museum at Cambridge but a further small collection from this site made by C. L. Collenette is in the British Museum (Natural History). No angle-burins are represented in these collections but it is obvious that they must exist and their absence merely means that they have not been collected. A further site of Hargeisan form was found by Barrington Brown at Hobat in Hegligab (Higligab) district and consisted of blade cores, utilized blades and one typical angle-burin. Specimens from Hormo and Hobat are illustrated at Pl. 21, nos. 15–33.

HARGEISAN DERIVATIVE INDUSTRIES

Two sites have yielded assemblages which typologically suggest that they are later in date than the typical Hargeisan Culture described above, and which would appear to be derivatives of this culture, and in part transitional between the Hargeisan and the 'Neolithic' facies of the Somaliland Wilton.

The first, and most important of these sites is situated at Hargeisa and is a small site (site H. 2) at which the artifacts are eroding from the upper levels of a red-grey sand. This bed is the same as that yielding the true Hargeisan industries, but typologically the artifacts from site H. 2 must be assigned to an intermediate place between the Hargeisan and the Somaliland Wilton.

As can be seen from the tool lists the same forms occur as were found with the Hargeisan (though in somewhat different percentages) but in addition the small unifaced points are now present in greater numbers, and there occur also hollow-based arrowheads worked on both faces of the tool, and it would seem that we are dealing with a development towards the Wilton industries of 'Neolithic' facies that are found in the Mudugh plains of central Somalia.

The second site is at Eil approximately ¾ mile from the estuary mouth where a few artifacts were found eroding from the surface of the lowest terrace deposit, a red sandy alluvium which was accumulated during the First post-Pluvial Wet Phase. These tools are therefore presumed to postdate the aggradation of this alluvium.

[1] Burkitt, M. C. & Barrington Brown, C. (1931).

Points, unifaced—20. (Pl. 20, no. 30.) These specimens are nearly all small—under 2 in. in length, and show careful pressure technique. They are characteristic of the Magosian type of point.

Points, bifaced—1. (Pl. 20, no. 26.) One triangular specimen was found and it can be seen how this form gave rise to the hollow-based forms which are clearly derived from it.

Points, hollow-based—4. (Pl. 20, nos. 27–29.) Four of these points were found, two represent the butt ends of points and the other two were nearly complete specimens. The simplest form is where a large flake has been removed from the centre of the base of a triangular point. The base was then trimmed until it was slightly concave as in no. 27. Accentuation of the concave base produced a form of which no. 28 is an example: two specimens of this type were found. The most developed specimen shows incipient barbs (no. 29) and is finely made, with lenticular section.

It is of interest to note that while the hollow-based point or arrowhead is not uncommon in the later industries in the Horn, not a single instance of a tanged arrowhead has yet been recovered and it is believed that this type is entirely foreign to our area. The hollow-based arrowhead on the other hand may be of autochthonous growth and be derived, by way of the triangular point, from the Stillbay subtriangular or leaf-shaped specimens.

Microliths—16. (Pl. 20, nos. 31, 32.) The typical specimens are straight backed with the backing directed from one or both sides as with the Hargeisan proper.

Scrapers—21. (Pl. 20, nos. 34, 35.) These comprise typical end of blade (no. 34), short end, and diminutive round scrapers (no. 35). In addition the single and double side-scraper on end-flakes and one hollow scraper occur.

Backed flakes—5. Atypical.

Burins—6. (Pl. 20, nos. 36, 37.) While no angle-burins were found, this assemblage is clearly consistent with their presence here. Good *bec-de-flute* burins, both single and double, are found and no. 37 is entirely typical.

Blades—18. (Pl. 20, no. 33.) Typical and mainly microlithic though some large examples occur.

Flakes: prepared—6. Represented.

Flakes: unprepared—17. Primary flakes only.

Cores: prepared—7. Seven disc cores exist—all are flattish and of Magosian type with one exception which is oval and high-backed.

Cores, blade—9. Typical.

Pottery—1. (Pl. 20, no. 38.) One small rim sherd in a light brown to buff well-fired paste was found in circumstances suggesting its association with this assemblage. The specimen is heavily coated in one place with concretionary lime, as are several of the artifacts.[1] The specimen shows a simple rounded rim and would appear to have formed part of a shallow bowl. There are signs of two impressions on the outer surface immediately below the rim such as might have been made by impressing a stick on the clay and dragging it sideways.

Various—11. Under miscellaneous tools may be included one piece of haematite, four serrated flakes, four redirecting flakes, one small crude pick-like tool from the Hargeisan site, and one plano-convex trimming stone, or core scraper, from Eil.

On account of the intermediate position that these two assemblages occupy between the Hargeisan proper and the later 'Neolithic' variants of the Somaliland Wilton, they are of considerable interest particularly in regard to the appearance of the hollow-based point or arrowhead. These industries would therefore seem to be contemporary with the Magosio/Doian transitional industries of southern Somalia, and may be tentatively dated to the end of the First post-Pluvial Wet Phase.

[1] It is of interest to note that the Somaliland Wilton tools in no instances show this concretionary lime deposit, which would indicate that the climatic conditions which gave rise to its formation had ceased to exist by Somaliland Wilton times.

CHAPTER 11

THE DOIAN CULTURE

I. INTRODUCTION

In the southern and central parts of the Horn, the Somaliland Magosian Culture develops, as has been shown, through transitional stages into a local, rather specialized series of industries. Localization is more apparent in the Late Stone Age cultures than at any preceding time and necessitates the employment of new regional terms. The name Doian[1] has, therefore, been applied to those industries of the south and centre which are considered to have some special cohesion. This name is derived from the Somali word *doi* which is used to describe the flat, scrub-covered, red sand country of southern Somalia in which so many of the factory and living sites of these Late Stone Age hunting and food-gathering peoples have been found.

The distribution of the Doian can be seen from the Map (Text-fig. 22) to be focused on the southern plateau and coastal plain between the Webi Shebeli and Juba rivers. The culture is also well represented in the southern Haud, but thins out rapidly towards the north. No typical Doian tools have been found within British Somaliland, and only a very few mixed sites exist in the southern Mijertein. As yet no detailed work has been done in the more eastern parts of the Benadir plain or in Trans-Jubaland (between the Juba and Tana rivers) but the writer believes that Doian sites will be found in these areas also. In the west the Doian undoubtedly stretches into the Ogaden, but here again, as in the north, its place appears to be taken by a regional form of Wilton. It is just possible that a variant of the Doian may be found to exist in the Danakil Rift, but from the scanty evidence so far available from that area it is more likely that the Magosian here developed into a form of Wilton B. The distribution of the Doian with its characteristic pressure-flaked points would seem to be fairly well defined being confined within those parts of the Horn bounded by the Northern Frontier District of Kenya in the south, the Ogaden in the west and

[1] This term was applied to these Late Stone Age industries by the writer before seeing Graziosi's final report on his work in Somalia (Graziosi, P. (1940)). Graziosi has used the term 'Eibian' to describe the industry found by him in the Gure Makeke at Bur Eibe. The writer had considered using this name but rejected it on account of the rather specialized or localized nature of the Bur Eibe industry, as it was thought that a more general term, which would embrace all the variants of this culture would be preferable. The term 'Doian' in preference to 'Eibian' has been retained therefore.

the northern Haud and Mijertein in the north. In origin it is a direct derivative of the Somaliland Magosian, but certain elements of its material culture, notably the high degree of pressure flaking and the hollow-based arrow-

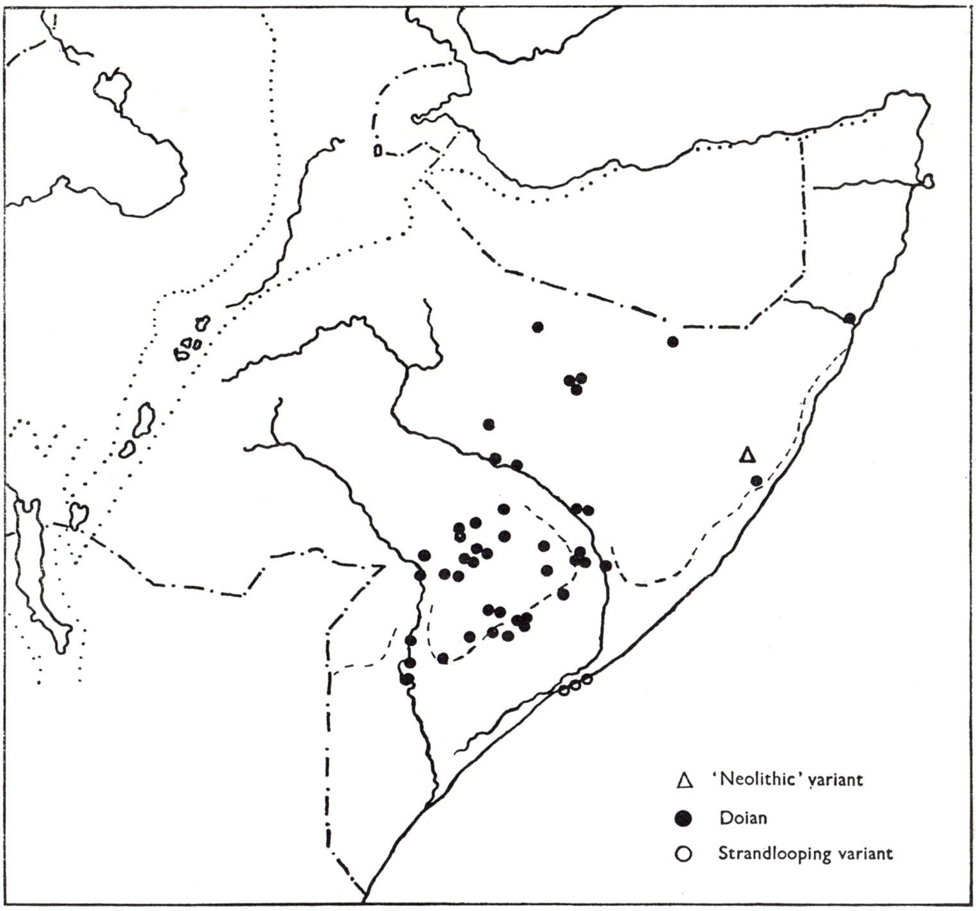

Text-fig. 22. Distribution of Doian industries.

heads, almost certainly owe their origin to extraneous influences which were possibly introduced by way of the Harar plateau from the south-eastern Sahara.

The most important sites are those at Bur Eibe, notably the Gure Warbei, and the Rifle Range site, at Bur Hakaba. The excavations at both these sites are described in detail below. In addition to the typical form of this culture as represented at these two sites, there are a few coastal sites where a strand-looping form existed, and one site which has yielded an industry in which the stone technique shows more marked 'neolithic' characteristics.[1] These variants have been described separately. (See pp. 252, 254.)

Stratigraphy. The stratigraphical position of the Doian is fairly well established. A large number of surface sites exist as might be expected—God'dere (35), Domo (62), Sijjia (41), Wejit (28) and El Dubbo (40) are examples, but in addition the climatic and geological horizon is established by several well-stratified sites correlated below. The more important of these are the Gure Warbei (18), Bur Hakaba (20), Bur Degis (17) (typical of the rock-shelter sites), Bulo Burti (86), Mirsale (45), Duldir Dune (82), Far Abou Dunes (33) between Ato (36) and Yet and Gawan Wein (22) (typical of the open station sites in the red sand areas); Ted (38),[2] Morogavi (31) and Burdo (39), for the 'sahas' or diminutive lake areas, and Mogadishu (89), Gesira (87) and Danane (88) for the coastal dune sites. The stratigraphy at the main sites is correlated below.

It will be seen that correlation between each of these areas is close. At the open station sites the Somaliland Magosian is separated from the Doian by wind-deposited sands, and similarly deposited sands overlie the latter culture. The major break in the rock-shelter deposits also occurs at the end of Magosian times. The succession at the inland open station sites such as Bulo Burti and those on the coastal dunes is identical, and in both cases the deposits containing the Late Stone Age rest on the Upper Pleistocene beds.

[1] The word 'neolithic' is used in inverted commas to emphasize that no implication is intended of a neolithic mode of existence (that is, permanent settlements, domestication of animals other than the dog, etc.).

[2] The sequence at Ted is typical of that seen in almost all the southern plateau lakelets. Ted is a depression sunk some 50 ft. or more below the plateau surface. The bottom of this depression must have been filled by a small lake about 1 mile long and $\frac{1}{4}$ mile wide: today it still contains water all the year round except during times of drought. Here three former lake levels can be seen: (*a*) a well-marked beach ± 4 ft. above the existing level, on which are three circles of stones (Galla hut circles?) but no Late Stone Age material; (*b*) an ill-defined level ± 7 ft. above the existing 'lake' level down to which the minor erosion of certain earlier tufa deposits has been graded; and (*c*) tufas, representing the highest lake level. These start at ± 12 ft. above present level and extend up the slope of the depression to some 20–30 ft. The tufas are equated with the El Dubbo and Saha Gurral tufas and yielded one primary flake of Levalloisian affinities; they are presumed to be of Upper Pleistocene Age. The ± 7 ft. level—to which erosion of the tufas has in most cases been graded—is presumed to represent the lake level during the first of the post-pluvial minor wet phases, though this requires substantiation. Lying on the eroded tufa slopes and on the ± 7 ft. level are numerous Doian artifacts and factory debris. Much of the Doian material is abraded, but whether this is the result of eluvial action following from long exposure on the surface or whether it represents contemporaneity of the Doian and ± 7 ft. level is not certain. The first alternative is considered to be the most likely, however, as the ± 7 ft. level is a degradation terrace. The lowest lake level at $\pm$ 4ft., which is very well defined, is therefore either contemporary with the Doian, the most likely possibility, or else is later.

Correlation table of Doian stratified sites in southern Somalia

PERIOD	CLIMATE	GURE WARBEI (BUR EIBE)	BUR HAKABA (Rifle Range)	BUR DEGIS ('War' excavations)	BULO BURTI
Present phase	Drier	Loose yellow-brown sandy earth	Rahanwein and Eile occupation pits Loose yellow-brown sandy earth	Fine grey sand and dust	Minor erosion Loose red-orange sand
Second post-Pluvial minor Wet Phase	Wetter	Grey earth passing into calcareous grey earth	Calcareous grey earth	Calcareous grey sand and grit with some clay	Land surface Minor erosion
Post-Pluvial Dry Phase	Drier		Compact white sand		Compact red sand ↓
First post-Pluvial minor Wet Phase	Wetter	Gritty brown earth	Ferruginous black earth	Ferruginous sand and ironcrete lenses	↑ Red bed ferruginized in places
		Yellow sand: decomposed granite	Eluvial gravel: decomposed granite	Eluvial gravel: decomposed granite	Buff-grey calcareous alluvium

PERIOD	CLIMATE	DULDIR DUNE GAWAN WEIN	MOGADISHU; DANANE; GESIRA	TED	CULTURE
Present phase	Drier	Minor erosion Loose red sand	Minor erosion Loose orange sand	Existing lake level	
Second post-Pluvial minor Wet Phase	Wetter	Land surface	Land surface Minor erosion	Lake level ±4 ft.	Doian
Post-Pluvial Dry Phase	Drier	Compact red sand	Compact red sand	Fall in lake level	Transitional industries
First post-Pluvial minor Wet Phase	Wetter	Red land surface	Red land surface Kunkar	Lake level ±7 ft.	Somaliland Magosian
		Consolidated gypsum dune	Consolidated white calcareous dune sands	Erosion of tufas: ±12 ft. Tufa deposits	

Geological horizon. Post-Pluvial dry phase and Second post-Pluvial Wet Phase.

Climate. Arid to semi-arid.

II. THE TYPE DOIAN INDUSTRIES

THE STRATIGRAPHY OF THE TYPE SITES

The type Doian as shown by the Bur Eibe and Bur Hakaba sites is considered first. The strandlooping and 'Neolithic' variants are described separately.

THE GURE WARBEI: BUR EIBE

The Gure Warbei[1] is a rock-shelter situated at the foot of Bur Eibe on its east side, about ¼ mile north of Berdale village. (See Sketch-map, Text-fig. 20.) The shelter is formed by a massive, overhanging granite rock supported at the back at the south end by a smaller boulder, and although it faces east it is protected from the weather by the outlying burs on the opposite side of the valley. The centre part of the shelter at the back is open, but the lower part of this opening has been blocked by several smaller boulders from the scree at the foot of the bur. The shelter thus formed is now some 60 ft. wide, 30 ft. deep, and 12 ft. high at the drip-line, but the roof slopes fairly regularly back to the rear part of the shelter where it is only a foot or so above the present floor level. The shelter is presumed to be in the first instance the result of negative spheroidal weathering of the granite, the formation of which long antedates the earliest occupation of the shelter. During the rains, water running off the bur collects in the scree at its foot and from there percolates through the opening at the back of the shelter and saturates the lower deposits (i.e. layers D and E).

Excavations were carried out at this shelter by the writer from 13 to 24 November 1944, inclusive. With the limited time available it was not possible to do more than excavate the two areas running on lines from front to rear and from side to side of the shelter (see Ground Plan, Text-fig. 23). These sections were known as T and M sections respectively. T trench was situated just within the drip-line on the right but owing to lack of time only the front part of this trench was excavated. M trench lay immediately in front of the first section under the drip-line and extended towards the left side of the shelter.

The deposits in the shelter slope very gently down to meet the floor of the valley and there is no scree or talus in front of the cave. Small shrubs and

[1] *Gure* = cave; *Warbei* = arrow poison (a species of *Acocanthera*), i.e. 'The cave of the arrow poison'; so called as it is reputed to have been the cave of Geda Babo the traditional hero of the Eile renowned for his prowess with the bow and arrow. The Eile are said to have brought their arrows to Geda Babo to this cave where he applied the poison and bound it on to the foreshaft of the arrowhead with gut.

trees grow in front of the shelter and effectively mask it from the path running round the bur. The shelter itself is perfectly dry during the rains. There are no paintings nor is there any evidence that such ever existed.

T section. This was subdivided into four sections, A–D, and dug in 2 in. layers down to the 1 ft. level after which it was dug in 6 in. layers except where change of deposit dictated otherwise.

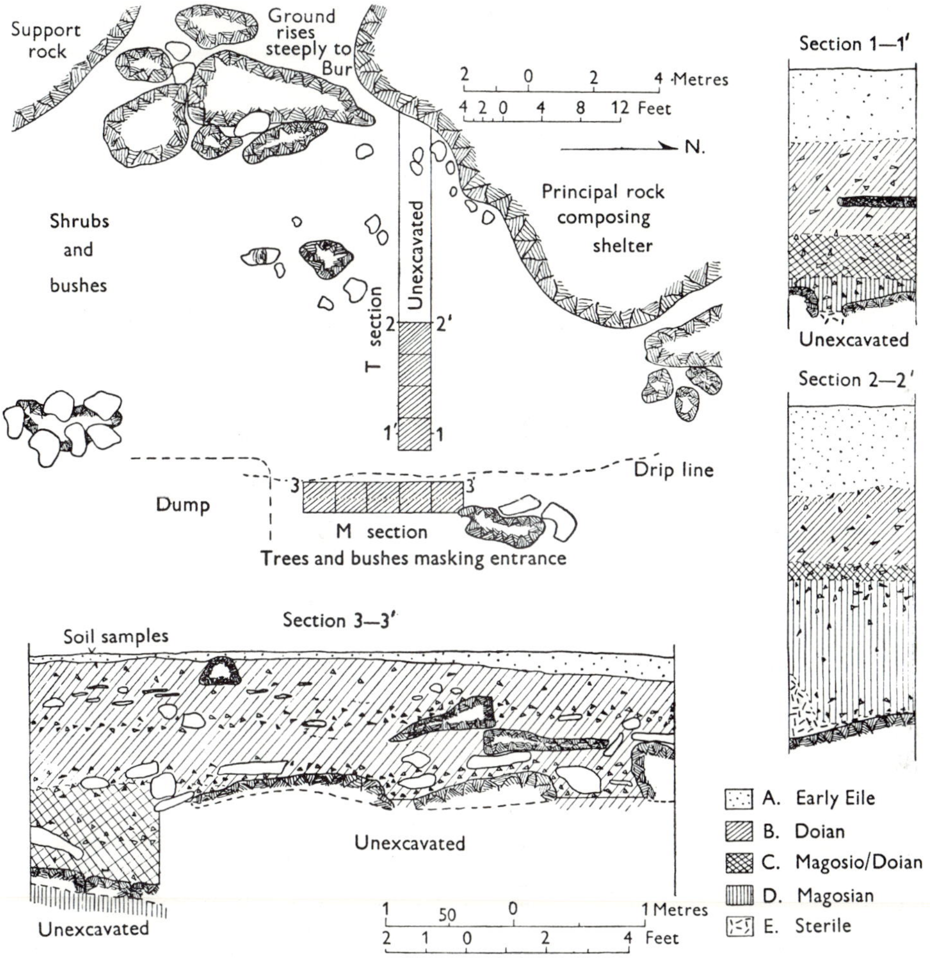

Text-fig. 23. Plan and section, Gure Warbei, Bur Eibe.

A number of nuts and sticks lay on the surface of the shelter together with two stones showing grinding surfaces. These rested on a fine yellow to grey sandy dust so loose that it was only with difficulty that it was possible to get it to stand in a vertical section. Outside the drip-line the greater part of this deposit had been eroded and that part of it which remained had been compacted into a yellowish grey earth. Down to the 3 in. level this deposit (layer A) contained many potsherds the majority of which could have come from pots which are similar to modern Eile pots, but there were in addition a few sherds, for the most part weathered and with rounded edges, from thin, burnished pots made from a fine dark brown paste. The sherds were very small and it was not possible to determine accurately the form of the pots from which they were derived; no rim sherds from this pottery were discovered.[1] Some indeterminate stone factory debris, a little unfossilized and some derived fragments of fossil bone were associated with the pottery. At a depth of 4 in. the earth became a little more compact and contained sherds similar to those above, as well as charcoal, mollusc fragments and some whole shells of *Achatina* and ? *Ampularia* sp.; bone was partially fossilized, and three curved-backed microliths were found with the factory debris. At this level there appeared in the left half of subsection C a shallow oval pit, some 2 ft. wide along the long axis and 1 ft. 3 in. along the short axis. The pit was some 5–6 in. deep with gently sloping sides and contained much ash; its purpose is unknown and it contained little of interest. Layer A continued to a depth of from 1 ft. 9 in. at the front to 2 ft. 3 in. at the rear of the trench when the yellow colour of the soil changed very gradually to a grey, sandy earth of a somewhat more compact nature. Some lamination of layer A could be discerned which would appear to be the result of alternating layers of wash entering through the opening at the rear of the shelter and wind-accumulated dust. The absence of any quantity of occupation material with the exception of the potsherds in the upper levels, and the presence of small mammal bones and whole mollusc shells is considered to indicate that the shelter was only sporadically occupied at this time. In the lower half of layer A, between 1–2 ft.,

[1] The true Somali as distinct from the Sab tribes use little or no pottery, due presumably to the rarity of suitable raw materials and the need for less easily broken containers for a pastoral life. The production of pottery is confined to the coast (i.e. Danane and one or two other places in the Merca area) where tradition has it that the art was first imported by the Arabs, to the negroid Gosha of the lower Juba valley, probably to some of the riverine negroid tribes on the Webi Shebeli, and to the Eile at Bur Eibe. Some of the Rahanwein tribes, notably the Elai and Helleda, also manufacture pottery but the raw material is all derived from Bur Eibe. The writer found an Eile at Wejit, who transported all his pot clay from Bur Eibe about 100 miles away. The 'clay' at Bur Eibe is produced by pounding the red-yellow sand (from which the bur derives its name) in a mortar, and mixing this with old potsherds, until the necessary degree of fineness is obtained. The actual pot making is done by the Eile men on a wooden 'table' or proto-wheel which is turned by the feet. It is unusual to find men undertaking this work, particularly as at Danane on the coast the pottery was observed by the writer to be made by women following the more usual practice. The Eile pots are of the usual kinds found in southern Somalia and have a very wide distribution; no unusual forms were noted here. The standard of workmanship was high, and the walls of all except the larger pots were kept thin. Firing is done exclusively by women among the Eile. Revoil noted pottery made by the Midgan in the northern Mijertein. (Revoil, G. (1882, p. 231).)

pottery became increasingly more rare until by the time the base of the deposit was reached it had completely disappeared, except for one sherd from layer B 1. Factory debris increased in quantity in layer A 2 (1 ft. to 2 ft. 3 in.) and yielded amongst other tools two lanceolate, bifaced points, a few microlithic curved-backed blades, a thumbnail-scraper, as well as two ostrich egg-shell beads, and two further pieces of pierced shell. The bone was now partially fossilized and was often covered by an incrustation of lime as were also some of the artifacts.

Layer B became consistently more calcareous towards the base, the greater part of the bone recovered was now fossilized and covered by lime incrustation. The factory debris and implements became more numerous and while there were not many artifacts sufficient were obtained to show that layer B represents the main Doian occupation horizon. In the front half of the trench there occurred at a depth of approximately 3 ft. 4 in. a large flat horizontal slab of granite. Other but smaller slabs of rock were also found to occur at this level and it would seem that they had become detached from the roof near the entrance presumably during a period of somewhat higher seasonal rainfall than exists today. Numerous broken fragments of *Achatina* shell from fully mature specimens occurred throughout.

At a depth varying from between 4 ft. 6 in. at the rear and 5 ft. 6 in. at the front of the section the friable, grey earth became more compact and more highly calcareous with small pisolites of 'kunkar'. The change from the grey earth to the grey, calcareous earth is fairly gradual, and in fact the latter probably represents the base of the former. The two have been separated and described as layer B and layer C mainly on the change in the material culture, which shows certain variations from the Doian proper, and represents the transitional phase between the Doian and the Magosian. At the Gure Warbei this industry might in fact be termed Lower Doian. The calcareous grey earth (layer C) is thickest towards the front of the shelter where it attains a depth of 1 ft. 3 in. but this thins out and slopes upwards towards the rear where it is only some 3 or 4 in. thick. Bone is now entirely fossilized.

Layer C rests on a gritty, reddish brown, sandy earth containing numerous small fragments of granite (layer D). The change from C to D is well marked. The latter deposit contained a Somaliland Magosian industry. Bone is present in the upper levels only but not in as great a quantity as in the overlying deposits; lime incrustation is absent and the bone has been stained black by iron oxides. This deposit becomes increasingly more 'gritty' towards the base where a few pisolites of iron oxide occur and there is a suggestion of consolidation into an incipient ironcrete. Layer D slopes upwards from front to rear, and is thicker at the rear being some 6 in. thick at the front and 2 ft. 6 in. thick at the back. This latter thickness was not due to occupation material as after the upper 9 in. to 1 ft. the deposit at this end of the trench became much finer and was found to be sterile. It was perfectly clear that we were dealing with sand and silt washed into the shelter from the opening at the back. The laminated bedding provided confirmation. However, immediately overlying the sterile yellow sand and bed rock, under this rain-wash was found a further thin, occupation deposit with Magosian tools, bone, and charcoal; the two deposits merged at the front end of the trench. Boulders and fallen slabs of granite were also found to be present in this deposit. Layer D and the yellow sand (layer E) beneath it were found to be damp, and it would seem that they are kept almost perpetually so by 'run-off' water from the foot of the bur entering the shelter by the opening at the back.

Beneath the gritty Magosian deposit was found an immature yellow sand which in turn rested upon granite rock. This deposit was thicker towards the front end of the shelter attaining a depth of 1 ft. 6 in. in the part excavated. It comprised yellow quartz sand and decomposed granite and is considered to represent an accumulation resulting from rain-wash and the decomposition of the underlying bed rock. It was quite sterile except for a few boulders. Bed rock was reached at 8 ft. and was found to slope gently downwards from back to front of the shelter. Soil samples were taken from top to bottom of the deposit 3 ft. from the rear end of the trench. The section exposed in the right side of T trench can be seen in Text-fig. 24.

M section. This was situated 3 ft. to the front of T trench, in the drip-line along the width of the shelter. The greater part of layer A had been removed under the drip-line and only the lower levels remained. These were not more than 3–4 in. thick on an average and contained numerous sherds of pottery similar to that from T trench, except that almost all the specimens were weathered, with well-rounded edges. Layer B which was divided into B 1 and B 2 was here found to consist of a gritty black-brown earth, compact and containing numerous decomposed or partially decomposed fragments of granite. These are flat flakes or scalings from the roof and lie horizontally in the deposit. The deposit was damp when excavated. Pottery was present in the upper 9 in., but was commonest in the topmost 4 in.; those few sherds found below the 6 in. level are considered to have fallen from above down root or rodent holes. Bone, mollusc shell and ostrich egg-shell are practically non-existent having presumably been leached out. At a depth of 1 ft. were found several longish blocks or boulders of granite which had fallen during the accumulation of layer B. From the top of layer B to a depth of 1 ft. 6 in. (B 1) the deposit contains a prolific factory debris the characteristic tools being many lanceolate and limace points, both bifaced and unifaced. Below the 1 ft. 6 in. level the bifaced lanceolate and limace points become rare and the commonest types of tool are the microlithic backed blade and the unifaced lanceolate and limace points: well-made scrapers are conspicuous by their rarity. Between 2 ft. and 2 ft. 6 in. there is a marked falling off in the number of tools and in the amount of the factory debris though the latter continues down to the 3 ft. 6 in. level. Below 2 ft. cores are more common and the single-platformed and bipolar blade core is more noticeable. At the 3 ft. level the deposit changes to a friable, nodular granite soil which is sandy, but devoid of the rock scalings common above.

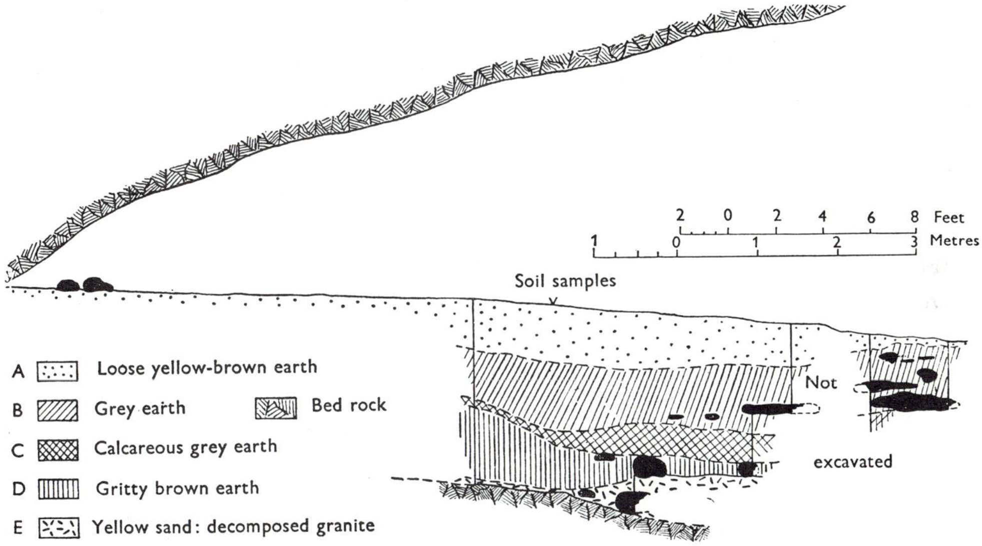

Text-fig. 24. Sections, Gure Warbei, Bur Eibe.

Large flat slabs of rock lying on the floor at this depth made excavation difficult and it was found impossible to move them in the majority of instances without widening the trench.

At the left-hand end of the trench excavation was continued to a depth of 4 ft. yielding very few tools, but at the 4 ft. level (layer C) was encountered an occupation floor yielding quantities of transitional or Lower Doian tools and debris. This floor continued for a depth of 9–12 in. Below this the deposit was again practically sterile for 6 in. when fallen granite blocks made further excavation impossible. Soil samples were taken from top to bottom of the deposits.

The deposits in the Gure Warbei can be interpreted as representing first, a long, rather dry period continuing to the present day (layer A) when the cave was only sporadically occupied, probably by the Eile[1] (judging from the pottery) though the similarity between sherds of the fine, thin ware with sherds from Wilton B sites in Kenya and southern Abyssinia must not be forgotten. Layer B, subdivisible into B 1 and B 2 shows evidence of wetter conditions. The Doian industry which was contemporary with this period can be subdivided into a later phase—when pressure-flaked bifaced and limace points were the predominant tool—and an earlier phase, when the unifaced limace point and the backed blade predominated. Layer C, which on the evidence of the 'kunkar' formation is considered to indicate somewhat drier conditions, contains the transitional Magosio/Doian, or Lower Doian, industry. The main break in the occupation deposits came between the layers C and D. The latter containing the Magosian is presumed to have formed under wetter conditions than exist in those parts of the Horn today. An exactly similar series of deposits was found by Graziosi[2] in the Gure Makeke nearby.

Photographs, plan and section-drawings of the deposits in the Gure Warbei will be found in Text-figs. 23, 24 and Pl. 44.

THE RIFLE RANGE SITE: BUR HAKABA.

(See Text-figs. 25, 26 and Pl. 45)

This site is situated at the foot of Bur Hakaba in a small embayment at the north end of the bur. This embayment is some 200 yd. wide and is protected on all except the north side by spurs of rock which project from the bur for a distance of 100 yd. (see sketch-plan). Within this sheltered bay the ground slopes gradually from the scree at the foot of the bur down to the level of the plain. The soil is here a white to grey quartz sand with some finer material. There is record that a small Elai family homestead was at one time situated in this bay, but it had disappeared before the coming of the Italian administration in 1913. During the early years of the Italian administration a rifle range was constructed in the bay and a trench some 70 yd. long, 5 yd. wide and 5 ft. deep was excavated at the foot of the 'bur'. The soil from this trench was used to form the bank for the butts. The bank is now about 3 ft. high and the ditch about 3 ft. deep. This excavation passed through several deposits containing occupation material. At the left-hand end of the butts nearest the observation

[1] Until recently the Eile certainly used rock-shelters as dwellings. Two such are situated at the south-east side of the 'bur' close to Berdale village. The floors are strewn with occupation material and ash, and the walls and roofs blackened by smoke. On the arrival of the Digil (Jiddu) in southern Somalia (about the end of the fourteenth century according to Puccioni) they found a people called the 'Loo Medo' inhabiting the 'Bur country'. These people were almost certainly negroids and not Galla, and probably survive today as the Eile, Bon and Ribi.

[2] Graziosi, P. (1940).

pit, the excavation reached to bed rock, which was only 4 ft. or so from the surface, but in the right-hand half the trench penetrated only as far as layer E or in some parts stopped in layer C. An examination of the spoil from the trench resulted in the discovery of numerous partially and wholly fossilized fragments

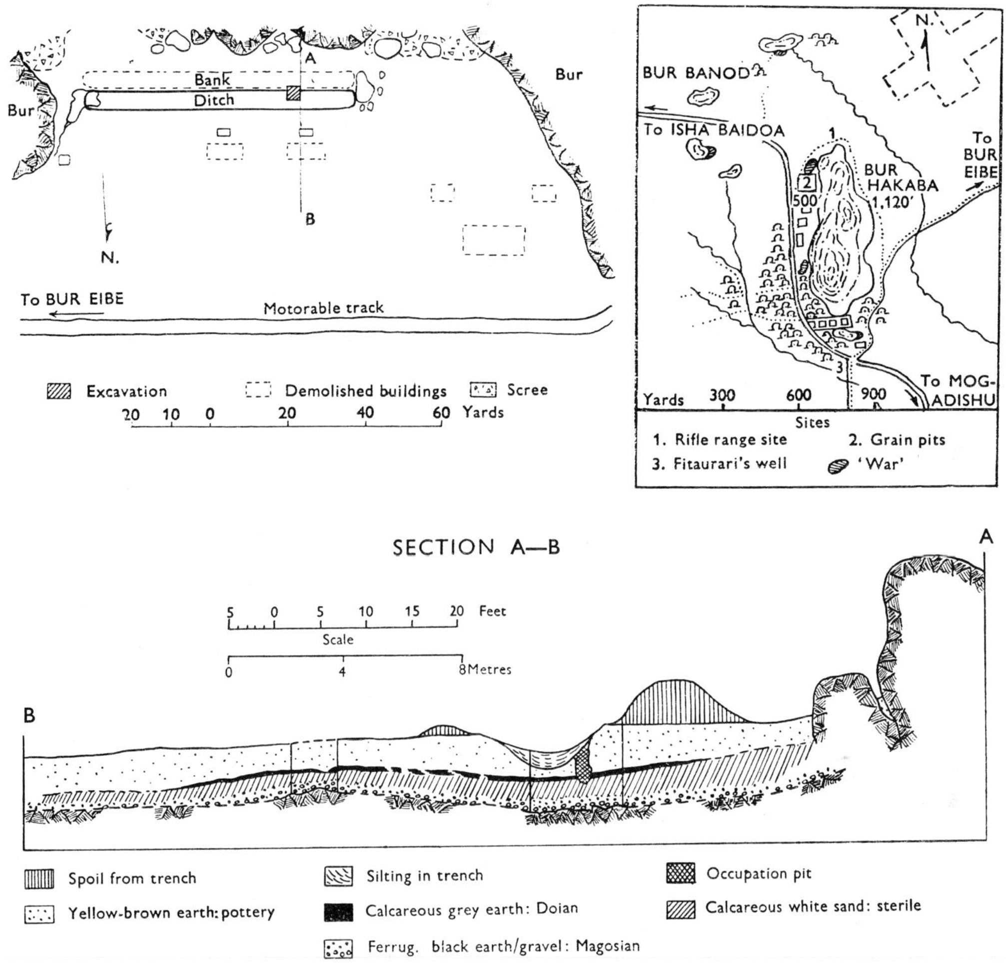

Text-fig. 25. Sketch and ground-plans and section, Rifle Range site, Bur Hakaba.

of bone, all covered by a coating of lime as well as numbers of flakes and other factory debris and a few microliths. In addition to the fossil animal bone, however, there were found a number of fragments of fossil human bone, these comprised pieces of skull, including parts of maxilla, fragments of mandible, and of long bones, vertebrae, ribs and phalanges. The majority of the human bones showed fracture in recent times and it became evident that the excavation of the trench had disturbed at least one burial which on account of the fossilized nature of the bone must have been of a reasonable antiquity. It was with a view to determining the age and association of the human skeletal material and the possibility of recovering further remains that excavation was undertaken at this site during April and May 1945.

The area chosen was, as far as could be ascertained, at the centre of that part of the bank where the greatest number of fragments of human bone had been recovered; this was situated approximately a quarter of the way along the trench from the right-hand end.

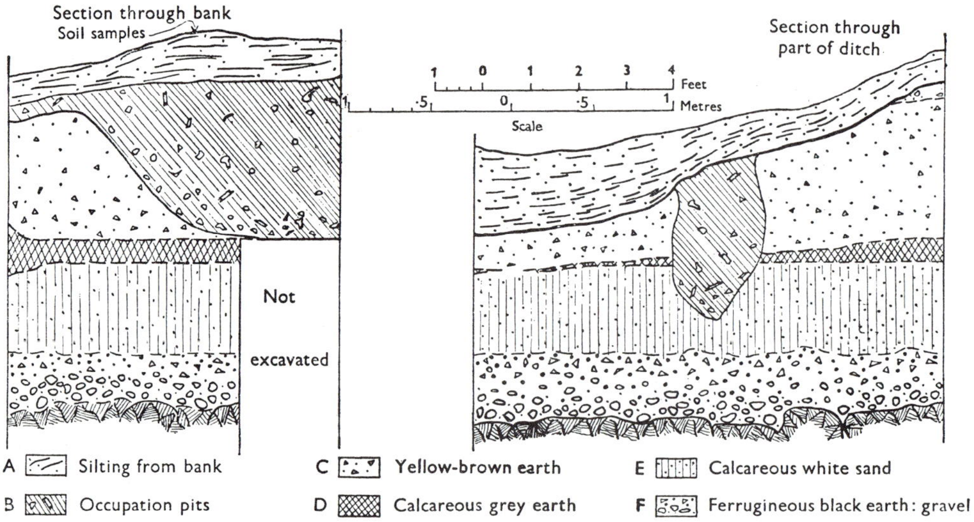

Text-fig. 26. Sections, Rifle Range site, Bur Hakaba.

A small area 6 ft. long by 3 ft. wide was first dug along the bottom of the Italian trench in order to determine whether fossil human bone occurred in the silting of the ditch, and the depth of bed rock. A depth of 2 ft. of silting (layer A) was found in the ditch, which was presumably derived from the bank. This silting was laminated and contained a mixture of potsherds, fossil bone and some Late Stone Age factory debris. The lower three-quarters of the deposit contained the greatest quantity of non-fossilized bone. The greater number of fossil-bone fragments, which included two small pieces of human skull, one molar tooth and a large piece of human temporal bone, were derived from the upper levels in the silt. This wash or silt rested on a consolidated, white to buff, calcareous sand (layer E) with occasional small pisolites and veins of 'kunkar'. This deposit was completely sterile and continued for a depth of 4 ft. when bed rock was reached. Immediately overlying bed rock, however, was a thin occupation floor with numerous small pieces of comminuted bone black in colour and mineralized, together with flakes and cores of a semi-microlithic industry.

This excavation was extended on the bank side of the trench and an area 10 ft. long by 7 ft. wide was dug cutting through the bank.

Layer A. This consisted of spoil and silting from the bank and contained the usual assortment of potsherds, fossil and non-fossil bone and some Stone Age factory debris. The deposit consisted of fine sand and dust laminated where it was washed down the sides of the bank into the trench.

Layer B. The top of this deposit represented the former ground-level prior to the excavation of the Italian rifle range. It consisted of a fine, loose, grey earth or sand containing some humic matter. This fine, grey earth was found filling pits which had been excavated in the underlying deposits. These pits were identical with the saucer-shaped storage pits dug by the present-day Elai inhabitants of the area for storing millet. The deposit contained numerous potsherds from water pots and incense burners closely similar to those still in local use; non-fossilized bone of, among other animals, sheep, goats and camels; cowries; round hammer-stones or grindstone 'sharpeners', as well as a small percentage of Stone Age factory debris and a few weathered and derived fragments of fossil bone. The main pit excavated was found to have been some 3½ ft. deep with an estimated diameter of 12 ft., and to have penetrated to the base of layer C, that is to say to the top of the hard, consolidated earth. On the side of the ditch was discovered a smaller oval pit the use of which is unknown. This pit was approximately 1 ft. 6 in. by 2 ft. 6 in. wide at the top and some 4 ft. deep. It contained a few fragments of unfossilized bone and two small potsherds near the base. It penetrated into but not below the sterile layer E.

Layer C. This had formed the most superficial deposit, prior to the occupation by the Elai, or Ma'adanle, who excavated the grain and other pits.[1] It attained originally an average depth of 3 ft. or so, and consisted of a rather light brown to yellow earth contrasting markedly with the grey colour of the occupation

[1] In connexion with a government grain buying scheme the writer excavated twenty large storage pits on the west side of Bur Hakaba, just north of the administrative buildings. This area had formerly been the site of an Elai village. A large number of filled-in grain pits were found to cover the whole area. Two, possibly three occupation periods were noted when pits had been dug into earlier, filled-in pits—the difference in the colour of the earth in-filling being clearly discernible. These pits yielded a similar assortment of occupation material as did those at the Rifle Range site, with the addition, however, of two pottery pipe bowls—probably for smoking hemp. The Elai state that they have never smoked, such practice being forbidden by their religion, so that if their information can be accepted it may be to the Galla Wardai (who were driven from Bur Hakaba by the Elai in the fifteenth century) or, more likely, to the Ma'adanle (an off-shoot of the Ajiuran, who are said to have been the first to excavate the 'wars' or catchment dams in this area) that the first agricultural settlement at Bur Hakaba should be ascribed.

pits. This deposit was for the most part of fine consistency, but contained some coarse and angular grit. This layer is somewhat more compact than the grey earth layer B and contained in its lower levels some semi-fossilized bone fragments, and a few specimens of microlithic debitage. A few small and weathered sherds came from the upper levels. As far as it was possible to tell, these sherds in texture showed no great variation from those recovered from the occupation pits: rims, bases and decoration were absent, however, for comparison. This layer has the appearance of a sub-aerial deposit accumulated during dry climatic conditions or such as persist at Bur Hakaba today. In consistency layer C became more compact towards the base and passed fairly rapidly into layer D.

Layer D. This consisted of a thin layer some 6 in. thick on an average, of consolidated, grey calcareous earth which contained Doian artifacts and factory debris as well as fossil bone with lime incrustation, identical with that recovered from the Italian bank. This layer is considered to represent a land surface on which the Doian people camped. No further human skeletal remains were found.

Layer E. This is a compact, hard, calcareous, white to buff, wind-blown sand with typical, rounded aeolian grains and was sterile. There was no bedding. The depth, on an average, was 2–3 ft.

Layer F. Layer E rested on a gritty, black 'earth' with iron pisolites and numerous comminuted fragments of mineralized bone, black in colour, and a fairly prolific factory debris and artifacts of Somaliland Magosian type. This deposit again clearly represented a former land surface—the surface and base were irregular following the undulating surface of granite bed rock. This horizon rested on and was in part mixed with a residual gravel of well-rounded quartz and quartzite pebbles. These pebbles on an average were small and not more than an inch in length, but occasional larger specimens occurred. They are presumably of the same age as the pebble 'banks' found scattered over the coastal plain and are thus of considerable antiquity. All the Magosian artifacts are quite fresh, but there were indications that a very few abraded Levalloisian flakes occurred. Exfoliated granite rubble filled pockets in the granite rock.

These deposits can be interpreted as representing a semi-arid climate (layer F) when conditions were wetter than they are today and the Somaliland Magosian was present at Bur Hakaba, followed by more arid conditions when wind-blown sand accumulated (layer E) over the Magosian occupation level. A return to wetter conditions (layer D) with which the Doian Culture was contemporary, was followed by a gradual drying up when detritus and sand (layer C) accumulated over the Late Stone Age horizon. At some time subsequent to the accumulation of layer C the area was occupied by a pottery-using, agricultural community storing their grain in saucer-shaped pits.

Inventory of tools from excavations at the Gure Warbei (Bur Eibe) and the Rifle Range site, Bur Hakaba, type sites of the Doian Culture

	GURE WARBEI, BUR EIBE									RIFLE RANGE: BUR HAKABA			
	Early Eile (?)		Doian	Magosio/ Doian transition	Magosian	Early Eile (?)	Doian	Doian	Magosio/ Doian transition	Early Elai (?)	Doian	Magosian	Totals
	Section T Layer A1	Section T Layer A2	Section T Layers B1, 2	Section T Layer C	Section T Layer D	Section M Layer A	Section M Layer B1	Section M Layer B2	Section M Layer C	Layers B and C	Layer D	Layer F	
Lanceolate and foliate points—bifacial	—	3	2	1	—	1	57	2	8	—	7	1	82
Triangles—bifacial	—	—	—	—	—	—	2	2	1	—	2	—	7
Subtriangular points—bifacial	—	—	1	—	2	—	1	—	—	—	2	5	11
Trihedral rods	—	—	—	—	—	—	2	—	—	—	1	—	3
Hollow-based arrowheads	—	—	—	—	—	—	—	—	2	—	1	—	3
Lanceolate and limace points—unifacial	—	4	1	—	1	—	56	5	29	—	4	—	100
Subtriangular points—unifacial	1	1	—	—	8	—	4	4	7	—	—	2	27
Asymmetric (crescent) points—unifacial	—	—	4	—	—	—	11	1	6	—	2	—	24
Crude foliate points (broken)	—	1	—	—	—	—	2	—	—	—	1	—	4
Large backed blades	—	3	4	—	—	—	12	5	36	—	8	1	69
Microliths—straight-backed blades	—	5	6	—	—	1	37	12	50	—	15	20	146
Microliths—straight-backed blades with basal retouch	—	—	1	—	—	—	17	—	10	1	—	—	29
Microliths—curved-back blades	5	14	18	1	3	15	147	34	142	21	38	9	447
Microliths—crescents	—	—	—	1	—	4	13	1	10	—	12	—	41
Microliths—double-backed blades	—	—	—	—	—	—	8	8	6	—	2	1	25
Blades with Heluan-type retouch	6	—	1	—	—	—	—	1	1	—	12	—	15
Tranchets (?)	—	—	—	—	—	—	3	—	—	—	—	—	3
Blades with Kasouga retouch	—	5	1	—	—	—	7	3	2	—	2	—	20
Blades worked down one or both edges	2	—	3	—	—	—	6	3	6	—	7	8	35
Scrapers—end-	—	1	1	—	1	—	4	6	15	—	3	—	31
Scrapers—side- and end-	2	5	—	—	2	—	12	1	8	—	1	2	33
Scrapers—small round and thumb-nail	—	2	—	—	1	—	1	4	2	1	7	1	19
Scrapers—hollow and notched	—	1	—	1	—	—	2	3	3	—	14	1	25
Burins—*bec-de-flute*	—	—	—	—	1	—	—	5	6	—	—	2	14
Burins—single-blow	—	1	—	—	—	2	1	—	3	—	2	—	9
Burins—gouge	—	—	—	—	—	—	1	2	1	—	—	—	4
Diminutive hand-axe	—	—	1	—	—	—	—	—	—	—	—	—	1
Pottery	210	29	1	—	—	188	17	—	—	Very numerous	—	—	445
Ostrich egg beads and fragments	16	53	50	2	5	1	—	—	—	—	5	1	133
Mollusc and marine shell beads	—	1	1 (?)	—	—	—	—	—	—	—	2	—	4
Utilized bone	1	2 (?)	1	—	—	1	—	—	—	—	1	2	8
Outils écaillés	3	8	3	—	4	4	32	7	13	4	15	6	99
Hammer and pestle stones	3	2	—	—	—	1	3	1	—	—	2	2	14
Rubbers and flakes from rubbers	—	10	8	—	1	—	1	—	2	—	—	—	22
Redirecting flakes	—	2	4	—	1	—	35	10	47	—	3	—	102
Haematite	20	8	11	3	3	9	23	9	21	1	2	5	115
Blades—some utilized	19	49	76	10	77	7	482	149	516	14	73	109	1581
Flakes—some utilized	124	151	228	33	355	32	777	205	545	117	120	195	2882
Blade cores—single-platformed	—	3	10	1	—	3	31	8	25	3	17	7	108
Blade cores—multi-platformed	—	6	2	1	—	—	19	8	17	2	6	5	66
Blade cores—bipolar	2	3	1	1	—	—	12	9	36	—	8	13	85
Disc and pseudo-disc cores	—	—	3	—	14	2	13	16	29	2	5	11	95
Totals	408	373	443	55	479	271	1851	524	1605	166	402	409	6986

THE CULTURE

The Gure Warbei, as we have seen, has given a most valuable, uninterrupted sequence from the Somaliland Magosian through a transitional industry (which at this site might be termed Lower Doian) to the typical Doian. At Bur Hakaba, the transitional industry is absent but the Magosian and Doian are separated by a deposit of sterile wind-blown sand. The Magosian has already been described but the transitional industry, or Lower Doian, at the Gure Warbei must first be considered before examining the Doian proper. A full inventory of all tools is given in the Tool Lists (p. 241).

THE MAGOSIO/DOIAN TRANSITIONAL INDUSTRY (LOWER DOIAN) AT THE GURE WARBEI

This industry is particularly well represented both by implements and factory debris and is found in layer C.

Points. The characteristic point is a small, delicately made, pressure-flaked limace, or lanceolate, unifaced example (Pl. 23, nos. 69–73); asymmetric specimens are also found (Pl. 23, no. 64). Bifaced points do occur but are rare at this level. The Magosian subtriangular, unifaced point is also found (Pl. 23, nos. 49, 68) with asymmetric variants of this form (Pl. 23, no. 61).

An important form is the hollow-based point, or arrowhead, worked over all or part of both faces (Pl. 23, nos. 65, 67). No 65 is a rather poorly made example in quartz with a more or less spherical section, but no. 67 is a well made specimen closely comparable to those from the contemporary industry at Jesomma. One bifaced triangle (or asymmetric triangle) was also found; these tools are presumed to be arrowheads (Pl. 23, no. 66).

Microliths. These are the commonest artifacts and 214 specimens were recovered; of these the greater number are curved-backed blades of usual form with the backing carried down the whole or part of one edge (Pl. 23, no. 50). Straight-backed microliths are also an important group; specimens both with (Pl. 23, nos. 51, 54, 55) and without (Pl. 23, nos. 52, 53) basal retouch occur. The backing on these tools is usually from both sides giving the back a straight rather than a convex section. Crescents are represented (Pl. 23, nos. 56, 57), in the case of no. 57 the cutting edge shows a nibbling retouch. Three eared-crescents occur.

A few double-backed blades are also found and two backed blades with the suggestion of a shoulder towards the butt end. Of particular note are two blades with Heluan-type retouch—that is to say where the backing is not flat but has been worked to a blunt edge by flaking directed from both sides of the tool;

this is a specialized form which hitherto has been restricted in its known distribution to Egypt and Palestine. It is surprising, therefore, to find it occurring in southern Somalia (Pl. 23, no. 58).

Large backed blades. This is a group which grades into the microliths. No hard and fast distinction is possible but those specimens of approximately 2 in. in length and over have been described as 'large'. The types found are similar to the microliths, both curved- and straight-backed examples occurring, usually with some secondary retouch on the cutting-edge (Pl. 23, nos. 47, 48).

Scrapers. The commonest form is a small end-of-blade scraper, which is usually poorly made though a few more carefully finished specimens occur (Pl. 23, no. 76). The scraper worked on the side and end of a flake or blade is also typical (Pl. 23, no. 74). Two small atypical round scrapers, and a few hollow and notched scrapers also occur (Pl. 23, no. 75).

Burins. These are poor and doubtful specimens for the most part. The usual *bec-de-flute* and single-blow burins occur as well as one small gouge-burin—a re-utilized core resultant from bipolar technique.

Blades worked down one or both edges. Six specimens of blades utilized in this way were found. Usually the working is restricted to the edges of the tool, but in one specimen (Pl. 23, no. 59) the whole of the upper face has been worked, and in another (Pl. 23, no. 60) the greater part of both faces has been carefully pressure flaked.

Ostrich egg-shell beads. One bead (Pl. 23, no. 62) was recovered from T section, showing polish on one side. A second irregular fragment of pierced shell was also found in the same section. The bore was of hour-glass type in both cases.

Outils écaillés, rubbers, redirecting flakes, Kasouga flakes and haematite are similar to those found with the Doian.

Technique. Blades were produced either from single-platformed cores in which case the platform may be either straight or circular, in the latter case resulting in a small, conical core (Pl. 23, no. 79), or else from multi-platformed (nobbly) cores, or cores with two platforms at opposite ends—the outcome of bipolar technique. This latter type of core is fairly common, thirty-seven specimens occurring against forty-four of the other types. Diminutive, prepared disc cores also occur (Pl. 23, no. 78), from which were struck small Levallois-type, faceted flakes, presumably for manufacture into subtriangular Magosian-type points. The pseudo-disc core is also present. The greater majority of flakes recovered show plain inclined striking-platforms and represent primary flakes removed in the preparation of blade cores. Only some 8 per cent of these flakes show faceted platforms and were derived from prepared cores. A typical example is figured (Pl. 23, no. 77) and would be entirely in place in a Magosian *milieu.* Secondary retouch is executed by a pressure technique. Thus it can be seen that there has taken place a great increase in blade production since the Magosian but that the latter tradition is still fairly strong.

THE TYPE DOIAN INDUSTRIES AT THE GURE WARBEI AND BUR HAKABA

(Pl. 23, nos. 1–46; Pl. 24, nos. 1–32)

This culture shows the development of pressure flaking in its highest form as found in the Horn; this is best seen on the foliate and lanceolate points. The characteristic pressure-flaked points show a fine degree of control which is executed in such a way that alternate flaking from both sides forms a wavy, almost a serrated edge on some of the specimens. This form approximates very closely to the type of pressure technique found on Neolithic artifacts from Egypt and the Sahara. The raw material used consisted of chert and semi-crystalline and crystalline quartz. The culture is in essence a blade culture depending on the manufacture of miniature blades from unprepared cores, but occasional disc cores and rather poorly faceted flakes are still found and help to indicate that its origins lie in the Stillbay/Magosian complex. At the Gure Warbei the culture is found in layer B, which in M section, for technological reasons, has been subdivided into layer B 1 (the upper half of the deposit) and layer B 2 (the lower half). The material recovered from T section was insufficient to warrant subdivision here. The tool lists give an inventory of all artifacts recovered. It will be seen from these lists that cultural material is most prolific in the upper part of the deposit and that there is a marked falling off in tools and debris in the lower half. Tools do not become prolific again until the Magosio/Doian layer is reached. At Bur Hakaba the Doian is found in the thin occupation layer D.

Bifacial lanceolate and foliate points. (Pl. 23, nos. 1–6; Pl. 24, nos. 1, 2.) These vary in shape from the foliate to the finely pointed form and in length from $2\frac{1}{2}$ to under 1 in. In section they are either biconvex, lenticular, or more rarely subtriangular. With the unifaced points they constitute the most characteristic tools of the culture. They usually show a fine degree of secondary retouch and the wavy edges due to alternate flaking are entirely typical. There can be little doubt that these tools were used as arrowheads. The two bifaced points in layer B 2 at Gure Warbei are both foliate.

Triangles, bifacial. (Pl. 23, nos. 18, 19, 36; Pl. 24, no. 17.) These tools are rare but are of interest. They are not equilateral triangles but asymmetric, such being probably intentional rather than fortuitous. They are believed to have been used as arrowheads and to represent a variant on the hollow-based form.

Subtriangular bifaced or partially bifaced points. (Pl. 23, no. 10.) These are rare and represent a survival of the Magosian type of point.

Trihedral rods. (Pl. 23, no. 7; Pl. 24, no. 3.) A rare form, yet sufficiently distinctive. They have been named after similar tools as described by Caton-Thompson[1] from the Fayum B Culture. In section these tools are triangular and each side has been pressure-flaked over the entire surface. Several of the triangular-sectioned, unifaced points approximate to these tools. They do not occur in layer B2 at Gure Warbei.

Hollow-based arrowheads. (Pl. 23, no. 21; Pl. 24, no. 18.) None of these tools was found in the Doian level in the Gure Warbei, but from a Doian site adjacent to the Gure Makeke was recovered the unifaced example with serrated edges illustrated. Other specimens from the surface were recovered from the Bur Eibe sites. The Bur Hakaba specimen is made from crystalline quartz. It is of interest to note that the triangular and hollow-based arrowheads have their origin in the Magosio-Doian transitional industries.

Lanceolate and limace unifaced points. (Pl. 23, nos. 8, 9.) These are a characteristic tool and are finely finished often with serrated or wavy edges. Some show varying degrees of retouch on the under surface; they were used presumably as arrow tips. In section they are either plano-convex (usually the case with the limaces) or triangular (with sides of equivalent length) or else wedge-shaped (similar to a backed blade in section).

Asymmetric (crescent) points, unifaced. (Pl. 23, no. 32.) A few examples occur of a type which has persisted from the Magosian.

Crude foliate points (broken). Four examples of large, foliate points, flaked on both sides, probably by wood technique, were found. These are all broken across the mid-line but when complete must have been some 3–4 in. in length. They are identical with those tools illustrated at Pl. 25, nos. 98–100. The technique in the production of these tools is in marked contrast to the fine pressure flaking of the foliate arrowheads.

Large backed blades. These are crude for the most part and grade into the microlithic forms. Specimens of 2 in. in length and over have been classed as large.

Microliths

i. *Straight-backed blades.* The backing on these tools is nearly always directed from both sides of the tool to form a flat back. The usual form shows no basal retouch (Pl. 23, nos. 38, 40; Pl. 24, no. 36) but specimens with this retouch are also typical (Pl. 23, nos. 11, 12)—in the case of the latter example the retouch has been carried up most of the length of the cutting edge.

[1] Caton-Thompson, G. & Gardner, E. W. (1934, p. 59).

ii. *Curved-backed blades.* (Pl. 23, no. 39; Pl. 24, nos. 26–30.) This is the commonest form of microlith, the backing being directed from one or both sides. Three specimens show an incipient shoulder or constriction of the backing at the base. Kasouga retouch is present on some of these microliths.

Crescents. (Pl. 23, nos. 14, 15; Pl. 24, nos. 19–23.) A small group, some of which show notching and nibbling of the cutting edge. Three complete and one broken specimen are eared and more triangular in shape (Pl. 23, no. 15).

Double-backed blades. (Pl. 23, no. 13; Pl. 24, no. 24.) Fairly typical.

Blades with Heluan-type retouch. (Pl. 23, no. 17; Pl. 24, nos. 4–10.) These were rare at the Gure Warbei, the specimen illustrated was recovered from a Doian site nearby. At Bur Hakaba they were more numerous, and were on the whole more developed. Whereas in the usual form the retouch is confined to the edge, in some instances at the latter site the working had been carried over the whole or the greater part of both sides to form a type of large crescent. The most finished specimen is illustrated at Pl. 24, no. 4. The writer believes that this type of tool develops from the simple crescent form (Pl. 23, no. 17) through the form illustrated at Pl. 23, no. 4 to the tranchet (?) type of tool illustrated at Pl. 23, no. 16. These specimens may therefore have been used as transverse arrowheads. This form of the Heluan technique is more developed than is the case in Egypt or Palestine, though it also occurs on some of the Sahara Neolithic sites.

Tranchets. (Pl. 23, no. 16.) The unique pressure-flaked tool illustrated is thought to have been used as a tranchet. Two small blades with opposed concave backing may also be placed in this class.

Blades with Kasouga retouch. (Pl. 23, no. 27.) The large blade illustrated shows typical retouch of this kind along the underside of most of the cutting edge. Usually blades utilized in this way are smaller.

Blades worked down one or both edges. (Pl. 23, nos. 25, 26.) Pl. 23, nos. 25 and 26 are examples worked down both edges; Pl. 24, no. 11 is an example worked down one edge only. Occasionally the working may extend over the whole of the upper surface of the tool (Pl. 23, no. 37).

End-scrapers. (Pl. 23, no. 31.) These are usually poorly made but occasionally better specimens are found. The specimen illustrated is on the bulbar end of a broken faceted flake.

Side- and end-scrapers. Scrapers formed by working down one side and round the end of an end-flake are an early type and the form has persisted from Stillbay times. The specimens with the Doian are usually fairly small.

Small round and thumbnail scrapers. These are made usually on short end-flakes—well-made specimens are very rare. Pl. 23, no. 24 is a small round scraper made on a core. One thumbnail or small fan-shaped scraper (Pl. 23, no. 46) was recovered from the Gure Warbei.

Hollow and notched scrapers. The usual forms occur.

Burins. All poor for the most part, no angle-burins occur, but occasional *bec-de-flute*, single-blow and gouge examples are found. Pl. 24, no. 31 is an example of a single-blow burin.

Diminutive hand-axe. One small core biface was found with the Doian from layer B 2 at the Gure Warbei (Pl. 23, no. 30).

Pottery. This is rare with the Doian but it seems that some at least of the pottery recovered should be associated with the final phase of this culture. For example at the Gure Warbei 427 sherds were found above the Doian layer, seventeen sherds were found in the upper levels of layer B 1, and one sherd was found in layer B 2 in T trench: the latter is presumed to have been *in situ* as no artificial disturbance of the deposit was noted. While it is just possible that the sherds in layer B 1 could have found their way there from the overlying deposit the writer does not believe that this altogether satisfactorily explains their presence particularly in regard to the one sherd from layer B 2 (Pl. 23, no. 42). This is a rim-sherd of thin pottery made of a fine dark brown paste, with a plain tapered rim rounded on both sides. A faint drawn line runs round the outside just below the rim, and there is the suggestion of decoration with a wedge-shaped tool. It seems probable, therefore, that pottery had been introduced at the very end of Doian times, though its use may have been known somewhat earlier.[1] The bulk of the sherds are divisible into four groups:

(i) A red ware of medium thickness in a fine paste and evenly fired. This ware shows inverted rims flattened horizontally, with projecting edges usually on the outside but sometimes on the inside also. The decoration is usually incised and consists of oblique or crossed lines interplaced with deeply drawn lines. These sherds have come from globular pots or bowls presumably with rounded bases though one sherd was found which formed part of a flat base.

(ii) A grey ware oxidized on the outer surface thus giving it a red colour. These sherds are believed to have come from shallow bowls. Two rim sherds with decoration were found. The rims are flattened horizontally one with slight rounded projections on both sides, the other with rounded projection on the inside. Decoration on the first specimen consists of deeply incised, oblique lines just below the rim, and on the second of lightly incised crossed lines on the flattened top of the rim.

(iii) One undecorated sherd of grey to buff ware with rolled and flattened rim.

[1] Graziosi, P. (1940), states that rough pottery occurs in the upper layers of the Gure Makeke, Strata I–III, associated with a degenerate stone industry.

(iv) Thin sherds of buff to dark brown ware in a fine paste. These are probably derived from bowls. The most complete sherd of this ware was found on site 4*a* and was coated by oxides which have obscured the edges. The exterior is buff in colour while the interior surface is grey. This sherd shows a carination, accentuated by two lightly incised lines and the outer surface shows the light impress of fibre or (?) woven grass.

Ostrich egg-shell beads. (Pl. 23, nos. 20, 43–5; Pl. 24, nos. 15, 16.) While numerous fragments of ostrich egg-shell were found, beads were not common. The beads had been made by rubbing down on a grindstone, but the circumference still shows irregularities. The bore is of hour-glass type. Fragments of bored shell exist but these are more probably beads in process of reduction than parts of shells bored for suspension. The beads vary in size and the size of the bore varies also. Some have been polished on both sides, presumably from being threaded on a string, but two from the Gure Warbei (nos. 20, 45) and one from Bur Hakaba show a black to greyish polish on one side and a rough surface on the other. It is interesting to note that Leakey found similar beads with the upper Kenya Aurignacian (now Capsian) in Gamble's Cave[1] and suggested that they were used as ornaments sewn on to leather skins.

Mollusc and marine shell beads. (Pl. 24, nos. 13, 14.) One small bead made from *Achatina* shell (?) was found at the Gure Warbei from the lower part of layer A and two beads made from part of a marine shell were found at Bur Hakaba. It is possible that the two latter, which show direct perforation are later and should be assigned to the Elai/Ma'adanle occupation. They were found on the right-hand side of the section and could have been derived either from the base of the occupation pit or from the top of the Doian horizon. If they belong to the latter then they presuppose some contact with the sea coast at this time.

Utilized bone. (Pl. 23, no. 41; Pl. 24, no. 12.) A few pieces of bone that have been rubbed smooth at one end were found. One broken fragment from Bur Hakaba shows an oval section and possibly formed part of a bone point. The almost complete absence of worked bone in deposits where such tools would have been preserved had they occurred suggests that bone was only rarely employed for tools by the Doian peoples.

Outils écaillés. (Pl. 24, no. 32.) These fabricators are entirely typical. The specimen illustrated is double-ended and made on a flat rejected bipolar core.

Hammer and pestle stones. (Pl. 23, no. 28.) These are usually on the small side and made of quartz though two limestone specimens occur. They show typical battered surfaces at one or both ends of the long axis of the pebble.

[1] Leakey, L. S. B. (1931, p. 103).

Rubbers. The presence of these tools, usually made from limestone with one or more rubbing or grinding surfaces suggests that edible roots or grasses may have formed part of the diet of these people. One specimen shows dimple scarring on one rubbing surface perhaps indicating its use as an anvil stone for bipolar technique. Two other anvil stones were also found.

Redirecting flakes. The majority are resharpening flakes struck down the length of a blade core to remove part of an edge of the striking-platform and the upper end. Rejuvenating flakes struck at right-angles to the platform to remove part or the whole of the platform also occur.

Haematite. A number of pieces of haematite occur at these sites but few if any show signs of grinding surfaces.

Technique. Small, unprepared cores for the production of miniature blades are the usual forms and are the same as those found with the industry from layer C at the Gure Warbei. That is to say, the usual forms of single-platformed core, the single-platformed conical core (Pl. 23, no. 23), the bipolar core (Pl. 23, no. 22) and the multi-platformed core which may show two, three or sometimes more, platforms. Cores are rarely longer than 2 in. and are microlithic. The blades struck from these cores vary in length from 2 in. to $\frac{3}{4}$ in. or less.

In addition to the blade element, however, there are still found a few true disc cores. These number no more than half a dozen specimens, but there can be no doubt that they comprise an integral part of the culture, and were used to produce small, carefully faceted, pointed flakes. Some 3 per cent of the flakes recovered show preparation of the striking-platform, the remainder being primary flakes removed in the preparation of blade cores. Certain 'nobbly' cores have been classed as pseudo-disc cores. In appearance they are similar to the specimen illustrated at Pl. 20, no. 21. These cores do not show the finish found with the true disc cores, but the writer considers that they represent a degenerate form of this type of core: the possibility that they are an extreme form of multi-platformed core must not be lost sight of, however.

It is possible to observe, particularly from a study of the Gure Warbei, a gradual development of the Late Stone Age Doian Culture of southern Somalia from Magosian, or even Stillbay origins. The persistence of Magosian tool forms (points and scrapers) though rare, helps to substantiate this, as does also the development of the limace points from the subtriangular points as well as of the limace to the fully bifacial pressure-flaked lanceolate and foliate points.

The percentage increase of blade to flake cores from the Magosian to the Doian clearly shows the greater development in blade production and the attendant decrease in prepared core technique that has taken place. Of the total of flakes and blades recovered the Magosian layer yielded 17·8 per cent blades and 82·2 per cent flakes, the transitional Magosio/Doian layer yielded 47·7 per cent blades and 52·3 per cent flakes, while the Doian layer yielded 36·9 per cent blades and 63·1 per cent flakes, predominantly long.

The position is reflected more clearly if we consider the percentage of blade to disc cores. In the Magosian layer no blade cores were found so that 100 per cent of the cores (14) were prepared. In the transitional Magosio/Doian 73·6 per cent are blade cores and 26·4 per cent flake cores. In the Doian 73·1 per cent are blade cores and 26·9 per cent flake cores. Thus the great impetus to blade production appears to have come at the time of the transitional industries. The cultural material from layer A at the Gure Warbei and from layer C at Bur Hakaba suggests that there was no sudden replacement of the Doian Culture but rather that it persisted in this part of the country until recent times, changing its make-up as a result of fresh cultural contacts and by a gradual falling out of the stone culture.

All the cultural and faunal evidence recovered independently by Graziosi and the writer goes to prove that the Doian peoples were hunters and food-gatherers, owning no domestic animals, unless it were the dog, and planting no crops. All evidence of the presence of cattle was confined to the topmost layers in the rock-shelters, in association with pottery and a very degenerate stone industry, and in fact it would appear that this late industry was the work of the descendants of the Doian hunters who had by that time acquired domesticated cattle and (?) sheep. It is tempting to see in the Eile, the Bon and the Ribi, the present-day descendants of the Doian people, albeit much altered by racial admixture. It is of interest also to compare the two types of Doian arrowhead—the foliate point and the triangular or hollow-based arrowheads, with the iron forms in use today in southern Somalia. With the exception that both the iron heads have a short tang and the triangular head has two incipient barbs, the two forms are identical. It is most unfortunate that the human skeletal material collected and believed associated with the Doian is so fragmentary and does not allow of reconstruction. Any closer comparisons with present-day inhabitants must wait, therefore, until further skeletal remains are recovered.

NOTE ON THE HUMAN SKELETAL REMAINS FOUND AT THE RIFLE RANGE SITE: BUR HAKABA

The human skeletal material is all fossilized and covered with a thin coating or incrustation of lime, very difficult to remove without injuring the bone. The remains were found scattered over about 15 yd. of bank at the point where excavation was undertaken. The greater number of fragments comprise pieces of skull (seventy-nine pieces)—recently fractured by the workmen who excavated the Italian trench—fragments of a frontal bone, a temporal and an occipital were found, together with part of a left maxilla with the 1st and 2nd molars. Fragments of mandible and isolated teeth were also recovered as a result of sieving the spoil from the bank: and other remains were found in the silting of the ditch. Recently broken fragments of limb bones, vertebrae, ribs and phalanges were also present. The long bones were rather slender. It appeared that there were certainly two individuals and a fragment of a mastoid bone suggests that a third may be represented. It is certain that these individuals were buried and the burials later disturbed when the trench was dug. The bottom of the trench at this point was excavated into the upper half of layer E which was sterile, so that the fossils could not have come from the Magosian level (layer F) and in any case the bone from this horizon is mineralized and black in colour with no lime incrustation. The trench, therefore, penetrated layers B, C and D, and into E but not deeper. This was found to be the case for two-thirds of the length of the trench. Layer B represents the Elai/Ma'adanle occupation and the associated bone is quite unfossilized. The bone from layer C was unfossilized in the upper two-thirds but showed partial fossilization in the lowest one-third though it was not incrusted with lime. The bone from layer D was fossilized to exactly the same degree as the human remains from the bank and was similarly lime incrusted. In the opinion of the writer, there is, therefore, little doubt that the human skeletal remains represent the makers of the Doian industry here, and that they were buried most probably in a shallow grave dug into layer E.

It is greatly to be regretted that excavation yielded no skeletal remains *in situ* and that lack of time prevented further work, for with the exception of the fragment of mandible discovered in 1932 by the Abbé Breuil in the Porc Epic cave at Dire Dawa[1] the Bur Hakaba fossils are the only other remains of fossil man yet recovered from the Horn of Africa. The remains have been handed over to Dr J. C. Trevor of Cambridge University who has examined them. Dr Trevor states that while the degree of fossilization suggests a respectable

[1] Boule, M. & Vallois, H. V. (1946, p. 460). Vallois, H. V. (1951).

antiquity and the remains are all of persons of *homo sapiens* type, they are unfortunately too fragmentary to be able to make any accurate determinations.

The numerous types of tumulae which occur throughout the Horn are ascribed to the Galla and known colloquially as 'Galla Graves'. That the majority are the graves of these people there can be little doubt, but it is not impossible that some may be Late Stone Age, or 'Neolithic' on analogy with those recorded by Monod from the western Sahara, and a critical study of these mounds would be of value.[1]

III. THE STRANDLOOPING VARIANT

This is poorly represented by three sites at Mogadishu (89), one at Gesira (87), and one at Danane (88). All these sites are stratified, however, and the artifacts are found eroding from a land surface underlying loose, orange dune sands, and on the eroded surface of compact red sands (see Correlation Table for Doian sites, p. 229).

As might be expected on sites where the principal form of diet was shell-fish the material culture is not rich and the degree of workmanship is poor.

MOGADISHU

The first site is situated on top of the dunes just to the north of the Italian battery post on the left of the Via Imperiale before entering the town. Here under orange dunes were found one poorly made crescent and four flakes of chert associated with a debris of broken shells.

The second site is situated at the pig farm on the right of the road between Forte Cecchi and the Military Details camp. Here were found scattered flakes of chert patinated white, a micro-blade core and six much-weathered sherds of red to dark brown pottery, which were probably associated. The specimens come from a land surface under loose yellow sands resting on the reddened surface of the earlier fossil dunes.

The third site occurred at the top of the dunes west of the airport where the land surface between the orange and red sands is again exposed. Here were found two more crudely made crescents, two rubbing stones, two battered quartz pebbles, a chopping tool of silcrete and a chert flake. These were apparently associated with a curious ellipsoidal piece of glass much etched by wind action, and with discoloration by lime incrustation, as well as potsherds with lime incrustation. Modern Somali pottery is found on the surface.

[1] Monod, T. (1948).

The earlier pottery can be subdivided into:

(i) Sherds (comprising the greater number found) in a fine red paste. Rims are either rounded on both sides; bent inwards, rounded and everted; bent inwards and squared; rolled and undercut. With one exception all these sherds are undecorated. The decoration is formed by a band of crossed lines under the rim incised on the clay after drying but before firing.

(ii) One sherd in a fine reddish paste, red on the outer surface and dark grey on the inner. This sherd shows a thick, squared rim with decoration of what appears to be panels under the rim filled by crossed lines. The flat top of the rim being decorated by two parallel drawn lines with between them a series of impressed circles.

(iii) Two rim sherds in a thick grey paste with partially reddened outer surface. One shows a simple everted and rolled rim with a single incised line at the neck. The other shows an inverted, squared rim with a single incised line just below it. Above this line the outer surface is black to grey, below the line it is red.

The potsherds appear to have come from bowls or shallow wide-mouthed pots. It can hardly be supposed that the pottery which is well made and well fired can have been made by the strandloopers. It is more likely that they obtained their pottery—if indeed the pottery is associated—from a more advanced people with whom they were in contact.

GESIRA

From a site at the foot of the inland dunes west of the island there were found eroding from the land surface between the orange and red sands fourteen microflakes; two utilized blades; three thin sherds in reddish brown paste, red on the outer surface, grey on the inner surface; and numbers of shell fragments, the commonest being the crescent-shaped pieces from the base of the large cowrie shell.

DANANE

From a site just above the gully north-west of the Sanatorium were found eroding from the land surface between the orange and red dunes ten small flakes, some utilized; seven sherds with red exterior and black interior surfaces; one very weathered piece of haematite; three septa from large conus shells; numerous crescent-shaped fragments from large cowrie shells; and five pieces of semi-fossilized bone.

More evidence is required from the southern strandlooping sites and it is probable that the Kismayu/Gobwein area will furnish this, though it is obvious that the industries are very degenerate and late.

IV. THE 'NEOLITHIC' VARIANT

The site at Mirsale Wells (45) in the eastern Mudugh on the road from Galkayu to Obbia differs somewhat from the Doian of the type sites and approximates more to the Somaliland Wilton of 'Neolithic' facies which is found in the same locality. The Mirsale site is important though it is the only one of its kind that has yet been found. It is probable that similar sites will be found to exist to the west and south though the whole of this area is difficult of access and is very infrequently visited by Europeans.

Mirsale is a shallow well area lying in the Eocene gypsum beds and is situated just to the right of the Galkayu/Obbia road some 87 miles from Obbia. About $\frac{3}{4}$ to 1 mile south of the wells is a low line of hills composed of conglomerates and limestones of the Warandi Series. These hills provided most of the raw material for the Late Stone Age industry of the well area—the chert being derived from the limestones—and flaking sites were found scattered thickly round the foot of these hills.

The pan or well area itself is approximately $\frac{1}{4}$ mile long on a N.W./S.E. axis in which have been dug ten wells though the majority have now silted up and only one is producing water. The pan is surrounded by bare gypsum rock, rose red in colour and covered in places by a thin capping of red quartz sand. Within the pan itself the red sand assumes a depth of some 4 to 6 ft. and in the centre has been leached to a white or grey sand. The two main wells have been dug through irregular fissures or sink-holes in the gypsum. The present excavation passes through a grey sandy clay with which the sink-holes or cracks had, at some previous period, become filled.

The most prolific occupation area was found to be at the north side of the pan and the artifacts appeared to be deriving from the red sand as they were commonest where this had been eroded.

On the first visit to the wells a slit-trench 1×3 ft. was dug to establish the stratigraphy. This showed from the surface to a depth of 10 cm. an unconsolidated, leached, quartz sand, and from 10 to 50 cm. a consolidated red sand. In the upper 15 cm. of this sand very few flakes were found, but in the lower 25 cm. factory debris was common. At 50 cm. the red sand gave place to a sterile red to grey sandy clay with 'kunkar' nodules. This rested on a weathered, red gypsum.

A second visit was paid to the pan and an area was excavated approximately in the middle of the north side between the centre of the pan and the outcropping gypsum at the edge. The area was 2 m. long by 1 m. wide and was excavated in 10 cm. layers until bed rock (gypsum) was reached at between 70 and 65 cm.

The section through the deposit is seen in Text-fig. 27. The top 2 cm. or more is composed of loose, leached yellow to white quartz sand, not calcareous and not well sorted. This passes into a grey to buff consolidated earth also leached in its upper 10 cm. This layer passes gradually and almost imperceptibly into a compact and 'stiff', red to orange quartz sand which is more strongly calcareous. This red sand continues unaltered in consistency and colour until the grey, sandy clay is reached immediately overlying pink gypsum rock.

The 30–40 cm. level appears to be particularly lacking in factory debris, but at 40–45 cm. debitage begins to appear in quantity and continues to a depth of

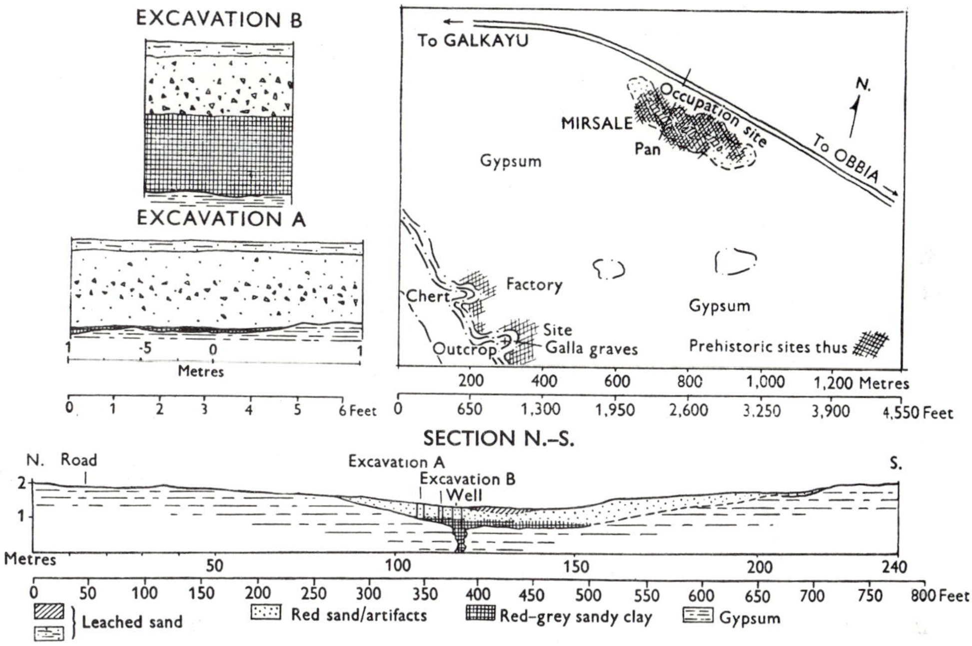

Text-fig. 27. Plan and sections, Mirsale Wells, Mudugh.

50–60 cm. For the most part the flakes lie horizontally in the sand. At the junction of the red sand with the grey, sandy clay implements and debitage cease. The gypsum rock floor had been eroded and weathered prior to the accumulation of the overlying deposits. There is no well-defined break in the stratigraphy here with the exception of that between the gypsum rock floor and the grey, sandy clay. The latter deposit may indicate somewhat wetter climatic conditions than exist in the area today, and is almost certainly to be equated with the grey, sandy clay filling of the sink-holes or cracks in which the wells have been excavated: it is correlated with the First post-Pluvial Wet Phase deposits.

The red sand may, with some degree of certainty, be correlated with the red, wind-blown sands which accumulated after the Magosian occupation of Somalia in the dry phase between the First and Second post-Pluvial Wet Phases. Only one industry is represented, which extends almost from top to bottom of the red sand. Occupation material is most prolific in the 40–50 cm. level, however, approximately half-way up in the deposit. Allowing for the tendency for debitage to work its way downwards in loose, shifting sands it would seem that a considerable part of the red sand had accumulated before the main settlement had begun around the pan and this industry would therefore be contemporary with the typical Doian of the 'Bur country'. That it extended over a longish period when wind continued to accumulate sands in the pan is clearly seen, but, as has been suggested, these sands need not necessarily, and most probably were not, accumulated under arid conditions only. Stronger monsoon winds, carrying an increase in rainfall must have caused sands to accumulate in most of the shallow pans in the Mudugh and Haud, and the writer believes that some of these red sands were accumulated under more humid conditions during the Second post-Pluvial Wet Phase. If this interpretation is correct, the Mirsale industry is therefore the contemporary of the typical Doian.

THE INDUSTRY (Pl. 25, nos. 98–115)

The greater number of artifacts were found in the surface exposures, but a sufficiently representative series was recovered from the excavation to show beyond any doubt that the surface material belongs to the industry in the red sand. The factory debris from the excavation provides the technical evidence which is lacking from the surface. The inventory of finds is given opposite.

Bifaced and foliate points. (Pl. 25, nos. 105–7, 109.) Eight of these are foliate, twenty-six pointed or lanceolate, seven lozenge-shaped and four are broken specimens. They vary in length between 1½–2 in. for the lanceolate and foliate examples and 1–¾ in. for the lozenge-shaped specimens. These tools have been carefully pressure flaked over both sides of the tool and are lenticular or sub-triangular in section. The type has already been noted from the typical Doian, with the exception of the lozenge-shaped examples which are probably a local form.

Limace points. (Pl. 25, no. 101.) These are similar to those found with the typical Doian. Sections are triangular.

Asymmetric (crescent) points. (Pl. 25, nos. 102, 103.) The Mirsale specimens are all well made and the bulb has been usually carefully reduced. Some show incipient Heluan-type retouch.

Inventory of tool types from Mirsale wells: eastern Mudugh, Somalia

	Excavations						Surface	Totals
	0–10 cm.	10–20 cm.	20–30 cm.	30–40 cm.	40–50 cm.	50–65 cm.		
Lanceolate and foliate bifaced points	—	—	—	1	2	2	40	45
Limace points	—	—	—	—	—	—	4	4
Lanceolate unifaced points	—	—	—	—	1	—	10	11
Asymmetric (crescent) points	—	—	—	—	—	—	11	11
Crude foliate points	—	—	—	—	2	—	75	77
Straight-backed microliths	—	—	—	—	—	—	3	3
Curved-backed microliths	—	—	—	—	2	—	1	3
Crescents	—	—	—	—	—	—	1	1
Blade with Heluan-type retouch	—	—	—	—	—	1	—	1
Burins—*bec-de-flute*	—	—	—	—	—	—	1	1
Burins—angle	—	—	—	—	—	—	1	1
Scrapers—end	—	—	—	—	1	—	7	8
Scrapers—short end- or round	—	—	—	—	1	2	11	14
Scrapers—thumbnail	—	—	—	—	1	1	9	11
Blades utilized down both edges	—	—	—	—	—	—	9	9
Flakes utilized down one edge	—	—	—	1	2	—	10	13
Diminutive hand-axe	—	—	—	—	—	—	1	1
Outils écaillés	—	—	—	—	2	2	—	4
Ostrich egg-shell	6	12	6	29	118	37	16	224
Blades	9	10	7	29	112	50	—	217
Flakes	16	26	15	39	185	76	67	424
Debitage	18	9	12	60	213	73	—	385
Cores—single-platformed	—	—	—	—	—	—	2	2
Cores—multi-platformed	—	—	—	—	1	—	—	1
Cores—bipolar	—	—	—	—	—	—	2	2
Cores—pseudo-disc	—	—	—	—	—	—	5	5
Various	—	—	—	1	—	—	11	12
Totals	49	57	40	160	643	244	297	1490

Crude foliate points. (Pl. 25, nos. 98–100, 108.) These are the most characteristic tools of the industry. Sixty-eight are the broken halves of such tools, but nine are complete specimens. The length of the longest complete specimen is $3\frac{1}{2}$ in., but others were undoubtedly longer, while the shortest was approximately 2 in. long. These tools appear to have been produced by wood technique—the edges and points are usually blunt and uneven and it is difficult to know what can have been their use, as they would have been clumsy weapons if employed as points to spears or arrows. At the Gure Warbei only two broken halves of such tools occurred whereas at Mirsale they are the commonest tool. One possible explanation is that they were used as picks for keeping the waterholes clear of silt which at this time was filling them.

Microliths. These are conspicuous by their almost total absence. They are poorly made and larger in comparison with the typical Doian or Wilton specimens.

Straight-backed. (Pl. 25, no. 110.) Three specimens were recovered from the surface exposures.

Curved-backed. (Pl. 25, no. 111.)

Crescents. One specimen only from the surface.

Blades with Heluan-type retouch. An unfinished specimen, but showing characteristic retouch.

Burins. One typical *bec-de-flute* and one oblique, concave, angle-burin were found. The latter presumably indicates contact with the contemporary Somaliland Wilton peoples.

End-scrapers. All made on shortish flakes and blades. One is double-ended, six are single-ended and one is diminutive.

Short end- or round. (Pl. 25, nos. 112, 113.) A typical form, carefully made, either with single (no. 112) or double (no. 113) scraping edges. Thirty-two broken scraping edges are thought to have been derived from these forms of scraper but have not been included in the tool lists.

Thumbnail. (Pl. 25, nos. 114, 115.) These are usually well made with either single or double edges. Three broken tips also probably come from thumbnail scrapers.

Blades utilized down both edges. As in the true Doian.

Flakes utilized down one edge. These are usually broadish, unprepared, primary flakes which have been worked with a steepish, coarse trimming down one straight or convex edge. They might perhaps be classed as side-scrapers on end-flakes.

Hand-axe. This specimen is very weathered and may be derived. It is, however, characteristic of those found with the Somaliland Stillbay and later cultures.

Outils écaillés. Typical.

Ostrich egg-shell. No pierced fragments or beads were found but the numerous pieces recovered indicate that ostrich eggs were in common use—presumably as water-containers. The present-day Somali specimens, now rarely found, are not pierced except at the top but are held in a leather cradle.[1]

Various. Four crude oval discs; three crude trimming stones; four redirecting flakes; four fragments of marine mollusc shell, one of which is a cowrie, and one a pierced half of a small bivalve. These molluscs (all from the surface) are more probably to be ascribed to Somali occupation.

Technique. The greatest part of the factory debris is microlithic but some larger flakes occur and seven of these show crude preparation of the striking-platform. This preparation is probably fortuitous as no prepared cores were found with the exception of five pseudo-disc cores. The typical single-platform and bipolar blade cores are represented. That a greater number of cores was not found may perhaps indicate that the majority were left at the factory site where the raw material was obtained. The percentage of flakes to blades is approximately double, but the greater part of these flakes are macrolithic. A number of the flakes show crackle due to fire. While, therefore, the predominant technique was directed towards the production of micro-blades and flakes, the large flakes indicate the persistence of macrolithic forms, presumably in connexion with the large, foliate points. The pressure flaking employed on the finished tools shows a high degree of workmanship.

This industry shows certain features in common with the true Doian which determines its associations; notably the pressure-flaked, lanceolate, limace and foliate points but the number of small end- and thumbnail-scrapers and the angle-burin indicate connexions with and borrowings from the contemporary Somaliland Wilton people. The great rarity of microliths and the profusion of crude foliate points are, however, characteristics which differentiate the Mirsale industry from both, though the microliths which do occur are similar to the Doian rather than to the Mudugh Wilton forms. The absence of hollow-based arrowheads, of rubbers, and of pestle stones is of interest, but the industry has a general 'Neolithic' appearance particularly about the pressure flaking.

[1] Not used as water-containers but as exterior decoration to the apex of the roof of a hut.

CHAPTER 12

THE SOMALILAND WILTON CULTURE AND ITS VARIANTS. CONCLUSIONS

THE SOMALILAND WILTON CULTURE

This culture has been so called because in its more usual form it approximates closely to those Late Stone Age industries and cultures which may be grouped together under the term Wilton Complex. There are individual and local differences which distinguish the Somaliland form of this culture from its colateral branches in East, Central and South Africa, but the stone culture of this group of industries from the northern and western part of the Horn is sufficiently close to this Late Stone Age complex to warrant the term Wilton being applied to it.

The distribution of the Wilton is shown in Text-fig. 28 and it will be seen that in its typical form it is restricted to the northern plateau and coastal plain, with outliers in the eastern Mudugh and the southern Danakil Rift. A somewhat different form of Wilton is found in the Ogaden and on the Webi Shebeli. This is considered to represent a variant of the main culture and has been tentatively termed the Ogaden Wilton. More information is required about this variant.

A strandlooping form of Wilton, somewhat affected by Doian and perhaps Levalloisian (?) influences is known from the mouth of the Nogal and one further form from the eastern Mudugh shows more developed 'Neolithic' characteristics in its stone industry. These variants will be considered separately.

I. THE TYPE SOMALILAND WILTON

Numerous surface sites exist throughout the northern part of the Horn but the type site is a small rock-shelter, the Gumbur Todoballa, situated at Mandera, near the foot of the escarpment leading from Berbera to Hargeisa. Another rock-shelter site also exists at the pass through the coastal hills on the road to Burao from Berbera. Afgudut, a well area in the eastern Mudugh has also provided a good series eroding from wind-blown sands.

Geological horizon. Second post-Pluvial Wet Phase and extending into the succeeding dry phase.

Climate. Semi-arid changing to arid.

THE GUMBUR TODOBALLA ROCK-SHELTER, MANDERA

Stratigraphy. This small shelter provides the type site for the Somaliland Wilton Culture. It is situated in an isolated, pointed, residual ridge of Archaean rock, from which it takes its name,[1] and which rises straight out of the peneplain

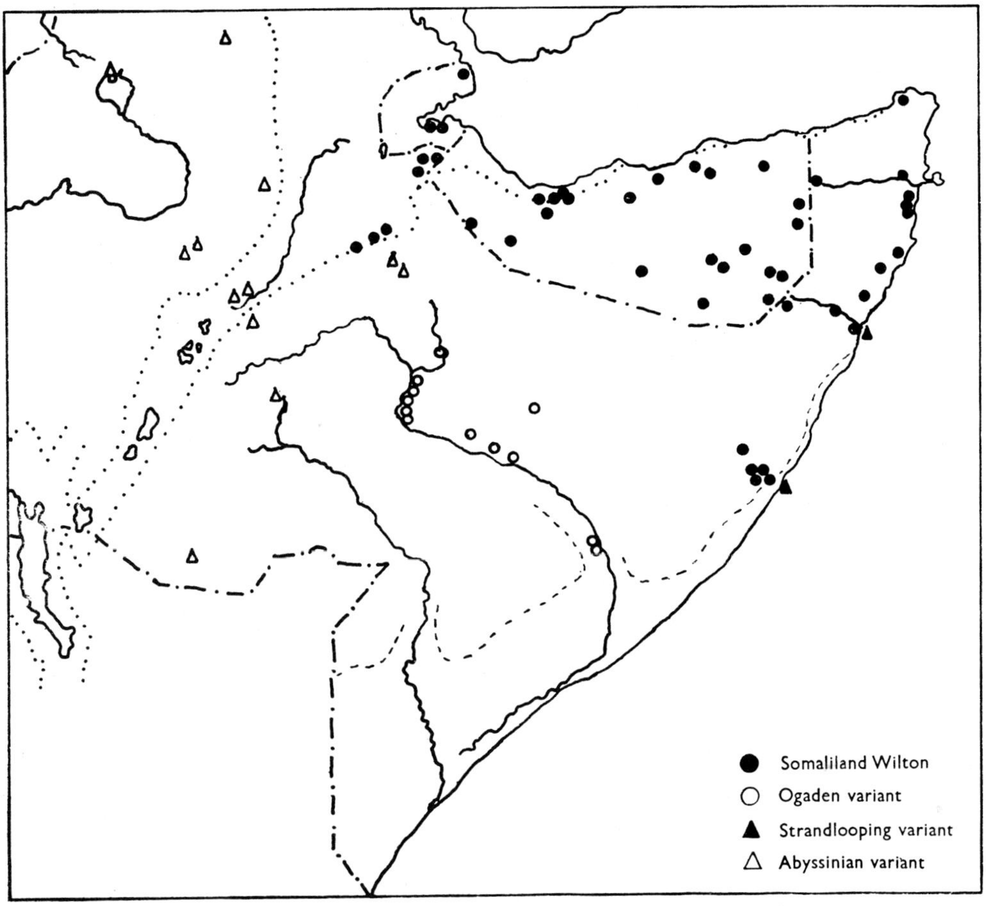

Text-fig. 28. Distribution of Somaliland Wilton industries.

surface about 1 mile to the east of Mandera Camp. The western side rises almost vertically from the Mandera Tug. When viewed from north to south this granite and gneiss residual is really a ridge of three points of rock which diminish in height from west to east. The highest rises some 250 ft. out of the plain and contains several small shelters. The largest of these is situated on the west side but does not appear to have been inhabited except by the present-day Somalis

[1] The hill derives its name from the Somali words 'Gumbur'—a small, isolated hill, and 'todoba'—the numeral 7, and was so called as it featured as the place of the last stand of seven Jibril Abokr robbers who were caught raiding stock by the local Esa Musa. (Swayne, H. G. C. (1895, p. 243).)

who use it as a shelter for sheep and goats. On the north side are two further shelters, which, although small, contained signs of occupation by a people practising a microlithic industry, the factory debris from which was found in the scree, which, owing to the steep nature of the slope below the shelters, had fallen and lay scattered down the hillside in several places. These shelters are situated at a height of about 50 ft. above the plain and face almost due north; there are signs that they featured in the fighting at the time of the temporary Italian occupation of the Protectorate, and they are still used today by the local Somalis as temporary living sites as is testified by the remains of hearths, bones and low thorn zeribas for stock. It is of interest to note that while these shelters would afford no protection from the north-east monsoon during the 'dair' rains, now blowing between October and December, they would provide complete protection from the south-west monsoon—the Karif—which usually carries with it the heaviest rains and blows between May and August.

It was decided to excavate a trench in the right-hand one of the two shelters and an area was accordingly marked out on the platform immediately in front of the shelter where it was considered that the greatest depth of deposit could be expected. The cave floor inside the shelter itself showed no evidence of Stone Age occupation, being composed mainly of a fine, light brown dust with a percentage of organic material derived from decayed vegetable matter. Details of the excavation can be seen from the plan and section-drawings of the site (Text-fig. 29), but a word of explanation may be given here. The rock floor was shallow, being covered by a few inches only of cave earth, and rose gradually towards the rear and the right-hand side. On the left the cliff falls almost sheer to the foot of the hill. On the right the floor rises steeply to the rear wall which meets the roof some 10 ft. above the floor in a long fissure. The platform outside the shelter dips gently to the right and the greatest depth of deposits occurred here. The deposits rise again steeply at the right edge of the platform. The excavation was divided into two main sections each subdivided into three subsections, and excavated separately.

From the surface were recovered a few flakes of crystalline quartz and chert which were found in a powdery and gritty brown earth or dust, containing also numerous angular rock fragments (granite, gneiss, etc.); two small fragments of fossilized bone also came from this level. This deposit continued for a depth of from 2 to 4 in. and then gave place to a more compact but still loose and gritty, red to brown earth, containing, in addition to the quartz flakes, a few microliths and a little bone. There occurred also several large, angular stones of local rock which may have been used as artifacts but were so decomposed that it was not possible to be certain on this point. The brown earth was sandy in texture and appeared to represent an accumulation of loose sand and grit with angular lumps of granite and gneiss fallen from the rock above. Some of the sand is probably wind-blown and the deposit most resembles a detritus accumulated during a dry phase such as exists in the district today. At a depth of from 5 to 10 in. this brown earth passes suddenly into a compact, black earth. This deposit is deepest at the right, rear, part of the excavation and slopes

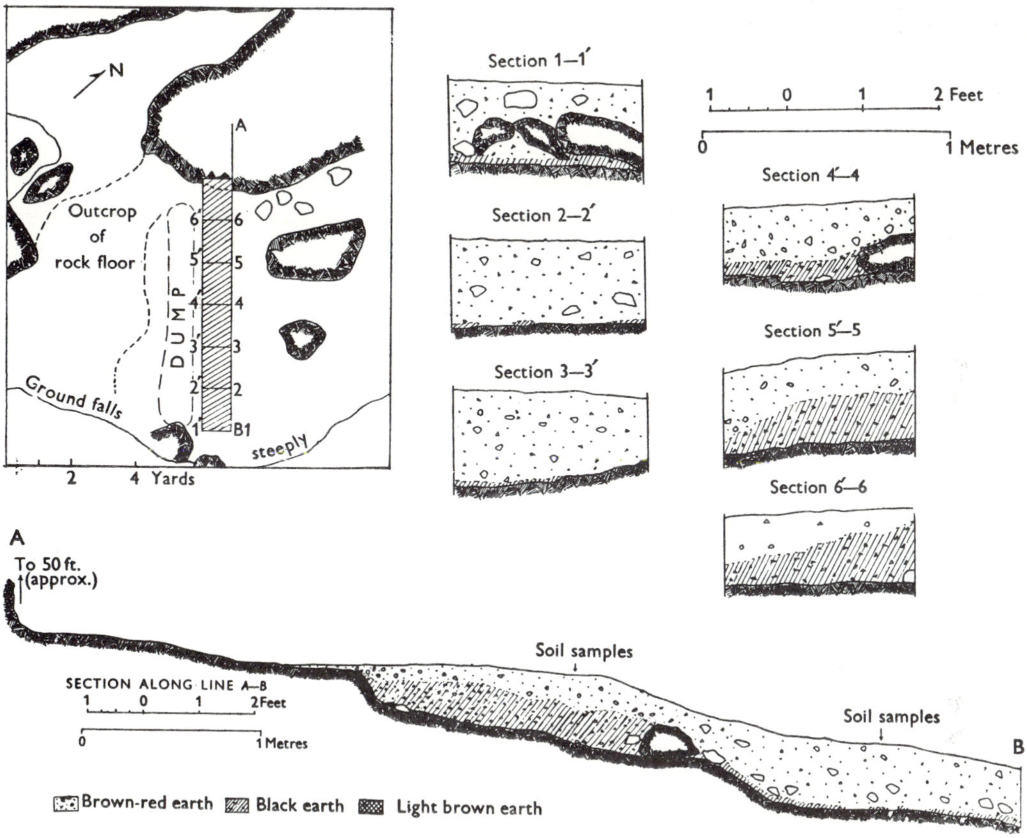

Text-fig. 29. Plan and sections, Gumbur Todoballa, Mandera.

steeply away towards the left and front of the platform. The deposit consists of a fine, grey to black earth which contains a number of small fragments of granite rock and an occasional larger, weathered fragment. It is a typical occupation deposit and contains the Somaliland Wilton industry. The implements are derived in the main from the junction of the brown and black earth and in the topmost one-third of the lower deposit. While they continue to the base of the black earth they are not as numerous here as at the top, although factory debris appears to be equally common throughout. No bone or charcoal was found but a high percentage of the factory debris and implements showed 'crackle' due to heating by fire. The almost complete absence of bone and charcoal may probably be explained by the fact that the deposit lying fully exposed to the weather, only imperishable material such as stone, has survived. The disintegration of the charcoal has possibly given this deposit its characteristic, black appearance. In places below the black earth are found a few lenses, in hollows in the underlying rock, of a deposit similar to the brown earth overlying the occupation level. This should perhaps be interpreted as a detritus which had accumulated prior to the occupation by the makers of the Somaliland Wilton. This basal deposit was found to be sterile.

The excavation showed a natural division into two parts due to the presence of a large boulder which either fell from the cliff above or was placed intentionally at a time almost coincident with the occupation of the shelter. This rock effectively retains the black earth which is found between it and the rear of the shelter. On the opposite and outer side of the rock the black earth is found only as an intermittent, thin layer immediately above bed rock and is overlain by the brown earth which in its lower levels becomes almost red in colour. At an average depth of 7 in., that is to say about the middle of the red earth, implements and factory debris occurred and continued sporadically to the rock-floor; artifacts were not nearly as common here as in the black earth, however. Some small fragments of bone and a small piece of enamel from the tooth of a large antelope as well as three sherds of pottery were recovered from this deposit. While the black earth contained a preponderance of microliths, the red earth on the opposite side of the retaining boulder yielded practically no microliths and artifacts consisted mainly of blades and scrapers.

The question to be decided is whether the lower part of the brown to red earth in front of the boulder is contemporary with the black earth and whether the industry contained within it is likewise contemporary. The fact that in places the black earth occurs as thin lenses underlying the red earth seems to suggest that the latter is in the main a more recent deposit and the presence of small, mineralized fragments of bone and of pottery (though there appears to be no difference in the stone industry), seems to bear this out. The red earth appears to have been contemporary with the later stages of deposition of the black earth. There is no doubt that only one industry is represented here which was in the main coincident with a period of increased humidity correlated with the Second post-Pluvial Wet Phase.

Excavation ceased just before reaching the steep slope marking the front edge of the platform. There was found a series of rocks and boulders which had fallen or been washed down from the cliff above and had become wedged or rolled into position on the edge of the platform and overhung the slope. These boulders were firmly fixed in position, the interstices being filled by smaller debris and brown earth. Soil samples were taken at two points in the excavation.

THE INDUSTRY. A complete inventory of tool types is given in the table on p. 267.

Crescents. (Pl. 25, nos. 10–15.) These are typical, and some quite small specimens occur. They are backed from the under side only.

Double-backed blades. (Pl. 25, nos. 16, 17.) These are the only two specimens found and one of them shows a double patina.

Short end- or round scrapers. (Pl. 25, nos. 22, 23.) These are entirely characteristic of the culture. They are smaller than those found with the Hargeisan, but there can be little doubt that they are developed from them.

Thumbnail-scrapers. (Pl. 25, nos. 24–26.) These tend to the 'fan-shaped' form and are usually single-ended though one good, double-ended specimen was found. Both forms are typical.

Outils écaillés. (Pl. 25, no. 27.) The typical fabricator.

Burins. (Pl. 25, no. 18.) No micro-burins were found and no by-products of the micro-burin technique occurred. The usual forms of burin are angle (Pl. 25, no. 18), of which one good example was found, or single-blow.

Utilized blades. (Pl. 25, nos. 19, 20.) Blades showing signs of utilization but with no intentional retouch on either edge are a typical form.

Chopping tools. Two flat, weathered pieces of granite were probably used as crude choppers.

Pottery. (Pl. 25, no. 28.) Three sherds were found close together at a depth of $10\frac{1}{2}$ in. in the brown-red earth in section 3/4 against the left-hand wall of the trench and in association with the double-ended, thumbnail-scraper (no. 24). The sherds are derived from two pots, and two of the pieces join and probably represent a break due to the weight of the overburden of the deposit. None of the pieces are rim sherds. The paste is darkish brown, felspathic and well fired. There is no evidence of burnish or decoration of any kind. Average thickness 5 and 6 mm. respectively. Edges rubbed and weathered except at joins. There seems to be no reason to dissociate this pottery from the Wilton industry, though the fact that only three small sherds were found suggests that it was a rare item of material culture, or possibly that these people merely collected sherds as the Bushmen in South Africa used at one time to do.[1]

Pigment. Some pieces of haematite, red ochre and one piece of white earth (calcrete?), were found. None showed any definite signs of striation or rubbing planes, however.

[1] Schofield, J. F. (1948, p. 36).

Ostrich egg-shell. One small fragment only from the upper deposit. It is possible that, as with the bone, exposure to the weather has destroyed other shell remains.

Fossil bone. Only small weathered fragments of bone and two pieces of enamel from antelope teeth were recovered, but these fragments are fully fossilized.

Utilized flakes. A few largish, primary flakes with plain striking-platforms were found utilized down one or part of one edge.

Hammer and pestle stones. Two hammer stones in quartz were recovered and one pestle stone in limestone and five polyhedral stones in quartz—these latter may be detaching hammers or possibly worked out, multi-platformed (nobbly) cores.

Various. Included in this category is one piece of weathered aplite or gneiss.

Technique. The micro-burin technique was unknown. All artificially fractured stone was retained and there is not a single piece which is either a micro-burin or shows any evidence of micro-burin technique. The industry is essentially microlithic specializing in the production of small blades from single-platformed, conical or bipolar cores (Pl. 25, no. 21).

As every piece of debitage and waste material was retained the percentage of tools (16 per cent) to debris can be accurately computed. Out of a total of 374 flakes, only fifteen were large, primary flakes, the remainder being microlithic and struck from diminutive cores. No prepared striking-platforms occurred nor were any disc cores found. A number of specimens show 'crackle' from fires.

This culture is the logical development of the Hargeisan industries and its distribution covers approximately the same area.

AFGUDUT

Situation and stratigraphy. This is a water-hole in the eastern Mudugh some 38 to 40 miles south-south-west of Mirsale wells. The site consists of the usual shallow pan, in the Eocene anhydrite series surrounded by low hills of the same rock. A well has been dug near the centre of the pan through a long irregular sink or fissure in the red gypsum from which the well derives its name.[1] Weathered gypsum rock is exposed in the centre of the pan which is surrounded by orange to yellow wind-blown sands, similar to those found at Mirsale and Rakon and some 3 ft. or so in thickness.

The area was first recorded as yielding Stone Age material by Stefanini who found there a few indeterminate artifacts comprising two discs, one blade without retouch, one end-scraper on a blade and two atypical flakes. The only evidence

[1] From *Af* meaning 'mouth' and *gudut* meaning 'red'.

Inventory of tools from type sites of the Somaliland Wilton Culture

	Microliths, straight-backed	Microliths, curved-backed	Microliths, crescents	Microliths, double-backed	Short end- or round scrapers	Thumbnail-scrapers	*Outils écaillés*	Burins—angle	Burins—single-blow	Burins—*bec-de-flute*	Burins—gouge	Utilized blades	Utilized flakes	Chopping tools	Hammer and pestle stones	Blades	Flakes	Cores—single-platformed	Cores—bipolar	Cores—multi-platformed	Debitage	Pottery	Pigment	Ostrich egg-shell	Fossil bone	Various	Totals
GUMBUR TODO-BALLA																											
Brown earth.	4	2	1	—	2	3	2	—	2	—	—	7	5	—	8	40	97	1	—	—	68	—	1	1	10	1	255
Layer A brown/red earth outside boulder.	—	2	1	1	5	2	—	—	—	—	—	4	—	—	—	16	54	3	5	2	17	3	1	—	8	3	127
Layer B black earth.	1	13	7	1	14	1	1	1	—	—	—	2	2	2	1	49	223	1	2	1	64	—	7	—	—	1	394
Totals	5	17	9	2	21	6	3	1	2	—	—	13	7	2	9	105	374	5	7	3	149	3	9	1	18	5	776
AFGUDUT WELL	28	46	14	2	9	4	—	2	1	1	4	—	—	—	—	68	15	6	2	4	—	—	—	14	—	10	230

of prehistoric occupation found by the writer belonged to the Late Stone Age. From the east and south sides of the pan was recovered a good series of Somaliland Wilton artifacts in a small area some 100 yd. from the centre of the pan. The tools are eroding from the orange to yellow sand in the same way as were those at Mirsale and Rakon. All were made from grey to brown chert and showed varying degrees of white or mottled patina. In some cases they also showed weathering due to wind action. The north and west sides of the pan appeared to be bare of occupation material.

The industry must be considered to be contemporary with those at Mirsale and Rakon and is assigned to the Second post-Pluvial Wet Phase, continuing probably into the succeeding drier period.

The industry. (Pl. 25, nos. 29–60.) The inventory of tools is given below and is sufficiently close to that from the Gumbur Todoballa not to require detailed description. The more important tools are the burins, angle (Pl. 25, nos. 54, 55), *bec-de-flute* and single-blow varieties, as well as single and double ended gouge-burins on small bipolar cores. The microliths are of two kinds, the usual curved-backed variety (Pl. 25, nos. 29–36), including crescents (nos. 39–43) and one triangle (no. 44) (the backing being directed from the under side only), and straight-backed microliths (nos. 45–52), usually smaller than the first group with the backing worked from both faces; these latter sometimes show basal retouch. The double-backed microlith (no. 38) is present, and one atypical shouldered specimen was found (no. 37). Scrapers are of the usual short end, or round forms (nos. 56–58) and thumbnail (nos. 59–60). Not included in the inventory are seven broken ends from short end-scrapers. Cores are of the usual single- and multi-platformed and bipolar (no. 53) types. The blades, some of which show signs of utilization, vary between 2 to $\frac{3}{4}$ in. in length. Ten redirecting flakes struck from the length of a core occur. There is again no evidence of the micro-burin technique. Fragments of ostrich egg-shell testify to the probable use of these shells as containers. The absence of pottery is not surprising in an area where no pottery exists even today.

The value of the Afgudut industry lies in the fact that it represents a pure form of the Somaliland Wilton in a datable horizon on an open station site. The interpretation of the deposit in which it is found corroborates that of the Gumbur Todoballa sequence, and the site can be dated in relation to the Doian and 'Neolithic' form of the Wilton, with which it has been shown to have been contemporary.

ADDITIONAL SITES

A complete list of sites will be found in Appendix E, but the more important may be noted here. They include the Berbera rock-shelter, the lower Nogal sites at Eil, the eastern British Somaliland sites found by Gilliland, the painted rock-shelter sites of Gerbakele and Dukokolol-yero, found by Glover on the northern plateau, the Danakil Rift sites of Lago Oda, Porc Epic and Aisha, and the painted rock-shelter of Errer Kimiet I found by Björn von Rosen in the region of Kondoddo on the Harar plateau.[1]

Satisfactory corroborative evidence for the Gumbur Todoballa sequence is provided by the two rock-shelters in the barrier hills on the Burao road some 12 and 13 miles respectively from Berbera. Both shelters are developed in Eocene limestone. The first is situated some 50 ft. up in the cliff on the right-hand side of the entrance to the pass through the barrier hills. It is a small rock-shelter some 20 ft. wide across the drip-line and 15 ft. deep. The roof is some 10 ft. above the floor at the drip-line and slopes gradually to meet the floor at the back of the shelter. A small trial trench was dug in this cave just inside the entrance, and the following sequence obtained, from the surface downwards:

(i) A bed of scalings from the roof, some 4 ft. thick at the back of the shelter, thinning out gradually towards the entrance, where it was found to be a foot or more thick.

(ii) A sandy deposit with numerous roof scalings, containing occupation material of Somaliland Wilton type. This bed was approximately 6–8 in. thick.

(iii) A further deposit of rock scalings retained at the entrance by rocks which had fallen from the roof of the shelter. This bed was thickest apparently at the entrance where it is some 2 ft. thick.

The cultural material from layer (ii) included two *bec-de-flute* burins, three end-scrapers on short and long blades, one straight-backed microlith, one curved-backed microlith, twenty-eight blades of which approximately half show evidence of utilization, five small primary flakes, one redirecting flake, one small bipolar core, two fragments of ostrich egg-shell and one fragment of mollusc shell. The stone tools are all in chert and are unpatinated as is the case with the material from Gumbur Todoballa.

The second shelter was at ground-level, approximately 1 mile further on and on the left of the road. The floor sloped downwards so that the surface of the

[1] Von Rosen, B. (1949, pp. 382–4).

deposits within the shelter is now at ground-level, and the lower deposits are below this. The shelter stands about 4 ft. above the level of the tug approximately 15 yd. distant. In extent the shelter is some 30 ft. wide at the drip-line, 10–20 ft. deep and 8 ft. high. A pit had been dug in the deposits for road metal and showed the following sequence from top to bottom:

(i) Bed of scalings from the roof mixed with grey to yellow, wind-blown sand; 1–2 ft. Sterile.

(ii) Grey to black sand and scalings from the roof, with charcoals and unfossilized, animal bones, artificially fractured. 3 ft. No artifacts.

(iii) River-deposited grit, sand and fine gravel, probably representing a flood deposit filling the front half of the shelter. 1–2 ft. thick, thinning out to nothing at the rear, where it is banked against the underlying deposit. Sterile.

(iv) Thickish bed of scalings from the roof and walls. 4 ft. Sterile.

(v) Uneven bed rock—Eocene limestone.

The deposits in these two shelters are considered to indicate (*a*) a dry phase (layer (iii) in the first shelter and layer (iv) in the second shelter), followed by (*b*) a minor wet phase gradually becoming drier (layer (ii) in the first shelter and layers (ii) and (iii) in the second shelter); followed by (*c*) a further drier phase which continues to the present day (layer (i) at both shelters). The wet phase is tentatively equated with the Second post-Pluvial Wet Phase.

II. THE OGADEN VARIANT AND DERIVATIVES

The evidence warranting a distinction being made between the Wilton found in the Ogaden and that on the northern plateau and Rift areas comes in the main from one site situated at Gabredarre on the Tug Fafan. This was the only stratified site found by the writer which yielded typical Ogaden Wilton material, but other sites, notably those in the neighbourhood of Giglei have yielded industries which are tentatively considered to be derived from this Wilton variant. The main differences distinguishing the Wilton in the Ogaden from that on the northern plateau are in certain elements of technique, secondary retouch and the apparent absence of the angle-burin and the end-scraper.

GABREDARRE, TUG FAFAN

The site is found approximately 1 mile upstream from the Administrative Headquarters, near the water-point. Here the Fafan runs through banks some 10–15 ft. high, eroded to the top of the kunkar band marking the top of the lower calcareous alluvium (see Text-fig. 12). Small tributary gullies drain into the main tug and have dissected the upper deposits. Towards the centre of the valley the topmost deposit, a friable, brown alluvium resting on the eroded surface of the upper calcareous alluvium, is compact and deepest, attaining a depth of some 3 ft. On the south-west slopes of the valley (the north-east bank was not investigated), the deposit thins out, following the contour of the valley, is much more friable and consists of a fine, brown dust, 1 ft. or so in depth. This stratum appears to be partly water-deposited (towards the centre of the tug), partly wind-deposited, and is presumed to indicate semi-arid conditions.

On the land surface between the eroded upper calcareous alluvium and the brown alluvium, is found in an unrolled, but slightly weathered state, the Ogaden Wilton assemblage. It is fairly prolific over the whole of the area examined, but consists mainly of factory debris.

The age of this assemblage depends on the age of the brown alluvium and on whether this deposit followed directly upon the minor degradation period when the upper calcareous alluvium was eroded, or whether this erosion surface is earlier, and there was a time interval between the erosion phase and the accumulation of the brown alluvium.

'Kunkar' nodules are fairly numerous on this surface and suggest that the one event did not follow immediately upon the other. The writer would tentatively equate the brown alluvium with the deposits accumulated during the Second post-Pluvial Wet Phase.[1] If, on the other hand, the brown alluvium represents the First post-Pluvial Wet Phase deposits, then the Wilton in the Ogaden must be considered to have appeared here earlier than in other parts of the Horn.

Geological horizon. Second post-Pluvial Wet Phase (?).

Climate. Semi-arid to arid.

The industry. (Pl. 24, nos. 46–56.) Tools consist of backed blades and crescents and some utilized blades. No pressure-flaked points of any form occur. Angle-burins are absent and no true bipolar cores are found.

Microliths. These comprise thirteen ordinary, curved-backed blades (no. 52), five crescents (nos. 53, 54), three triangles (no. 55), one with a shoulder or concave retouch (no. 56); also four straight-backed blades with no basal retouch

[1] The industry beneath it must therefore antedate, probably not by long, the accumulation of this alluvium. The First post-Pluvial Wet Phase deposits are either not represented in this particular part of the valley or else the upper calcareous alluvium, which contains at its base an Upper Levalloisian industry of more developed form than is found elsewhere, represents them. If the latter should prove to be the case, then the Levalloisian must be held to have continued in the Ogaden into post-Pluvial times.

(no. 51), and one with basal retouch. One curved-backed blade shows a notch on the cutting edge, and two others show Kasouga retouch. The backing on all these microliths has been executed from both faces of the blade to give a uniform convex back to the tool. Working from one edge only does not occur, and the microliths show a markedly wedge-shaped section.

Utilized blades. (Pl. 24, nos. 49, 50.) Eight specimens were found. These are typical and show signs of utilization but no secondary retouch.

Burins. Two poor *bec-de-flute* burins were found.

Scrapers. The complete absence of scrapers of the short end and thumbnail varieties may be fortuitous, but is more likely to be a true indication that such tools do not occur in this variant. The only scraper, if such it can be called, is a micro-disc which shows evidence of marginal retouch.

Technique. The factory debris collected comprised eighty microlithic blades and micro-flakes; forty-eight broad, primary flakes with unprepared striking-platform, one redirecting flake, one *outil écaillé*, twenty-three broken flakes and other debitage and four small pebbles in quartzite and lava which showed signs of utilization, probably as hammer stones.

In addition to the above, however, there were recovered ten single-platformed blade cores (Pl. 24, no. 46), two multi-platformed cores (in this instance with two platforms at right-angles to each other), and one flat, bipolar or pseudo-disc core. The above cores are typical of those found with the Late Stone Age industries in the Horn, but there were also found types which did not occur at any other site. These comprised five diminutive, disc cores and eight flat, alternate-ended cores (nos. 47, 48). These diminutive, prepared cores are of considerable interest. Apart from their size they are identical with the prepared cores found at the base of the upper calcareous alluvium, with a developed form of the Upper Levalloisian industry, and there can be no doubt that they represent the final development of this type of core. The alternate-cores are flat and subrectangular, worked only on the upper face, and were employed for the removal of short, broadish blades.

Although the range of tools and the quantity of material yielded by this site is not large, it was derived from a sealed deposit and the writer believes that it is of considerable interest as there can be little doubt that the industry represents the ultimate development of the Levalloisian, or the epi-Levalloisian, in the Horn. By a gradual process of reduction in size the flake-blade core has become the single-platformed micro-blade core, and the alternate cores have developed into a kind of pseudo-bipolar core. The finished tools, divorced from the factory debris, might pass as Somaliland Wilton or Doian, but it is in the basic technique and the absence of the 'neanthropic' element that this industry differs from both.

It is more closely akin to the Doian in technique but is effectively distinguished from it by the complete absence of pressure-flaked points.

It would appear that in this industry and that from the base of the upper calcareous alluvium we have the Somaliland counterpart of the Egyptian Sebilian Culture and that the local form of the Levalloisian continued in some areas to develop to a large degree unaffected by the Stillbay-Magosian complex or by the intrusive blade and burin industries in the north.

ADDITIONAL SITES

A surface site at Kaptinag (66) can probably be assigned to the Ogaden Wilton, as well as some of the sites found by the du Bourg de Bozas Expedition in the western Ogaden, notably stations 13 and 15 on the left bank of the Webi Shebeli.

THE GIGLEI INDUSTRIES

Stratigraphy. On the right bank of the Webi Shebeli, 3½ to 3 miles on the Beletwein side of Giglei, at sites named by Somali informants Habr Fordisso and Durdur Heime, respectively, are found living sites yielding industries which appear to show connexions with the Gabredarre industry, though the latter is presumed to be earlier.

These sites are situated on low dunes which overlie the alluvium and gravel of the First post-Pluvial Wet Phase. In this gravel occur shells of the freshwater oyster. At Habr Fordisso (80) the dunes are some 10–12 ft. high and are composed of white to yellow, compact sands turning to a loose grey sand at the surface. They are approximately ¼ mile from the river bank. The artifacts are eroding from the top levels and surface of the dunes, which at this site are two in number, situated about 300 yd. apart and just to the right of the road. The artifacts are found on the summit of the dunes and at intervals down the slopes where erosion has carried them. Numbers of mussel shells (*Unio*) litter the site as well as fragments of ostrich egg-shell and there can be little doubt that the makers of this industry were a riverine people living on fish and shell-fish. The dunes themselves appeared to be sterile and the excavations at Durdur Heime established that this was in fact the case. The inventory of tools recovered from these dune sites is given below (p. 275).

DURDUR HEIME (81). This was an interesting home-site found on a small, low dune overlying the brown alluvium and gravels, about 25 yd. to the left of the road and 3 miles from Giglei. The dune or mound was only some 200 yd. from the river which here flowed over rapids, the rock being limestone. The mound is about 6 ft. high and 40 yd. long (see Sketch-map, Text-fig. 30). The surface was littered with artifacts, in fact this was one of the richest open station sites for its size anywhere in the Horn. The mound is being slowly eroded and is undercutting the occupation level, which is on the top of the mound. There is no occupation material found on the slopes unless it be an isolated specimen that has fallen down from above, and there is not a single specimen on the surface of the brown alluvium on which the dune or mound rests. At one end of the mound is a 'Galla Grave' which contains no artifacts and is presumably later in age than the Stone Age occupation.

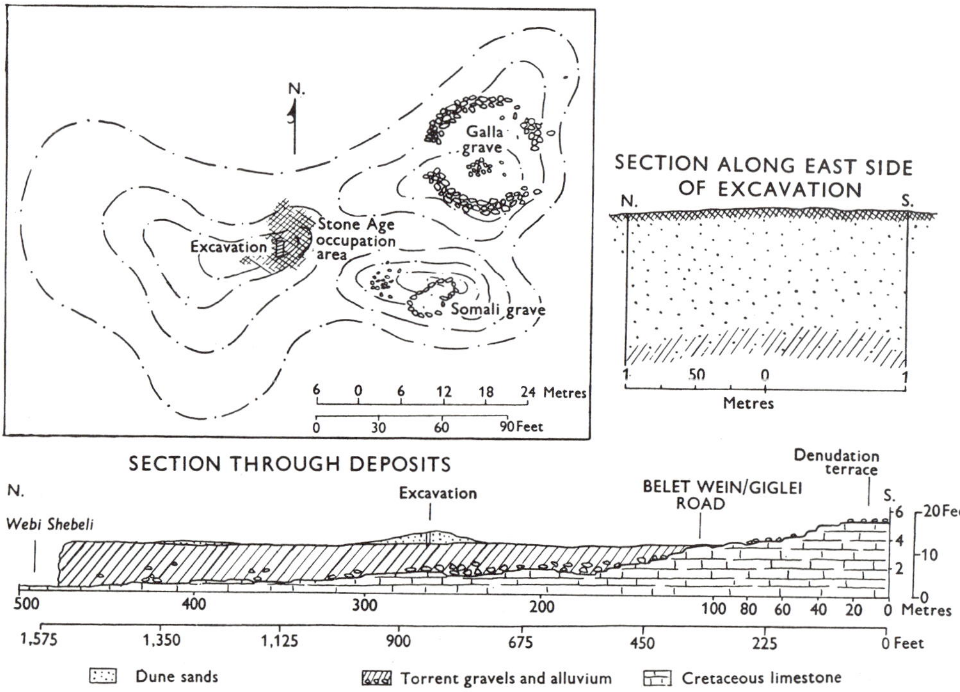

Text-fig. 30. Plan and sections, occupation dune, Durdur Heime.

In order to establish the horizon from which the artifacts were deriving a trench was excavated across the centre of the mound, 6 ft. long by 3 ft. wide. This excavation showed that the occupation material occurred only in the topmost 6 in. and that the great majority of tools were in the upper 4 in. of this. The deposit was a loose, white to grey sand on the surface and passed imperceptibly into a compact, white to yellow dune sand which was quite sterile. The amount of gypsum in the sand appears to decrease with depth. This sand rested in turn on the brown alluvium, which in a tributary gully just to the south of the mound was seen to be 4 ft. thick, and resting on 2 ft. of coarse, angular gravel which was also sterile.

Inventory of tool types Giglei Industries: Ogaden Wilton

	Durdur Heime	Habr Fordisso	Totals
Large backed blades	12	—	12
Microliths:			
Curved-backed blades	141	4	145
Straight-backed blades	71	8	79
Straight-backed blades with basal retouch	12	1	13
Backed blades with Kasouga retouch	14	—	14
Crescents	19	—	19
Double-backed blades	3	—	3
Backed blades with concave retouch	7	—	7
Burins	3	1	4
Scrapers:			
Hollow	1	—	1
Thumbnail	—	5	5
Utilized flakes	9	5	14
Ostrich egg-shell	6	16	22
Fossil bone	42	—	42
Hammer-stone	1	—	1
Large flakes	1	3	4
Points	1	1	2
Rubbers	2	—	2
Outils écaillés	1	2	3
Blades, some utilized	147	24	171
Flakes	73	14	87
Cores:			
Single-platformed	5	3	8
Multi-platformed	2	—	2
Micro-disc	—	1	1
Alternate ended	8	—	8
Redirecting flakes	3	2	5
Various: snapped blades and flakes	52	—	52
Totals	636	90	726

These white to buff dunes on the river banks are considered to be the counterpart of the compact, red sands on the higher slopes of the valley sides. This was proved to be the case at similar sites yielding a comparable industry at Bulo Burti at the back of the Residente's house. The industries which rest on these dunes are therefore later in date than the dry period which follows the First post-Pluvial Wet Phase and they are believed to belong to the wet phase which followed this. They are therefore the contemporary of the Doian and Somaliland Wilton, but are probably somewhat later than the Gabredarre industry.

Geological horizon. Second post-Pluvial Wet Phase.

Climate. Semi-arid to arid.

The industries. (Pl. 24, nos. 57–75.) These are rather monotonous and consist almost entirely of backed blades and crescents and utilized flakes. End-scrapers and points do not occur, with the exception of two large, crude unifaced specimens. All the tools are made from chert with the exception of one microlith in silcrete, one piece of obsidian and one large, lava flake from Durdur Heime, two large flakes of lava and one of limestone from Habr Fordisso.

The only abundant form found in these industries which are monotonous in the paucity of tool types is the rather large microlith. The characteristic type is curved backed (Pl. 24, nos. 68–70, 72, 75), but the straight-backed microliths, with (no. 71), and without (nos. 67, 73), basal retouch are also common. A small number of crescents also occur (nos. 63–65). A percentage of microliths show Kasouga retouch down the length of the cutting edge (no. 66), or nibbled notches due to utilization. A rare type is a microlith where an incipient shoulder has been formed by concave retouch at the butt-end (no. 74). A few large backed blades occur and grade into the microlithic forms: the length of the former is usually over 2 in. (nos. 61, 62).

Other tools include primary flakes which show signs of utilization down one edge—possibly as scrapers, and a few poor thumbnail or diminutive end-scrapers from Habr Fordisso. The only well-made scraper that was found was a hollow scraper (no. 59). A few burins also occur. These are *bec-de-flute* (no. 60), but the burin facets tend to curve over from the edge on to the face of the blade. One good example of the oblique, straight angle-burin was picked up at Habr Fordisso, but it seems so strangely out of place in such a milieu that it is possible that it may not belong to this industry.

Two further types of tools are worthy of note; the one is a rare, unifaced point, of which two crude, thick specimens were found. The working is of a resolved nature and is confined to the edges. They might be referred to as hand-picks rather than as points (Pl. 24, no. 58). Similar tools were found with the strandlooping industry from Biede at the mouth of the Nogal. The second type of tool consists merely of a large, primary flake of lava or limestone showing a plain striking-platform and no secondary retouch (Pl. 24, no. 57).

Technique. Cores are similar to those from Gabredarre. The alternate-ended is again present, though it now approximates a little more closely to the true bipolar blade core. The blades require no description, but the flakes are of two kinds, the broad, primary flake a small number of which show incipient preparation of the striking-platform—possibly fortuitous, and small micro-flakes struck from blade and alternate-ended cores. Some of the factory debris shows 'crackle' from fires.

The backing on the microliths varies considerably in workmanship. Many of the straight-backed forms (nos. 61, 73), are merely backed flakes or broad blades, others are better made, however, and compare with the Doian forms.

The backing is usually from both sides but there are some specimens with flat backing from the underside. A number of broken specimens occur.

The Giglei industries must be presumed to reflect a food-gathering economy based on fish and shell-fish, though the small fragments of fossil bone seem to indicate that such a diet was eked out with a more substantial food. The crude points or hand-picks and the large flakes could have been used for opening shell-fish of which the freshwater mussel and oyster are the commonest forms. Quite how the microliths were used is difficult to determine. The straight-backed form may possibly have been used as a gorge though it is more likely that the fish were caught in reed traps and weirs and by spearing than by line or net. That fish were eaten is indicated by the occasional presence of vertebrae. It would appear that these nomadic hunters and fishers formed temporary camps on the low sand hills and mounds at a time when seasonal flooding of the river swamped the alluvial flats and conditions were most favourable for the trapping and spearing of fish. Apparently similar industries but with Doian features occur at other sites in the Webi Shebeli valley.

III. THE 'NEOLITHIC' VARIANT

This variant is known as yet from only one site in the eastern Mudugh. This is situated at Rakon some 10 miles north of Mirsale and 63 miles south-east of Galkayu, on the road to Obbia.

Here are two shallow, circular pans about 300 yd. apart, which have formed in the flat, monotonous gypsum plain. The pans are closely similar to the one at Mirsale, and two of the wells in the southern pan are re-excavated sinkholes or cracks. Except in the pans themselves the gypsum rock is overlain by an orange to red sand identical with that at Mirsale (see Text-fig. 31).

Stratigraphy. The main occupation area is found on the left of the road going towards Obbia, between the two pans—that is to say on the north-west side of the southern pan. This occupation seems to have been limited to the one area, as the remaining sides, as well as the northern pan, are devoid of material other than a sparse factory debris. The orange and red sands which appeared to be some 3–4 ft. thick on an average were being eroded by wind action in their upper levels and in the shallow depressions so formed was found a prolific industry of considerable interest. A small excavation was accordingly undertaken at this site in order to establish the sequence and the horizon from which the industry was deriving.

An area of 1 m. square was excavated near the centre of the occupation zone, 80 yd. from the edge of the pan and just to the left of a Somali tomb. The excavation was 125 yd. from the road. Here were found 10–15 cm. of orange to red sand which showed some stratification owing to wind deposition or re-deposition. From this sand irregular shrinkage cracks run down into the main body of the sand. 316 specimens were found in this horizon, all showing a characteristic whitish patina and quite fresh. The sand also yielded five pebbles of tuffaceous limestone and one of grey vesicular lava.

Below this stratified upper level the sand became more compact and redder in colour, though the change-over was imperceptible. The greater number of specimens (317) in this compact red sand, were found between the 10–20 cm. levels. Artifacts and debitage continued to be plentiful down to 30 cm. but below the 30 cm. level only rare factory debris and ostrich egg-shell were found. This compact sand continued to a depth of from 40 to 50 cm. when it was replaced by a consolidated red sand cemented by gypsum which was difficult to excavate. This 'hardened complex' rested on weathered gypsum rock. Bed rock was uneven and occurred between 35 and 60 cm. below the surface of the orange to red sands.

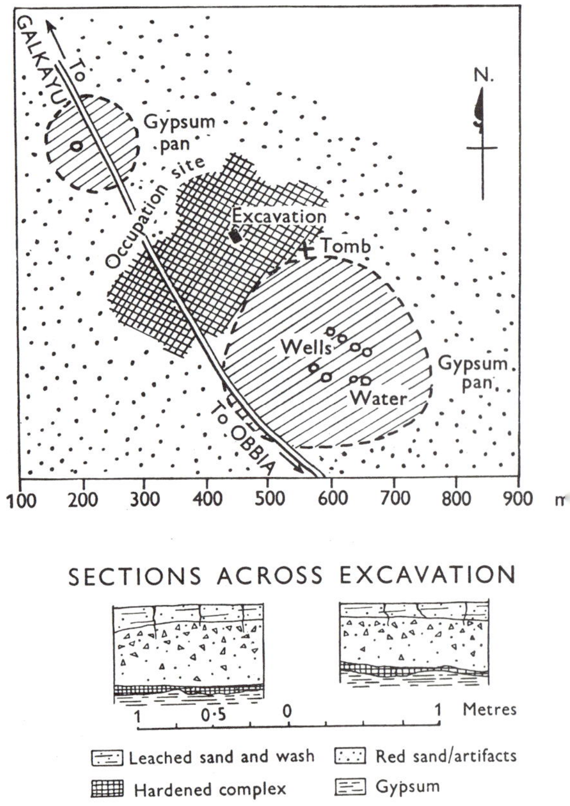

Text-fig. 31. Plan and sections, Rakon Wells, Mudugh.

The writer believes that the stratigraphy at Rakon closely parallels that at Mirsale. The 'hardened complex' is, Professor Zeuner suggests, probably a secondary formation at the base of the sand due to weathering processes. It will be noted that the industry which occurs only in the upper half of the sands must postdate the major accumulation of these sands. The Rakon industry is therefore the contemporary of, or possibly somewhat later than, the Mirsale and Afgudut industries.

Geological horizon. Second post-Pluvial Wet Phase.

Climate. Semi-arid to arid.

The industry. The tool lists show a rich industry with a number of interesting groups of tools. Those recovered from the excavation are identical with those that had been eroded and were lying on the surface. It was not possible to note any development in the industry between the top and bottom of the occupation layer. It would seem as if these wells formed a regular camping ground for the makers of this Late Stone Age industry for a number of years, but that their occupation was terminated by the drying-up of the pans. While the industry is basically of Somaliland Wilton origin, the pressure-flaked tools—notably the points and the arrowheads, indicate contact with and borrowing from the Doian peoples to the south. Such an industry is in fact what we might expect to find at the meeting place of the two culture complexes.

Inventory of tools from Rakon, eastern Mudugh

	Excavation				Surface	Totals
	0–10 cm.	10–20 cm.	20–30 cm.	30–50 cm.		
Lanceolate and foliate bi-faced points	—	1	—	—	12	13
Limace points	—	—	—	—	10	10
Lanceolate unifaced points	—	—	—	—	13	13
Hollow-based arrowheads	—	1	—	—	11	12
Asymmetric (crescent) points	—	—	—	—	5	5
Crude foliate points (broken)	—	—	—	—	7	7
Straight-backed microliths	5	7	3	—	75	90
Curved-back microliths	4	4	4	—	66	78
Crescents	4	1	1	—	22	28
Burins—*bec-de-flute*	—	—	—	—	9	9
Burins—angle	—	—	—	—	4	4
Scrapers—end	—	2	—	—	7	9
Scrapers—short end- or round	2	9	3	—	23	37
Scrapers—thumbnail	2	1	—	—	17	20
Utilized blades	5	—	—	—	9	14
Outils écaillés	—	—	—	—	4	4
Ostrich egg-shell—beads and pierced fragments	1	2	—	—	6	9
Ostrich egg-shell	49	57	42	7	—	155
Blades	82	68	59	6	82	297
Flakes	84	84	76	11	38	293
Debitage	65	51	25	—	8	149
Cores—single-platformed	5	4	—	—	8	17
Cores—multi-platformed	2	5	6	—	8	21
Cores—bipolar	2	1	5	—	—	8
Various	5	11	—	—	32	48
Large-backed blades	—	3	7	—	19	29
Fossil bone	—	4	—	—	—	4
Cores—pseudo-disc	—	—	—	—	8	8
Truncated blades	—	—	—	—	15	15
Blades with concave retouch	—	—	—	—	15	15
Double-backed microliths	—	—	—	—	4	4
Totals	317	316	231	24	537	1425

All artifacts are made from chert almost certainly derived from the local Eocene beds, and a few artificially fractured pieces of quartz—from small iron-stained quartz pebbles—also occurred. These comprised seven artificially fractured pieces, four flakes and two single-platformed micro-cores. Also two fragments of shell from marine molluscs were found on the surface. They may be associated with the Stone Age material and if so would indicate traffic with the coast at this time, but they are more likely to have been carried to the wells and dropped by Somalis—though the species of shells are not such as they usually carry about with them.

Points, bifaced. (Pl. 25, nos. 61, 64.) These show the 'Neolithic' pressure technique that was noted on the Doian specimens. The majority are broken. In shape they are either foliate or lanceolate; no. 61 is an unfinished example of the foliate arrowhead.

Limace points. (Pl. 25, nos. 62, 67, 68.) These are again typical forms with the Doian, and as can be seen vary considerably in size. The main flake surface is usually unworked, but occasionally basal retouch is present.

Lanceolate unifaced points. (Pl. 25, no. 66.) These are similar to those found with the Doian and usually show triangular sections.

Hollow-based arrowheads. (Pl. 25, nos. 87–91.) These are bifacial or parti-bifacial and are with one exception small and triangular in general shape One good example was recovered from the 10–20 cm. level in the excavation. They vary in form from the triangular type with only slightly concave base to the type with barbs formed by accentuation of the basal concavity. One specimen (no. 87), differs from the above forms in being larger and more pointed and triangular in shape. It approximates to the forms found with the Hargeisan derivative industry from site H.2 at Hargeisa (see p. 224).

Asymmetric (crescent) points. (Pl. 25, no. 63.) These are unifaced and typical of such tools described above under the Doian.

Crude foliate points. (Pl. 25, no. 93.) Five are bifaced and two unifaced—all are broken. They are typical of the specimens from Mirsale, but the rarity of these tools at Rakon, only 10 miles distant from the former site, is of interest.

Microliths

i. *Straight-backed.* (Pl. 25, nos. 69, 70, 73, 75, 77.) These are the characteristic form and all are for the most part small. The backing on all except the smallest examples is from both sides of the tool.

ii. *Curved-backed.* (Pl. 25, nos. 71, 72, 74, 76.) Small and usually elongated.

iii. *Crescents.* (Pl. 25, nos. 84–86.) Typical, backing is from one or sometimes from both sides.

iv. *Truncated blades.* A few specimens occur which have been truncated by oblique backing.

v. *Blades with concave retouch.* (Pl. 25, nos. 82, 83.) These belong to a type already described. The concavity is often slight and is usually found on crescents.

vi. *Double-backed blades.* Typical.

Large backed blades. Poorly made and atypical.

Burins. (Pl. 25, no. 92.) Good angle and *bec-de-flute* burins occur. The angle-burins are either transverse-straight (two specimens, one of which has been worked into a *bec-de-flute* burin at the opposite end), or oblique-concave (two specimens). Three broken backed blades show the *coup du micro-burin*, but there is little doubt that this is fortuitous and is the result of accidental fracture of a finished artifact. The facet is in each instance on a backed microlith not on the butt end of a micro-blade.

Scrapers

End-scrapers. (Pl. 25, nos. 94, 95.) Usually on small blades they are typical of the forms found with the Hargeisan and Somaliland Wilton.

Short end- or round scrapers. Typical of those from Afgudut, Mirsale and elsewhere.

Thumbnail-scrapers. (Pl. 25, nos. 96, 97.) These show usually one convex scraping edge, but one small specimen is rhomboidal in shape and has been worked by pressure over the whole of the upper side.

Utilized blades. (Pl. 25, no. 65.) Specimens are micro-blades worked down one or both edges. Occasional, more developed forms are found, such as the specimen illustrated, in which the whole of the upper face has been pressure flaked.

Outils écaillés. Typical.

Ostrich egg-shell beads and pierced fragments. (Pl. 25, nos. 78–80.) Six beads were found in the surface exposures two of them showing irregular edges which have not been ground smooth. A fragment showing part of a perforation was recovered from the 0–10 cm. level in the excavation and two further irregular, pierced fragments came from the 10–20 cm. level. The perforation in all cases shows an hour-glass section. Those specimens which have had the circumference ground smooth are still irregular, however,

Ostrich egg-shell. (Pl. 25, no. 81.) A quantity of broken fragments were recovered from all levels in the excavation, but none showed any evidence of engraved decoration. Only one piece had been worked (the specimen illustrated) and this had been reduced to an approximately circular disc, presumably preparatory to piercing to form a bead. The quantity of shell again suggests the use of ostrich egg-shell containers. A few burnt fragments occurred.

Various. Under this category are included seven backed blades with Kasouga retouch on the cutting edge, one triangle, a double-ended pestle stone in quartz and a fragment of a second, redirecting flakes, one crude awl, five utilized flakes, two with serrations, and one trimming stone or core scraper.

Fossil bone. Two small fragments and two fragments of enamel from teeth were recovered from the 10–20 cm. level in the excavation.

Technique. All artificially fractured stone from the excavation was kept. In all 297 blades—the majority microlithic—and 293 flakes—a large proportion microlithic—were recovered. None of the flakes show preparation of the striking-platform. A number of pieces show crackle from fire.

The usual types of Late Stone Age core are found—the single-platformed, the circular-platformed conical, the bipolar and the multi-platformed cores are all present. In addition eight pseudo-disc cores occur though these may be worked-out multi-platformed cores which have been re-used as detaching hammers.

The microliths are all made from small blades which have been backed down one edge with pressure from an *outil écaillé* fabricator from one, though usually both, sides of the blade. The back is therefore flat rather than rounded.

There can be little doubt that this industry with its burins and scrapers and true blade-cores is basically of Somaliland Wilton affinities, but the pressure-flaked tools are certainly more developed than anything that has been found with the typical Somaliland Wilton and give the industry a 'Neolithic' appearance though it is fairly certain that the makers of the Rakon industry were still little more than hunters and food-gatherers.

IV. THE STRANDLOOPING VARIANT

This is represented by material from two areas on the east coast—the mouth of the Nogal and Obbia. It is certain, however, that examination would bring to light many other sites up and down this coast and on the north coast also.

THE BIEDE ROCK-SHELTER

Situation. The type site is a rock-shelter at Biede at the mouth of the Nogal. A number of small rock-shelters have been eroded in the limestone and sandstone walls of the lower part of the Nogal Gorge, and similar shelters occur at intervals up and down the coast from the estuary mouth. A large shelter exists about $\frac{1}{4}$ mile down the coast across the estuary from the fishing village of Biede. This shelter is situated approximately 12 m. above sea-level. It contains a very thin deposit of strandlooper occupation material. Below this shelter is a smaller one facing south and eroded in an isolated deposit of cream-coloured limestone with sand lenses. This deposit does not appear to belong to the Miocene (?) sandstones and limestones which compose the cliffs and it is possible that it is an eroded and altered remnant of consolidated fossil dune. The erosion of the shelter was effected in part by the sea of the 12 m. beach—with which level

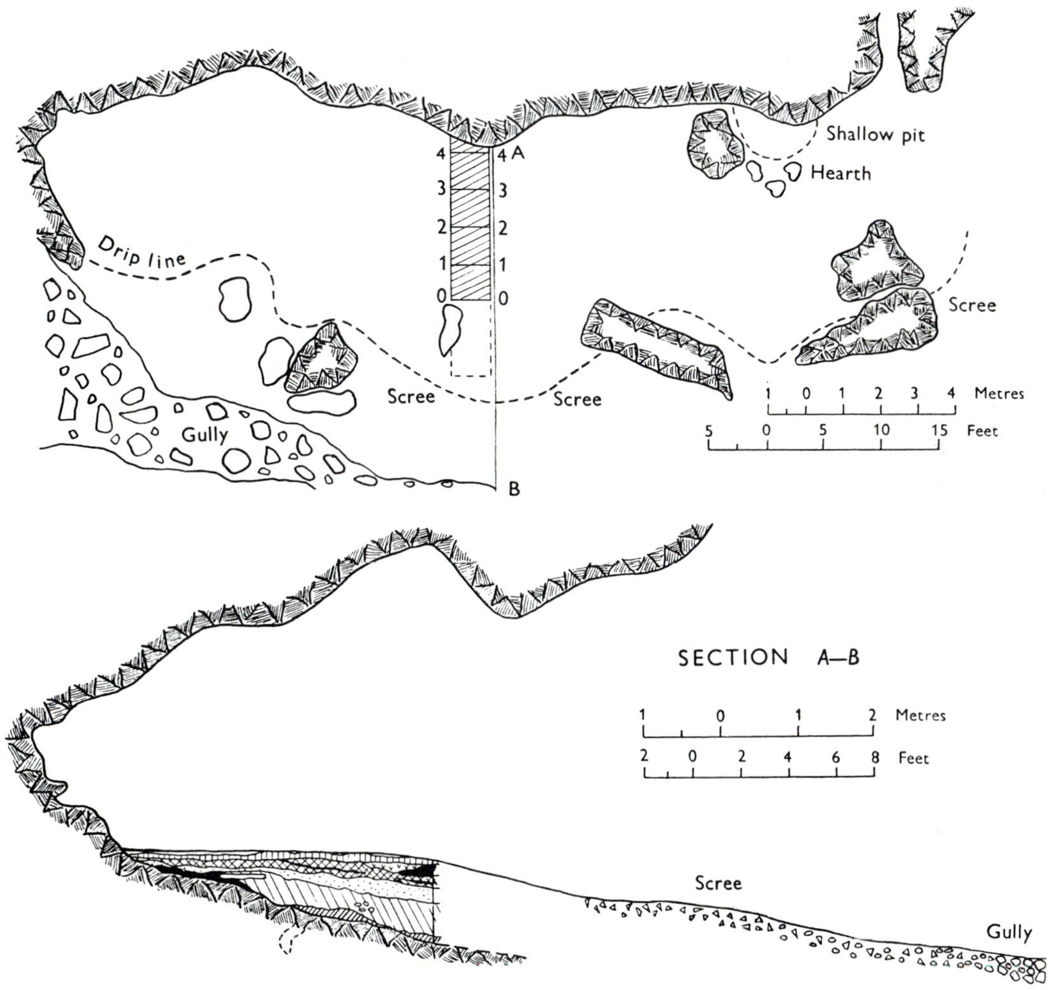

Text-fig. 32. Plan and section, Biede rock-shelter, Eil.

the floor of the shelter coincides—and in part by wind action. It is certain, therefore, that all the occupation deposits within the shelter and overlying the beach in its immediate vicinity are later in date than the 12 m. beach.

Several other shelters with occupation material were noted both up the coast from Biede and in the gorge of the Nogal. One such was on the immediate right of the road half-way between the estuary mouth and Eil village. Here a hole had been dug in the centre of a low shelter and showed in its upper levels a mussel-shell midden with a matrix of black earth and yielding a crude industry of flakes and debitage and one large pebble chopper. The sequence in the other shelters appeared to be identical.

Stratigraphy. For the sections and plan of the Biede shelter excavation see Text-figs. 32, 33, and Pl. 46.

A trench 1 m. wide and 4 m. long, subdivided into four sections A–D, was excavated from back to front of the shelter. The stratification was well marked.

The surface of the shelter consisted of a fine, loose, grey sand, flat over the greater part of the floor but dipping gently at the entrance. At the end nearest the sea was evidence of temporary occupation by Rer Magno fishermen. The deposits can be divided into three main groups:

(i) An upper, Rer Magno (?), midden.
(ii) A sterile, yellow sand.
(iii) A lower, Late Stone Age, midden.

The upper midden is of comparatively recent date and is separated from the lower midden by a most distinct deposit of yellow sand. This sand was practically sterile of occupation material. It is composed of fine grains and shows some lamination. It is clearly an accumulation of sand that has been blown into the shelter from the beach outside and scalings from the roof and walls; it effectively seals the lower midden with its degenerate, strandlooping industry. The sand is correlated with the orange sands which overlie the Doian strandlooper sites on the southern half of the east coast, and the true Doian further inland. It is considered, therefore, to mark the end of the Second post-Pluvial Wet Phase, and the beginning of the present climatic phase. The lower, Late Stone Age, midden is therefore dated to the Second post-Pluvial Wet Phase and the upper midden to recent times.

THE UPPER, RER MAGNO, MIDDEN

This deposit extended to an average depth of 40 cm. from the surface and at the outer edge of the shelter dipped gently downwards. It consisted of a series of midden layers separated by thin layers of hearth and floors of compressed vegetable matter and dung. The uppermost midden deposit contains numerous mussel-shells and charcoals set in a gritty, black earth matrix. In the lower layers the shells, again almost always mussel, are more comminuted and the earth is browner. The hearths are lenticular and contain ash and charcoals. The floors, often only an inch or so thick, consist of compressed dung, vegetable matter and some wood and twig fragments and are reddish brown in colour.

This midden yielded very little occupation material but that which was found places it without any doubt in recent times. The few specimens of worked stone from this deposit can be considered to be derived.

The upper midden rests on the yellow sand and the stratified break between the two levels is clearly defined.

Yellow sand. This deposit, some 10 cm. thick on an average, consists of clean sand, minute scalings from the roof and walls of the shelter and a few comminuted shell fragments. It forms a uniform deposit over the whole of the area excavated and was quite undisturbed by the upper occupation level except in the central

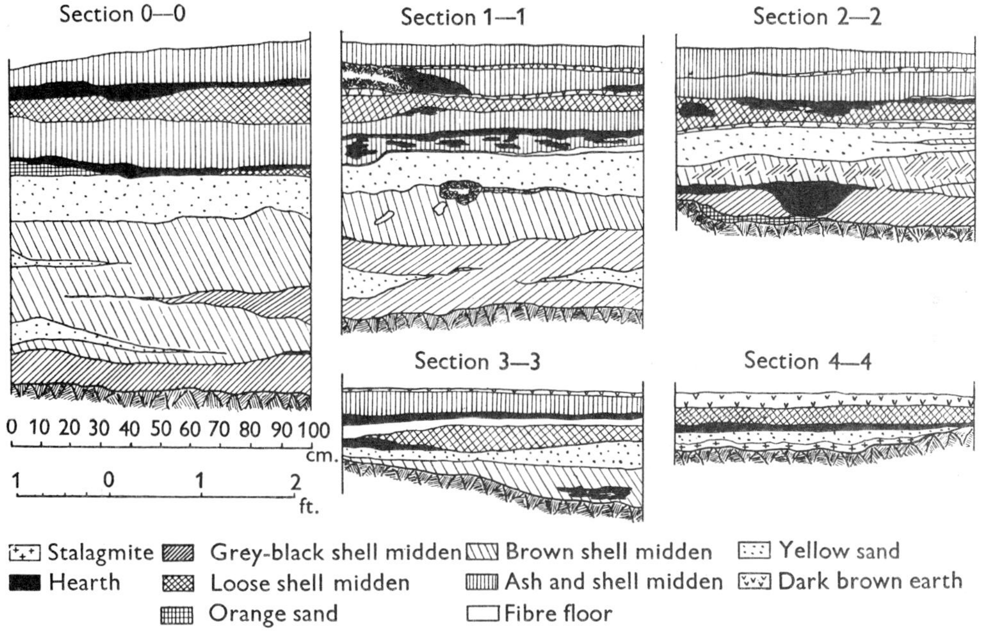

Text-fig. 33. Sections, Biede rock-shelter, Eil.

part of area C, where was found a shallow depression dug in the yellow sand and filled with ash and charcoal, which probably represented a hearth.

The yellow sand passes gradually into the underlying midden and in places lenses of the latter are interbedded with the base of the sand.

LOWER, LATE STONE AGE, MIDDEN

With the exception of the shallow pit in area C the whole of the lower midden was sealed by the sterile, yellow sand. The lower midden is laminated like the upper midden and in its upper levels is brown in colour, gradually changing to grey in the lower levels, which contain hearths and charcoals, until immediately above bed rock a thin deposit of black earth was found. This brown to grey earth is loose and gritty and appears to contain flakings from the cave roof. In places in area B were found thin, lenticular 'crusts' of comminuted shell and calcium carbonate; these crusts appeared to centre round the hearths, but this was not always the case. The hearths are usually circular, shallow depressions rather than flat, lenticular hearths as are found in the upper midden.

In bed rock in the centre of area B was found a small, tapering hole, probably eroded by the sea of the 12 m. beach. This hole was narrow at the top and wider at the base and reached a depth of 50 cm. slanting back towards the rear of the shelter. It was filled with a fine, black earth similar to that occurring immediately above bed rock, and yielded factory debris and a few implements. In fact this area yielded almost as many specimens as did the whole of the remainder of the lower midden. Three smaller holes of this type were also found similarly filled with black earth.

In places were found lenses of yellow or orange sand interbedded with the brown and grey earth.

The shells in the lower midden were generally broken, unlike those in the upper midden which were usually whole. The mussel is still the commonest shell but the limpet and a marine gastropod are also found in the lower midden. See Appendix G (p. 366).

The industry contained within this lower midden is essentially degenerate and poor. No occupation floor was present but the tools and debitage were found distributed sparsely throughout the deposit.

The floor of the shelter slopes gradually down from back to front and the deposits of the lower midden conform to this angle of slope.

These deposits are interpreted as indicating a period when the climatic conditions were wetter than they are today, but with sufficient dry intervals to allow of the accumulation of lenses of wind-blown sand. The 'crusts' of lime bear out this interpretation. During this time the Late Stone Age inhabitants lived a strandlooping existence in the shelter.

The climate then became drier until the shelter ceased to be occupied and wind-blown sands accumulated over the lower midden.

The upper midden represents a reoccupation of the shelter in fairly recent times, probably by the Rer Magno fishers. These people represent a 'caste' of fishermen who are found along the whole of the east coast of Somalia. They are said by one authority to be of Arab origin, but while this may perhaps be the case in regard to the chief centre of these people at Merca, the writer considers that such an origin is most unlikely for the Mijertein group who show no Semitic features whatsoever. A family of such fisherfolk was found by the writer inhabiting a small shelter a few hundred yards to the south of the excavation. These people were 'shegats' (i.e. tenant-holders) of the Isa Mahmoud and had recently moved to the shelter which they then occupied from another shelter a few miles up the coast where they had lived for the previous ten years, remaining as long as the fishing was good. This particular family kept dogs and a few goats which were herded in a pen of thorn bushes at one end of the shelter. The family lived in the other half with only a poorly constructed windbreak of mats for protection. It is tempting to see in these Rer Magno people the descendants of the 'troglodytae' and 'ichthyophagi'[1] recorded by the classical writers as living along the Gulf of Aden coast and indeed the mode of life of the Late Stone Age strandloopers can have differed in no major respects from that of the present-day, Rer Magno, occupants of these shelters.

The industry. While the blades and cores point to a Wilton or mixed Wilton/Doian origin for this industry the rare flakes with prepared striking-platforms and the crude points suggest the possible persistence on this coast of Levalloisian elements which had long since disappeared in the interior. The pebble picks and choppers may possibly be taken as a further indication of this persistence, though the association of these crude tools with most strandlooping industries throughout the world renders this latter suggestion a doubtful one.

UPPER MIDDEN. (Pl. 26, nos. 17–23)

The most interesting objects from the upper midden are the seven potsherds. One is a sherd from a wheel-turned pot and comes from a depth of 10 cm. (no. 19). The remaining sherds are from hand-made pottery and probably of local origin. Other objects worthy of note are two approximately circular pieces of shell with ground edges (the use of which is not known), one fragment of pierced mussel-shell and a crude bone point. The bone from this midden is quite unfossilized. Pebbles and fragments of pebbles broken by fire also occurred.

Yellow sand. This deposit yielded nothing of note with the exception of one small sherd of a thin black ware similar to that found in the upper midden. This sherd was found 3 cm. below the top of the sand. The deposit also contained four large pieces of thick, flat bone, probably from some large fish.

[1] Captain H. J. Courtney informs me that whilst he was Civil Affairs Officer at Galkayu and Obbia in 1945–6 he employed a fisherman at the latter place who claimed to be a member of the Osman Mahmoud tribe from Bender Kassim but who had lived for many years among the Rer Magno of Obbia district, and obtained his living entirely by fishing. On one occasion when preparing to cut up bait this man found that he had left his knife behind. Captain Courtney states that the man picked up a piece of broken bottle glass from the ground and holding this in his left hand, with an edge resting on a rock, he knapped it along one side of the upper edge with a pebble hammer-stone. Moving the glass round he similarly removed flakes down the other side of the same edge, thus producing a sharp wavy edge with which he cut up the bait.

Inventory of tools from the Biede rock-shelter

Tool types	Biede rock-shelter excavation			Open station site deposit outside shelter	Totals
	Upper midden	Yellow sand	Lower midden		
Microliths:					
(1) Straight-backed	—	—	2	1	3
(2) Curved-backed	—	—	1	2	3
Awl	—	—	—	1	1
Large backed blades	—	—	—	1	1
Crude unifaced points	—	—	—	3	3
Pebble picks	—	—	1	14	15
Side choppers	—	—	—	4	4
Bilaterally notched pebbles	—	—	—	5	5
Potsherds	7	1	—	—	8
Shell objects	3	—	4	—	7
Utilized bone	1	—	2	—	3
Hammer-stones	—	—	3	—	3
Blades, some utilized	—	3	39	55	97
Flake-blades, some utilized	1	—	8	—	9
Flakes, some utilized	2	9	257	26	294
Cores:					
(1) Single-platformed	—	1	4	6	11
(2) Multi-platformed	—	—	3	3	6
(3) Bipolar	—	—	1	3	4
Scrapers	—	—	8	16	24
Ostrich egg-shell	—	—	1	—	1
Bone fragments	5	5	34	—	44
Debitage	—	—	42	—	42
Burins	—	—	1	1	2
Iron slag	—	—	1	—	1
Rubbers	—	—	—	1	1
Wood	2	—	—	—	2
Redirecting flakes	—	—	6	—	6
Utilized pebbles	—	—	2	3	5
Totals	21	19	420	145	605

LOWER MIDDEN. (Pl. 26, nos. 24–35)

Of interest is the fact that very few finished implements were found. The microliths (nos. 28–31) are poorly made and atypical; so also are the scrapers (nos. 25–27), and for the most part the makers of this industry contented themselves with using primary flakes with little or no secondary retouch (no. 24). The cowrie shells are of interest as they have been pierced for decoration in a manner identical with that in use today. The most interesting object is a lump of iron slag which was found at a depth of 55 cm. in area B. This specimen was submitted to Dr A. F. Hallimond who replied as follows: 'An iron slag in a fairly good state of preservation. This is a typical magnetite-fayalite slag, the fayalite having a refractive index somewhat lower than that of the pure iron mineral. It probably contains a eutectic admixture of finely divided silica. Slags of this type result from small native processes and might be produced up to the present day or might be very early.'

It is clearly of interest that such a specimen should occur in a sealed, albeit degenerate, Late Stone Age deposit and would suggest that although the makers of this industry probably did not manufacture iron tools, they were already in contact with an iron-manufacturing and using people.

While the greatest use was made of short, broad flakes, micro-cores for the production of micro-blades, are also present. A few of the flakes show crude preparation of the striking-platform.

OPEN STATION SITE OUTSIDE BIEDE ROCK-SHELTER. (Pl. 26, nos. 1–16)

Immediately outside the shelter is the ±12 m. raised beach. Lying on this beach are detrital deposits on the surface of which, but underlying a foot or so of wind-blown sand, is further occupation material of this strandlooping site. In addition to the forms already recovered from the shelter the artifacts included pebble picks and choppers (nos. 1, 3, 4), a rubber (no. 2) as well as crude, unifaced points or picks (nos. 6, 7, 16), one of which showed a prepared striking-platform (no. 16), also several large, crude scrapers on thick, broad flakes (no. 15).

The artifacts from this site have been eroded by wind action and the majority of specimens show typical scouring and polishing.

OBBIA. (For the stratigraphy at the lagoon site at Obbia, see p. 137)

Although this is an open site the sequence is identical with that at Biede. The Late Stone Age occupation horizon rests on the surface of a bed, which may represent a lagoon deposit of a low-level ±6 ft. raised beach, and it is sealed by dune sands. Overlying the dune sands is the upper or recent midden—several of which occur in the locality, comprised of mussel-shell in a black earth matrix. This midden is itself covered by shifting dune sands. It is of interest to note that the Late Stone Age horizon on the Gowein sands rests on the eroded surface of the compact red sands, and is covered by a varying thickness of orange sands which have themselves been subjected to some erosion.

The material from the Late Stone Age occupation level at Obbia is rather poor. All specimens are in chert with a characteristic white or mottled patina similar to those from the open site at Biede. They are usually unweathered. Specimens comprise four fragments of ostrich egg-shell, five micro-blades, twenty small, primary flakes, one worked into a crude, short end-scraper, one *outil écaillé* worked on a bipolar core and two double-platformed micro-cores. While the material is not as conclusive as could be wished, it is nevertheless sufficient to provide interesting corroborative evidence for that from Biede.

Other important corroborative evidence for the Biede excavation comes from Graziosi's shelter at Eil. The deposits in this shelter can be divided into two main series:

(1) The upper levels (strata 1–3), where a detrital deposit with sparse evidence of occupation covers a midden occupation with numerous molluscan remains and in which is also noted the occurrence of three molars of sheep, presumably domesticated. The midden contained numerous charcoals and a crude industry entirely comparable with that from the lower midden at the Biede shelter.

(2) The lower levels (strata 4, 5), which consisted of a yellow, sandy earth resting on a deposit of large boulders and smaller rocks, yielded an industry of Hargeisan or Magosian affinities with long, utilized blades, blade cores, one unifaced point of the type previously noted as occurring with the Hargeisan industries, microlithic backed blades, a burin and a crude, pebble chopper. There can be little doubt that the industry from the upper levels is developed immediately from that in the lower and yellow cave earth, the latter industry being a derivative of the Hargeisan such as we have already noted as occurring in the Nogal Gorge.[1]

V. MINING AND FACTORY SITES

In the greater part of the area covered by the Jurassic limestones, Cretaceous limestones and sandstones and the Basement Complex rocks the raw material consists primarily of chert, silcrete, quartz and quartzite—the chert occurring in the form of nodules and pebbles, the silcrete and quartzite in bands or pebbles, and the quartz mainly as pebbles.

In the area where the Tertiary rocks predominate, that is to say in most of the central and north-eastern parts of the Horn, the chert occurs in the form of bands, interstratified with limestone or gypsum rock. In the case of the former the chert is usually creamy white or translucent brown in colour and in the case of the latter, black or grey.

In the Mudugh the raw material occurs as outcrops on low hills sometimes protected by a capping of limestone. The people who made use of these outcrops did no actual mining here but broke off large blocks of the tabular chert, which varies in thickness from an inch to a foot, and then roughly quartered them. These

[1] Graziosi, P. (1940, pp. 57–66 and Pls. 23–7).

quarters were then worked into crude, formless cores varying greatly in size—some of the largest having a diameter of 2 ft., while the smallest are of microlithic proportions. Numerous, rough flakes, with and without crude preparation of the striking-platform, littered the slopes of these outcrops and combined with the cores and other debitage to form a scree of variable thickness, though usually not more than a foot, round the base. A few of the flakes forming the scree had been worked up into roughouts for artifacts and presented a primitive appearance. The technique employed was invariably direct, percussion flaking. Those specimens which lie upon the surface show some degree of weathering though not as much as might be expected, but those contained within the scree are usually quite unweathered and unpatinated.

At first glance it appeared that the crude artifacts represented an early culture, but it later became apparent that this was not in fact the case, but that the sites were mines and workshop sites of comparatively recent date. They are considered by the writer to represent in the main the workshop sites where the Late Stone Age population obtained its raw material, though it is possible, on typological grounds, that some of the material may be as early as the Middle Stone Age, although no home sites of this cultural stage have been found anywhere in the eastern Mudugh, unless Puccioni's site at Warandi may be quoted as an example.

The three main workshop sites found by the writer in the Mudugh are at Mirsale (already noted), Daba Gallo and Kourou or Kurou.

DABA GALLO (46)

A small isolated bur lying 6 miles to the right of the Obbia/Galkayu road and 14 miles from Mirsale. In addition to the crude quarterings in the scree were found crude *coups-de-poing* and large 'side-scrapers', as well as some large, crude 'tortoise cores'. Specimens from the surface are weathered and patinated, but those from within the scree are quite fresh. Specimens retained include three *coups-de-poing* or picks with approximately parallel sides, two crude concavo-convex side-scrapers, one ovate *coup-de-poing* and one chopper.

KOUROU (50)

This is a low outcrop on the left of the road, 56 miles from Obbia. It takes its name from a pan about 4 miles distant to the right of the road. The tabular chert here is thick and forms a protective capping to the gypsum rocks on which it rests. Very little of the material from the scree has been weathered. Specimens include one large, crude disc or pseudo-disc core, one circular-platformed blade core, one multi-platformed core, one bipolar core, one trimming stone or core scraper, one *coup-de-poing*, one broken pick, one broken, crude, foliate point, one large flake-blade, four broad, large primary flakes (two of which are concavo-convex flakes), five smaller flake-blades and one small flake.

UPPER NOGAL

In the upper Nogal valley tabular chert is exposed in the sides of the numerous isolated hills between the central plain and the southern scarp. Here the chert occurred in three or four different levels, but it was usually the lowest band that was worked. From the Middle Stone Age onwards these chert bands were 'tapped' for raw material, as is evidenced by the thickness and character of the scree. Two sites were examined in the upper Nogal valley, one at Laba Higlo (53), 14 miles from Garoe on the road from Donkukoq, and the other some 6 miles from Bihen (60) on the Boundary Cut road. These are burs with steepish sides and near their base is a band of tabular chert some 2 in. or less thick. At the first site it seems probable that the Stone Age inhabitants 'mined' the chert, by digging away the soft gypsum rock above and below the chert band, besides merely detaching those parts of the band which had been exposed by natural weathering.

The Middle Stone Age factory material is always weathered and patinated an ochreous or chestnut brown, while the Late Stone Age debitage is always quite fresh and unpatinated. The Middle Stone Age material includes flakes and flake-blades with faceted striking-platforms, and discoidal and subtriangular prepared cores. The Late Stone Age specimens comprise blades, primary flakes, blade cores, choppers, occasional artifacts and the usual rough quarterings.

The Somaliland Wilton material from Laba Higlo is typical. It comprises seven single-platformed blade cores, and one with circular platform, nineteen blades ranging from $2\frac{1}{2}$ in. to $\frac{1}{2}$ in. in length, some with retouch, six short, broad, primary flakes, a core chopper, one concavo-convex side-flake, one end-scraper on a blade and one typical, double-ended, oblique-concave angle-burin. Some of the blades show notches but these are usually close to the tip end and are in no way typical of the micro-burin technique.

CONCLUSIONS. THE END OF THE STONE AGE SEQUENCE

The evolution of culture in the Horn has been tabulated in the Correlation Table (p. 152). The earliest culture of which there is definite evidence is the Acheulio-Levalloisian, but it is very probable that an Upper Acheulian preceded it in the north-west. From the Acheulio-Levalloisian evolved the long sequence of pure Levalloisian cultures during the Upper Pleistocene. In the north and south the Levalloisian is followed by the Somaliland Stillbay, which develops into the Magosian and so in Late Stone Age times into the Doian. In the west, however, it would appear that the Stillbay forms are not so common and the Levalloisian may have followed a local line of development, emerging finally into the Ogaden variant of the Somaliland Wilton. In the north the end of the Pleistocene is marked by the introduction of blade and burin elements (the Hargeisan), which appear to have no antecedents in the Horn and which develop into the Somaliland Wilton. The remains of fossil man himself are meagre in the extreme. The Porc-Epic mandibular fragment points to the possibility of a Palaeoanthropoid stock (Neanderthaloid) being present in the Horn, associated we may suggest with a Levalloisian Culture, but the remains of Late Stone Age man are too fragmentary to be able to determine his physical make-up.

The writer was unable to adduce any single piece of evidence which pointed to the existence of a true Neolithic Culture in the Horn. The Late Stone Age industries of 'Neolithic facies' may be so termed only in regard to certain elements of their stone culture, and it can be confidently stated that the makers of those industries were still living a predominantly hunting and food-gathering, in other words a Mesolithic, existence. From a study of their material culture and of the fauna found with this, it seems unlikely that they had any domesticated animals, with the possible exception of the dog, but if it can later be proved that they were the authors of the paintings and engravings occurring in some parts of the Horn, they must either have had some domesticated cattle themselves or have been in contact with a pastoral people using spears and shields and owning cattle, and later sheep, goats and probably camels. That the greater part of these paintings and engravings are late there can be little doubt and if they should in fact prove to be the work of the Late Stone Age hunters then they must postdate the main and best period of their cultures.

The evidence from the excavated sites suggests that the Late Stone Age was not suddenly replaced by a metal-using culture, but that it lingered on and its material culture gradually degenerated and underwent a change by reason of contact and possibly conflict with new and alien influences. In this connexion it is of interest to note that D'Errico quotes Conti Rossini as saying: 'Valentissimi nella fonda, i Cusciti lo furono altresi nell' arco. . . . Le frecce munivansi di punte di scheggie d' osso o di pietra. Come arma da getto, primeggiava il giavellatto, in origine interamente di legno, al piu con punta d' osso o di pietra.' (Conti Rossini, *Storia d' Ethiopia*, p. 77.)[1]

No ground or polished stone tools that can be associated with the Late Stone Age have been found in the Somalilands. Discounting the *pseudo-morphs*

[1] D'Errico, L. (1937*b*, p. 194).

described by de Morgan,[1] not a single polished stone axe has been found. The nearest specimens recorded come from Yubdo, on the western part of the Abyssinian high plateau east of the Omo river, where they were associated with ancient gold-mining,[2] and from the Tuli Kapi plateau[3] and from Garba Tula near Isiolo, in the Northern Frontier District of Kenya, where the writer found the cutting edge of a polished axe.

The bored stone is nowhere found in a prehistoric context in the Somalilands. It occurs only on the north-western plateau at Hargeisa, to which place it was introduced from Harar in the second half of the last century. The weighted digging-stick is used today by the Galla on the Harar plateau, but rarely, if ever, occurs off the high plateau.[4]

There are also a few artifacts which, as yet, cannot be fitted into the cultural succession. That they are all late there is no doubt, and in fact they almost certainly postdate the Stone Age proper. Such specimens are the T and U pieces that occur as isolated finds in the northern parts of the plateau and in Danakil. They are made on rough flakes or lumps of any convenient material. The T pieces are formed by two opposed notches or hollows, while the U pieces have the appearance of a rather crude hollow scraper.

Site H.6 at Hargeisa yielded one such T piece to the writer, and the Garasleh Tug site yielded a second, and also a U piece. Other U pieces are found in the collections of Trevor from Damero, in Barrington Brown's collection from Buran, in Phillips' collection from Hargeisa, in Glover's collection from mile 45 on the Hargeisa/Wardere road, and Breuil's and Wernert's collections from Djebel Djinn[5] contained further T pieces. The U pieces are reminiscent of the crescent drills found by Caton-Thompson at the gypsum quarries in the Fayum.[6] A carved T piece of soapstone or schist from Borama in British Somaliland recovered by the writer may be connected with the ruined town of 'Amud' close by: its use is unknown though it may have been some kind of amulet. Similar T pieces are found in the Sahara where they are known as *coches doubles* and are considered to be camel charms.

Other tools that cannot be fitted into the sequence are the picks described by Wernert from Lake Assal in French Somaliland,[7] and a few crude *coups-de-poing* or choppers in limestone from Isha Baidoa and Deileb on the edge of the southern plateau and from Taffe (222) on the Harar Plateau.

[1] de Morgan, J. (1921).
[2] Graziosi, P. (1938).
[3] Leakey, M. D. (1943, pp. 182–95).
[4] Clark, J. D. (1944).
[5] Teilhard de Chardin, P. *et al.* (1940, pp. 497–522).
[6] Caton-Thompson, G. & Gardner, E. W. (1934).
[7] Teilhard de Chardin, P. *et al.* (1940).

CHAPTER 13

PREHISTORIC ART IN THE HORN OF AFRICA

PAINTINGS AND ENGRAVINGS

Both paintings and engravings have been found in our area but as yet under twenty sites are known and considerably more work is necessary before anything more than tentative results can be obtained from a study of this art.

The paintings are all found (with two exceptions) either on the southern edge of the Danakil Rift, or on the northern part of the plateau and occur in rock-shelters or on protected rock faces. A number of different styles exist.

The engravings also occur in small rock-shelters and on exposed rock surfaces.

The sketch-map (Text-fig. 36) shows the distribution of painted and engraved sites in the Horn and the granite areas where further painted sites may be expected to exist.

PAINTINGS

The first paintings to be found occur in the neighbourhood of Dire Dawa and have been described by Breuil.[1] They consists of the painted cave of Porc Epic, at Dire Dawa, and the painted rock face of Genda-Biftou, at Sourré,[2] 60 km. W.S.W. of Dire Dawa. We may summarize briefly the points which arise from a study of these paintings.

PORC EPIC

(i) Rock—limestone.

(ii) This shelter was occupied by the makers of the Somaliland Wilton Culture, or a variant of this. The occupation layer containing this industry overlay a Magosian and possibly a late Stillbay industry.

(iii) The paintings were partially covered in places by stalagmitic formations, which is considered by Breuil to indicate that they are earlier than the most recent of the stalagmitic layers in the floor of the cave.

(iv) The style is conventionalized and semi-schematic.

(v) The great majority of the paintings are in red but a few reddish brown and yellow, and one doubtful black painting, are considered to be earlier.

[1] Breuil, H. (1934). [2] Italian spelling, Surè: R.G.II spelling, Sureh.

(vi) Twenty human figures are represented. The figures illustrated (Breuil (1934), Fig. 2) are drawn in profile. Steatopygia is shown, and one figure wears a skirt.[1]

(vii) The fauna represented is apparently all wild (i.e. elephant, lion, jackal, antelope, buffalo, bubalis?, stag?, and ostrich?), there being no domesticated species.

It is reasonable to suppose that the only colour here was red or red-brown, and that the earliest paintings would have faded to yellow.

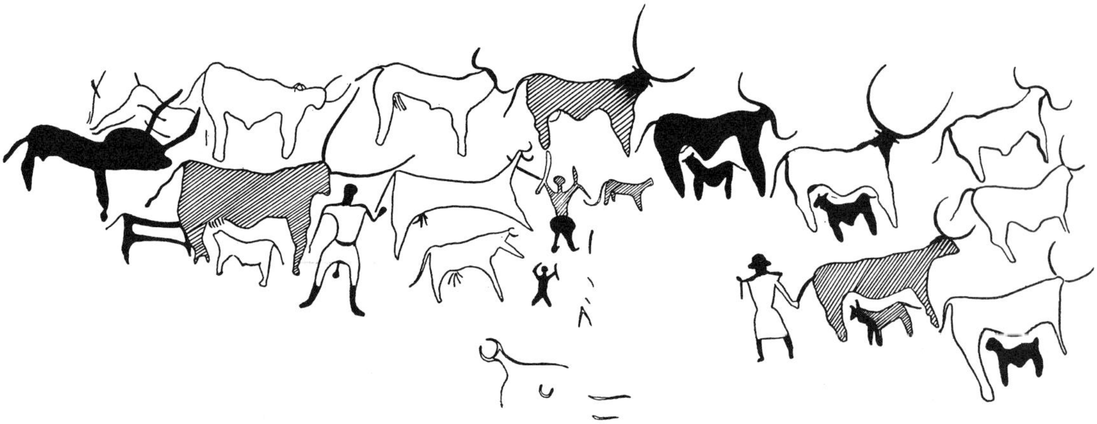

Text-fig. 34. Herdsmen and cattle, Genda-Biftou, Sourré.

Breuil, H. (1934). 'Peintures rupestres préhistoriques du Harrar (Abyssinie).' Fig. 7. *L'Anthropologie, Paris,* vol. XLIV, nos. 5–6, pp. 473–83.

GENDA-BIFTOU (Text-fig. 34)

(i) Rock—limestone.

(ii) Sparse and indeterminate occupational material in the vicinity.

(iii) The paintings have been preserved only where protected by the overhang of the rock.

(iv) Two main series are found, an earlier naturalistic series comprising Breuil's styles 1–5 and style 7 and a later schematic series comprising his styles 6 and 8.

(v) The earliest paintings are in yellow, and are followed by those in black, pale red, red-brown (most numerous and subdivided by Breuil into styles 4*a* and 4*b*), yellow, red and black in that order.

[1] Steatopygia is common in mature Somali women, especially among the true pastoral tribes.

(vi) In the styles later than 4*b* the human figures are all schematic. The earliest naturalistic figures are drawn full face, clothes are shown, one figure wears a belt and another wears a hair-comb and carries a bow.[1] In the schematic groups the human figure has been reduced to a capital H with various appendages.

(vii) The majority of the animals represented are domesticated and represent cattle with long horns and no humps. These are probably *Bos africanus* and not *Bos brachyceros*. Breuil identifies buffaloes amongst the herd scenes and the illustrations leave little doubt that the animals represented are indeed buffalo. More rarely carnivores are depicted.

(viii) Characteristics of the animal drawings are the manner in which the horns are represented (as if seen full face); the careful sexual differentiation, notably the indication of the udders of the cows; method of representing variegated colouring (i.e. the partitioned, or *cloisonné* technique); the inverted U technique for the underbelly and tapered legs; the representation of only one front and one rear leg.

LAGO ODA

Two more rock-shelters are situated at Lago Oda some 20 km. from Genda Biftou. They were discovered by Père Azais in 1934. A note of this discovery was recorded by Breuil,[2] and they were visited and many of the paintings photographed by W. B. Fagg in 1941, to whom the writer is indebted for the following information and copies of the paintings:

(i) Rock—limestone.

(ii) There are indications from the surface material collected that the shelters were inhabited by people practising a Somaliland Wilton industry.

(iii) Many hundreds of paintings in red, yellow, white and black occur, with many superpositions. Motifs include pastoral scenes and the protection of a herd against wild animals. An earlier, naturalistic, and a later, schematic, series are represented.

(iv) Animals represented in the naturalistic styles appear to be predominantly domestic cattle. Some are drawn in outline except for the head, some have the whole body coloured, while in others the body is filled in with dots. Elephants also appear to be represented. One drawing shows what may be a human figure riding an animal which is probably an ox.

[1] Both bow and comb are similar to those found in use among the Somali today. The former is a simple reversed bow, two of which are in the writer's possession. Others are illustrated by Puccioni (1936).

[2] Breuil, H. (1934).

(v) The characteristics of style noted at Genda-Biftou are also present at Lago Oda.

(vi) Such human figures as can be seen in the photographs belong to the schematic paintings and are shown either full face—with a shield, or side face—with bow and arrow.

A second group of painted shelters exists in the Mount Kondoddo area, close to and north of Harar. They have been copied and photographed by Count Björn von Rosen to whom I am much indebted for his description of the sites and for permission to reproduce six of his photographs.[1]

Von Rosen found three sites. The first of these was at Saka Sharifa, a high ridge or scarp not far from Kondoddo mountain.

SAKA SHARIFA

The shelter was a large one situated immediately below the crest of this ridge facing south and commanding a magnificent view of the surrounding country. The country rock is probably granite (as it probably is for all these Harar sites) and the paintings are usually on the protected sloping back walls of subaerially weathered masses of rock such as are commonly found in granite country. At Saka Sharifa the painted surface is some 4×3 m. in area in the centre of the wall, well protected against weathering. The paintings are all executed in a reddish to dark brown pigment and vary in length from a centimetre to half a metre. Both outline and flat-wash paintings occur. They consist of:

(1) Heavily built and long-horned humpless bovids—either buffalo or domestic cattle, most probably the latter are intended. The complete body is filled in with colour. Rather poorly drawn in comparison with those from Genda-Biftou and Lago Oda, they approximate more to the Porc Epic paintings. (See Pl. 50, no. 1.)

(2) A few human figures with the arms outstretched, flat-washed. (See Pl. 50, no. 2.) The attitude of these figures suggests a characteristic stance of Somali and Galla herdsmen today, the arms resting over a stick or spear lying across the shoulders.

(3) Possibly a hunting scene in which a hunter aims with his bow at an ostrich. (See Pl. 51, no. 1.)

(4) Low down near the floor was a good naturalistic drawing of a jackal again in flat-wash.

[1] Von Rosen, B. (1949, pp. 201–26, 380–2).

(5) Two poorish paintings, one an outline drawing of what may be intended to represent an elephant, the other perhaps a snake with other motifs.

(6) Rather fine line drawings of animals, almost certainly in view of the Sourré paintings, representations of cattle, one with a calf. (Pl. 50, no. 2.) The heads have in each case been purposely left undrawn, however. Of particular interest is the fact that one of these animals shows a clear branding mark, presumably contemporary, in the form of an inverted 'S'. The style of these beasts recalls similar paintings from Lago Oda. (See Text-fig. 35, nos. 3, 4.)

No superposition was found but it is apparent that at least three different styles occur.

A few feet away was found a flat stone on which was an elliptical ground-out hollow which showed evidence that it was here that the pigment had been ground; in association there also occurred several small cup-shaped depressions.

Some hundred metres higher, close to the rocky summit of the ridge, was found another protected rock face with a single rectangular geometric drawing in the same reddish brown pigment.

Von Rosen visited several other rock-shelters in this neighbourhood but found nothing more, other than daubed lines and blotches which were obviously modern; the pigment having been mixed with a large amount of animal fat could easily be rubbed off. It is, however, of some interest when looking for the reason behind the 'prehistoric paintings' that villagers in the nearby Galla village told von Rosen that it was customary each year after the slaughtering to make these 'daubings' to show how fat the animals had been.

BAKE KHALLO

About 20 miles to the south near the village of Bake Khallo was found a small kopje with another typical rock-shelter. The painted area was again on the sloping and protected back wall, but this time the shelter was facing west. The paintings are not so well preserved as those at Saka Sharifa. The pigment was probably the same though this is not actually stated. Here the paintings are essentially of a schematic nature and are drawn in outline only.

Motifs consist of:

(1) Paired 'soles of feet' likened by von Rosen to those depicted in the Bronze Age engravings at Bohuslän in Sweden.

(2) 'Curving lines partly joined together and meeting at a single point.'

(3) Rectangular motifs like a child's drawing of a window and similar to the isolated painting at Saka Sharifa.

(4) Kidney- or heart-shaped designs.

(5) What is considered by von Rosen to be a naturalistic painting of a butterfly.

These paintings do not as yet bear comparison with any of the other groups in the Horn. Near this shelter was found an occupation deposit from which von Rosen recovered two worked flakes.

ERRER KIMIET

The third painted area was again found to be situated near the crest of a mountain ridge, this time at the mountain of Errer Kimiet south of Kondoddo. Here von Rosen found two rock-shelters some hundreds of metres apart. He has named them Errer Kimiet I and Errer Kimiet II. The first of these is visible from a distance and is found just below the top of the ridge and commanding a clear view towards Kondoddo.

ERRER KIMIET I. Superposition and three styles and colours exist here:

(1) The latest (presumably) is in thick white paint which has been carelessly and unevenly smeared on the rock. These consist of single paintings—a cross and an outline drawing of what may be a stylized human with an hour-glass body can be seen from von Rosen's photographs. These overlie paintings of the third and earliest series.

(2) Paintings in black 'like writing on a notice board' on a small area of rock face in a protected niche or recess. These are said to be better drawn than those of series (1) but no further details are given.

(3) The most numerous series are flat-wash paintings in red-brown pigment as at Saka Sharifa. Represented are filiform humans and human figures with circles in outline on or just over the head. (Pl. 51, no. 2.) These may represent a headdress or bundle and are reminiscent of some of the Sahara paintings, particularly those of the Hoggar region.[1] At the western end of the shelter is a group of long-horned and humpless cattle behind a large leader. These paintings with their sometimes angular profile and 'saddle' backs are the nearest that have been recorded to the Genda-Biftou group but are not so well drawn. (Pl. 52, no. 1.)

Also on the rock wall facing south is a row of rather systemless wavy lines.

ERRER KIMIET II. The second and more westerly shelter lies in a ravine, somewhat difficult of access, on the slope of the mountain facing south. The paintings are here all drawn in a white paint which is nevertheless of some antiquity as

[1] Lelubre, M. & Cousin, J. (1951).

it has become absorbed into the rock in the same way as have the red paintings of Saka Sharifa and Bake Khallo; von Rosen, therefore, considers these white paintings to be earlier than the white series at Errer Kimiet I.

One-half of the painted surface has been badly weathered and has largely flaked away. Von Rosen found here some of the largest figures that occurred in this Harar group of sites. Most of these represent domestic cattle with large spreading horns. Also depicted is what the discoverer calls a running lion or an old male baboon, but it could also, and I think more probably, represent a cheetah, these animals being commonly met with in the Horn. Another painting, one of the largest, von Rosen considers probably represents a horizontally lying human being with the arms outstretched. Between these flat-wash paintings is a row of outline drawings which can most probably be interpreted as stylized representations of warriors with shields and spears or bows. (See Pl. 52, no. 2.) They can be compared with similar capital H paintings found at Genda-Biftou.

Both these sites were occupied and from the eastern one (Errer Kimiet I) von Rosen recovered a microlithic factory debris and typical crescents and straight-backed bladelets of Somaliland Wilton form which he illustrates.

A third group of paintings is found in small shelters in the Gan Libah range on the edge of the Gulf of Aden fault-scarp, in British Somaliland. These have been described by Burkitt & Glover.[1] They are all of a stylized, semi-schematic nature, and the salient points from each shelter have been summarized below.

TUG GERBAKELE SHELTERS (9° 53′ N., 44° 54′ 30″ E.)

(i) Rock—limestone.

(ii) Typical Somaliland Wilton artifacts were recovered from the scree of no. 3 shelter.

(iii) The earliest paintings are in black, and are covered by irregular, horizontal and perpendicular scratches. Over these are rock engravings of animals (cattle or camels), executed by a pecking technique. Over the engravings are red paintings of animals, the marks of fingers and schematic motifs. The most recent are probably two schematic paintings in white.

(iv) The animals depicted are said to represent elephants, giraffe, lion, kudu, oryx, and probably cattle.

TUG KHABOBA SHELTER (9° 52′ 40″ N., 44° 49′ 40″ E.)

(i) Rock—limestone.

(ii) No evidence of occupation recovered.

[1] Burkitt, M. C. & Glover, P. E. (1946).

(iii) Two styles of paintings, in red, the earlier granular, and the later streaky, in appearance.

(iv) Schematic representations of what are thought to be the wild ass, baboon, and a hornbill.

DOMBOSLEH SHELTER (9° 54′ 2″ N., 44° 51′ E.)

(i) Rock—limestone.

(ii) No prehistoric occupation material recovered.

(iii) Much flaking and scaling from the walls had taken place, rendering many of the paintings incomplete or unrecognizable.

(iv) Two styles of paintings in black, the earlier granular, the later streaky in appearance—semi-naturalistic animals and symbols.

GALA-AD SHELTERS (9° 57′ 30″ N., 45° 1′ 30″ E.)

(i) Rock—limestone.

(ii) No occupation material recovered.

(iii) Paintings in shelter 1 are on the roof, and well preserved.

(iv) Sequence of colours is probably red (earliest) then black (possibly subdivisible into two series), white, or dirty grey (most recent).

(v) The authors say of the dirty white paintings that they 'occurred on a clean chipped fawn rock surface'.

(vi) Subjects include paintings of cattle and/or elephants, stylized humans and ?frogs.

BALLEH SHELTERS, NR. SKUTAR (9° 55′ 3″ N., 44° 56′ 30″ E.)

(i) Rock—limestone.

(ii) No occupation material recovered.

(iii) The paintings occur on partially protected rock faces.

(iv) Two colours are used, black and white, the paintings in the latter being more numerous: no superposition of styles was seen.

(v) Motifs consist of stylized human figures in profile (one showing steatopygia) and stylized animal figures.

A note on a series of Somaliland Wilton and earlier specimens collected by Glover from Dukokolol yero, between El Afwein and Las Dureh, states that a new type of cave drawing was found here, but no details are given.

Near the eastern end of this range of high country marking the Gulf of Aden fault-scarp an officer of the British Somaliland Government, named

Stephen Stock, found in 1947 a further painted rock-shelter. I am indebted to Dr Macfadyen for a description of this shelter. This is situated at Karin Heganeh (48° 38′ E., 11° 03′ N.), an isolated granite hill about $1\frac{3}{4}$ miles south-east of Las Bar wells. Enormous boulders or blocks of granite lay at the base of the hill and it is in the space between two such boulders that the paintings occur. They are found on the wall facing the entrance and from the description given by Stock are polychrome paintings in red, white and black paint. The scene depicted represents a herdsman with a herd of cattle. The cows are hornless, but the bulls, which are drawn about twice the size of the cows, are painted in red and white paint with long white back-sweeping horns. The herdsman is naked and painted in red. In one corner is a small painting of a camel though from its poorer style it may perhaps be later than the main scene. The paintings, like our other Somaliland groups, are all of the flat-wash or silhouette type and lack detail and perspective. The floor of the shelter is said to be filled with dust to the depth of several feet and would obviously repay careful excavation. As these paintings have not yet been copied it is not possible to make any close comparison with our other sites but it would seem that they are similar to the main group found by Glover further west along the Jir. It is considered most probable that all the way along this escarpment country where the Basement Complex rocks are exposed paintings will be found to occur.

A small rock-shelter on the south-west side of Bur Eibe in southern Somalia had the only paintings of any age found in the area investigated by the writer.

This shelter was at ground-level in granite some 12 ft. wide across the drip-line, 10 ft. deep, and 7 ft. high. There was no deposit on the floor except towards the back, where occurred a few inches of black, sandy earth with broken and burnt bone and potsherds, but no worked stone. The paintings are all in red, usually fairly well preserved, and are on the roof of the shelter. They consist for the most part of groups of parallel lines, 4–8 in. long, but two motifs, which may possibly be highly conventionalized representations of oxen also occur. (Pl. 48.)

In bed rock at the rear of the shelter were eight conical holes, approximately 3 in. deep and 3 in. in diameter at the top. These had been formed by grinding and were unevenly spaced along the floor near the back wall.

The Eile to whom the paintings were shown declared that they had never seen them, or any like them before. The 'cattle motifs' (?) were put down as a representation of an Eile hunter with his bow and arrow, and the holes in the floor as 'some kind of game or small places to crush millet before the present-day grindstones were introduced'. Pressed further my informants considered that the 'cattle motifs' might be an early Eile tribal sign which had now gone out of use.

The large rock-shelter, the Gure Makeke, partially excavated by Graziosi, contained some modern paintings. Besides Italian monograms and initials, there occur a few schematic paintings in red, black and white. These are mostly representations of Eile and Rahanwein tribal marks, similar to those commonly carved on the *garas* trees over the whole of the Rahanwein country. The most numerous are representations of the capital H, one of the Eile and also Shekal tribal marks. In addition there occur three drawings of the *bilawa*, or knife, in red, and one highly conventionalized red painting of what may be a quadruped; there are also three tectiform motifs.

These paintings have now no ritual significance and are scribbled at will by anyone who so wishes.

CONCLUSIONS

There can be little doubt that the earliest paintings are those in the naturalistic and semi-naturalistic styles from Genda-Biftou, Lago Oda, and Errer Kimiet, where the animals and humans are for the most part carefully drawn. As the schematic and conventionalized series overlie the naturalistic paintings at both these two sites, they are presumed to be of later date. It is this later conventionalized series which is represented at Porc Epic and in the British Somaliland paintings. Those from Bur Eibe may perhaps represent the extreme schematic form. The earlier, naturalistic painters specialized in pastoral scenes with long-horned, humpless cattle, and buffalo; the cattle shown in the later conventionalized series are zebu, and the camel is, also, rather doubtfully depicted.

Where an industry is found in relationship to paintings it is always of Somaliland Wilton form; there is reason to infer that a thin horizon containing Somaliland Wilton tools overlay the Magosian industry at Porc Epic.

Of considerable importance is the discovery at Gerbekele of pecked engravings overlying and in turn overlain by paintings of the schematic group. This suggests very strongly that the paintings and engravings in these styles are more or less contemporary.

Text-fig. 35 has been drawn to demonstrate the gradual degeneration in technique in the representation of cattle, and the gradual development of the conventionalized series out of the naturalistic series.

The dress and weapons carried by the naturalistic humans at Genda-Biftou are, as far as can be determined, consistent with those found among the present-day Hamitic inhabitants of our area.

The associations and age of these paintings will be discussed later.

Text-fig. 35. To show the degeneration in style and technique of cattle paintings in the Horn.

1. From the main frieze (Third Series) at Genda-Biftou, Sourré, Abyssinia. (After Breuil, H., 1934.)

2. Fourth series from Genda-Biftou, Sourré, Abyssinia. (After Breuil, H., 1934.)

3, 4. From Lago Oda, Abyssinia. (After photographs by W. B. Fagg.)

5, 6. From Cave 1, above Tug Gala-ad, British Somaliland. (After Burkitt, M. C. & Glover, P. E., 1946.)

7. From rock-shelter at Balleh, near Skutar, British Somaliland. (After Burkitt, M. C. & Glover, P. E., 1946.)

8. Possible conventionalized cattle motif from rock-shelter at Bur Eibe, Somalia. (From photograph by the writer.)

ENGRAVINGS

BUR DAHIR SHELTER (Pls. 27 and 47)

Two small shelters were found in an isolated hill rising steeply from the edge of the Webi Shebeli: the name of this hill was said to be Bur Dahir by my informants. It lies approximately ½ mile to the right of the road from Callafo to Mustahil, and 16 miles from the latter place. Here the river opens out into a wide flood plain which is swampy during the rains, but at one time the river must have spread almost to the foot of the hill. The rock is white gypsum and represents presumably the lower beds of the Cretaceous Series. The shelters are 25 ft. above the surrounding plain and face north-west. No other shelters were found.

The shelters are now some 5–6 ft. deep, 15–20 ft. long, and the roof rises gradually but steeply to overhang the shelters to a height of 12 ft. or so at the drip-line. A scree falls steeply to the base of the hill. These shelters are undoubtedly the result of subaerial weathering of the soft gypsum rock. Weathering is gradually eating the shelter deeper into the hill, but at the same time causes the outer edge of the roof to fall. The first such collapse formed the basis of the present scree. At a later time, two large tabular blocks of rock had fallen from the roof of the second shelter and lay propped at an angle to each other, it is on these blocks and on the back wall of this shelter that the engravings occur. There are no engravings in the first shelter except rectangular and linear motifs incised on the wall in recent times—presumably by the local Somalis.

Three series of engravings were found:

Series I

These engravings occur only on the fallen blocks and the majority are on the larger of the two blocks presenting a flat surface some 5 × 6 ft. which is inclined at an acute angle towards the rear wall of the shelter.

The natural colour of the rock is white or creamy white, but these blocks, together with those parts of the walls which have resisted recent weathering, are patinated a dark brick or chestnut red in colour. The Series I engravings, on the flat upperside of the larger slab are, with one exception, equally patinated with the rock so much so that it was quite impossible to photograph or trace them without first outlining them in white. At the outer edge of this block the engravings had been removed by weathering, the patina here being of the same brick-red colour.

Other engravings in Series I are found on the side of the main block, and by reason of their position have been subject to the least degree of weathering.

Here the engravings have patinated to a creamy red, while the rock itself shows the deeper chestnut patina.

The figures on the second rock are not so numerous but are similarly patinated to those on the upper side of the main slab.

It would seem, therefore, that these slabs had been in their present position sufficiently long before the engravings were made to have become well patinated. When first made the engravings would have shown up white on a red background; if on the other hand the rock had not been patinated they would have been very difficult to discern.

All the engravings were made by a pecking technique. Presumably any convenient hard stone served as a hammer and with this the whole of the figure was pecked out. No outline figures were found. There was evidence, however, that the outline of a few of the figures had first been incised into the soft rock with a sharp tool (perhaps a metal knife) and that the whole figure had then been pecked out.

The engravings were usually evenly concave in section, the deepest part in the centre being sunk some 5 or 6 mm. in the rock.

The figures are all conventionalized and represent:

(i) Filiform humans with shield and spear, drawn full face.

(ii) Cattle with and without humps: a characteristic is that the horns and tails are always carefully drawn.

(iii) Animals with pointed muzzles and ears which may represent dogs or possibly goats.

(iv) Animals which may be intended to represent horses but are more probably goats or calves.

(v) Camels; these show a flatter pecked surface than the other engravings of this series. The figure near the left-hand bottom edge of the main slab is somewhat less patinated than the other engravings, and approximates in patina to those on the side of the slab.

(vi) One drawing of an animal, possibly representing a rhinoceros, also occurs.

These engravings depict, therefore, a pastoral people using spears and shields, who have domesticated humped and humpless cattle, camels, goats (?), dogs (?) and very doubtfully horses.

Series II

Crude engravings with very irregular outline unpatinated are found on the back wall and on the side of the main slab. They represent camels and elephants.

The workmanship is poor and the outlines of the animals are not clear-cut. They have been executed by pecking. They appear to represent fairly recent copies of the earlier engravings which they overlie.

Of interest is one inscription in Amharic. This was translated for the writer by Ras W. B. Akalou, as the word *Quelelett*, meaning 'melting'—possibly an observation by the engraver on the prevailing climatic conditions! It would seem as if the camels underneath this inscription were drawn by someone lying on the flat, upper surface of the rock and leaning over the side, as they are upside down to anyone seated on the shelter floor.

Series III

A few rectangular and linear motifs incised on the rock wall with a metal knife. These are the most recent.

Series II and III engravings are obviously of fairly recent date and the only early ones belong to Series I.

At the foot of the hill below the shelter were found a few pieces of Late Stone Age factory debris and one or two pieces were also recovered from the scree of the second shelter. A small trial excavation was made in the floor. This showed at the surface 4 in. of fine dust, grey-brown in colour, with occasional flakes. Under this occurred a more compact layer of an approximately similar thickness, containing charcoals and a factory debris in somewhat greater quantity. The earth was a grey-brown powdery sand, as in the overlying level. This deposit rested on the rock floor of the shelter.

The excavation and surroundings yielded the following: one bifacially flaked side-chopper on a piece of tabular chert, one triangular piece of chert worked down both sides of one edge, four micro-blades, seventeen micro-flakes, one straight-backed microlith, one curved-backed microlith, one crude thumbnail scraper and five worked nodules. Specimens are all in black or grey chert either unpatinated or with mottled patina.

EL GORAN SHELTER (Pls. 28 and 48)

The shelter is situated at ground-level in the left wall of the gorge immediately above the upper water-hole at El Goran. It is a shallow shelter formed in the gypsum rock, some 15 ft. wide, 4–5 ft. deep and 6–8 ft. high, and about 8 ft. above and 15 yd. from the water-hole. The back wall slopes inwards and upwards to meet the roof and it is on this wall that the engravings are found. The floor contains no occupation material other than the droppings of stock, which make use of the shade afforded by the shelter. Late Stone Age factory debris is found on the upper slopes of the gorge.

Series I

At the top of the back wall just below its junction with the roof, where the maximum protection is afforded and on the roof, where not affected by later weathering, the rock has been patinated to a dark brick-red, or chestnut colour, identical with that found at Bur Dahir. On the former areas was found the

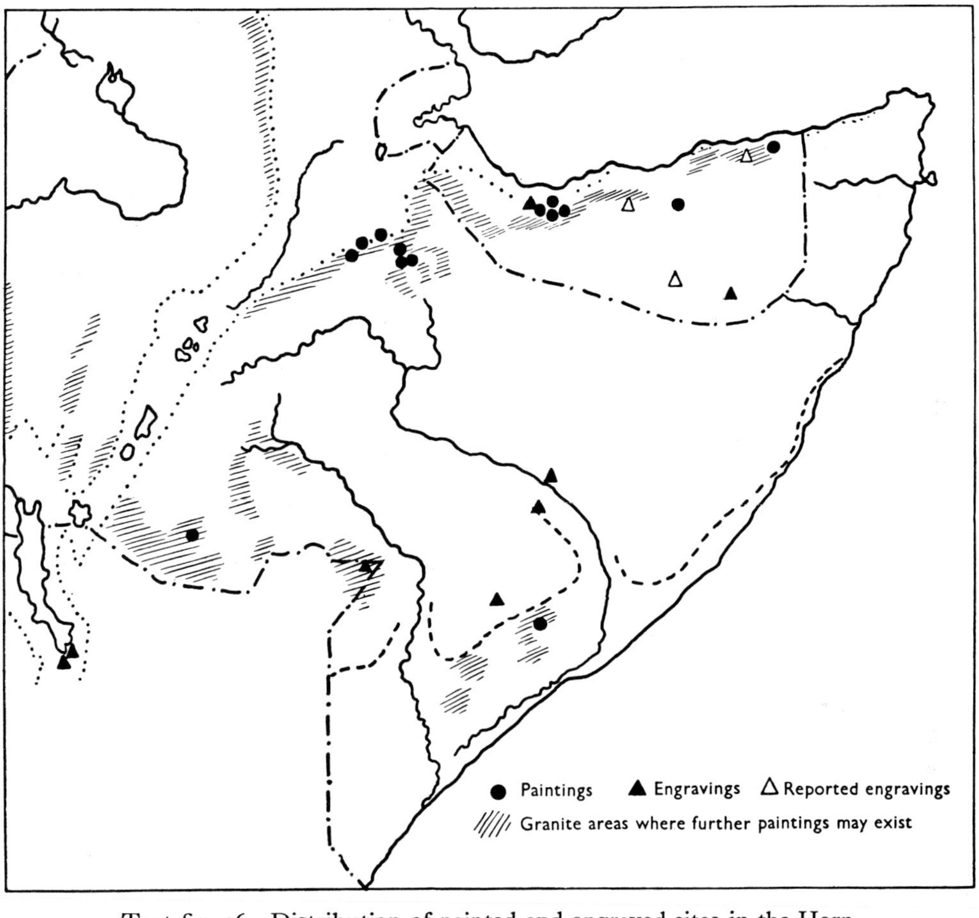

Text-fig. 36. Distribution of painted and engraved sites in the Horn.

remains of what appears to have been a frieze of animals in the Series I style of Bur Dahir.

Those figures are executed by a similar incised outline and pecked technique and are equally patinated with the rock though the pecking is not so deep as at Bur Dahir. The same care is taken to represent the tail of the animals. Much weathering has taken place and has rendered copying of the engravings very difficult, but a few were sufficiently well preserved for tracing and identification.

There are representations of:

(i) Cattle.
(ii) Camels—one fairly well drawn.
(iii) One elephant.
(iv) One calf or possibly horse.
(v) One portion of a human figure.

Possibly of the same age, as they were equally weathered with the semi-naturalistic drawings, were a few schematic motifs.

Series II

The commonest figures are tectiforms and other schematic signs which occur over the whole of the sloping part of the back wall of the shelter. This wall has been considerably weathered and the surface is glossy and polished, presumably by the bodies of animals rubbing against it. Numerous superpositions can be seen here, and all but the latest motifs on the lower half of this wall are quite unrecognizable.

Figures that are recognizable are parallel lines, circles, curved lines, tectiforms and twinned tectiforms, and rectilinear motifs and *garre* boards (i.e. the almost universal African game played with two, four or more rows of shallow holes).

The H sign appears, as well as the arabic letter *min* (Pl. 28, no. 10) which is now the tribal sign of the Rahanwein Harau.[1] The inverted V—now the sign of the Elai Warasili—is also present.

Series III

Recent copies of the *garre* boards, hands and finely incised geometric and linear motifs.

At El Goran, therefore, the same semi-naturalistic forms are present as were seen at Bur Dahir, but the schematic, geometric motifs are most probably later.

JID BANAN (THE BARREN PASS)

This is a pass leading from El Hedit in the northern Haud to Birtableh in the Nogal depression. The pass is situated near the south-east corner of British Somaliland and on the edge of the Nogal escarpment, in what is Dolbahanta territory. The rock is Eocene limestone.

[1] A single instance of this sign also occurs, about 3 miles due west of Bur Eschunle. It was pecked on an isolated slab of Jurassic limestone, and is not greatly weathered. It occurs in the Harau tribal area. No other engravings were found associated. Record may be made here of the tradition that the Elai 'marked the stones' at Bur Eschunle after their defeat of the 'Ma'adanle (Ajiuran)' sometime during the fifteenth century.

The writer was informed of the reputed existence of these engravings by Brigadier Smith,[1] then the Military Governor of British Somaliland.

The engravings are all found near the head of the pass on worn and weathered slabs of limestone. The great majority are in the actual pass itself and have been worn smooth by the passage of stock. Some of the engravings are covered by bushes and a little earthwash, but the majority are quite exposed. They all represent geometric and other schematic motifs and are executed by a pecking technique. They are of several different ages and it would appear that the original drawings, which are now quite unrecognizable, were copied on later occasions up to the present day.

The Dolbahanta state that the first engravings here were the work of the Galla whom they drove from the country and who emigrated to Abyssinia by this pass. They say that the Galla drawings have now been worn away but that the Dolbahanta themselves saw the drawings and copied them. The majority of the more recent engravings represent Dolbahanta tribal marks and Arabic letters; the marks of the Rer Baharasami, the Rer Hassan Ugaz and the Rer Gamasiya are all represented.

Pl. 49 illustrates some of these engravings. Those which are white and quite unpatinated must be very recent.

There can be little doubt that the Jid Banan engravings represent some of the latest to be found in the Horn and are perhaps of Galla, certainly of Somali origin.

ASSOCIATIONS AND AGE OF THE ROCK ART

To summarize briefly there would appear to exist three main styles in the Horn: first, naturalistic and semi-naturalistic paintings of pastoral scenes; second, semi-naturalistic and conventionalized paintings and engravings (the approximate contemporaneity of which is proved by the Gerbekele shelter), mainly individual figures, and schematic, geometric motifs; the third group need not concern us greatly—it is undoubtedly very late and tribal marks of the present-day Somali tribes are represented.

The second group is really a degeneration of the first and is later in age, as is shown by its superposition at Genda-Biftou and Lago Oda, and by the presence of camels (certainly in the engravings, doubtfully in the paintings): it is significant that the camel is entirely lacking from the first series, as is also the horse.

[1] Brigadier R. N. Smith informs the writer that he has noted similar engravings at the pass to Las Khoreh from Mogor to Yaffa, at Duberin pass east of Burao, and on rocks at well-heads at Horu Fadi south of Ainabo.

Breuil[1] has drawn attention to the affinities which exist between our first series paintings and the paintings and engravings at Djebel Uweinat, in the Fezzan, and at other places in the Sahara. A comparison between the Sahara art groups and those from the Horn shows that a close connexion undoubtedly exists between the two. Overhanging rock faces and small shelters are the places where this art was practised in both areas. The earliest Sahara group—the incised and polished engravings of southern Oran, the Fezzan and elsewhere are not found in the Horn, and the similarity is with the 'later prehistoric' groups in the Sahara. Cattle (*Bos africanus*), are the commonest animals represented and pastoral scenes predominate in both areas. The *cloisonné* character of the paintings in our own area, the prominence given to udders, horns and the inverted U underbelly and pointed legs are paralleled by those at Djebel Uweinat,[2] the Gilf Kebir[3] and the Hoggar,[4] the Fezzan,[5] Tibesti[6] and the Nile Valley in Upper Egypt.[7] Whereas, however, sometimes two, sometimes four legs are represented in the Sahara groups, only two are ever shown in the groups from the Horn. Humans are drawn side and full face in both groups and there is also a suggestion of the 'hour-glass' body in the Horn. The garments and weapons shown are the same in both areas, and Breuil has drawn attention to the similarity between the Egyptian pre-dynastic bow and that seen at Genda Biftou. The whole technique appears to be closely similar. Certain engravings at Onib, 130 miles W.S.W. of Haleib in the Nubian desert closely parallel those from Somalia: the camel is represented in both.[8]

The writer is of the opinion therefore that the earliest painters in the Horn were of a similar race and practised a similar pastoral economy to that of the painters of cattle scenes and the later engravers of the Sahara.

Vaufrey[9] and others have shown that the prehistoric paintings were the work of the makers of the Neolithic of Capsian tradition, and that their culture combines Egyptian elements of various dates. He considers that these cultural elements reached the Sahara and North-west Africa one at a time over a long period, and gradually transformed the Upper Palaeolithic Culture into a Neolithic civilization. Now if we can accept, as the writer maintains, that there were close racial connexions between the Sahara and the Horn then we should expect to

[1] Breuil, H. (1934).
[2] Breuil, H. (1928). Caporiacco, L. di & Graziosi, P. (1934). Shaw, W. B. K. (1934).
[3] Shaw, W. B. K. (1936).
[4] Wulsin, F. R. (1941). de Chasseloup Laubat, F. (1939).
[5] Graziosi, P. (1938*b*).
[6] Monod, T. (1947).
[7] Sandford, K. S. & Arkell, W. J. (1933, chap. 8). Myers, O. H. (1948).
[8] Parker, O. F. & Burkitt, M. C. (1932).
[9] Vaufrey, R. (1938).

find Egyptian cultural elements in both areas, unless special circumstances should have prevented them from reaching the Horn.

As has been pointed out there is absolutely no evidence as yet of the existence of Neolithic elements such as polished axes, conical based pottery, tanged arrowheads, etc., in the Horn.

The only true Neolithic form is the isosceles triangle hollow-based arrowhead.[1]

The Neolithic of Capsian tradition and its art groups, has been dated to between the fourth or fifth and the second or first millennia B.C. The later art groups approximate presumably more nearly to the latter date when the culture had become enriched by most or all of the Egyptian elements.

The absence of these Egyptian Neolithic elements in the Horn (with the exception noted above), might be explained by the fact that connexion between the two areas had been severed before these cultural elements had had time to penetrate to the Sahara. However, while a true Neolithic is absent in the Horn we have noted certain intrusive neanthropic industries (the Hargeisan) which introduce into our area new cultural elements—notably the angle-burin, the true end-scraper, and the true blade-core—and which develops into the Somaliland Wilton Culture. Much more research is necessary before it can be determined whether there exists any connexion between the naturalistic and semi-naturalistic paintings and the intrusive Hargcisan and Wilton elements. It is, perhaps, significant that wherever an industry is associated with this art it is of Somaliland Wilton form. If a connexion indeed exists between the art and the Somaliland Wilton complex then the Somaliland Wilton folk must have been Hamitic pastoralists and hunters and not merely hunters and food-gatherers as a study of the material culture would lead one to believe. The excavation of a painted site where bone is preserved would presumably solve this problem, for as yet we have no faunal remains associated with the Somaliland Wilton.[2] It would seem more probable, however, that the Hamitic cattle-owners depicted in the paintings should not be confused with the makers of the Late Stone Age microlithic cultures and that the paintings were either the work of the Somaliland Wilton hunters and depicted the strangers they saw, or else, more likely, that they were the work of the cattle-owners themselves.

With the meagre evidence at present at our disposal it is difficult, if not impossible, to arrive at an approximate date for the Series I paintings. On a comparison with the Sahara evidence the apparent absence of the camel, horse

[1] Vaufrey, R. (1938). It is of interest to note that Vaufrey places the hollow-based arrowheads of isosceles triangle form as one of the earliest traits to be diffused from pre-dynastic Egypt to the Sahara.

[2] But see the occurrence of sheep with a late strandlooping Wilton industry at Eil, cf. Graziosi, P. (1940, p. 60).

and zebu cattle, place them in the prehistoric group. The absence of any evidence of the 'helmeted ram cult' suggests they are early in this group, and Breuil's observation that the Sourré paintings (with their domesticated buffalo, bubalis (?), and antelope) are appreciably earlier than those at Djebel Uweinat (Ain Doua) tends to confirm this. However, the complete absence of the earlier type of Sahara engraving, the close comparison that exists with the Sahara paintings and later engraving groups and the absence of extinct forms with the doubtful exception of bubalis and a cervid suggests to the writer that the Sourré paintings are in time late prehistoric.[1] On the negative evidence mentioned above they may be considered to be earlier than 600 B.C. (when the riding horse was introduced into the Sahara) but whether they can be dated to the first or second millennium B.C. or earlier—to an age comparable with the late Sahara groups—still remains to be proved. An important pointer, however, is the fact that the climate was apparently wetter in both areas, than it correspondingly is today, and the apparent absence of *Bos brachyceros* from the Sourre paintings suggests that they may date prior to the Middle Kingdom in Egypt (2200–1788 B.C.) when short-horned cattle were introduced; this does not take into account any appreciable time-lag, however, before this breed was introduced into Ethiopia.[2]

The Deir-el-Bahari bas-reliefs of Hatshepsut's expedition to Punt, undertaken between 1501 and 1479 B.C. depict the cattle of Punt as belonging to two different breeds—long-horned (*Bos africanus* (?)) and short-horned (*Bos brachyceros* (?)).[3] It would seem, therefore, that short-horned cattle had already penetrated to the Ethiopian region, whether we suppose the land of Punt to have been between Suakim and Assab, or further east along the Somaliland coast, by the middle of the second millennium B.C. None of the Ethiopian paintings are as well drawn as those from the Hoggar, Tassili-n-Ajjer, or Uweinet, but on the other hand they are not as conventionalized as the paintings and engravings found by Mrs Drew[4] in the Akele Guzai division of Eritrea and the northern part of the Tigré province of Ethiopia, which are considered to be of 'Early Ethiopian' age and generally 'Christian'. The purely symbolic motifs of the Eritrean sites are quite absent from our Series I paintings but there is some similarity in the treatment of the human figure in both groups. Weighing all the evidence available it would seem that the earliest art-group in the Horn probably dates to about the beginning of the Christian era or shortly before.

[1] They also clearly antedate the conventionalized engravings of Gazien in northern Abyssinia which are dated by a southern Arabian inscription to between the second century B.C. and the second century A.D.; cf. Mordini, A. (1947, pp. 321–3 and Plate).

[2] Sandford, K. S. & Arkell, W. J. (1933).

[3] Naville, E. (1898).

[4] Drew, S. (1951, 1952).

The semi-naturalistic and conventionalized paintings and engravings from the Horn must be appreciably later. They show the camel, zebu cattle, as well as *Bos africanus*, and perhaps the horse. The engravings are much more akin to those of the Sahara Libyco-Berber engravings than they are to any of the prehistoric groups. They cannot be earlier than the fourth century A.D. and are probably appreciably later, though the weathering and patination to which some of them have been subjected assigns to them a reasonable antiquity.

I can find no reference to the probable date of introduction of the zebu cattle into the Horn. It appears to be in origin an Asiatic breed as it is represented in the Indus civilization at Mohenjodaro[1] and it seems not unlikely that it may have been introduced into the Horn during the Semitic migrations from south-west Arabia which are said to have taken place between the first millennium B.C. and the first half of the first century A.D. The Galla and Somali cattle are today all of zebu type.

The schematic and geometric motifs from El Goran (and perhaps also the earlier ones from Jid Banan) recall the engraved monoliths and grave-stones in the country of the Galla lakes on both sides of the Rift.[2]

Analogous, pecked engravings have been found at Mandera on the Juba on the Kenya/Somalia border, and in northern Turkana. Paintings which can be ascribed to the same conventionalized group have been described by the writer from southern Abyssinia, associated with a degenerate Wilton industry.[3] It will be for future investigators to decide whether in fact the earlier art group in the Horn was the work of early Hamitic immigrants (pastoralists and hunters) bringing with them the domesticated *Bos africanus* or whether they were painted by the Somaliland Wilton peoples, and whether again the later paintings and engravings point to a later migration of Hamites or Semites, bringing the camel[4] and zebu cattle.

[1] Childe, V. G. (1935).

[2] (*a*) Kammerer, A. (1926). (*b*) Instituto per gli Studi di Politica Internazionale, 1936, *L'Africa orientale*, pp. 275–6.

[3] Clark, J. D. (1945*b*).

[4] That the camel was already present in Egypt as early as Old Kingdom times, however, is proved by the discovery of a camel-hair rope in the Fayum in a third or early fourth dynasty context. Caton-Thompson, G. (1934).

PART IV

A TENTATIVE CORRELATION OF CULTURES AND CLIMATES

CHAPTER 14

A TENTATIVE CORRELATION OF CULTURES AND CLIMATES IN THE HORN WITH OTHER AREAS OF THE AFRICAN CONTINENT AND WITH SOUTHERN ARABIA

The Ethiopian region lies at the northern end of the tropical zone and is thus in a most favourable position for correlation of the climatic sequence observed there with that of the Sahara/Arabian dry belt and so with the sequence in the Mediterranean region. As Zeuner has pointed out[1] there is no evidence that the tropical pluvials were contemporary with those of the Mediterranean area, but the links afforded by the cultural and climatic succession in the Horn can be used to relate more closely the two zones, and give a strong indication (unless we must suppose cultural time-lags) that the pluvials of these two zones will most probably be found to be contemporary.

It must be emphasized once again here, that in the application of the term 'pluvial' to those occasions when the climate of the Horn was wetter than it is today, it is not implied that great increases in precipitation took place and that any thick covering of forest spread throughout the whole area. The increases in rainfall (except for the earliest) do not appear ever to have been on as large a scale as occurred in Kenya and other parts of East Africa, though the Abyssinian high plateau with its specialized climate must necessarily be considered separately from the other regions of the Horn. In such areas as the Mudugh, the Mijertein and the Danakil Rift it is very doubtful whether the increases in precipitation in later Pleistocene and Recent times caused the climate to pass out of the arid or semi-arid classification of Twenhofel. Within these classificatory bounds, however, it has been shown that fluctuations in rainfall intensity did in fact occur. Bearing in mind these limitations, therefore, the term 'pluvial' and 'wet phase' are applicable in the Horn as well as in tropical Africa.

In the Horn evidence has been adduced for one, possibly two major pluvials during the Lower and Middle Pleistocene, of long duration followed by a long and arid interpluvial. In the Upper Pleistocene there is evidence for a further pluvial, of two, perhaps three phases, and a further arid phase, marking the end of the Pleistocene. In Recent times two minor wet phases of decreasing intensity

[1] Zeuner, F. E. (1946, p. 245).

and separated by a period of arid conditions are recognized. It will now be of interest to see how this sequence agrees with that in East and North Africa, and other areas.

CORRELATION WITH EAST AFRICA (KENYA, UGANDA, AND NORTHERN TANGANYIKA)

The cultural and climatic succession in these areas is well known and documented from the work of Leakey[1] in the lakes of the Kenya Rift, northern Tanganyika and elsewhere; and of Nilsson[2] in the Kenya Rift lakes, Wayland[3] and O'Brien & Solomon[4] in Uganda, and Fuchs[5] in the Lake Rudolf basin.

The earlier cultures so well represented at Olduvai and in the Kenya and Uganda Rifts, have not yet been recorded from the Horn. The intense and long erosion period of the Kageran Pluvial can be correlated with the major erosion period in the Horn, as also presumably can the Kamasian deposits with the thick conglomerates and marls of the major aggradation phase of the Somali rivers. The major arid phase would then correspond to the Great Interpluvial in East Africa.

The cultural sequence begins with the Acheulio-Levalloisian (perhaps with an Upper Acheulian stage) which belongs to the beginning of the last, or Gamblian, Pluvial, but may in part belong to the end of the preceding Interpluvial. Its counterpart in East Africa is provided by the Kenya Fauresmith with its hand-axes, U-cleavers and associated flake tools.[6] The comparison is a particularly close one, identical forms occurring in both areas. The Kenya Fauresmith is dated by the lake beds of the Kinangop plateau, to the very end of the Kamasian Pluvial, which is probably a little anterior in date to its counterpart in the Horn. The correlation is a sufficiently close one, to make it possible that subsequent work may prove that the two cultures were in fact contemporary. In Uganda the place of the Fauresmith is taken by the Sangoan, but in the Lake Rudolf basin Fuchs[7] found certain surface industries which unfortunately cannot be dated but which suggest the extension of the Fauresmith into this area, so linking up Kenya with northern Abyssinia and the Somalilands.

When we compare the cultures of the main Gamblian Pluvial we find contemporaneous Levalloisian cultures in both areas. O'Brien[8] extends the bounds of this culture into Uganda and it would seem that practically the whole of

[1] Leakey, L. S. B. (1931, 1936).
[2] Nilsson, E. (1940).
[3] Wayland, E. J. (1934).
[4] O'Brien, T. P. (1939).
[5] Fuchs, V. E. (1939, pp. 219–74).
[6] Leakey, L. S. B. (1931, p. 38).
[7] Examined by kind permission of Dr V. E. Fuchs.
[8] O'Brien, T. P. (1939, p. 38).

eastern Africa during this time was a province of the Levalloisian Culture complex. Kenya and northern Uganda with their East African Capsian Culture alone show alien elements. It might be thought that the Kenya Capsian reached this area by way of the Somalilands, but there can be little doubt from the whole of the stratigraphical evidence that is so far available, that this was not the case. This leaves the possibility that western Abyssinia was the area through which this culture reached Kenya (if indeed it can be proved to be intrusive there and not of autochthonous growth) but as yet Abyssinia is a blank, or virtually a blank, on the prehistoric map.

The Developed Levalloisian of Kenya is very close in tool form and technique to that in the Horn, the chief differences being that in the former area, the Tabelbalat technique is rare, the parti-bifaced, foliate points do not appear to occur and the double-ended, subrectangular, prepared core is only rarely found, the almost universal form being discoidal or oval.[1] In fact the Developed Levalloisian culture of Kenya and Uganda is much more nearly the immediate ancestral form of the Stillbay, than is the Developed Levalloisian in the Horn. The sudden change-over here from the oval flake and the tendency towards developing the foliate point, at the end of the Levalloisian sequence, to the markedly subtriangular Stillbay flake and point may be indicative of a northward migration of Stillbay stock into the Horn. Whether this is in fact what occurred must be left for future investigation to decide, but this supposition does explain the differences which exist between the basic technology of both cultures, and is substantiated in some degree by the very close affinities that exist between the Somaliland Stillbay Culture and its counterpart in Kenya. These similarities might of course be explained by parallel evolution from a common form, but if this were, in fact, the case we might expect the Stillbay in the Horn to reflect and develop the most evolved forms of the final Levalloisian there. This is not altogether the case, however.[2]

Whatever the origin of the Stillbay in the Horn, its distribution is almost identical with that of the Levalloisian, and just as our area was a province of the Levalloisian culture, so was it also of the Stillbay.

The Stillbay in Kenya is found in the deposits laid down during the closing stages of the Gamblian Pluvial, but it extends into the dry period marking the end of the Pleistocene. This corresponds very closely to its chronological position in the Horn, though it is there in the main confined to the dry end-Gamblian period.

[1] Small double-ended cores are noted by O'Brien from Walasi in eastern Uganda, and are ascribed by him to a variant of the Uganda Stillbay. O'Brien, T. P. (1939, p. 197).

[2] See, however, the foliate points associated with the lower Somaliland Stillbay at Hargeisa.

The Magosian in the Horn is a logical development from the Stillbay in the same way as it is in East Africa. Regional differences exist, but they can be explained by local specialization.

Neanthropic elements (blade and burin) do not appear until post-Pluvial times in the Horn and after their appearance the cultural connexions can be seen to be much more closely allied to the northern part of the Continent than to East Africa. Whatever the origin of the Late Stone Age cultures of Kenya, they bear no detailed resemblance to those from the Horn (with the exception of the Somaliland Wilton) and it is certain that the blade and burin elements reached the latter area from the north and not from the south. The specialized pressure flaking of the Doian has no counterpart in Kenya, the microlithic element is also in the main quite distinct, and the hollow-based arrowheads of isosceles triangle form are unknown south of Somalia. It would seem, therefore, that the neolithic elements found in Kenya reached that country by a route west of the Abyssinian plateau, or possibly from the plateau itself. It is certain that they were not derived from the Somalilands, and in Late Stone Age times the Northern Frontier District of Kenya seems to have formed just such a natural regional boundary as it does today.

While, therefore, the prehistoric cultural succession during the latter part of the Pleistocene shows that close connexions existed between East Africa and the Horn, from the end of the Pleistocene they began to diverge and developed independently of each other.

ABYSSINIAN HIGH PLATEAU

The fragmentary and scanty evidence at present available from the Abyssinian high plateau renders any correlation impossible, and at best only affords a very incomplete indication of the cultural associations. Much of the evidence has yet to be described.

A Kenya or Abyssinian Fauresmith site on the Dabat plateau near Gondar[1] and a somewhat similar but unpublished collection made by McInnes at Mega[2] in southern Abyssinia suggests that the Acheulio-Levallois/Fauresmith complex incorporated the high plateau also in its sphere of influence. Levalloisian cultural material is particularly meagre. The Dagha Dima beach of the Zwai/Shala Basin yielded Levalloisian type tools in quantity to Nilsson.[3] These have not yet been described though Dr Leakey informs the writer that they are of Developed Levalloisian type. D'Errico[4] describes material from Zergnat, a site discovered by him near Magdala. The material is chalcedony and chert and the

[1] Clark, J. D. (1945*a*).

[2] In the Coryndon Museum, Nairobi

[3] Nilsson, E. (1940, p. 39).

[4] D'Errico, L. (1937*a*).

specimens are of two ages. The later series shows a glossy patina and includes long blades and flake blades with subtrapezoid section, little or no retouch, but evidence of utilization on the edges. The material is ascribed by d'Errico to the Neolithic; it is reminiscent of the Elmenteitan of Kenya. The earlier specimens comprise red-patinated flakes one of which has been worked into a crude unifaced point. It is not clear whether the striking-platforms are faceted or not; this assemblage is presumably of Stillbay or earlier affinities. Moysey's collections have added considerably to our knowledge of this area. Amba Alagi yielded Developed Levalloisian or Stillbay material, and his excavation of a rock-shelter at Gorgora has provided most important evidence of the apparently long development of a Stillbay industry on the northern plateau.[1] The developed nature of this later Stillbay with its triangular, bifaced points, and high degree of secondary retouch is of particular interest.

South-east of the Abyssinian Rift, Teilhard[2] and Breuil[3] found Levalloisian material in the alluvial deposits of the Errer valley, and on the plateau near Harar. Blanc[4] records Stillbay artifacts from Moggio.

The evidence suggests, therefore, that the cultural sequence on the high plateau for the greater part of the Upper Pleistocene followed that in the Somalilands and Kenya, the Acheulio-Levalloisian or Fauresmith giving place in turn to the Levalloisian and a local form of the Stillbay.

A greater number of Late Stone Age sites exist. Although the majority are surface collections they suggest that two distinct cultural elements are present. The Stillbay is the ancestral form of the one, and develops by way of a Magosian into a local Mesolithic or Neolithic Culture of Wilton form. The origin of the other, which approximates more nearly to the Elmenteitan and its derivative cultures than to any other East African culture, is unknown, though it may be found to have had a similar origin with the Elmenteitan in an ancestral 'Capsian' form. Besides the excavations at Porc Epic and Gorgora the Magosian is known from surface collections made by Moysey at Batie near Dessie, Wadelai Delanta and Bethor (?) between Debra Tabor and Dessie; and the Mille river crossing on the road to Assab. What is thought to be its final development, into a pottery-using microlithic culture, is seen in the uppermost levels at Gorgora and Porc Epic and again at Yavello.[5] At the last two sites it may perhaps be associated with conventionalized paintings. The distribution of this, or an allied culture is extended into the Galla highlands and into the Arba, Awash and Shoa valleys by the du Bourg de Bozas collections; on to the Harar

[1] Moysey, F. (1943). Leakey, L. S. B. (1943).
[2] Teilhard de Chardin, P. (1930).
[3] Information received verbally from the Abbé Breuil.
[4] Blanc, A. C. (1938).
[5] Clark, J. D. (1945*b*).

plateau by Breuil's and Teilhard's collections (notably from the shelter at Omerdin), and into the region of Addis Ababa by the collections described by Blanc from Moggio and Lake Biscioftu. It appears in fact to be widely distributed on the high plateau.

Basically distinct from the Wilton industries of Middle Stone Age ancestry is that from a rock-shelter at Quiha near Macalle, excavated by Moysey, but unpublished.[1] The material is predominantly obsidian but a small collection of chert specimens is also found. Tools include three large and twelve small, round or disc scrapers, one core scraper, thirteen *outils écaillés*, one redirecting flake, twenty-nine utilized blades ($2\frac{1}{2}$ to 1 in. in length, constricted bulb and platform, showing utilization of edges but no retouch), forty-three primary flakes, four burin spalls, eight single-blow and *bec-de-flute* burins, four poorly made, large, backed blades, forty-nine lunates (usually backed from one side only), thirty-five curved-backed microliths (rather poorly made), twenty fabricators (flakes and blades with battered edges) and a number of sherds of red-fired pottery, mainly undecorated—derived for the most part from deep bowls. There is no evidence of the micro-burin technique. This industry appears to be allied to that from Zergnat. A similar surface collection comes from Amba Sel, in approximately the same area.

It would appear, therefore, that on the northern part of the plateau there existed two cultural elements in Late Stone Age times, one allied to the Somaliland Wilton Culture, the other with affinities not found in the Somalilands and reminiscent of the Elmenteitan and derivative cultures of Kenya.

Much of the correlation work undertaken by Nilsson in Abyssinia has already been mentioned but a summary of his findings in regard to the glaciated areas of British East Africa and Abyssinia and his suggested origin for the Nile silt regime in Egypt must be given here.[2]

Nilsson correlates pluvials with glacials and considers that the climate, besides being wetter was also colder during the pluvials, as the glaciated areas on the higher mountains are found by him to have been much more extensive during these periods. Wright[3] considers, however, that Nilsson's evidence is not sufficient grounds for correlating the East African glacier extensions with any of the European glaciations. During the Kamasian Pluvial the summits of Mount Kenya, the highlands of Semien, and probably also the tops of the highest mountains on the eastern wall of the Rift, in the region of the Galla

[1] Collection in Coryndon Museum, Nairobi.

[2] Of interest also is Nilsson's correlation of the curves for the East African and Abyssinian lakes with the Fayum. The correlation is a close one. (Nilsson, E., 1940, p. 65.)

[3] Wright, W. B. (1939, p. 162).

lakes, were covered by glaciers. During the last Interpluvial volcanic activity on a vast scale caused the uplift and outward tilting of the Somaliland and Abyssinian plateaux. During the Gamblian Pluvial the snowline was appreciably lower than it is today, for instance in the Semien mountains, the glaciated zone stood at 4200 m. (there is no permanent snow on these mountains today), and Nilsson was able to correlate the zones of moraine-ridges on each of the mountains he examined. He states—'this great similarity in situation and size, of the moraine-ridges demonstrates the synchronism of the whole recession of the glaciation through all its stages from its maximum extension, which also, of course, was simultaneous on all the mountains of East Africa.'

While direct cultural proof is lacking for corroboration of Nilsson's evidence for the contemporaneity of Pluvials and Glacials in East Africa, his findings are a very strong indication that this correlation will be found to be the correct one—particularly when considered in conjunction with and through the medium of correlations between eustatic movement, and river and lake terraces. It would seem, therefore, that the theory of Hume & Craig[1] that during the Glacial Period the Asiatic centre of high pressure was very much stronger, the N.E. monsoon dominant, and Abyssinia a practically dry country, cannot be substantiated. Of importance also when considering the correlation of the Nile terraces and those of the rivers in the Horn is Nilsson's explanation of the great silt phase in the Nile during the second half of the Upper Pleistocene. He has suggested that most of this silt may have come through the Blue Nile, from the Abyssinian plateau and that the tilting of the plateau during the last Inter-Pluvial and beginning of the last Pluvial, was the initial cause of this so that the rejuvenated Blue Nile cutting deeply into the Lake Yaya and Mesozoic sandstones and limestones, removed vast quantities of these sediments and deposited them in the Nile below the second cataract.

SOUTHERN ARABIA

Huzayyin[2] noted in the Yemen evidence of an earlier and major Pluvial when terraced gravels were deposited up to 30 m. thick. This was followed by an arid Interpluvial, accompanied by volcanic activity and the pouring out of extensive lava-flows. This Interpluvial was followed by a second Pluvial, not so intense as the first, when silts and fine gravels were laid down in three terraces. The deposits of this Pluvial have yielded material representative of a crude and relatively unprogressive facies of the Levalloisian. Huzayyin also found evidence for a final, or historic, wet phase which lasted well into the Christian era.

[1] Hume, W. F. & Craig, J. J. (1911, pp. 382–3).

[2] Huzayyin, S. A. (1941, p. 125).

Caton-Thompson's & Gardner's[1] work in the Hadhramaut corroborates Huzayyin's later sequence; only the later quaternary deposits are known in the Hadhramaut, however, as the earlier Pleistocene geology is buried in deeply incised channels. This major erosional phase resulted in the cutting of channels over 100 m. below the present floor and was followed by a major depositional phase when the valleys were filled with gravels and aeolian silts. A further erosional phase followed and resulted in the deposition of gravels in three terraces which presumably correlate with Huzayyin's second Pluvial deposits in the Yemen. They contained the same crude, Levalloisian industry. Tufa also formed during this phase. Huzayyin correlates these deposits with his second (Gamblian) Pluvial.

While as yet no detailed correlation can be attempted between the Horn and south-west Arabia, it is of interest to note that the same major erosional and major depositional phases followed by an Interpluvial with volcanic activity, are found in both areas, and that these events are similarly followed during the Gamblian Pluvial by aggradation deposits containing Levalloisian industries. The complete absence of *coups-de-poing* in the Hadhramaut and the crude nature of the Levalloisian Culture certainly suggests that no land-bridge can have existed at the straits of Bab-el-Mandeb during the Upper Pleistocene or more recent times. If a land-bridge had been present it is difficult to see why the developing Levalloisian of the Horn was not diffused to south-western Arabia.

CORRELATION WITH EGYPT AND THE SAHARA

Any correlation of the Egyptian sequence with that in the Horn is complicated by, and must be considered in relation to, the changes in sea-level in the Mediterranean, and no correlation can be considered as final until the age of the Monastirian II beach is established beyond further doubt. Zeuner[2] places this beach in the second part of the Last Interglacial (Riss-Würm) and there is much to support such a date. Caton-Thompson[3] has shown, however, that if this beach with its inferred Upper Levalloisian culture is to be assigned to this date it implies (correlating the Nile aggradation silts and the Fayum 34 m. lake with the Monastirian II beach) that the Egyptian Upper Levalloisian was typologically further advanced than its counterpart in Europe and moreover leaves nothing in Egypt other than the epi-Levalloisian cultures to fill the period occupied in Europe by the whole of the Upper Palaeolithic. As an alternative, therefore, Caton-Thompson suggests the correlation of the Monastirian II beach with a Würm I–II interstadial. This on the cultural evidence seems to be much more in keeping and is that which has been adopted in the second table.

[1] Caton-Thompson, G. & Gardner, E. W. (1939). [2] Zeuner, F. E. (1946).
[3] Caton-Thompson, G. (1946, 1947).

Comparing the Egyptian succession[1] with that in the Horn it would seem at first sight that the Nile aggradation silts, with their Upper Levalloisian culture, equate with the Tug Gravels and Alluvium of the Somali rivers which contain the local form of the Levalloisian. Thus the initial erosion of the Nile upper buried channel would correspond to the buried channel erosion in the Horn. If this correlation is the correct one, however, and bearing in mind the correlation of the Nile aggradation silts with the Monastirian II beach, it means that the Tug Gravels and Alluvium and the Somaliland Levalloisian must be assigned to the early part of the Gamblian, while the Acheulio-Levalloisian must be put back into the preceding Interpluvial. On the existing evidence while the latter might not be inconsistent, it is very difficult to believe that the Somaliland Levalloisian should be sandwiched into the first half of the Gamblian leaving little more than the Stillbay to fill the major part of this pluvial. Moreover should the Last Interpluvial (or Last Interglacial) age of the Monastirian II beach be confirmed this would put back the Levalloisian of the Horn still further and place it in the last Interpluvial.

An alternative correlation is to equate the Older Tug Gravels and Alluvium containing the Acheulio-Levalloisian in the Horn with the period of the Nile aggradation silts. This would mean, however, that there was a cultural time-lag with a locally persisting Acheulio-Levalloisian the contemporary of an Upper Levalloisian stage of culture in Egypt. It is of interest to note that these two series of aggradation deposits in Egypt and in the Horn stand in the same relationship to the tilting of the Abyssinian and Somaliland plateaux, which event, as has been shown, is considered by Nilsson to be the probable cause of the aggradation of the Nile silts.

The period of major Nile bed erosion which follows the aggradation silts and is contemporary with the 28 m. lake in the Fayum, would then equate with the phase of buried channel erosion in Somaliland and with the period of major marine regression in the Mediterranean and the Horn. On this basis the Lower and Upper Somaliland Levalloisian contained within the Tug Gravels and Alluvium (and believed contemporary with the eluvial deposits and fossil dunes at the coasts) would equate with the epi-Levalloisian I and II of Egypt.

A typological comparison between the Somaliland and Egyptian cultures shows that such a dating would not be inconsistent especially when the very conservative nature of the Levalloisian complex in both areas is taken into account. The Acheulio-Levalloisian in Egypt seems to have occupied a much shorter period of time than its counterpart in the Horn, so that it would appear

[1] Sandford, K. S. & Arkell, W. J. (1933; 1939*a*, *b*). Caton-Thompson, G. & Gardner, E. W. (1932). Caton-Thompson, G. (1946). Caton-Thompson, G. (1952).

that typological development may have been much more rapid in Egypt. It must also be recalled that a late form of culture with hand-axes, and Levalloisian-type cores and flakes—probably a local form of Sangoan—occurs in the Sudan.[1]

When we come to compare the Levalloisian cultures an even closer relationship is apparent. On the whole, however, the tools of the Upper and Lower Levalloisian from the Tug Gravels in the Horn are smaller than those of the Egyptian Levalloisian and comparison is probably closer with the epi-Levalloisian I and II of the Nile and Fayum and the Levalloiso-Khargan and Khargan from Kharga Oasis. If this correlation is substantiated it may well be that the industries from Eil and Dubar, which typologically are transitional between the Acheulio-Levalloisian and the Lower Levalloisian in the Horn, are the Somaliland equivalent of the Egyptian Lower and Upper Levalloisian.

A correlation such as the second alternative suggested above would seem to fit the available evidence better and is that preferred by the writer, but it must be admitted that it implies that the Tug Gravels and Alluvium in the Horn, which are aggradation deposits, equate with the epi-Levalloisian degradation gravels in the Nile, an apparent inconsistency.

Over the remaining parts of the Horn the Upper Levalloisian is replaced by the Stillbay, the southern and eastern counterpart of the Aterian. No Aterian influences have yet been found in the Horn. The tanged point does not occur, the subtriangular points almost invariably show a high degree of pressure and flat percussion flaking, and the presence of backed flakes and blades further sets apart the Stillbay of the Horn from the probability of any Aterian connexions.

That the cultures in the Horn in early post-Pleistocene times were quite independent of those of the Nile Valley, there can be no doubt, but intrusive elements now begin to make their first appearance. The Stillbay developed directly into the Magosian without any outside interference, but it is in post-Magosian times that the blade-and-burin elements which had penetrated to the northern part of the Horn during Stillbay and Magosian times began to make themselves felt.

The origin of the Hargeisan is a matter of considerable interest. It shows little or nothing in common with the Egyptian Sebilian nor, owing to the apparent complete absence of the micro-burin does it closely compare with the North African or Kenya Capsian. The true burins, however, are sufficient evidence to show that it had connexions with the Capsian complex.

The geometric microliths of the Capsian are entirely absent in the Horn and the large backed blade, though it does occur, is a rare tool. On the other hand, the straight-backed microliths with and without basal retouch, are common to both cultures, as is also the method of flaking from both faces to produce a flat back.

[1] Arkell, A. J. (1949).

While the Hargeisan shows typological similarities to the Upper Capsian of Tunisia it lacks the developed forms found with this culture and approximates more nearly to the culture known in French North Africa as 'Ibero-Maurusian' or 'Oranian'. The same absence or rarity of the micro-burin technique and the concentration on the production of non-geometric microliths, the straight-backed blade and lunate, and the utilized blades are common to both cultures. It would seem more likely, however, that this similarity is due to the existence of a similar ecology and to the use of a similar form of raw material rather than to any cultural connexions. Until further evidence is forthcoming, therefore, we must regard the Hargeisan as a regionally specialized culture of mixed descent but in the North African blade and burin tradition.

The full expression of the Late Stone Age in the Horn shows a number of traits in common with the Saharan cultures and there can be no doubt that fairly close contact, however sporadic it may have been, did exist between the two areas.

The 'Neolithic Wet Phase' in the Sahara and Egypt allowed the spread of the 'Neolithic of Capsian Tradition' throughout the Sahara, and its influence to extend into the Horn. Until detailed work is carried out on the Abyssinian high plateau it will not be known for certain whether the impetus for the spread of this culture came from the Abyssinian highlands or from North-west Africa, but the existing evidence seems to indicate that the spread was from north-west to south-east. Again we can point to no complete identification of the Late Stone Age cultures in the Horn with any of the Saharan, Egyptian or Sudanese forms of the Neolithic, but the distance of the Horn from the centre of diffusion of these cultural traits is most probably the reason for this distinction. It is probable that by independent development the triangles and crescents of the Ogaden Wilton had evolved from diminutive Levalloisian-type flakes, in the same way as did the Sebilian trapeze.

The culture which approximates most nearly to any of the Egyptian or Saharan cultures is the Doian and the industry from Rakon. This is clearly a similar development to the Neolithic B group of the Fayum[1] and the Beduin Neolithic of Kharga. The same delicate pressure flaking is seen on the lanceolate and foliate points, and the trihedral rods, and unifaced points. The shanked form of microlith has its counterpart in the eared microliths and those with concave retouch; the forms and backing of the microliths are also closely similar. Heluan has also yielded the triangular-sectioned point.[2] The specialized Heluan retouch also occurs, as we have seen, in the Doian. The microlithic

[1] Caton-Thompson, G. & Gardner, E. W. (1934, p. 55).
[2] Huzayyin, S. A. (1941, p. 260).

forms, foliate and lanceolate points are again reflected in the Ounanian[1] of the southern Sahara, though the tanged blades are missing from the Doian. The Neolithic from Ténéré[2] again shows the trihedral rod; but also examples of the hollow-based arrowhead almost identical with those from Doian transitional sites and from Rakon. This type of arrowhead is reflected again and again in the Saharan Neolithic and its isosceles triangle form suggests that it was diffused into the Horn, though it is possible that it developed independently inside our area from the triangular points of the Upper Somaliland Stillbay. It is of interest also to note that both Vaufrey[3] and Breuil[4] place this form as one of the earliest to reach the Sahara from Egypt. The tanged arrowhead is quite absent from the Horn but is common in the Sahara where the Aterian tradition must also have been present, as Aterian industries occur as far east as Kharga. It is possible, therefore, that the tanged form found with the Saharan Neolithic may have developed from the Aterian tanged point in a centre of Aterian influence. Its absence from the Horn would then be due to the fact that the Aterian failed to penetrate to this area. Again, the large crude, foliate point best seen at Mirsale, is found also with the Saharan and North African Neolithic. As far as I have been able to determine the tanged arrowhead does not occur or is very rare in the pre-dynastic cultures of Egypt, the hollow-based and winged forms being the characteristic ones. At Tamaya Mellet and Taferjit[5] in the central Sahara the earlier forms are all apparently hollow-based and leaf-shaped. The tanged forms appear to be most abundant in the northern and western parts of the Sahara and it is not improbable that they spread from there towards the centre and east. It would seem not at all unlikely therefore that the tanged arrowhead owes its origin to an Aterian ancestry while the hollow-based arrowheads may well have developed out of the Stillbay complex. The presence of paintings in the Horn, similar to those of the later prehistoric groups in the Sahara has already been noted. It becomes apparent, therefore, from the above brief recapitulation of the evidence that there occurred a definite migration of peoples into the Horn of Africa during Late Stone Age times, bringing with them traits which link them with the Saharan Neolithic cultures, and with the Neolithic B group people of the Fayum. This migration took place during a period when climatic conditions were wetter than at the present time—that is to say during the 'Neolithic Wet Phase' of North Africa and the Sahara, when, besides human cultures an intermingling of Mediterranean and Ethiopian forms of fauna and flora took place in the Saharan dry belt.[6]

[1] Huzayyin, S. A. (1941, p. 282).
[2] Joubert, G. & Vaufrey, R. (1946, pp. 325–30).
[3] Vaufrey, R. (1938).
[4] Breuil, H. (1930).
[5] Kelley, H. (1934, pp. 135–43).
[6] Huzayyin, S. A. (1941, p. 73).

CORRELATION OF STRAND-LINES

Owing to the paucity of the cultural evidence and the amount of isostatic movement that has occurred in the Horn during quaternary times any correlation of strand-lines both within the Horn itself and with other areas, must remain tentative in the extreme. However, it is worthy of note that the broad sequence of events on the Mediterranean coast during the later Pleistocene is reflected on the north and east coasts in the Horn. Sandford's evidence of raised beaches in the Gulf of Suez and the Red Sea will probably provide the link with those at the lower end of the Red Sea and on the Gulf of Aden coast.[1] Sandford's 26 ft. coral beach in the Gulf of Suez should correspond to the ±5–10 m. Monastirian II beach in the Mediterranean. The evidence from the lower Nogal suggests that the Acheulio-Levalloisian Culture was contemporary with a period of high sea-level at ±30 m. Working through Miss Caton-Thompson's correlation of the Nile sequence with that in the Mediterranean, this may well correspond to the ±18–20 m. Monastirian I raised beach. The ±30 m. raised beach was followed by a regression of sea-level to approximately 0 m. on the east coast of Somalia. This would correspond to the intra-Monastirian oscillation, when the Mediterranean shoreline stood at −30 m. Further marine transgression followed, the height of which, due to local isostatic movement (and to lack of definition of the beach element) varies between 5 and 25 m. above present sea-level. The average is about 12 m., though more than one beach may be represented. It is considered most probable that this beach corresponds to the ±5–10 m. Monastirian II beach which has been shown to have been contemporary with an Upper Levalloisian stage of culture. For the reasons given above, these beaches have been shown in the table as occupying the early part of the Gamblian Pluvial; their alternative position is in the last half of the preceding Interpluvial. Overlying this beach are fossil dunes and eluvial formations of sands and some clay, the origin of which was due to the action of running water; in the uppermost levels of these eluvial beds have been found Somaliland Stillbay artifacts. These deposits appear to follow a fall in sea-level on the Somali coasts. This drop in sea-level is well represented at Zeila on the north Somali coast where a fall in sea-level occurred to at least −78 m., and again to an unknown depth at the mouth of the Nogal on the east coast. This major regression phase is considered to correspond to the Mediterranean post-Monastirian regression to −100 m. or more. The fossil dunes and eluvial formations of the Somali coasts find their counterpart in the Mediterranean in the dunes and Red Beds of the north-west African coast, notably those of Berard and Pointe Pescard in Algeria,[2] and in the fossil dunes of the Cyrenaican coast.

[1] Sandford, K. S. & Arkell, W. J. (1939*b*, p. 60). [2] Caton-Thompson, G. (1947).

A field comparison of the Algerian beds and their contained culture with those in the Horn leaves the writer in little doubt that they are contemporary. If this correlation is confirmed by later work it will confirm also, at any rate for the northern half of the continent, the contemporaneity of the Last Glacial in Europe with the Last Pluvial in Africa. The late low-level, unconsolidated beach occurring in the Horn at ± 2 m. finds apparently no corresponding level in the Mediterranean.

An attempt has been made in the second table to correlate the cultural and climatic succession in the Horn with that from Kharga Oasis, the Nile Valley, East Africa and the Sahara; tentative correlation of the changes in sea-level with those of the North African littoral have also been attempted. It must be emphasized, however, that these correlations must not be considered to be other than tentative until they have been confirmed by further research. The time has now come when correlation of the Stone Age succession in the various regions of the Continent will no longer be the rather uncertain matter that it has been hitherto for, with the development of the C^{14} radioactive carbon method we have for the first time a method of dating the Middle and Late Stone Age Cultures which when it is perfected promises to go a long way towards providing the answer to the problem of correlation of the prehistoric cultural and climatic succession in the African Continent.

In our area the main succession, the attributes and broad associations of the different cultures has, we may claim, been largely established but it must remain for others to perfect our knowledge of them and to confirm the dating evidence. The areas likely to be most productive for future research can now be determined, and what is now needed is for a team of workers—geologist, palaeontologist and prehistorian—to undertake studies of specific problems in selected localities. Many problems remain to be elucidated. Did a land-bridge in fact exist for a time in the Pleistocene at the Straits of Bab-el-Mandeb, how many raised beaches are there and what are their cultural associations and their correlation with the North African littoral? The origin of the apparently intrusive blade and burin industries at the end of the Pleistocene? Can the art groups be linked conclusively with any of the Late Stone Age Cultures? What was the extent of the Pleistocene rifting and earth-movement in the Abyssinian Rift Valley and in Danakil? What cultures existed on the Abyssinian high plateau?—to cite but a few of the problems which await further research.

Another area on the prehistoric map of Africa has begun to take shape but here, as in not a few of the other regions the conclusions need supplementing and substantiating before the multitudinous problems confronting workers in African prehistory are even partially solved.

APPENDIX A

DETAILS OF COLLECTIONS (NOT PREVIOUSLY DESCRIBED) MADE BY CIVIL AND MILITARY PERSONNEL IN THE SOMALILANDS AND ABYSSINIA

C. Barrington Brown Collection

Area. Eastern British Somaliland. In the Univ. Mus. of Arch. and Eth. Cambridge. Cat. no. 30.

HOBAT: HEGLIGAB DISTRICT, no. 1537.

Comprises chert specimens of varying patina and *état physique.*

(1) Levallois forms—chestnut to black patina, usually weathered.
(2) Stillbay forms—unpatinated; flake-blades and alternate core.
(3) Hargeisan forms—long blades, blade-core, and angle-burin.

AL HILLS: HEGLIGAB DISTRICT, no. 1536. Levalloisian tortoise core, chert.

AL HILLS AND DISTRICT, no. 1535.

Material. Chert.
Mixed series showing different patinas and degrees of weathering.
Levalloisian—most patinated.
Stillbay—three good examples of subtriangular prepared cores.
Hargeisan—long blades.
Somaliland Wilton—microliths and thumbnail scrapers.

HORMO, no. 1533.

Material. Chert.
From eroded terrace deposits.
Levalloisian—a trace.
Hargeisan—typical collection of factory debris, straight-backed blades and end-scrapers.
Thirteen end-scrapers on blades (single-ended).
Two end-scrapers on blades (double-ended).
Three short end-scrapers (single-ended).
One short end-scraper (double-ended).
Seven straight-backed blades and bladelets.
One convergent blade-core.
Two double-ended blade-cores.
Also utilized blades and bladelets.

In addition there occur in the British Museum (Natural History) (Collection Collenette) the following:

Four end-scrapers on blades (single-ended).
One end-scraper on blade (double-ended).
Two short end-scrapers (single-ended).
Three straight-backed blades and bladelets; also utilized blades and bladelets.

SUNTO OR HORMO DISTRICT, no. 1530.

Large disc or chopping tool—Levalloisian (?) chert.

ERIGAVO, no. 1529.

Small disc, probably Late Stillbay—chert.

WAR EIGLEB DISTRICT, no. 1528.

Levalloisian (2), and Late Stone Age (1), tools—chert.

BURAN, no. 1527.

Surface finds. Heavily patinated Levalloisian tools in chert, also small series of later Levalloisian or Stillbay, and Hargeisan and Somaliland Wilton tools.

A. T. Curle Collection

Area. Western British Somaliland. In the Univ. Mus. of Arch. and Eth., Cambridge. Cat. no. 32.

NOTES ON THE OCCURRENCE OF STONE TOOLS IN WESTERN BRITISH SOMALILAND, *by* A. T. CURLE

Leaving Hargeisa along the main Berbera motor road traces of flint implements were observed at about every halt to Baderwanak and a few samples were collected; in every case these were found on the surface of the ground where it had been washed away by the erosion of wind and rain.

From Baderwanak to Adadleh flints are to be seen along the main track and at places where the ground is washed away—e.g. at Barad on Ononuf Plain.

At Sik samples were found to the south where a small stream had washed away the surface and to the west where the outcrop ceased and washaways in the soil began.

From Sik to Shilmaleh no samples were observed except in a washaway but once the latter area was reached the whole surface to the east of the hill appeared to be intermittently covered with them and many specimens were collected. A small pile of flint chippings was found in one area where recent erosion was most noticeable since dead trees were standing with their roots some 2 ft. above the present surface.

From Shilmaleh to Daba Gumbur occasional samples appeared in patches on the surface, at Daba Gumbur to the south of a low stony hill east of the cleared area the ground is littered with flints for several acres; many roughly made implements were found, flints worked on one side only, and lumps of stone worked as scrapers. Here and at Shilmaleh the same types and skill appear to have existed. The flints are scattered so thick that it is hard to believe that such quantities could have been deposited by human means yet examination shows signs of most of the pieces having been worked.

From Shilmaleh to Baderwanak only occasional samples were seen.

From Baderwanak to Au Bakadleh, a few samples seen in washaways but at Lugud many worked specimens were found and continued along the track to Au Bakadleh where many samples were found especially on the south bank of Tug Amoul among the stones and at the back of the mosque.

Flints from this area are much better worked and the arrowheads are finished on both sides, practically no scrapers appear in this area.

From Au Bakadleh to Haleya a few samples were seen in the tug bed and again at Haleya in washaways.

Hargeisa flints are found in the tug bed and on the golf course about 50 yd. on the fairway from the second tee.

Tesseh Tug close to the Hargeisa/Buramo road to the north of it some 5 miles before the Gebileh road junction is reached. Here the flints are found in the bank of the tug and in the washaways at the sides.

Traces are also seen in the Ijara Tug. Also in a small tug some 6 miles from Hargeisa on the Gedkedebke track.

Arabsiyo—many found to the north-west of the village across the tug and in the river bed.

Gebileh—on the banks of the Gebileh Tug where the surface has broken away.

Ijara Tug—on the banks where the surface has broken away.

Tesseh Tug—the spot where the majority of specimens were found was about 60 yd. to the south-west of where the Dagahhanyada/Arabsiyo road crosses the tug. The breakaway is particularly large and very fine perfect specimens were found often three or four together. It was impossible to locate any level. The area where the artifacts are to be found is very large and the specimens are all distinguished by their fine workmanship.

To the north lie the Dubburoh hills and round them flints are also picked up where the rush of water has washed away the surface.

In the whole of the area covered, artifacts were found especially in the riversides and in breakaways, but they are most abundant and perfect in the drainage area of the upper Tesseh Tug.

None of the tortoise-shaped cores found on the Shilmaleh site and sent to the Royal Scottish Museum were found on this trip at all.

TESSEH TUG. Lat. 9° 35′ N.; long. 43° 40′ E. Cat. no. 32.X 2.

A good series of Stillbay and Hargeisan or possibly Magosian tools. All is surface material but has been divided on a physical and typological basis.

Material. Chert and quartzite.

(1) Unpatinated or mottled blue patina (believed derived from tug gravels). Number of subtriangular unifaced points and some radially worked faceted flakes.

More rare are bifaced points sometimes of lanceolate form well made and with lenticular section.

(2) White patina with occasional concretion of lime. These comprise the main group. Numbers of unifaced and bifaced points triangular, foliate and limace in shape. Butts either predominantly unprepared or entirely removed by secondary trimming. No microliths.

EAST SIDE OF CENTRE DUBBUROH HILL CLOSE TO TESSEH TUG. Cat. no. 32.3.

Small series of Stillbay and Hargeisan-type tools.

ARABSIYO. Lat. 9° 42′ N.; long. 43° 46′ E. Cat. no. 32.4.

Developed Levallois or Lower Stillbay tools including one good unifaced foliate point. Cf. Las Anod and Jesomma types.

GEBILEH TUG. Lat. 9° 42′ N.; long. 43° 36′ E. Cat. no. 32.5.
Five specimens of Stillbay type including one diminutive *coup-de-poing*.

GEBILEH AREA. Cat. no. 32.6.
Fifteen specimens of Stillbay and Magosian form.

IJARA TUG. Lat. 9° 37′ N.; long. 43° 40′ E. Cat. no. 32.7.
Six specimens of Lower Stillbay or Developed Levalloisian type including diminutive hand-axe, subtriangular prepared core and flakes with Emireh or Tabelbalat retouch.

NO LOCALITY GIVEN. No Cat. no.
Good series of Lower Stillbay or Developed Levallois tools; points, cores and flakes.

Dr J. Parkinson Collection

Area. British Somaliland. In the Univ. Mus. of Arch. and Eth., Cambridge.
A collection from surface sites the exact provenance of the majority being unknown.
Material. Predominantly chert.

BOHOTLEH (183). British Somaliland/Somalia border.
Atypical flakes (2) of Stillbay or Magosian form.

BETWEEN GUMBURU AND AROR. (Perhaps between Dubburoh (9° 39′ N., 43° 32′ E.) and Aror (9° 14′ N., 43° 50′ E.))
One pressure-flaked 'slug' point and five Magosian (?) flakes.
Two epi-Levalloisian 'nibbled' flakes.

NEAR GUMBURU. (?Dubboroh (9° 39′ N., 43° 32′ E.))
Seven Magosio-Wilton specimens; three epi-Levallois-type tools.

DUMUK TUG.
One Levallois-type flake.

GORREN DABU.
One backed blade; eleven epi-Levallois-type specimens.

LAS ALLAN GOBBLEH.
Three Wilton-type specimens.

DUDWEIN. (?Perhaps 9° 19′ N., 44° 08′ E.)
Two epi-Levallois-type 'nibbled' tools.

ADEN KUDAN. (?Perhaps Adan Qudon (9° 46′ N., 47° 36′ E.))
One blade core.

MAGELLEN HUN.
One epi-Levallois tool.

NEAR DABBA GOB.
One Stillbay type of side-scraper.

HURAN HUR.

Two Levallois-type tools; one disc core and one prepared flake.

FROM BOTTOM OF HARGEISA TUG.

(Presumed to have been collected from the present tug bed as specimens similar to tools collected here by the writer in 1941.)

Fresh and unpatinated. Four specimens—one utilized faceted flake, one pointed flake, one pointed flake with faceted striking-platform, worked by pressure to form a crude point, one flake-blade utilized. Probably Somaliland Stillbay tools.

TUG AT SHARAB AD. (? = Garab Ad)

In situ 25 ft. below surface. Entirely typical example of a Developed Levallois flake with well-prepared side butt. White patina with concretions.

ABADLEH.

Nine unpatinated specimens probably of Stillbay age comprising two disc cores, one point (pressure-flaked) two small prepared flakes, two flake-blades.

AROR. (?Perhaps at 9° 14′ N., 43° 50′ E.)

Surface material which has been weathered and patinated. Specimens are of two ages (on physical grounds) the majority belong to Series I.

Group I. Notched and nibbled flakes and broken flakes of Levallois type suggesting a form of epi-Levalloisian industry. Tools consist of one bifacial and thirteen unifacial 'points' made on prepared flakes by notching in from both sides, nine flakes with nibbled retouch, three rough discoidal cores, three flakes (prepared) with opposing notches forming a waisted tool with a blunt flat end.

Group II. Four microlithic specimens of Wilton type.

Captain A. Phillips Collection

Area. Southern Somalia and British Somaliland. In the Coryndon Museum, Nairobi, Kenya.

LUGH FERRANDI: SOUTHERN SOMALIA. A number of specimens from a surface factory site the exact provenance of which is unknown. Variously weathered and patinated. Comprising small disc cores, faceted flakes and some unifaced points. A series referable to the Somaliland Magosian and Doian industries.

HARGEISA. epi-Levalloisian-type specimens weathered and patinated from the surface.

EL AF. epi-Levalloisian-type specimens.

AINABO. epi-Levalloisian-type specimens.

Captain H. B. Gilliland Collection

Area. Eastern British Somaliland. In the Archaeological Survey, Johannesburg, South Africa.

The material described below was collected by Capt. H. B. Gilliland during 1944–5 while serving with the Military Administration in British Somaliland.

The sites are situated in the eastern half of the Protectorate and are of particular interest in that they connect the better known sites in the western half of the country with sites recorded by the Italians in the Mijertein and with our own investigations in the lower part of the Nogal valley from Bihen to Eil.

A short description of the material from each site is given below and is followed by a general discussion of the finds and the conclusion that can be deduced from these.

GOVERNOR'S CAMP AREA, LAS ANOD. Cat. no. 2/46.

Situation. Las Anod is situated on the edge of the southern scarp of the Nogal depression on the gypsum/anhydrite beds of Eocene age which are exposed in the upper end of this valley.

The material from this site is derived from scree at the base of two hill slopes one at the side of the District Commissioner's house and the other about a mile to the north-east. That at the former site appears to be eroding from secondary gypsum beds which are being cut into by many small erosion gullies broken irregularly by ledges on which the implements have weathered out, and from the description and photograph it would seem that the deposit is a scree cemented in places by gypsum, from which factory debris and a few implements showing 'pink' and grey patina are derived. No thickness of deposit was noted by Gilliland and there would appear to be no connexion with alluvial beds of the Las Anod Tug.

On evidence of typology and physical condition the material can be divided into two groups and will be described accordingly.

Group A

Material. Brown chert from the upper levels of the Eocene gypsum beds.

Condition. Mainly fresh and unrolled, buff to 'pink' patina varying with degree of exposure and weathering.

Implements. Eight long blade-like flakes showing parallel flaking and small unprepared and inclined butts. Length of longest approximately 5 in., smallest 2 in. Some indication to suggest utilization as evidenced by small flake scars at intervals down edges; perhaps due to use.

Ten broader flakes with similar platform and butt the majority showing signs of use.

One round scraper and one point.

Three flakes showing faceted right-angled butts, one typical prepared flake with radial flaking (edge utilized), one crude point; these flakes show evidence of scratching of upper and lower surfaces (due to long inclusion in scree?).

One lanceolate point showing 'wood' technique (possibly pressure) over both sides. A deep reddish patina. Has been broken and resharpened at one end.

Two 'roughouts' for points, or poorly made pointed hand-axes. The flaking on the first specimen is not unlike that seen on the point recorded above but typical 'stone' technique of a primary nature predominates. Both specimens probably made from flakes.

One ovate hand-axe or chopper—primary work but a little secondary work round edges. Possibly a roughout for a prepared core.

One chopper made from a piece of tabular chert.

One subtriangular prepared core (so-called 'Tortoise point'), flaking of upper surface directed from opposing ends; good example, unstruck, with well-prepared platform.

One-half of weathered flake from a prepared core. In detaching the flake a hinge fracture resulted in the removal of the underside of the core at the end opposite to the butt; the butt was later snapped off.

Five flakes from which the butts have, in three instances, been removed by deliberate secondary retouch. Two show cortex on upper surfaces (one with scraping edge worked round two sides), remaining three show typical scars resulting from removal from prepared cores—two instances radial, one parallel flaking. Small intentionally worked notches are present. One possibly poor example of single-blow burin.

One good example of 'Levallois'-type point with radial primary flaking and secondary retouch round edges.

Three flakes with unprepared butts—two blade-like—one with later 'nibbling' retouch—one with notch.

One nodule—atypical.

Total. Thirty-eight specimens.

Group B

Material. Mainly black chert from lower levels of the Eocene gypsum series.

Condition. Weathered but unrolled. Patination varied between black, white speckled blue, and white.

Implements. Not necessarily a homogeneous assemblage.

Five discoidal to subtriangular prepared cores; small and well made—one struck.

One small cordiform *coup-de-poing*; milky blue patina.

Two small fresh flakes with white patina.

Five atypical flakes.

One discoidal core.

Four broadish semi-microlithic flakes.

Two blades, small.

One *bec-de-flute* burin.

Total. Twenty-one specimens.

KIRIT: BURAO/LAS ANOD ROAD. Cat. no. 3/46.

Situation. Material all from a surface site on right-hand side of the Burao/Las Anod road near the first wells. Just within the eastern half of the Protectorate and situate within the Haud to the west of the Nogal: approximately 55 miles from Burao.

Material. Mainly grey to brown cherts but some black specimens also occur. Some pebble material present due to situation in Ain valley presumably.

Condition. On typology, patina and physical condition tools can be divided into two, possibly three, groups.

Group A

Not a clearly defined group but showing a heavy white patina and a greater degree of weathering.

One large redirecting flake.

Six broad flakes, three suggesting Levallois technique on upper surface. Platforms where present are inclined and unprepared.

Group B

Shows a fairly uniform white or grey patina slightly thicker than Group C where patina is present at all in the third group. Unrolled and fresh condition.

Fifteen blades—unworked.

Six small microlithic blades—unworked.

Three microliths—two obliquely blunted, one straight-backed type with butt removed.

Three small typical end-scrapers.

One faceted flake showing radial flaking, small.

One blade denticulated down one edge.

One large crudely backed blade.

One eared or 'tap-borer' flake.

Nineteen broadish end-flakes small and unworked.

One broken blade showing oblique retouch at one end.

Five atypical nodules.

One subtriangular prepared core; good example.

Two poor and doubtful specimens of above.

One large flake with edge trimming in places.

One large end-scraper.

One small side-scraper.

Three burins—one single blow on redirecting flake; one double *bec-de-flute* on small core; one good beaked-burin.

Six side-flakes—unworked.

One typical small Magosian-type core.

One Sebilian-type core—not good.

Three blade cores, small.

Two atypical 'ridged' cores.

Group C

Condition fresh and unpatinated or with mottled blue/white patina. Evidence of surface weathering on some specimens. All microlithic forms with exception of the semicircular scrapers.

One flake from quartz pebble.

Thirty-five end-flakes—some with secondary retouch.
One largish side flake—unworked.
Nine atypical and one tabular, pieces of chert.
Four poor Sebilian-type cores; small, flat subrectangular to discoid in shape.
One microlithic multi-platformed core showing much battering.
One pseudo-prepared core.
Eighteen typical microlithic blade cores with one, two or more platforms.
Thirty-eight microlithic blades—unworked.
Two microlithic utilized flakes.
Two crudely denticulated blades.
One beaked burin on a 'rod' of tabular chert.
Four microlithic blades retouched—probably for use as microliths.
One broken flake retouched down one side and end.
Two small hollow scrapers.
Eleven round to semicircular scrapers of varying sizes.
Totals. *Group A*: 7 specimens. *Group B*: 78 specimens. *Group C*: 132 specimens.

MERKANWEIN BALLEH: BOHOTLEH. Cat. no. 4/46.

Situation. Slightly to the east of Bohotle and to the south of the Somalia/British Somaliland boundary road. From the earth thrown up by well diggers around the wells in the Merkanwein Balleh. Implements probably derived from an ochreous layer approximately 3 ft. below surface.

Material. The wells are in the so-called Plio-Pleistocene concretionary deposits which probably overlie the Cretaceous (Nubian) sandstones. Certainly most of the chert used is derived from these sandstones but may have been collected from these beds where they outcrop further to the south-west.

Condition. Fresh, unrolled and unweathered as newly derived from the well diggings. Yellowish patina due probably to staining from iron oxides from the red sand of the Haud.

Implements. Indications suggest this to be a homogeneous assemblage.

One large, thick primary flake.

One fragment of silcrete flake with *écaillé* scars at one edge.

One atypical flake.

One good example of a discoidal prepared core—unstruck—typical radial flaking: flattish; 3×3 in. approximately.

One flat cordiform prepared core with flake removed at butt end; good example of radial retouch on upper side; $1\frac{1}{2} \times 1\frac{1}{2}$ in. approximately.

One core for production of long flakes; double-platformed.

Six flakes with prepared butts; three good examples (two of which may be butt-scrapers); two straight butts; one *chapeau-de-gendarme*. Of these, one is a round scraper with fine resolved retouch, one similar to first but scraping edge down one edge and end only, one with slightly concave well-worked edge down one side—possibly a rudimentary point; one retouched flake suggesting use as a point. Primary flaking on upper surface is usually radial.

Two crude, subtriangular sectioned, denticulated tools—possibly fabricators; secondary retouch absent.

Four small blade-like flakes—no retouch.

One side-flake, no retouch.

Four short end-flakes with unprepared inclined platforms; two show use as crude scrapers.

One good *lame écaillée* on nodule.

One possible *lame écaillée* on part of broken flake.

Three flakes with butts removed; one with two small notches; one large but crude scraper; one flake showing pressure-flaking retouch—part of under surface on one edge shows Kasouga flaking, and small notch on upper surface on opposite edge.

SADDLE IN HILLS 12 MILES NORTH OF EL DUR ELAN. Cat. no. 5/46.

Situation. El Dur Elan is a Police Post lying about 25 miles east of Las Dureh at the head of the Hudma Tug which flows north into the sea at Ankhor.

Material. Grey and yellowish-white chert from the Eocene limestone series. Mainly nodular chert but one small flake has been derived from a pebble.

Condition. Two specimens show deep white patina and weathering and are probably earlier than the bulk of the material which is fresh and not greatly weathered.

Implements. One Levallois type flake with prepared butt and one other thickly patinated specimen probably belong to the Somaliland Stillbay complex and are earlier than the rest.

Three flakes struck from prepared cores, two with unfaceted inclined platform (one showing retouch near end on one edge), one with prepared butt.

One large broad flake with butt removed.

One side-flake; unworked.

Nine end-flakes—two with retouch down both edges due probably to utilization.

One large but poorly made backed blade—utilized.

Five small blade-like flakes.

One tap-borer (poor).

One double-ended scraper on blade (straight and convex working ends).

One crudely backed and thick end-flake with rough denticulation down both edges from opposing sides (fabricator ?).

RAGUDA TUG: ERIGAVO DISTRICT. Cat. no. 6/46.

Situation. The Raguda Tug flows into the sea between Ankhor in the west and Heis in the east. It flows north from the region of the Asseh gap. The implements were collected during a camel safari along the coast about 10 miles inland in a very desert region, and were found 30 ft. above the bed of the tug.

Material. All of honey-coloured chert probably from the Eocene limestones. There are four specimens of Eocene black chert however.

Implements. Are of two main groups which can be divided on technique and physical condition. Two-thirds of the specimens show a thin 'desert varnish' while the remainder are fresh and unpatinated. The division is accentuated in some instances where implements showing the 'varnish' and honey-coloured patination have been reflaked at a later date and show an unpatinated surface; the black chert is always unpatinated.

Group A

Six broad primary flakes, with inclined unfaceted platforms and not very typical radial flaking of upper surface.

Three prepared cores—small and ellipsoid in shape; one has been struck.

Three small broad flakes with unfaceted inclined butts.

Two blade-like flakes with butt missing.

Two good examples of flakes with unprepared butts but typical radial flaking of upper surface; long flakes.

Four prepared flakes with prepared butts and radial flaking.

One thick flake with Emireh retouch at butt.

One small point with Emireh retouch.

One unfaceted flake, showing semi-cone, with good resolved retouch down one edge.

Two long flakes, one with retouch on one edge, unfaceted and small butts with parallel flaking on upper surface.

Eleven indeterminate nodules and pieces of chert.

Group B

Two reflaked prepared cores of Group A, one now forming a flat disc.

One retouched flake forming a crude point by peripheral secondary working.

One *bec-de-flute* burin.

One end-scraper on small flake with *bec-de-flute* burin at opposite end.

Five smallish blade-like flakes.

Two large blade-like flakes with unprepared and inclined butts.

Four broad flakes of black chert with cortex adhering.

One flake, perhaps rolled, but now much weathered, shows steep almost vertical 'nibbling' trimming round irregular sides possibly due to natural agencies.

5 MILES FROM HUDIN ON HUDIN/AINABO ROAD. Cat. no. 7/46.

Situation. From the Nogal valley; this collection comes from the surface near the bank of a tug.

Material. Derived from Eocene gypsum beds probably; black, grey, and brown chert.

Condition. Seven specimens show a white patina and greater degree of weathering than the remainder: typology suggests these are earlier. The main group are unrolled, fresh, but show evidence of abrasion and weathering due to surface exposure.

Group A

One small prepared core—struck.

Two flakes with faceted prepared butts.

One polyhedral stone, i.e. much battered multi-platformed core.

One small core with prepared platform.

One poorly made point of Magosian type with pronounced butt, and secondary retouch round edges done at a later date.

One backed end-flake with partial retouch of the opposite edge.

Group B

Eight blades—small.
Fourteen microlithic flakes and blades.
Four side-flakes.
Five broadish end-flakes.
Three utilized blades.
One broken straight-backed microlith.
One backed blade with butt present.
One battered core—possibly a fabricator.
One unprepared, blade core showing 'crackle'.
One micro-core with three platforms.
Two micro-cores, typical.
One broken flake with *écaillé* retouch.
Two burins; one *bec-de-flute*, one single-blow on redirecting flake.
Two poor hollow scrapers.
Four end-of-blade scrapers.
Two scrapers on short end-flakes (poor).
Two thumbnail (poor) scrapers, one worked at both ends.

HEADWATERS OF TUG AT BUR KHANSA NEAR BIHEN. Cat. no. 8/46.

Situation. On the edges of the Hubera valley; Erigavo District surface finds.

Material. Black and buff chert, but two specimens of pink chert also occur.

Condition. Unrolled and for the most part unweathered but some show evidence of lengthy surface exposure.

Specimens. Probably a homogeneous series.

Seven side-flakes, unworked.
Twenty-three end-flakes and blades, unworked and with unfaceted, inclined platforms.
One 'prepared' end-flake with scraping edge at end opposite from butt.
One redirecting flake.
One unprepared blade core.
Four atypical nodules.
One thick primary flake with worked notch; fabricator ?
One oblique-convex angle burin.
One transverse-straight angle burin (doubtful).

SIDE OF ROAD AT HALIN. Cat. no. 9/46.

Situation. A surface site near the eastern boundary with the Mijertein, situated on the edge of the northern side of the Nogal valley on the Sol plateau.

Material. All black Eocene chert (from the gypsum series?). Two pieces of brown and honey-coloured chert, however, suggest an origin in the limestone series.

Condition. Possibly a homogeneous assemblage, not greatly weathered by surface exposure. Probably a factory and 'mine' site.

Implements. One piece nodular black chert with cortex.

Three micro-cores for production of blades.

One end-scraper on flake (poor).

One end-scraper (good) showing steep secondary retouch on part of large flake of brown chert.

One keeled or ridged core of brown chert.

Five micro-blades.

Four small blades, one with poor backing.

One side flake.

Five primary directing flakes.

Two atypical flakes.

One thick flake with some secondary retouch but much weathered and probably earlier.

(Note by H. B. Gilliland: 'Flakes very numerous indeed'.)

JUNIPER FOREST AT ERIGAVO. Cat. no. 10/46.

Situation. Surface finds in open glades on the upland side of the juniper forest at Erigavo, one of the highest points in British Somaliland, approximately 7000 ft.

Material. Grey chert—from Eocene limestones?.

Condition. Unrolled but considerably weathered due to long exposure on the surface.

Implements. Seventeen atypical lumps and nodules of chert.

Twelve long flakes with unfaceted platforms, three retouched.

One flake with prepared platform.

One large broad flake.

Three micro-cores, blade producing.

Three micro-blades or portions of same.

One core scraper.

Two end-scrapers.

Eleven atypical short end-scrapers.

One small hollow scraper.

One utilized nodule.

GRAVEL RIDGE AT HADED: BALAD AD. Cat. no. 11/46.

Situation. Collection derived from a surface 'gravel' on this plain about half-way between Taleh and Erigavo. The Haded Balad Ad plain extends from the Al hills in the north to the Nogal scarp in the south and is characterized today by a complete absence of tugs.

Material. Seven specimens of black chert and twelve examples of the grey or brown types; from Eocene beds.

Condition. Some show varying degrees of weathering but the majority are fresh and unpatinated.

Implements. Six small blade-like flakes—no secondary trimming.

Three large primary flakes.

Four worked nodules one with suggestion of scraping edge.

One primary blade.

One redirecting flake.

One transverse-concave angle burin.

Three 'round' scrapers.

SHIMBIR BERIS IN DALO FOREST: ERIGAVO. Cat. no. 12/46.

Situation. A surface collection from the summit ridge of Shimber Beris in Dalo forest, Erigavo.

Material. Grey chert nodules from Eocene limestone beds.

Condition. Considerably weathered with rounding and natural 'nibbling' of edges. An indeterminate collection showing signs of long surface exposure.

Implements. Nine broad primary flakes with unfaceted and inclined butts and pronounced bulbs.

Seven small flakes showing bulbs completely trimmed away or platform reduced to minimum.

Two larger flakes as above but with notches.

One pseudo-Kasouga flake.

Six largish blades.

Eight indeterminate nodules.

One redirecting flakc from micro-core, much weathered and iron-stained.

One flake with trimmed edges.

One 'round' scraper.

One long flake with *écaillé* flaking down one edge.

One utilized flake with rudiments of scraping edge at end.

One broad flake with nibbling retouch down one side.

One single-blow burin.

HILL SLOPES AT GOVERNOR'S CAMP AREA: LAS ANOD. Cat. no. 13/46.

Situation. Surface assemblage; see Cat. no. 2/46 above (p. 338).

Material. Predominantly black chert from Eocene gypsum series.

Condition. With one exception all unpatinated and fresh; presumably from a factory site.

Implements. Group A: Represented by one specimen with pink patina. A long blade, approximately 3½ in., with small unfaceted butt and incipient hinge fracture.

Group B: Seventeen small but broadish end-flakes.

Six short side-flakes.

One broken lower half of flake with prepared butt.

Fifteen blades—primary and unworked.

Seven atypical nodules.

Five pieces of tabular chert utilized as cores; flaking from both sides produces a kind of pseudo-prepared core due to shape of chert.

One pebble core.

One triple-platformed core.

Three crude tap-borer tools, possibly fabricators.
Two small blade-like flakes with notches.
Two blades retouched down edges.
One retouched micro-flake.
One oblique-straight angle burin (poor).
Eight small end-scrapers; convex edge.
Three small end-scrapers; straight edge.
Three doubtful scrapers on side-flakes.
Two hollow scrapers.

ABOUT 12 MILES FROM HUDIN ON HUDIN/HALIN ROAD. Cat. no. 16/46.

Situation. Surface collection in mouths of caves and on the slopes of hills parallel with the road on left side when travelling towards Halin.

Material. Grey, black and pink chert.

Condition. Varies, but mainly fresh and unpatinated.

Implements. Sixteen specimens amongst which can be recognized one thumbnail scraper and one broken small end-scraper, one micro-core, one piece of specularite (?), four microlithic flakes one with crude denticulation, one large flake with prepared platform from which a small flake has been removed.

BALLEH SIRRAU: NEAR WARDERE. Cat. no. 17/46.

Situation. From surface of balleh approximately 25 miles north of Wardere on Burao road.

Material. Variously coloured cherts from the Cretaceous sandstones.

Condition. Much weathered due to trampling by stock and annual immersion in the balleh, thus producing many small broken fragments of blades and flakes which have been reflaked by natural agencies.

Implements. From the material collected the following are recognizable.

Five small flakes with faceted butts, more weathered than remainder of collection, one showing a later secondary retouch at end to form a small end-scraper.
Fifteen end-flakes or fragments with unprepared butts.
Four utilized blades.
One utilized flake with faceted butt.
One micro-core.
One flake showing utilization round edges to form a crude type of point.
One notched scraper.
One microlith—doubtful.

DISCUSSION AND CONCLUSION

It will be apparent from the above lists of specimens that the greater part of the material from these sites is composed of factory debris, so that, especially where this has been derived from sites, for example 2/46, 13/46, 14/46, 15/46 and 16/46, where the tabular chert must occur as veins and nodules within the Eocene gypsum and limestone series and has been 'mined' it will be appreciated we are dealing with typical mine and factory sites. Sites of this nature we found to be particularly common along the south side of the Nogal valley and from Garoe to Bihen to Donkukoq, from the valley floor to the top of the scarp, a distance of approximately 30 miles, was one, almost continuous, factory site of this type, the lower slopes and talus from the hills comprising one large scree of nodules of chert and factory debris showing varying degrees of patination.

Although finished end-products are rare it is possible nevertheless from a study of technique, typology, and physical condition to divide the finds described above into three main groups; such a division is corroborated by comparative material from stratified and datable deposits.

Group I

An industry of Upper Levalloisian type. The Group A material from Las Anod listed under sites 2/46 and 13/46 and 15/46; weathered, with fairly uniform, thick, pinkish buff patina, comprising long blade-like flakes, faceted flakes, discoidal to subtriangular and cordiform prepared cores, *coup-de-poing*?, chopper and bifaced lanceolate point.

Group II

Industries of Somaliland Stillbay and Magosian type. Comprising the Group B material from Las Anod site 2/46, Group A from Kirit 3/46, Merkanwein Balleh 4/46, Group A from the Raguda Tug 6/46, and possibly the more weathered specimens with faceted butts from the Balleh Sirrau 17/46. Specimens of small prepared cores with radial flaking, prepared flakes with faceted butts; resolved, and probably pressure, secondary retouch; points and scrapers.

Group III

Industries of Somaliland Wilton and Hargeisan type. This is the largest series and includes the Groups B and C material from Kirit 3/46, El Dur Elan 5/46, Group B from the Raguda Tug 6/46, Groups A and B from the Hudin/Ainabo road 7/46, Bur Khansa 8/46, Halin 9/46, Erigavo 10/46 (?), Haded: Balad Ad 11/46, Shimbir Beris 12/46 (?), Group B from Las Anod site 13/46, Halin/Las Anod road 14/46, Group B from Las Anod 15/46, Hudin/Halin road 16/46, majority of specimens from the Balleh Sirrau 17/46; distinguished by their predominantly microlithic flavour, blades, burins and large round and end-scrapers.

Major P. E. Glover Collection

Area. Western British Somaliland and northern Haud. In the Univ. Mus. of Arch. and Eth., Cambridge.

BRITISH SOMALILAND

THIRD CAVE AT GERBAKELE. About 10 miles from Dudubo and on the same ridge as Gan Libah mountain. All specimens collected from the surface of this cave.

One fragment of partially fossilized antelope molar.

One good *bec-de-flute* burin.

One double faceted single-blow burin.

One end-scraper on a blade.
One curved-backed microlith.
One micro-core with two platforms.
One *outil écaillé.*
One large poorly made end-scraper.
One sherd of black pottery (probably of later age).
Material. Grey or brown chert and two flakes of obsidian. An industry of Somaliland Wilton type.

GED DOBO. 13 miles from Manjasseh. Specimens collected at random on the Old Oil road.
Material. Siliceous sandstone and quartzite.
Group I. Ovate hand-axe, one U-cleaver, one broken pick-like tool: Acheulio-Levalloisian specimens.
Group II. Two bifaced pressure-flaked points—subtriangular and foliate.
One unifaced high-backed point.
One flake with faceted striking-platform.
One atypical quartz fragment.
Specimens suggest Somaliland Magosian or Hargeisan affinities.

LAS ANOD. About 12 miles from Hudin on the Hudin/Halin road. Specimens collected in the mouths of caves and shelters and on the slopes of hills running parallel with the road on the left-hand side going towards Hudin. Divisible into two groups on condition and patina.

Group I. One long blade.
One bifaced foliate point—flat percussion technique.
One fragment of broken foliate point similar to above.
One side-scraper on a broken flake.
Somaliland Stillbay-type specimens.

Group II. Micro-blade cores, one a possible core burin.
One good transverse straight angle-burin.
One small end-scraper.
One large straight-backed blade.
Hargeisan- or Somaliland Wilton-type specimens.

DUDUBO COFFEE SHOP. Specimens from surface collections at a factory site on the Burao/Hargeisa road, 73 miles from Hargeisa and 200 yd. off the road on the left-hand side.
Material. Chert. Divisible on condition and patina into two groups.

Group I. One multi-platformed blade core.
Four faceted flakes with convergent flaking, two worked into crude unifaced points.
Late Stillbay or Magosian type specimens.

Group II. One radially flaked prepared flake with faceted striking-platform.

One pointed flake with convergent flaking and faceted striking-platform.

Upper Levalloisian type specimens.

DUKUKOLOL YERO. Found in and outside caves between El Afwein and Las Dureh where new type of cave paintings were observed.

One weathered prepared flake in sandstone and one flake-blade with white patina are probably of Levalloisian Age.

One *bec-de-flute* burin; one end-scraper on a blade; two diminutive prepared flakes; one square-ended scraper; one flake of obsidian. Somaliland Wilton assemblage.

HALEYA. 5 miles east of Hargeisa on the Tug Marodijeh.

Group I. Collected from 'Boulder bed' layer along tug wall.

Material. Chert. Rolled.

One subtriangular hand-axe on a flake.

Two crude prepared cores.

One heavily rolled prepared flake.

Acheulio-Levalloisian tools.

Group II. Collected from eroded superficial deposits in tug bank.

Material. Chert. Unrolled.

One unifaced point; one redirecting flake; one flake-blade.

Somaliland Stillbay specimens.

SKUTAR. On the Tug Der 20 miles north of Burao. Specimens found on a 'boulder bed' exposed by recent erosion. Some of the specimens were cemented into the deposit by concretions of lime. Apparently a mixed assemblage divisible on physical and typological grounds into three, possibly four groups.

Material. Predominantly chert.

Group I. Two 'formless' nodular cores.

Eight broad primary flakes with unprepared platforms.

Five flake-blades.

One subtriangular Micoque-type hand-axe made on a flake.

One roughout for a similar hand-axe.

One poor cordiform hand-axe.

One U-cleaver.

One side-scraper on a large flake.

Six flakes with faceted striking-platforms.

Six discoidal prepared cores.

Acheulio-Levalloisian tools.

Group II. Five flake-blades with unprepared butts.

One radially flaked prepared flake.

One bifaced foliate point.

One discoidal core—struck.

Three poorly made unifaced points, two with faceted butts.

One tranchet-type tool.

Nine atypical flakes and nodules.

Somaliland Stillbay-type assemblage.

Group III. Two round scrapers.

One formless core.

Three diminutive faceted flakes.

Hargeisan-type tools.

BURAO/HARGEISA ROAD. Surface finds.

Material. Chert.

Three triangular flakes two with prepared butts; one crude, disc core. Somaliland Stillbay type.

Two small blades; two small flake-blades; one small triangular flake with faceted platform. Somaliland Magosian and Hargeisan forms.

JIFA URE. Collected on the surface at the foot of the granite hill thus named, 20 miles from Borama on the Hargeisa/Borama road.

Material. Obsidian and some rare quartz and chert. The obsidian specimens are divisible into two groups, those patinated and those unpatinated.

Group I. Patinated. Broken fragments of flakes with faceted striking-platforms.

Group II. Unpatinated. Microlithic assemblage containing numerous broken fragments of flakes and blades and utilized nodules. Also six *outils écaillés*; two micro-cores; two broken straight-backed microliths; one thumbnail scraper; six poor scraper-like tools on flakes; one end-scraper on a diminutive flake; one crudely backed blade with opposite edge retouch; one Magosian-type diminutive point.

Assemblage suggests affinities with the Magosian and Somaliland Wilton of the southern Danakil Rift.

SOMALIA

ALABLAH BALLEH. Specimens collected from 'root excavations' on the edge of the balleh. See p. 217.

SIRRAU BALLEH. Well area near Somalia/British Somaliland border, 25 miles north of Wardere on the Burao road. Specimens collected on the surface near edge of pan. Material much broken up by passage of stock. Besides numerous fragments of small flakes and blades can be recognized—one broken curved-backed blade, diminutive and atypical scraperoid tools, fragment of a triangular-sectioned Magosian-type point, and a small eared scraper.

Material. Chert. Somaliland Magosian-type specimens.

45 MILES FROM HARGEISA ON THE HARGEISA/WARDERE ROAD. Found on the surface and in sides of ditch, 25 yd. from left side of road; one single-platformed blade core; one lanceolate double-backed microlith or rod, end-scraper, atypical factory debris. Somaliland Magosian-type industry.

BARDERA. From the surface of gravels at the Staging Post 3 miles north of Bardera. Small collection of Late Stone Age Magosian or Doian factory debris comprising microlithic flakes, blades and cores, and one crescent.

Material. Chert, and quartz pebbles.

Major P. Jackson Collection

Area. Tug Gerer at Dagahbur, Ogaden. In the Coryndon Museum, Nairobi, Kenya. Small series of Somaliland Stillbay and Magosian tools and Acheulio-Levalloisian limande hand-axe.

Major J. C. Trevor and Captain D. McInnes Collections

Area. British Somaliland. In the Coryndon Museum, Nairobi, Kenya.

UPPER SHEIK. For *Group I* see p. 163, and Inventory of Acheulio-Levalloisian Tool Types (p. 165).

Group II. Two small prepared cores; one small ovate hand-axe; three side-scrapers, unifaced points and flakes with resolved retouch; one broken large backed blade; one Magosian-type prepared core; one fragment of what must have been a large and long bifaced point carefully worked on both sides; one subtriangular prepared core; one triangular bifaced point; three triangular unifaced points; one foliate unifaced point; two radially flaked faceted flakes. Somaliland Stillbay assemblage. Also small Magosian or Hargeisan-type factory debris.

GEDEIS. Somaliland Stillbay and Hargeisan assemblages.
Seven flakes from prepared cores with faceted striking-platforms.
One plunging flake from a radially flaked prepared core.
One small disc core with high back.
One broken large straight-backed blade.
Three subtriangular and pointed unifaced points with marginal retouch.
Five blunt-ended unifaced points on prepared flakes with steep marginal retouch.
One small end-scraper; one broken point and two rhomboidal cores.

MANJASSEH. Developed Levalloisian or Somaliland Stillbay specimens.
Material. Chert and quartzite.
Four small disc cores.
Three flakes with faceted striking-platforms one trimmed to form an end-scraper.

DARBRUK. Specimens of Somaliland Stillbay and Levalloisian form.
Material. Chert, quartz and quartzite.
One unifaced point on a broad flake.
Two small oval prepared cores.
Three side-scrapers on end-flakes.
One prepared flake with parallel flaking.
Two broken tips of bifaced points.
One broken subtriangular point with steep marginal retouch.

TUG ARGAN GAP.
Material. Chert. Somaliland Stillbay-type specimens.
One unifaced point with steep secondary flaking down one edge.
Six subtriangular flakes with faceted striking-platforms.

BORAMA. Developed Levalloisian and Stillbay-type specimens.
Material. Quartz and quartzite.
Group I. Three slightly rolled oval prepared cores.
One faceted flake utilized as a side-scraper.
Group II. Three utilized flakes with faceted striking-platforms.
Two unifaced points.
One subtriangular prepared core.
One crude subtriangular point with steep marginal retouch round three edges.

DUBATO. See details on p. 200.

DAMERO. Exact location not known, but situated between Burao and Erigavo. Surface collection from scree at foot of low hills.

Material. Chert. Owing to the nature of the site (similar to the factory and mine sites in the Upper Nogal) many of the earliest tools have been reworked in later times. On patina and physical condition it is possible to subdivide these specimens into three groups.

Group I. Weathered and showing an iron-stained and polished patina. Specimens consist of tabular nodules and flakes of chert from which the bulb and platform have in most cases been removed. The flakes have often been snapped across and reworked by steep, almost vertical retouch round all or part of the irregular circumference. Notches, eared-scrapers and the pseudo-burin occur and are reminiscent of the Hope Fountain industry from Rhodesia.

Group II. Many of the earlier tools have been reworked. Besides the forms noted above there is present a tortoise core and a few flakes with unifaceted striking-platform.

Group III. Unpatinated. Several notched and U-shaped scrapers recalling the Egyptian crescent drills from the Fayum.

ERIGAVO. The same forms are present as are found at Damero, with the addition to the Group II forms, of definite prepared flakes with faceted striking-platforms. The Group III scrapers are absent, but one backed-blade and a few flakes suggest Late Stone Age Wilton affinities.

Colonel F. Moysey Collection

Area. Northern and Eastern Abyssinia. In the Coryndon Museum, Nairobi, Kenya.

GORGORA. For an account of the excavation of the rock-shelter here, see Moysey, F. *J. E. Afr. and Ug. Nat. Hist. Soc.* XVII, pp. 196–8; Leakey, L. S. B. *J. E. Afr. and Ug. Nat. Hist. Soc.* XVII, pp. 199–203. (12° 15′ N., 37° 15′ E.)

QUIHA. See details on p. 324. (13° 27′ N., 39° 33′ E.)

MILLE RIVER CROSSING. (11° 18′ N., 40° 50′ E.) On road from Dessie to Assab. Fresh and unweathered specimens in chert, agate and obsidian of Magosian form.

One diminutive prepared core.

Seven small utilized flakes some with faceted striking-platforms and marginal retouch to form diminutive points.

WADELAI DELANTA. (11° 29′ N., 39° 15′ E.) Between Debra Tabor and Dessie.

Material. Jasperite, chert, agate and obsidian. Late Stillbay or Magosian type assemblage.

Two prepared disc cores.

One blade core.

Two broad, utilized and notched, flakes.

Three flakes from which the bulbs and platforms have been removed.

Striking-platforms do not show preparation.

UAHANI (greenstone site). Between Debra Tabor and Dessie.

One scraper of jasperite on a broken blade.

DEBRA TABOR. (11° 51′ N., 38° 2′ E.) One utilized flake.

DESSIE. (11° 7′ N., 39° 37′ E.)

Material. Chert and obsidian. Late Stone Age specimens.

One microlithic blade.

Two small subtriangular scrapers.

AMBA SEL. (11° 28′ N., 39° 38′ E.) Between Dessie and the Eritrean boundary.

Material. Obsidian. Specimens resemble those from Quiha.

One square *outil écaillé* worked round the four sides.

Five ordinary *outils écaillés.*

Three small end-scrapers.

One side-scraper on a broken flake.

Eight micro-flakes and blades some utilized.

BATIE. (11° 2′ N., 40° 0′ E.) Near Dessie. Small series suggesting Magosian affinities—obsidian.

Two small faceted flakes with retouch.

Four small broad flakes without retouch.

Two large primary flakes.

GADARET. Near Dessie.

Material. Jasperite, chert and obsidian. Late Stone Age assemblage.

One hollow scraper similar to the Group III scrapers from Damero.

One single-blow burin.

One *outil écaillé.*

One diminutive faceted flake.

One utilized blade.

BETHOR. (11° 29′ N., 39° 15′ E.) Between Debra Tabor and Dessie.

Material. Jasperite and chert. Late Middle, or Late Stone Age utilized flakes.

GIDDA. Between Debra Tabor and Dessie.

Material. Jasperite and ? indurated shale. Late Stone Age (Wilton) affinities.

Five utilized flakes.

Two end-scrapers.

Two small hollow scrapers.

One small flake with vertical trimming.

Three small flakes.

AMBA ALAGI. (12° 59′ N., 39° 30′ E.)

Material. Chert and jasperite. Developed Levallois or Somaliland Stillbay surface collection.

Two small flake-blade cores with parallel flaking.

One nodular core.

Twenty short flakes with faceted striking-platforms and radial and converging flaking on the upper surface. Some show secondary retouch to form unifaced points.

Seven flake-blades some with secondary retouch. Platforms rarely faceted.

AXUM. (14° 7′ N., 38° 28′ E.)

Material. Chert.

One flake with shallow notch.

APPENDIX B

TOOL LISTS FOR ADDITIONAL SOMALILAND MAGOSIAN CULTURE SITES

Tool types / *Sites*	Points bifaced, subtriangular	Points, bifaced, foliate and lanceolate	Points, unifaced, subtriangular	Points, unifaced, foliate and lanceolate	Scrapers, thumbnail and diminutive round	Scrapers, end	Scrapers, side and various	Burins, single-blow and *bec-de-flute*	Backed-blades large	Microliths, curved-backed	Microliths, straight-backed	Microliths, crescents	*Outils écaillés*	Blades, some utilized	Flakes, some utilized	Prepared flake cores	Unprepared blade cores	Various	Totals
ABYSSINIA																			
Jigjiga (4)	4	—	3	1	—	—	5	—	1	—	—	—	—	4	13	2	—	1	34
Porc Epic (152) (Fagg Collection)	2	—	—	3	2	—	—	—	—	—	—	—	—	—	—	—	—	—	7
Roukatcha Wells (219)					See Breuil, H. & Kelley, H. (1936, p. 135)														
Sheik Hussein (178)					See Breuil, H. & Kelley, H. (1936, pp. 139–40)														
BRITISH SOMALILAND																			
Haud surface/Aware (10)	—	—	1	—	—	—	—	—	—	—	—	—	—	—	—	—	—	—	1
Jifa Ure (205)					See Appendix A, Glover Collection														
Merkanwein Balleh (183)					See Appendix A, Gilliland Collection														
Khansaha Au Plain (13)	2	—	1	—	—	—	—	—	—	—	—	1	—	—	12	1	—	1	18
FRENCH SOMALILAND																			
Mount Mabla (149)					See Teilhard de Chardin, P., Breuil, H. & Wernert, P. (1940, pp. 497–522)														
Obok (150)					See Teilhard de Chardin, P., Breuil, H. & Wernert, P. (1940, pp. 497–522)														
SOMALIA																			
Bun Mess (95)					See Graziosi, P. (1940, pp. 69–71)														
Bur Dauleh (18)	2	2	3	4	—	6	—	1	—	2	—	—	2	10	25	7	5	7	76
Bur Degis (17)	2	—	1	3	—	—	—	—	—	—	—	—	—	5	14	3	—	1	29
Dagahbur (116)					See Appendix A, Jackson Collection														
Eil (56)	—	—	—	—	—	8	2	—	1	—	7	—	—	24	29	4	4	2	81
Guretca (99)					See Puccioni, N. (1936, p. 119)														
Hargeisa/Wardere road (200)					See Appendix A, Glover Collection														
Lugh Ferrandi (100)					See Appendix A, Phillips Collection														
Segag (153)					See Breuil, H. & Kelley, H. (1936, pp. 116–21)														
Sheik Gure (83)	—	—	—	—	—	1	—	—	—	—	—	—	—	2	9	12	3	—	27
Tug Ensiral (157)					See Breuil, H. & Kelley, H. (1936, p. 116)														
Webi Shebeli (161)					See Breuil, H. & Kelley, H. (1936, pp. 121–2)														
Webi Shebeli (162)					See Breuil, H. & Kelley, H. (1936, pp. 123–4)														
Webi Shebeli (163)					See Breuil, H. & Kelley, H. (1936, pp. 123, 125–6)														
Webi Shebeli (164)					See Breuil, H. & Kelley, H. (1936, pp. 125, 127–30)														

APPENDIX C

TOOL LISTS FOR ADDITIONAL HARGEISAN CULTURE SITES

BRITISH SOMALILAND	
Berbera Plain (15)	One flake-blade with parallel flaking One blade One crescent
Borama (3)	One gouge burin Two unifacial points One *outil écaillé* Four small end-scrapers Two large crudely backed blades
Buran (137)	See Appendix A, Barrington Brown Collection
Dudubo Coffee Shop (201)	See Appendix A, Glover Collection
Ged Dobo (202)	See Appendix A, Glover Collection
Gedeis (232)	See Appendix A, Trevor & McInnes Collection
Hormo (139)	See Appendix A, Barrington Brown Collection
Kirit (182)	See Appendix A, Gilliland Collection
Raguda Tug (185)	See Appendix A, Gilliland Collection
Sheik (141)	See Appendix A, Trevor & McInnes Collection
SOMALIA	
Garoe Tug (54)	See Graziosi, P. (1940, pp. 74–6)

APPENDIX D

TOOL LISTS FOR ADDITIONAL DOIAN CULTURE SITES

TOOL TYPES

Sites	Lanceolate and foliate bifaced points	Triangles and sub-triangular bifaced points	Lanceolate and limace unifaced points	Subtriangular uni-faced points	Scrapers, thumbnail and small round	Scrapers—end-	Scrapers—various	Burins, single-blow and *bec-de-flute*	Backed-blades, large	Microliths, curved-backed	Microliths, straight-backed	Crescents, some with Heluan retouch	Double-backed microliths	*Outils écaillés*	Blades, some utilized	Flakes, some utilized	Prepared flake cores	Blade cores	Rubbers, hammer and pestle stones	Pottery	Various	Ostrich egg-shell	Haematite	Totals
										SOMALIA														
Alamo (32)	—	—	1	—	—	—	1	—	—	—	1	—	—	—	2	5	—	—	—	—	—	—	—	10
Allengo (98)										See Puccioni, N. (1936, p. 118)														
Ato (36)	—	1	—	1	—	3	2	—	1	3	—	—	—	2	2	3	4	1	—	—	1	—	—	24
Bardera (90)	—	2	—	3	—	—	—	—	—	2	—	1	—	2	—	6	1	—	—	—	—	—	—	17
Bio Ado (115)										See Graziosi, P. (1938)														
Bio Mudo Balleh (12)	—	—	—	—	3	1	—	—	—	1	—	—	—	—	16	36	2	3	—	—	1	—	3	66
Bugda Acable (84)	—	—	—	1	—	—	—	2	—	—	—	—	—	—	2	5	1	4	—	1	—	—	—	16
Bun Mess (95)										See Graziosi, P. (1940, p. 69)														
Bur Dauleh (18)	2	—	1	1	—	—	1	—	—	—	—	—	—	—	3	5	2	—	—	1	—	1	—	17
Bur Dinsor (97)										See Graziosi, P. (1938)														
Burdo (39)	—	—	—	—	—	—	—	—	1	—	—	1	—	—	—	1	—	—	—	—	—	—	—	3
Bur Eighei (74)	—	—	1	—	—	—	—	—	—	—	—	—	1	—	2	1	—	—	—	—	—	—	—	5
Bur Maha-mudo (19)	—	1	—	3	—	1	1	—	—	1	2	3	1	—	4	3	2	3	—	3	—	—	—	28
Bur Rari										See Puccioni, N. (1936, p. 118)														
Bur Yassin (18)	3	—	1	2	2	1	5	1	—	—	5	—	—	1	15	35	1	11	—	2	—	—	—	85

Domo (62)	—	—	—	—	7	—	—	—	—	—	—	—	—	—	4	10	3	2	—	—	—	—	—	26
Duldir Dune (82)	—	—	1	—	1	—	—	—	—	—	—	—	—	—	—	5	—	1	—	—	—	—	—	8
Dusa Mareb (51)	—	—	—	—	2	—	1	—	—	2	1	—	—	—	5	7	—	1	—	—	—	—	—	19
Eil (56)	2	—	5	2	5	4	1	3	—	5	—	—	—	—	29	16	1	1	—	—	1	—	—	75
El Colcolo (91)	—	—	—	—	—	—	—	—	—	—	1	—	—	—	2	—	—	1	—	—	—	—	—	4
El Dubbo (40)	—	—	—	2	—	2	9	—	—	3	—	—	1	—	7	22	3	4	—	—	8	—	—	61
El Goran (77)	—	—	—	—	—	1	1	—	—	1	3	—	—	2	—	8	2	8	—	—	1	—	1	28
El Ure (103)	See Puccioni, N. (1936, p. 118)																							
Far Abou (33)	—	—	—	—	—	1	—	—	—	3	1	—	—	—	8	—	—	2	—	—	4	—	—	19
Goffado (43)	—	—	—	—	—	—	—	—	—	1	—	—	—	—	3	9	—	1	—	—	1	—	—	15
Gawan wein (22)	—	—	—	—	1	—	—	—	—	1	—	—	—	—	—	1	—	—	—	—	—	—	—	3
Guriagale (Ohole area) (42)	See Graziosi, P. (1938)																							
Legullei (29)	—	—	—	—	5	2	1	1	—	7	1	—	—	1	25	10	1	5	—	—	2	3	—	64
Lugh Ferrandi (100)	See Puccioni, N. (1936, p. 118); Graziosi, P. (1940, p. 72); and Appendix A, Phillips Collection																							
Madeida (29)	—	—	—	—	—	—	2	—	—	—	—	—	—	—	11	26	8	9	—	—	8	—	—	64
Manaas (26)	—	—	—	—	—	—	—	—	—	—	—	—	—	—	1	—	—	1	—	—	—	—	—	2
Morogavi (31)	—	—	—	—	—	—	—	—	—	—	—	—	—	1	17	—	—	3	—	—	—	—	—	21
Sheik Gure (83)	—	—	1	—	—	1	—	—	—	—	—	—	—	—	2	10	6	2	—	—	—	—	—	22
Sijjia (41)	—	—	—	—	7	—	7	1	—	6	—	—	—	1	19	23	6	10	—	—	15	—	—	95
Sirrau Balleh (199)	See Appendix A, Gilliland and Glover Collections																							
Ted (38)	—	—	—	1	—	—	—	—	—	9	3	—	1	—	2	—	2	—	—	—	18	—	—	36
Walwal (114)	See Graziosi, P. (1938)																							
Wanei (24)	—	—	—	—	—	2	—	—	—	—	2	—	—	—	3	3	—	2	—	—	2	—	—	14
Warishe (21)	—	1	—	—	—	—	—	—	—	—	—	—	—	—	—	—	—	—	—	—	—	—	—	1
Washaga Guran (102)	See Puccioni, N. (1936, p. 118)																							
Wejit (28)	—	—	—	1	—	3	—	—	—	5	2	—	—	1	11	24	1	6	1	—	15	—	—	70
SITES OF MIXED DOIAN AND OGADEN WILTON AFFINITIES																								
Bulo Burti (86)	—	—	—	—	—	—	—	—	1	—	2	—	—	—	—	12	—	—	—	—	1	—	—	16
Bur Dahir (214)	See details in text, p. 308																							
Damburweini (67)	—	—	1	—	—	—	—	—	—	3	2	—	—	—	7	5	—	4	—	—	—	—	—	22
God'dere (35)	—	—	—	—	4	—	—	3	—	8	4	—	—	—	8	4	—	6	—	—	1	—	—	38

APPENDIX E

TOOL LISTS FOR ADDITIONAL SOMALILAND WILTON CULTURE SITES

TOOL TYPES

SITES	Microliths, curved-backed	Microliths, straight-backed	Microliths, crescents	Microliths, double-backed	Scrapers, thumbnail and small round	Scrapers, end-	Scrapers, various	Burins, angle, *bec-de-flute* and single-blow	Backed blades, large	*Outils écaillés*	Blades, some utilized	Flakes, some utilized	Blade cores	Pottery	Ostrich egg-shell	Various	Hammer and pestle stones	Bifaced points, foliate	Totals
ABYSSINIA																			
Adele (7)	1	—	4	—	1	1	—	1	—	—	3	5	1	—	—	—	—	—	17
Adjin (9)	5	—	2	—	3	—	2	—	—	1	6	—	1	—	—	—	—	—	20
Aisha (6)	5	—	6	—	4	—	—	1	—	—	7	19	7	—	—	2	—	—	51
Garasleh Tug (8)	4	—	—	—	10	—	6	1	—	4	11	19	7	—	1	—	—	—	63
Lago Oda (221)	—	—	3	—	2	—	—	2	—	—	—	—	—	—	—	—	—	—	7
Omerdin (220)	See Breuil, H., Teilhard de Chardin, P. & Wernert, P. (1951), p. 230																		
Porc Epic (152)																			
BRITISH SOMALILAND																			
Berbera rock-shelter (1)	1	1	—	—	—	3	—	2	—	—	28	5	1	—	2	2	—	—	45
Bihen (60)	—	—	—	—	—	2	1	1	—	—	20	3	—	—	—	—	—	—	27
Buran (137)	See Appendix A, Barrington Brown Collection																		
Bur Khansa (187)	See Appendix A, Gilliland Collection																		
Dubar (14)	1	—	—	—	—	—	—	—	—	—	—	3	—	—	—	—	—	—	4
Dukokolol yero (213)	See Appendix A, Glover Collection																		
El Dur Elan (184)	See Appendix A, Gilliland Collection																		
Erigavo (140)	See Appendix A, Trevor & McInnes and Gilliland Collections																		
Gerbakele (209)	See Appendix A, Glover Collection																		
Haded: Balad Ad (189)	} See Appendix A, Gilliland Collection																		
Halin (188)	}																		
Hamas (16)	—	—	—	—	1	—	—	1	—	—	—	1	—	—	—	1	—	—	4
Hargeisa (3)	1	—	—	—	—	1	5	—	—	—	12	16	4	—	—	2	—	—	41
Hudin/Ainabo road (186	} See Appendix A, Gilliland Collection																		
Hudin/Halin road (192)	}																		
Jifa Ure (205)	See Appendix A, Glover Collection																		

Kirit (182)	See Appendix A, Gilliland Collection																		
Las Anod (181)	See Appendix A, Gilliland Collection																		
Laferug (2)	1	—	—	—	—	—	—	—	—	—	2	2	1	—	—	—	1	—	7
Mandera North (11)	4	—	1	—	3	2	—	1	—	2	5	9	2	(2)	2	3	1	—	37
Raguda Tug (185)	See Appendix A, Gilliland Collection																		
Shimbir Beris (190)	See Appendix A, Gilliland Collection																		
FRENCH SOMALILAND																			
Djebel Djinn (151)	See Teilhard de Chardin, P., Breuil, H. & Wernert, P. (1940)																		
Gulf of Tajura (147)	See Teilhard de Chardin, P., Breuil, H. & Wernert, P. (1940)																		
Kor Ali (148)	See Teilhard de Chardin, P., Breuil, H. & Wernert, P. (1940)																		
Ras Kiro (145)	See Teilhard de Chardin, P., Breuil, H. & Wernert, P. (1940)																		
SOMALIA																			
Adad (107)	See Puccioni, N. (1936, p. 118)																		
Adani (110)	See Puccioni, N. (1936, p. 118)																		
Badi Ahmud (129)	See Puccioni, N. (1936, p. 118)																		
Between Bakbah and Bio Ado (124)	See Puccioni, N. (1936, p. 118)																		
Biede (57)	2	1	—	—	1	3	1	2	—	—	12	24	3	—	—	1	—	—	50
Bihen (Mijertein Coast) (126)	See Puccioni, N. (1936, p. 118)																		
Between Bihen and Ariri (125)	See Puccioni, N. (1936, p. 118)																		
Eil (56)	—	5	—	—	—	5	—	2	1	—	15	16	1	—	—	1	—	1	47
Garoe (54)	1	2	—	—	13	—	3	2	—	—	16	27	7	—	—	3	—	—	74
Gowein (48)	2	—	—	—	—	—	—	—	—	—	3	11	1	—	—	—	—	—	17
Hararile (55)	—	1	—	—	—	—	—	—	—	—	24	6	3	—	—	8	—	—	42
Laba Higlo (53)	—	—	—	—	—	1	4	1	—	—	19	6	12	—	—	3	—	—	46
Lower Darror (133)	See Puccioni, N. (1936, p. 118)																		
Between Martissor Dunti and Scerer Hor (131)	See Puccioni, N. (1936, p. 118)																		
Taleh (136)	See Puccioni, N. (1936, p. 118)																		
Upper Darror (134)	See Puccioni, N. (1936, p. 118)																		
Between Warassimoghe and Bender Beila (128)	See Puccioni, N. (1936, p. 118)																		
OGADEN VARIANT OF THE WILTON																			
Cura Ba'ad (72)	—	—	—	—	1	—	—	—	—	—	4	6	—	—	—	—	—	—	11
Dagahbur/Webi confluence (163)	See Breuil, H. & Kelley, H. (1936)																		
Kaptinag (66)	3	2	—	—	—	—	3	—	—	—	1	4	—	—	—	—	—	—	13
Left bank of the Webi Shebeli (160–162, 164)	See Breuil, H. & Kelley, H. (1936)																		
Melka Taka (70, 71)	2	—	—	—	—	—	1	—	—	3	—	7	—	—	—	2	—	—	15
Orahaut (155)	See Breuil, H. & Kelley, H. (1936)																		
Souri (165)	See Breuil, H. & Kelley, H. (1936)																		

APPENDIX F

REPORT ON THE FAUNAL REMAINS COLLECTED BY J. D. CLARK FROM SITES IN THE SOMALILANDS

BY D. M. A. BATE

The collection of animal remains obtained by Mr Desmond Clark from rock-shelters in Somaliland includes a considerable number of specimens, but these are unfortunately in an extremely fragmentary state of preservation, with hardly a complete bone present. It seems as if the broken condition is due to the animals having been used for human food, though it is impossible to suggest a reason for the very small size of most of the fragments. Nearly all the bones are not only broken transversely but are also split longitudinally, so that complete sections are rare. A single piece of bone is 15 cm. in length, but apart from this fragments seldom reach a length of 10 cm.

Identifiable remains form only a small proportion of the collection, but a close scrutiny suggests that very few unidentified specimens represent animals other than those which have been included in the list. Even when the presence of certain animals is demonstrated in no case are specimens sufficiently well preserved for specific identifications to be made with complete confidence. This seems to make it impossible for the collection to be of assistance in suggesting whether any change has taken place in the fauna since the time when the deposits were accumulated.

The comparatively great geographical and ecological differences in a small country ensures that the fauna is very varied in type and composition, and this adds to the difficulty of gaining a picture of past conditions from scattered fragments. However, at this early stage of our knowledge of the early history of the country every fact may be of possible importance.

The following is a list of the eleven mammals it has been possible to distinguish, with the addition of a monitor lizard, and a tortoise, whose remains were kindly examined by Dr W. E. Swinton. A point of interest is that the animals enumerated are all forms which might be, and probably were, used for food.

1. *Hystrix* sp.
2. *Potamochoerus* sp.
3. *Phacochoerus* sp.
4. ? *Giraffa* sp.
5. *Bos* or *Syncerus*
6. Cf. *Oryx* sp.
7. ? *Oreotragus* sp.
8. *Rhyncotragus* sp.
9. *Antilope* spp.
10. *Equus* cf. (*Asinus*) sp.
11. *Equus* cf. (*Dolichohippus*) *grevyi.*

Varanus sp.
Testudo sp.

1. *Hystrix* sp. Porcupine

A Porcupine is represented by only three isolated cheek-teeth from the Doian Levels of Bur Hakaba and Gure Warbei, both in the south of the country.

The common North African species, *H. cristata*, is recorded from several parts of the country at the present day (Peel, 1900, p. 505).

2. *Potamochoerus* sp. Bush Pig

A fragment of skull, and two foot-bones believed to be those of a Bush Pig come from Gure Warbei, and an imperfect cheek-tooth from Bur Hakaba, Magosian Level.

It seems probable that the distribution of the Bush Pig in Somaliland is very local, for several well-known books on the country omit it from their lists of fauna. This may be partly due to the nocturnal habits of these animals and also to the aversion to pigs evinced by the Mahommedan population. There is, however, no doubt that it does occur, and de Beaux (1924, p. 490) has separated it as a distinct subspecies, *somaliensis*, basing his description on specimens from the upper Webi Shebeli.

3. *Phacochoerus* sp. Wart Hog

Ten imperfect third molars prove the presence of Wart Hog in the collection; two are from Bur Dauleh, with Doian Culture, the remainder from Bur Hakaba, seven being from the Dump, and one from the Doian level.

Wart Hogs are widely distributed, and common in some parts of Somaliland; Prince Nicholas Ghika (1898, p. 139) records encountering a party of more than twenty individuals in the Ogaden country.

4. ? *Giraffa* sp. ? Giraffe

A single imperfect lower premolar is believed to be that of a Giraffe, though the evidence is not quite conclusive. It was obtained from a midden near Obbia on the central Somalia coast.

The reticulated Giraffe of Somaliland is a very rare animal found at the present day only on the upper reaches of the Webi Shebeli.

5. *Bos* or *Syncerus*. Ox or Buffalo

Four cheek-teeth in the collection may be those of either Ox or Buffalo, but it is not possible to be certain which they represent. The fact that three of the specimens come from Iron Age levels at Bur Hakaba, and the other from an Iron Age/Doian level at Gure Warbei suggests that domesticated cattle were probably present in the country at this time. Wild Buffalo are found in Somaliland, but are not common.

6. ? *Oryx* sp. ? Oryx

Only a single lower cheek-tooth is present resembling that of this species, which it probably represents. This specimen is from Bur Hakaba, Rifle Range site, Magosian level.

This beautiful large antelope was until recently both plentiful and widely distributed in the country.

7. ? *Oreotragus* sp. ? Klipspringer.

A single cheek-tooth from Bur Hakaba, dump, resembles that of the Klipspringer, while three others, also from Bur Hakaba, one Iron Age, are more doubtfully referred to this small Antelope.

8. *Rhyncotragus* sp. Dik-Dik

Remains of these tiny Antelopes are fairly plentiful, portions of seven lower, and two upper jaws, as well as a horn core in a fragment of skull having been identified. These remains come from Gure Warbei, Doian and Magosian levels; and from Bur Hakaba, Rifle Range site, from Doian, Magosian and Iron Age levels.

These tiny Antelopes, together with the closely allied genus *Madoqua*, may be described as typical of the Somaliland fauna, and are found in great numbers and variety, particularly in the acacia country.

9. *Antilope* spp. Antelopes.

Besides the remains of small antelopes mentioned above there is a small number of isolated cheek-teeth representing several large species, but it is not possible to determine the species to which they belong, owing to the similarity of the cheek-teeth in this group.

10. *Equus* (*Asinus*) sp. ? Somali Wild Ass

Nearly all the sites have yielded remains believed to be of the Somali Wild Ass, in the form of isolated cheek-teeth, which may be listed as follows:

One cheek-tooth from Gure Warbei. Magosio-Doian transition/Magosian level.

Seven cheek-teeth from Bur Hakaba Rifle Range site. Doian level.

Four cheek-teeth from Bur Dauleh. Doian level.

Six cheek-teeth from Hargeisa, site H. 7 Somaliland Stillbay.

One cheek-tooth from Obbia, central Somalia Coast, Historic Midden.

In recent years this handsome species was found in some numbers, chiefly on the great coastal plain in the north of Somalia.

11. *Equus* cf. (*Dolichohippus*) *grevyi*. Oustalet

A single equine tooth considerably larger than those ascribed to the Wild Ass may perhaps represent Grevy's Zebra, which is still found in Somalia. This specimen comes from Bur Dauleh west, and was associated with a Stillbay industry. Three portions of teeth in a hard limestone matrix from El Dubbo, south Somalia, also represent an *Equus*, either Ass or Zebra.

Varanus sp. Varanus Lizard.

A number of vertebrae of a varanus lizard, as well as a few jaws with teeth were found at each of the levels of Gure Warbei; two other vertebrae come from Bur Hakaba, Rifle Range site, Doian level.

Testudo sp. Tortoise

A fragment of the edge of the carapace of a tortoise from Gure Warbei suggests an animal about 2 ft. in length.

The following lists give the species found at the different sites:

GURE WARBEI, BUR EIBE, SOUTH SOMALIA

Hystrix sp.	Doian level
Potamochoerus sp.	Doian/Iron Age level
Bos or *Syncerus*	Iron Age/Doian level
Rhyncotragus sp.	Magosio/Doian transition and Doian level
Equus cf. (*Asinus*) sp.	Magosian level

Varanus sp.	Doian, Magosio/Doian transition and Iron Age levels
Testudo sp.	Doian level

RIFLE RANGE SITE, BUR HAKABA, SOUTH SOMALIA

Hystrix sp.	Derived Doian level
Potamochoerus sp.	Magosian level
Phacochoerus sp.	Derived Doian level
Bos or *Syncerus*	Iron Age level
Cf. *Oryx* sp.	Magosian level
Cf. *Oreotragus* sp.	Derived Doian level
Rhyncotragus sp.	Magosian, Doian, Iron Age level
Antilope sp.	Derived Doian level
Equus cf. (*Asinus*) sp.	Doian level
Varanus sp.	Doian level

EL DUBBO, SOUTH SOMALIA

Equus sp. *Asinus* or *Equus grevyi*	Levalloisian industry

BUR DAULEH, BUR EIBE, SOUTH SOMALIA

Phacochoerus sp.	Doian level
Small antelope	Doian level
Equus cf. (*Asinus*) sp.	Doian level
Equus cf. (*Dolichohippus*) *grevyi*	? Somaliland Stillbay

SITE H.7. HARGEISA, BRITISH SOMALILAND

Equus cf. (*Asinus*) sp.	? Somaliland Stillbay

OBBIA, CENTRAL SOMALIA COAST

Equus cf. (*Asinus*) sp.	Historic (Modern) Midden
? *Giraffa* sp.	Historic (Modern) Midden

LIST OF WORKS CONSULTED

de Beaux, O. (1924). Beitrag zur Kenntnis der Gattung Potamochoerus. Gray. *Zool. Jahrb. Syst.* **47**, pp. 379–504, 2 pls. 3 text-figs.

Drake-Brockman, R. E. (1910). *The Mammals of Somaliland.* London.

Ghika, N. D. (1898). *Cinq Mois au Pays des Somalis.* Geneva.

Graziosi, P. (1940). L'Eta della Pietra in Somalia. *R. Universitas. degli Studi di Firenze——Centro di Studi Coloniale*, xviii, pp. 89. 34 pls. (A small fauna was obtained from Bur Eibe by Professor Graziosi.)

Parker, H. W. (1932). Two Collections of Reptiles and Amphibians from British Somaliland. *Proc. Zool. Soc. London*, i, pp. 335–67, 3 figs.

Peel, C. V. A. (1900). *Somaliland.* London.

Swayne, H. G. C. (1895). *Seventeen Trips through Somaliland.* London.

APPENDIX G

LIST OF MARINE MOLLUSCS COLLECTED BY J. D. CLARK FROM RAISED BEACHES AND OTHER SITES IN THE SOMALILANDS

By L. R. COX

1. OBBIA: CENTRAL SOMALIA COAST; HISTORIC (MODERN) MIDDEN

Mytilus perna Linné

2. BIEDE ROCK-SHELTER, NOGAL ESTUARY MOUTH, NORTHERN SOMALIA COAST

Upper Midden. Iron Age. Strandlooper

(0–10 cm.)

Columbella sp.
Thais rudolphi (Lamarck)
Mytilus perna Linné

(10–20 cm.)

Thais rudolphi (Lamarck)
Mytilus perna Linné
Ostrea sp.

(20–30 cm.)

Thais rudolphi (Lamarck)
Nerita textilis Gmelin
Cypraea sp.
Mytilus perna Linné
Ostrea sp.

(30–40 cm.)

Thais rudolphi (Lamarck)
T. bufo (Lamarck)
Patella umbella Gmelin
Trochus erythraeus Brocchi
Mytilus perna Linné
Barbatia decussata (G. B. Sowerby)
Ostrea sp.
Plicatula sp.

Yellow Sand. Sterile

Fissurella cf. *spreta* Smith
Patella sp.

Lower Midden. Late Stone Age, Strandlooper

Thais rudolphi (Lamarck)
T. bufo (Lamarck)
Cymatium (Lampusia) pileare (Linné)
Oliva sanguinolenta Lamarck
Oliva cf. *ispidula* Linné
Terebralia palustris (Linné)
Nerita textilis Gmelin
Conus sp.
Mytilus perna Linné
Barbatia decussata (G. B. Sowerby)

3. NOGAL ESTUARY MOUTH, MIJERTEIN

Late Stone Age, Strandlooper. Midden outside Biede Rock-shelter

Thais rudolphi (Lamarck)
T. bufo (Lamarck)
Conus sp.
Chlamys senatoria (Gmelin)
Ostrea sp.

APPENDIX G

4. RAKON WELLS. CENTRAL MUDUGH

Surface find in association with tools of Somaliland Wilton of 'Neolithic' facies

Patella cf. *Umbella* Gmelin
Conus sp.
Cardium sp.

5. MIRSALE WELLS. CENTRAL MUDUGH

Surface find in association with tools of Doian of 'Neolithic' facies

Cypraea turdus Lamarck

6. MOGADISHU, LIDO

Low-level (0–2 *m.*) *beach under fossil dunes*

Conus cf. *lividus* Hwass
Cypraea cf. *arabica* Linné
Cymatium (*Lampusia*) *pileare* (Linné)
Nerita textilis Gmelin
Georgia guillainopsis Bourguignat
Barbatia fusca (Bruguière)
Spondylus cf. *fauroti* Jousseaume
Spondylus sp.
Cardium (*Trachycardium*) *flavum* Linné
Turbo marmoratus Linné
Mytilus perna Linné

APPENDIX H

LIST OF CHARCOALS COLLECTED BY J. D. CLARK FROM PREHISTORIC SITES IN THE SOMALILANDS

IDENTIFIED BY THE ROYAL BOTANIC GARDEN, KEW.

All of the charcoals appear to have been prepared from woody genera which are known to occur in Somaliland at the present time. The very miscellaneous assortment of woods suggests that those who were responsible for preparing the charcoal must have used almost any kind of wood which happened to be readily available.

Unfortunately, the material is so completely carbonized that repeated attempts to cut sections of it have all failed. It has therefore been necessary to base our identifications on features which can be seen on newly fractured surfaces of the charcoal when examined as opaque objects under the microscope. The disadvantage of this technique lies in the fact that some of the most important diagnostic microscopical features cannot be clearly seen. I would therefore emphasize that, although every care has been taken in making the identifications, they should be accepted with a certain amount of reserve. The identifications are as follows:

UPPER MIDDEN (IRON AGE) BIEDE ROCK-SHELTER,
NOGAL ESTUARY MOUTH, MIJERTEIN

1. *Sample B.* (0–10 cm.). This contained charcoal prepared from the wood of *Balanites aegyptiaca*, *Acacia* sp. resembling *Acacia arabica* and what appeared to be a species of *Commiphora*.

2. *Sample A.* (20–30 cm.) This contained charcoal prepared from the wood of *Balanites aegyptiaca* and some from what appeared to be a species of *Commiphora.* Brown fragments included in the sample are of animal origin.

3. *Sample F.* (30–40 cm.) This included charcoal which was probably prepared from a number of the Salvadoraceae, and strongly resembled species of *Dobera* and *Salvadora* received from Somaliland. Also a member of the Sapotaceae resembling *Mimusops*, similar to some of the material in Sample C.

LOWER MIDDEN (LATE STONE AGE STRANDLOOPER)

4. *Sample C.* This included charcoal prepared from the wood of a member of the family Sapotaceae in many respects resembling *Mimusops*; an unidentified member of the Leguminosae; *Acacia* sp.

5. *Sample D.* This included charcoal prepared from the wood of *Ficus* sp., possibly *F. sycomorus*, and *Acacia* sp. (Probably *not A. arabica.*)

6. *Sample G.* This sample included some carbonized fruits which could not be identified. The charcoal was of a different kind from that in the other samples and in some respects resembled *Pistacia*, but this identification is by no means certain.

GURE WARBEI ROCK-SHELTER, BUR EIBE, SOUTH SOMALIA

7. *Sample E.* This included charcoal prepared from *Ficus* sp., possibly *F. sycomorus*; an unidentified member of the Leguminosae (similar to material in sample C); *Buxus* sp., resembling *B. hildebrandtii.*

[*Note by J. D. C.* All the species and genera identified belong to the flora of a dry region. At first it might appear strange that both *Ficus* and *Acacia* should occur in the same deposit as the water requirements of these trees are different, but it must be remembered that the Nogal Gorge contained freshwater springs since Upper Pleistocene times which would thus allow of the continued existence in the gorge of trees requiring a fresher type of soil and high water-table, as is indeed the case today. In the anhydrite country, sink holes, some of which contain water, others being dry, are now commonly marked by very large wild fig-trees which grow in them. *Acacia* trees are often found growing quite close by on drier sites. These charcoals have therefore only a limited value as indicators of climatic conditions and there would therefore seem to be no difficulty in accounting for a mixture of charcoal from these two types of trees, at least in the vicinity of springs and sink holes.]

BIBLIOGRAPHY

ABRUZZI, DUC D' (1932). *La esplorazione dello Uabi-Webi Scebeli.* Milan.

ALOISI, P. & DE ANGELIS, A. M. (1938). *Le Rocce della Somalia.* Bologna.

DE ANGELIS, A. M. (1938*a*). Le Rocce sedimentare e le Sabbie. *Geol. della Somalia.* Bologna.

DE ANGELIS, A. M. (1938*b*). *See* Aloisi, P. & de Angelis, A. M.

ARAMBOURG, C. (1933). Découverte d'un gisement de Mammifères burdigaliens dans le bassin du Lac Rudolphe (Afrique Orientale). *C.R. Soc. Géol. France*, no. 14. Paris.

ARKELL, A. J. (1949). *The Old Stone Age in the Anglo-Egyptian Sudan.* Sudan Antiquities Service, Occ. Paper, no. 1.

ARKELL, W. J. (1933). *See* Sandford, K. S. & Arkell, W. J.

ARKELL, W. J. (1939*a*). *See* Sandford, K. S. & Arkell, W. J.

ARKELL, W. J. (1939*b*). *See* Sandford, K. S. & Arkell, W. J.

ARSANDAUX, H. (1905). Résultats pétrographiques d'un Voyage dans le pays Somali-Dankali et en Abyssinie. *C.R. Congrès des Soc. Sav. Paris* (1904).

ARSANDAUX, H. (1906). Contribution a l'étude des roches alkalines de l'Est Africain. *C.R. Scient. de la Miss. Duchesne-Fournet.* 4to. Paris.

ARSANDAUX, H. & NEUVILLE, H. (1905). Résultats petrographiques du Voyage de M. Maurice de Rothschild dans les Pays Somali-Dankali et en Abyssinie. *Bull. Mus. Hist. Nat.* XI, no. 3. Paris.

ARTINI, E. (1915). *Intorno alla composizione mineralogica di alcune sabbie ed arenarie raccolte dalla missione scientifica Stefanini-Paoli nella Somalia Italiana.* Pavia.

BALDACCI, L. (1891). Osservazioni fatte nella Colonia Eritrea. *Mem. Carta geol. d'Italia*, vol. VI. Rome.

BALL, J. (1939). *Contributions to the Geography of Egypt.* Survey and Mines Dept. Cairo.

BARRINGTON BROWN, C. (1931*a*). The geology of N.E. British Somaliland. *Q.J. Geol. Soc.* vol. LXXXVII.

BARRINGTON BROWN, C. (1931*b*). *See* Burkitt, M. C. & Barrington-Brown, C.

BASE, E. (1930). *See* Teilhard de Chardin, P. *et al.*

BATE, D. M. A. (1937). *See* Garrod, D. A. E. & Bate, D. M. A.

DE BEAUX, O. (1924). Beitrag zur Kenntnis der Gattung Potamochoerus Gray. *Zool.-Jahrb. Syst.* 47, pp. 379–504.

BLANC, A. C. (1938). Industrie palaeolitiche e mesolitiche del Moggio presso Addis-Abeba. *Rivista di Anthrop.* vol. XXXII.

BLANC, A. C. & TAVANI, G. (1938). Contributo all conoscenza delle industrie litiche della Somalia. *Atti Soc. Toscana Sci. Nat.*

BLANC, A. C. (1952). Industries lithiques des Iles Dahlac (Mer Rouge). *C.R. 2nd Pan-Afr. Congress on Prehist.* (In the Press.)

BLANDFORD, W. T. (1869). On the geology of a portion of Abyssinia. *Q.J. Geol. Soc.* vol. XXV. London.

BLANDFORD, W. T. (1870). *Observations on the Geology and Zoology of Abyssinia made during the progress of the British Expedition to that Country in 1867–68.* London.

BOULE, M. & VALLOIS, H. V. (1946). *Les Hommes Fossiles.* Paris.

BREUIL, H. (1928). Les gravures rupestres du Djebel Ouenat (Lybie). *Revue Scient.* vol. XXV, p. 11.

BREUIL, H. (1930). *L'Afrique préhistorique.* Cahiers d'Art. Paris.

BREUIL, H. (1931). *Afrique.* Cahiers d'Art. Paris.

BREUIL, H. (1934). Peintures rupestres préhistoriques du Harrar (Abyssinie). *L'Anthropologie, Paris*, vol. XLIV.

BREUIL, H. (1940). *See* Teilhard de Chardin, P., Breuil, H. & Wernert, P.

BREUIL, H. & KELLEY, H. (1936). Documents préhistoriques recueillis par la Mission du Bourg de Bozas en Abyssinie. *J. Soc. Africanistes*, vol. VI.

BREUIL, H., TEILHARD DE CHARDIN, P. & WERNERT, P. (1951). Le Paléolithique du Harar. *L'Anthropologie, Paris*, vol. LV, nos. 3–4, pp. 219–30.

BRITISH MUSEUM (1926). *A Guide to the Antiquities of the Stone Age.*

BURKITT, M. C. & BARRINGTON BROWN, C. (1931). Stone Implements from British Somaliland. *Man.*

BURKITT, M. C. (1932*a*). *See* Parker, O. F. & Burkitt, M. C.

BURKITT, M. C. (1932*b*). *See* Wayland, E. J. & Burkitt, M. C.

BURKITT, M. C. & GLOVER, P. E. (1946). Prehistoric investigations in British Somaliland, paper no. 3. *Proc. Prehist. Soc.*, pp. 49–56.

CAPORIACCO, L. DI & GRAZIOSI, P. (1934). *Le pitture rupestri di Ain Doua (El Auenat).* Centro di Studi Coloniali. Florence.

CATON-THOMPSON, G. (1934). The Camel in Dynastic Egypt. *Man*, Feb. no. 24.

CATON-THOMPSON, G. (1946). The Levalloisian Industries of Egypt, paper no. 4. *Proc. Prehist. Soc.*, pp. 57–120.

CATON-THOMPSON, G. (1947). *The Aterian Industry, its place and significance in the Palaeolithic world.* R.A.I. Huxley Memorial Lecture for 1946.

CATON-THOMPSON, G. (1952). *Kharga Oasis in Prehistory.* London.

CATON-THOMPSON, G. & GARDNER, E. W. (1932). The prehistoric geography of Kharga Oasis. *Geogr. J.* vol. LXXX, no. 5.

CATON-THOMPSON, G. & GARDNER, E. W. (1934). *The Desert Fayum.* Royal Anthropological Institute.

CATON-THOMPSON, G. & GARDNER, E. W. (1939). Climate, irrigation and Early Man in the Hadhramaut. *Geogr. J.* vol. XCIII, no. 1.

DE CHASSELOUP LAUBAT, F. (1939). Saharan influences on early Egyptian art? *Illus. London News.* 14 Jan.

CHILDE, V. G. (1935). *New Light on the Most Ancient East.* London.

CIPRIANI, L. (1932). *In Africa dal Capo al Cairo.* Florence.

CLARK, J. D. (1944). The use of the bored stone in Abyssinia. *Man*, vol. XLIV, nos. 1–25.

CLARK, J. D. (1945*a*). A Kenya Fauresmith Factory and Home Site at Gondar, N. Abyssinia. *Trans. Roy. Soc. S. Afr.* vol. XXXI, part I.

CLARK, J. D. (1945*b*). Short notes on Stone Age Sites at Yavello, Southern Abyssinia. *Trans. Roy. Soc. S. Afr.* vol. XXXI, part I.

CLARK, J. D. (1950). The Stone Age Cultures of Northern Rhodesia. *S. Afr. Arch. Soc.* Cape Town.

CLARK, J. D. (1952). *Recent Prehistoric research in the Somalilands.* C.R. Pan-Afr. Congress on Prehist. (1947).

COUSIN, J. (1951). *See* Lelubre, M. & Cousin, J.

CRAIG, J. J. (1911). *See* Hume, W. F. & Craig, J. J.

DAINELLI, G. (1936). *L'Africa Orientale.* Part II. Le condizioni climatiche and Part III, Costituzione e Storia Geologica. Bologna.

DAINELLI, G. (1939). *La Regione del Lago Tana*, p. 22. Verona.

DAINELLI, G. & MARINELLI, O. (1912). *Resultati scientifici di una viaggio nella Colonia Eritrea.* Florence.

EGOUTIN (1922). Description d'un gisement de sel actuellement en formation. *Ann. des Mines* (12) *Memoirs*, vol. II.

DIETRICH, W. O. (1921). *See* Reck, H. & Dietrich, W. O.

DIXEY, F. (1946). Erosion and Tectonics in the East African Rift System. *Q. J. Geol. Soc.* vol. CII, Part III.

DRAKE-BROCKMAN, R. E. (1910). *The Mammals of Somaliland.* London.

DREW, MRS S. (1951). Two Eritrean rock sculptures probably of Coptic origin. *Man*, vol. LI, p. 155.

DREW, MRS S. (1952). The mysterious rock carvings of the Eritrean Highlands. *Illus. London News*, 9 Aug., pp. 225–7.

DREYFUSS, M. (1931). Itinéraire Géologique de Dire Daua à Harrar (Abyssinie). *Bull. Soc. Hist. Nat. Doubs.*

DREYFUSS, M. (1932). Études de géologie et de géographie physique sur la Côte française des Somalis. In *La Géologie et les Mines de la France d'outre Mer.* Paris.

DREYFUSS, M. (1930). *See* Teilhard de Chardin, P. *et al.*

EVANS, SIR J. (1896). On some Palaeolithic Implements found in Somaliland by Mr H. W. Seton-Karr. *Proc. Roy. Soc. London.*

D'ERRICO, L. (1937*a*). Una Stazione litica nell Amara. *L'Africa Italiana.* Jan.-Dec.

D'ERRICO, L. (1937*b*). *Cenni sulla preistoria ethiopia.* Opere per L'Organizzazione civile in Africa orientale Italiana. Anno XVII.

FUCHS, V. E. (1939). The geological history of the Lake Rudolf Basin, Kenya Colony. *Trans. Roy. Soc.* B, vol. CCXXIX, pp. 219–74.

GARROD, D. A. E. (1938). The Upper Palaeolithic in the light of recent discovery. *Proc. Prehist. Soc.* New Series, vol. IV, part I.

GARROD, D. A. E. & BATE, D. M. A. (1937). *The Stone Age of Mount Carmel*, vol. I. Oxford.

GARDNER, E. W. (1932). *See* Caton-Thompson, G. & Gardner, E. W.

GARDNER, E. W. (1934). *See* Caton-Thompson, G. & Gardner, E. W.

GARDNER, E. W. (1939). *See* Caton-Thompson, G. & Gardner, E. W.

GHIKA, PRINCE N. D. (1898). *Cinq Mois au Pays des Somalis.* Geneva.

GILLILAND, H. B. (1947). An approach to the problem of the government of Nomadic Peoples. *S. Afr. Geog. J.* vol. XXIX.

GLOVER, P. E. (1946). *See* Burkitt, M. C. & Glover, P. E.

GRAZIOSI, P. (1938*a*). L'Età della Pietra in Ethiopia e in Somalia. *Rivista delle Colonie*, no. 6.

GRAZIOSI, P. (1938*b*). Graffiti rupestri e stazioni prehistoriche del Fezzan. *Gli Annali dell'Africa Italiana.* Anno I, vols. III and IV.

GRAZIOSI, P. (1940). *L'Età della Pietra in Somalia.* R. Universitas degli Studi di Firenze XVIII. Centro di Studi Coloniali. Florence.

GREGORY, J. W. (1921). *The Rift Valleys and Geology of East Africa.* London.

HEY, R. W. (1948). *See* McBurney, C. B. M., Hey, R. W. & Watson, W.

HUME, W. F. & CRAIG, J. J. (1911). *The glacial period and climatic changes in North-east Africa.* Rep. B.A.A.S., no. 8.

HUNT, J. A. (1943–4). Genealogies of the Tribes of British Somaliland and Mijertein. *Gen. Surv. Brit. Som.*, 1943–4, p. 32.

HUNT, J. A. (1944). *Report on General Survey of British Somaliland.* Govt. Press. B.M.A. Somalia.

HUNT, J. A. (1946). The geology of the Zeila Plain. *Q. J. Geol. Soc.*

HUNT, J. A. & VINEY, N. (1946). *Gazetteer of British Somaliland and Grazing Areas.* Colonial Office.

HUZAYYIN, S. A. (1941). The place of Egypt in prehistory. *Mem. Inst. d'Egypte*, vol. XLIII.

INSTITUTO PER GLI STUDI DI POLITICA INTERNAZIONALE (1936). *L'Africa orientale.* Milan.

JAMES, P. L. (1888). *The Unknown Horn of Africa.* London.

JOUBERT, G. & VAUFREY, R. (1946). Le Neolithique du Ténéré. *L'Anthropologie, Paris*, vol. L.

Jousseaume, F. (1895). Réflexions anthropologiques à propos des tumulus et silex taillés des Çomalis et des Danakils. *L'Anthropologie, Paris*, vol. vi.

Kammerer, A. (1926). Appendix to *Essai sur l'histoire antique d'Abyssinie, le Royaume d'Aksum et des voisins d'Arabie et de Meroë*. Paris.

Kelley, H. (1934). Harpons, objets en os travaillé et silex taillés de Taferjit et Tamaya Mellet (Sahara Nigérien). *J. Soc. Africanistes*, iv, pp. 135–43.

Kelley, H. (1936). *See* Breuil, H. & Kelley, H.

Krenkel, E. (1925). *Geologie Afrikas*, part i. Berlin.

Lacroix, A. (1930). *See* Teilhard de Chardin *et al.*

Lamare, P. (1930). *See* Teilhard de Chardin *et al.*

Lamare, P. (1935). Sur l'Age des coulées du fosse de l'Aouache (Ethiopie). *C.R. Soc. Géol. France.*

Leakey, L. S. B. (1931). *The Stone Age Cultures of Kenya Colony*. Cambridge.

Leakey, L. S. B. (1936). *Stone Age Africa.* Oxford.

Leakey, L. S. B. (1943). The industries of the Gorgora rock shelter, Lake Tana. *J. E. Afr. and Ug. Nat. Hist. Soc.* vol. xvii, pp. 199–203.

Leakey, M. D. (1943). Note on the ground and polished stone axes of East Africa. *J. E. Afr. and Ug. Nat. Hist. Soc.* vol. xvii, pp. 182–95.

Lelubre, M. & Cousin, J. (1951). Peintures et Gravures Rupestres nouvelles de la Téfedest Meridionale (Ahaggar, Sahara Central). *L'Anthropologie, Paris*, vol. lv, nos. 1–2, pp. 24–49.

Lowe, C. van Riet (1945). The evolution of the Levallois Technique in South Africa. *Man*, vol. xlv, p. 37.

Lowe, C. van Riet (1952). The Pleistocene geology and prehistory of Uganda. Part II. Prehistory. *Mem. VI Geol. Surv. Uganda.*

McBurney, C. B. M., Hey, R. W. & Watson, W. (1948). First Report of the Cambridge Archaeological Expedition to Cyrenaica (1947). *Proc. Prehist. Soc.* (N.S.), vol. xiv, pp. 33–45.

Macfadyen, W. A. (1932). The late geological history of British Somaliland. *Nature*, vol. cxxx (Sept.).

Macfadyen, W. A. (1933). *The Geology of British Somaliland.* London.

Marinelli, O. (1912). *See* Dainelli, G. & Marinelli, O.

Monod, T. (1947). Sur quelques gravures rupestres de la region d'Aozou (Tibesti). *Rivista di Sci. Preist.* vol. ii, no. 1. Florence.

Monod, T. (1948). *Sur quelques monuments lithiques du Sahara occidental.* Homenje a J. N. Santa Olalla, vol. iii. *Act. Soc. esp. Antrop. Etnogr. Prehist.* vol. xxiii. Madrid.

Mordini, A. (1947). Le incisioni rupestri di Gazien. *Rivista di Sci. Preist.* vol. ii, no. 4.

de Morgan, J. (1921). Sur l'Industrie de la pierre au pays des Çomalis. *L'Anthropologie, Paris*, vol. xxxi, p. 46.

Moysey, F. (1943). Excavation of a rock shelter at Gorgora, Lake Tana, Ethiopia. *J. E. Afr. and Ug. Nat. Hist. Soc.* vol. xvii, pp. 196–8.

Myers, O. H. (1948). *Illus. London News*, 13 Nov.

Naville, E. (1898). *The Temple of Deir-el-Bahari*, part iii. Egypt. Explor. Soc. Pub. London.

Negri, G. (1913). *Appunti di una escursione botanica nell' Ethiopia Meridionale.* Min. delle Colonie. Monogr. e Rapp. Col. no. 4, Rome.

Nesbitt, L. M. (1935). *Desert and Forest.* London.

Neuville, H. (1905). *See* Arsandaux, H. & Neuville, H.

Nilsson, E. (1940). *Ancient changes of climate in British East Africa and Abyssinia.* Stockholms Hogskolas Geolig. Inst. no. 50. Sweden.

O'Brien, T. P. (1939). *The prehistory of Uganda Protectorate.* Cambridge.

PARKER, H. W. (1932). Two collections of Reptiles and Amphibians from British Somaliland. *Proc. Zool. Soc. London*, vol. I, pp. 335–67.

PARKER, O. F. & BURKITT, M. C. (1932). Rock engravings from Onib, Wadi Allaki, Nubia. *Man*, vol. XXXII, no. 297.

PARKINSON, J. (1932). Climatic changes in British Somaliland. *Nature*, 30 April.

PARONA, C. F. (1932). Qualche considerazione geomorfologica sul corso dell' alto Uebi Scebeli. In Abruzzi, Duc d' (1932).

PATERSON, T. T. (1940). A world correlation of the Pleistocene. *Trans. Roy. Soc. Edin.* vol. LX, part II, no. 11.

PATERSON, T. T. (1945). Core, Culture and Complex in the Old Stone Age. *Proc. Prehist. Soc.* (N.S.), vol. XI.

PAULITSCHKE, P. (1898). Praehistorische Funde aus dem Somaliland. *Mitt. Anthrop. Gesellsch. Wien*, vol. XXXIII.

PEEL, C. V. A. (1900). *Somaliland*. London.

PUCCIONI, N. (1919). *Studi sui materiali antropologici ed etnografici raccolte della Missione Stefanini-Paolo nella Somalia Italiana Meridionale*. Archivo, per l'Antrop. e l'Etnol. vol. XLIX.

PUCCIONI, N. (1936). *Antropologici e Etnografici delle gente della Somalia*, vol. III. Bologna.

RECK, H. & DIETRICH, W. O. (1923). *Ein Streiflicht auf die geologische Verhältnisse der Kreide-Juragrenze in Abessinien*. Centralbl. f. Min.

REVOIL, G. (1882). *La vallée du Darror*. Paris.

ROBECCHI-BRICHETTI, L. (1899). *Somalia e Benadir*. Milan.

VON ROSEN, B. (1949). *Berget och Solen*. Stockholm.

SANDFORD, K. S. & ARKELL, W. J. (1933). *Palaeolithic Man and the Nile Valley in Nubia and Upper Egypt*. Univ. Chicago Orient. Inst. Pub. vol. XVII.

SANDFORD, K. S. & ARKELL, W. J. (1939*a*). *Palaeolithic Man and the Nile Valley in Upper and Middle Egypt*. Univ. Chicago Orient. Inst. Pub. vol. XVIII.

SANDFORD, K. S. & ARKELL, W. J. (1939*b*). *Palaeolithic Man and the Nile Valley in Lower Egypt*. Univ. Chicago Orient. Inst. Pub. vol. XLVI.

SCHOFIELD, J. F. (1948). *Primitive Pottery*. S. Afr. Arch. Soc. Handbook Series, no. 3.

SETON-KARR, H. W. (1896). Discovery of evidences of the Palaeolithic Stone Age in Somaliland. (Tropical Africa). *J.R.A.I.* vol. XXV, no. 4.

SETON-KARR, H. W. (1898). Further discovery of ancient stone implements in Somaliland. *J.R.A.I.* vol. XXVII.

SEWELL, R. B. S. (1935). Studies on coral and coral-formations in Indian Waters. *Mem. Asiatic Soc. Bengal*, vol. IX. Calcutta.

SHAW, W. B. K. (1934). The Mountain of 'Uweinat'. *Antiquity*, vol. VIII, no. 29.

SHAW, W. B. K. (1936). Rock paintings in the Lybian Desert. *Antiquity*, vol. VIII, no. 38.

SMITH, R. A. (1937). *The Sturge Collection*. British Museum.

SOLOMON, J. D. (1931). Appendix A, *The Stone Age Cultures of Kenya Colony*, by L. S. B. Leakey. Cambridge.

STEFANINI, G. (1916). *Ricerche idrogeologiche, botaniche, ed entomologiche falte nella Somalia Italiana Meridionale*. Florence.

STEFANINI, G. (1925). Primi risultati geologici della missione della R. Societa Geografica in Somalia (1924). *Rendic. R. Acc. Naz. Lincei* (Rome), ser. 6, 1.

STEFANINI, G. (1929). Successione ed éta della 'Serie di Lugh' nella Somalia. *C.R. Int. Geol. Congr.* vol. XV.

STEFANINI, G. (1933). Cenni sulle localita fossilifere del Pliocene e Pleistocene della Somalia. Palaeontologia della Somalia. *Palaeont. Ital.* vol. XXXII, suppl. I. Siena.

STEFANINI, G. (1936). *Saggio di una Carta Geologica dell' Eritrea, della Somalia, e dell' Etiopia* (2nd ed.). Florence.

SWAYNE, H. G. C. (1895). *Seventeen Trips through Somaliland.* London.
TAVANI, G. (1938). *See* Blanc, A. C. & Tavani, G.
TEILHARD DE CHARDIN, P. (1930*a*). La Paléolithique en Somalie française et en Abyssinie. *L'Anthropologie, Paris,* vol. XL.
TEILHARD DE CHARDIN, P., LAMARE, P., DREYFUSS, M., LACROIX, A. & BASE, E. (1930*b*). Etudes géologiques en Ethiopie, Somalie et Arabie Méridionale. *Mem. Soc. Géol. Fr.* (N.S.), Mem. no. 14, vol. IV, pp. 3–4.
TEILHARD DE CHARDIN, P. (1932). Observations sur les changements de niveau marin dans les régions d'Obok. *C.R. Soc. Geol. Fr.* Paris.
TEILHARD DE CHARDIN, P., BREUIL, H. & WERNERT, P. (1940). Les Industries lithiques de Somalie française. *L'Anthropologie, Paris,* vol. XLIX, no. 5.
TEILHARD DE CHARDIN, P. (1951). *See* Breuil, H., Teilhard de Chardin, P. & Wernert, P.
VALLOIS, H. V. (1946). *See* Boule, M. & Vallois, H. V.
VALLOIS, H. V. (1951). La mandibule humaine fossile de la Grotte du Porc-Epic près Dire Daoua (Abyssinie). *L'Anthropologie, Paris,* vol. LV, nos. 3–4.
VAUFREY, R. (1938). *L'Age de l'Art rupestre Nord-Africain.* Ipek.
VAUFREY, R. (1946). *See* Joubert, G. & Vaufrey, R.
VINASSA DE REGNY, P. (1924). *Dancalia.* Rome.
VINEY, N. (1946). *See* Hunt, J. A. & Viney, N.
WATSON, W. (1948). *See* McBurney, C. B. M., Hey, R. W. & Watson, W.
WAYLAND, E. J. & BURKITT, M. C. (1932). The Magosian Culture of Uganda. *J.R.A.I.* vol. LXII, pp. 369–90.
WAYLAND, E. J. (1934). Rifts, Rivers, Rains and Early Man in Uganda. *J.R.A.I.* vol. LXIV.
WEBER, M. (1906). Die petrographische Ausbeute der Expedition O. Neumann v. Erlanger nach Ostafrika und Abessinien, 1900–1901. *Mitteil. Geogr. Ges.* vol. I, 4. Munich.
WERNERT, P. (1940). *See* Teilhard de Chardin, P., Breuil, H. & Wernert, P.
WERNERT, P. (1951). *See* Breuil, H., Teilhard de Chardin, P. & Wernert, P.
WRIGHT, W. B. (1939). *Tools and the Man.* London.
WULSIN, F. R. (1941). The prehistoric archaeology of north-west Africa. *Pap. Peabody Mus.* no. XIX, no. I.
ZEUNER, F. E. (1946). *Dating the Past.* London.

PLATES

PLATE I

Implements from Older Tug Gravels I at Hargeisa

Lower Acheulio-Levalloisian Culture

1. Lower half of a well-made hand-axe; quartzite, slightly rolled. Site H. 12.

2. Small high-backed oval prepared core; quartzite, slightly rolled. Site H. 12.

3. Ovate or subtriangular hand-axe; quartzite, slightly rolled. Site H. 8.

4. Discoidal chopper; quartzite, slightly rolled. Site H. 12.

5. Subtriangular hand-axe with flake removed at butt; quartzite, slightly rolled. Site H. 12.

6. U-cleaver on side flake with converging edges; quartzite, slightly rolled. Site H. 12.

7. U-cleaver on side flake, broken at butt, quartzite, slightly rolled. Site H. 12.

8. Unfinished ovate on side flake; quartzite, slightly rolled. Site H. 12.

PLATE I

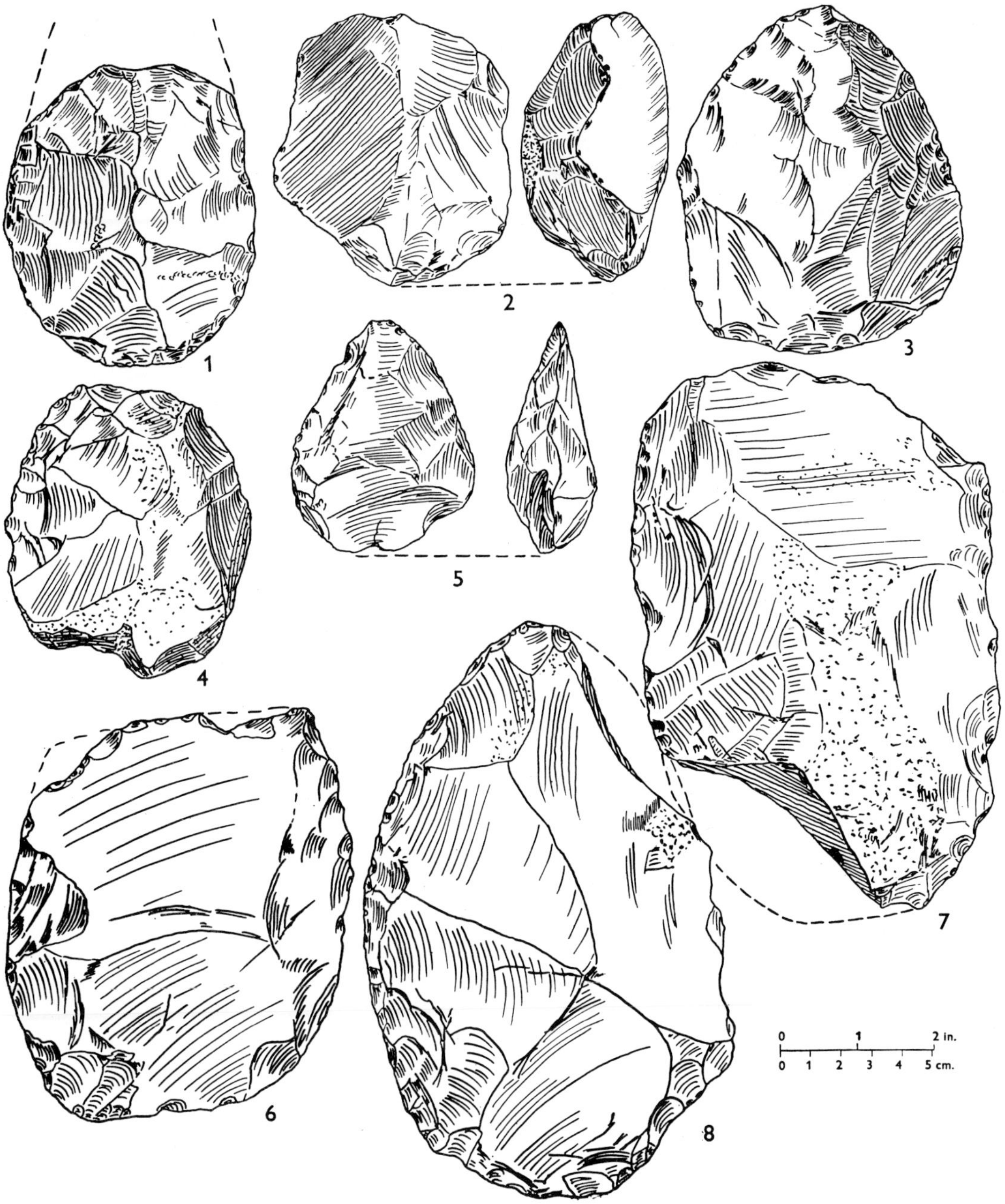

PLATE 2

Implements from Older Tug Gravels I at Hargeisa

Lower Acheulio-Levalloisian Culture

1. High-backed circular prepared core; quartzite, slightly rolled. Site H.8.
2. Flake with unfaceted platform struck from prepared core; quartzite, unrolled. Site H.8.
3. Large flake-blade or long flake with unprepared platform; quartzite, unrolled. Site H.12.
4. Flake with faceted striking-platform; quartzite, unrolled. Site H.8.

PLATE 2

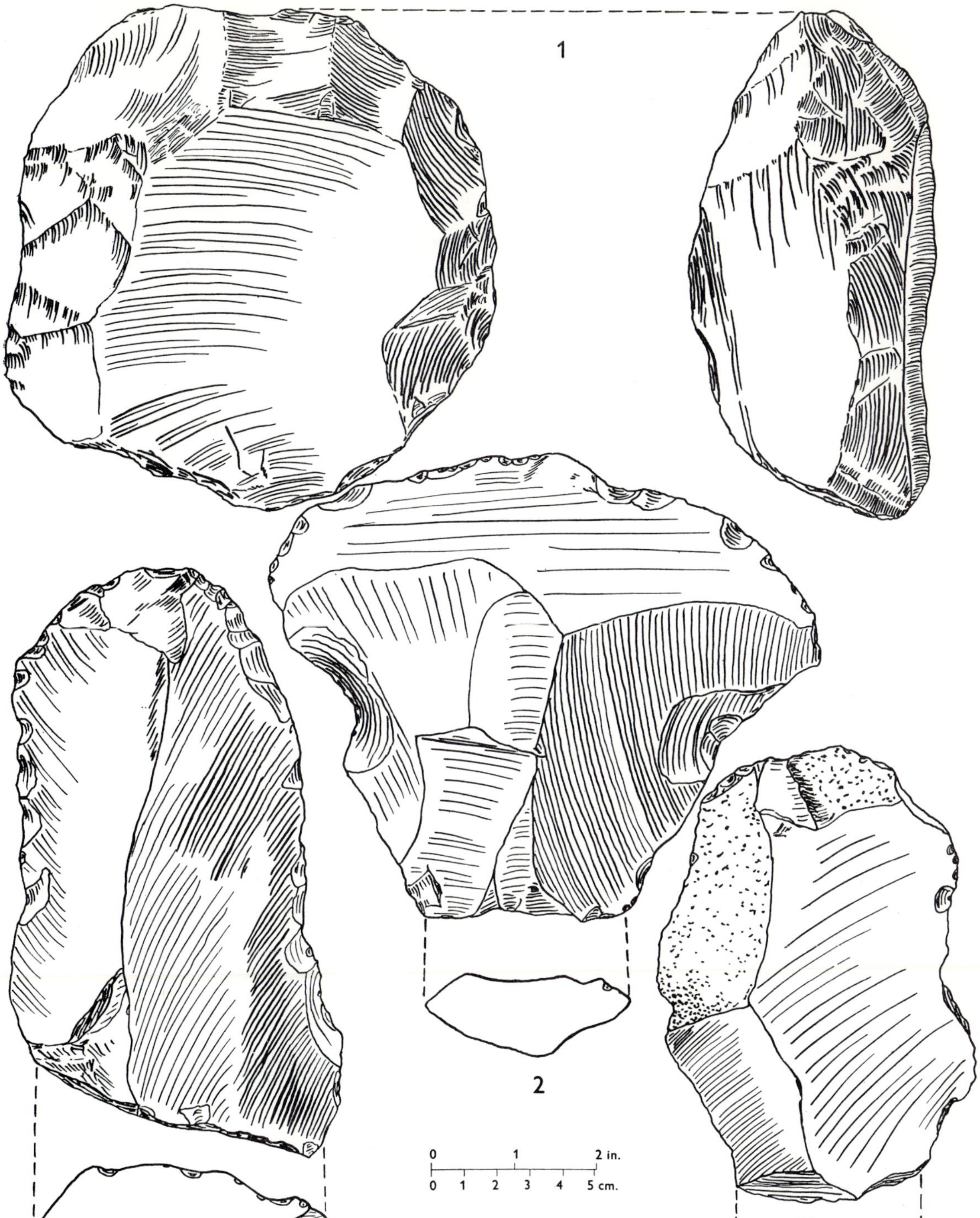

PLATE 3

Implements from Sheik (Trevor Collection)

Acheulio-Levalloisian Culture (undifferentiated)

1. Pointed pear-shaped hand-axe with thick butt; chert, unrolled.
2. Ovate hand-axe with Z-twist; chert, unrolled.

PLATE 3

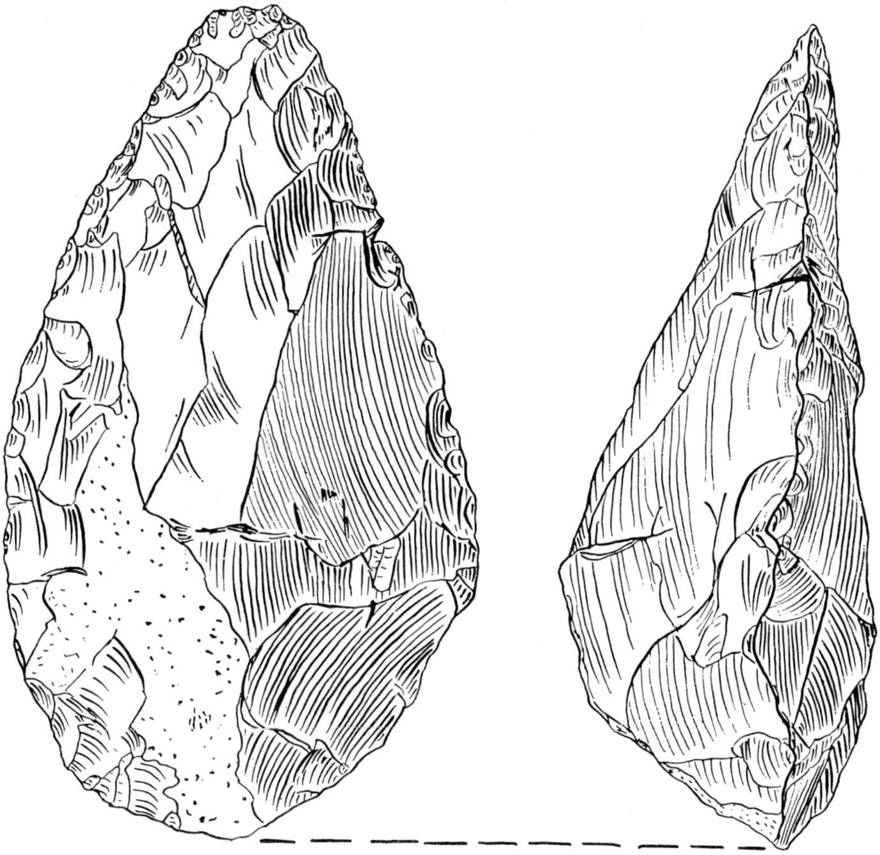

1

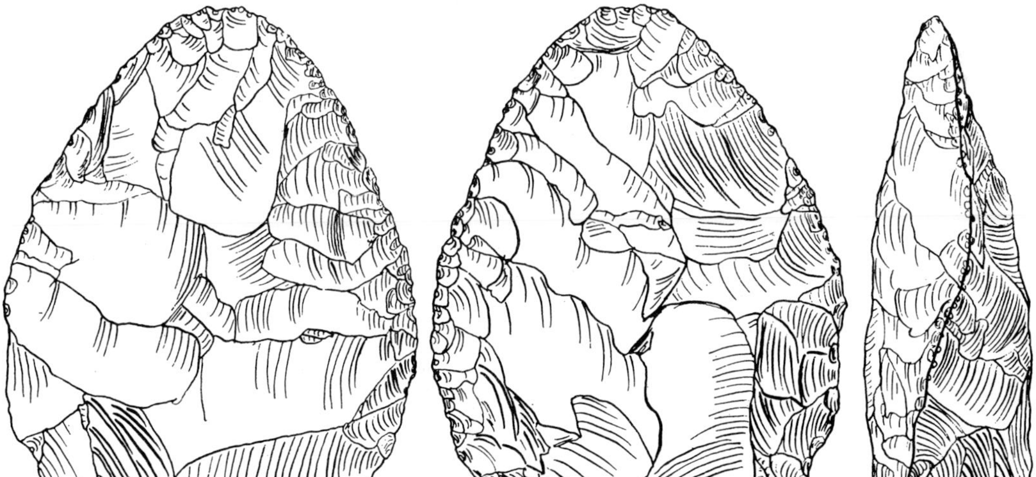

2

PLATE 4

Implements from Sheik (Trevor Collection)
Acheulio-Levalloisian Culture (undifferentiated)

1. U-cleaver on side-flake; quartzite, unrolled.
2. Micoque-type hand-axe; chert, unrolled.

PLATE 4

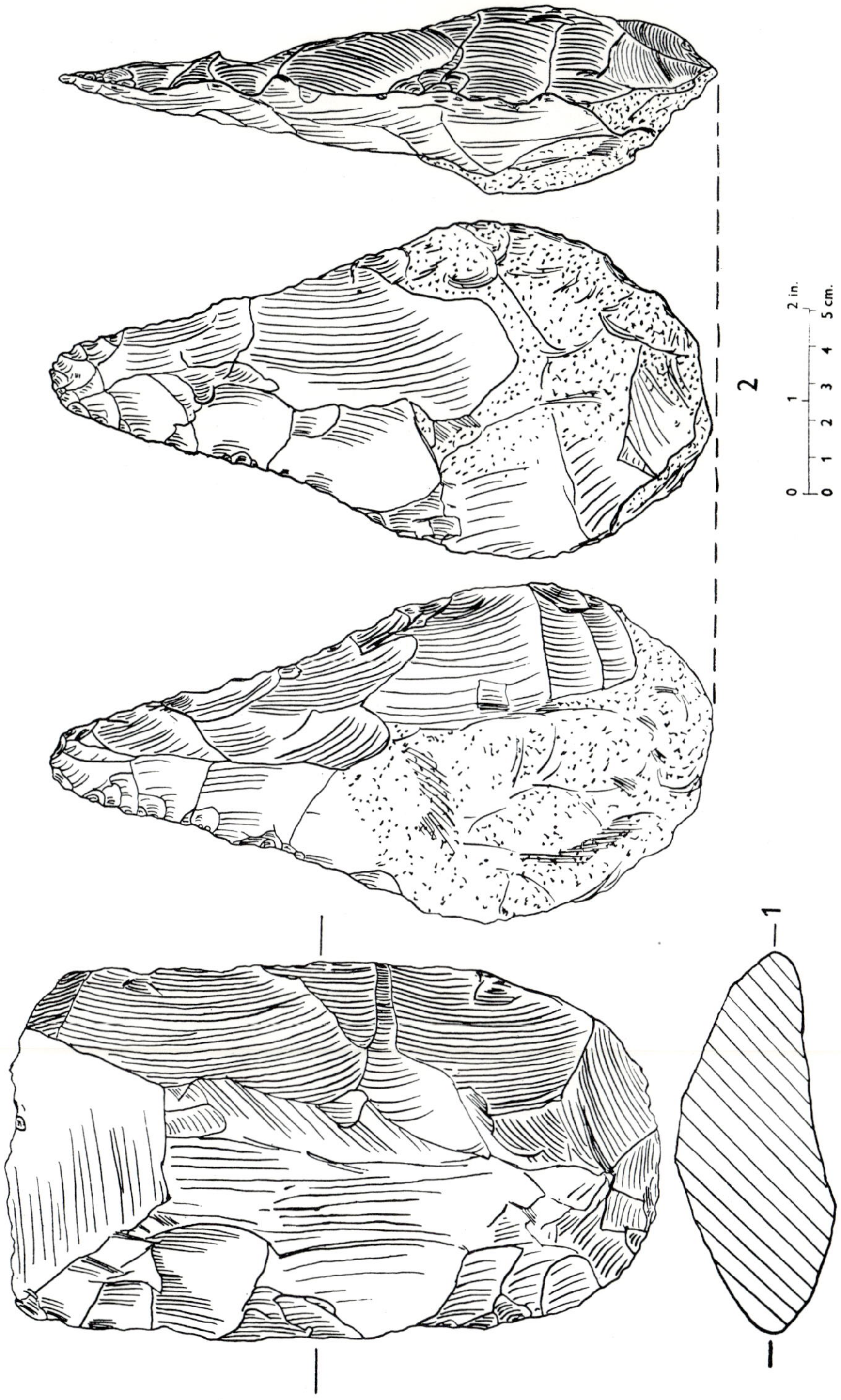

PLATE 5

Implements from Older Tug Gravels II at Hargeisa

Upper Acheulio-Levalloisian Culture

1. Subtriangular hand-axe with thick butt, made on a flake; quartzite, unrolled. Site H. 12.
2. Pointed pear-shaped hand-axe; quartzite, unrolled. Site H. 12.
3. Waste flake resulting from Acheul wood technique; quartzite, unrolled. Site H. 12.
4. Pear-shaped hand-axe; quartzite, slightly rolled. Site H. 12.
5. Ovate hand-axe; chert, unrolled. Site H. 12.
6. Ovate hand-axe; quartzite, slightly rolled. Site H. 12.
7. U-cleaver on side-flake, butt broken; quartzite, unrolled. Site H. 12.
8. Micoque-type hand-axe; quartzite, slightly rolled. Site H. 12.
9. 'Cleaver-flake' on an end-flake; quartzite, unrolled. Site H. 12.

PLATE 5

PLATE 6

Implements from Older Tug Gravels II at Hargeisa

Upper Acheulio-Levalloisian Culture

1. End-scraper on flake with faceted striking-platform; quartzite, unrolled. Site H. 12.

2. End-scraper on flake with faceted striking-platform; quartzite, unrolled. Site H. 12.

3. Broken hand-axe re-utilized as a core; chert, unrolled. Site H. 12.

4. Double-edged scraper on flake-blade with faceted striking-platform; quartzite, slightly rolled. Site H. 12.

5. Flake-blade with unfaceted striking-platform; quartzite, unrolled. Site H. 12.

6. Single-edged scraper on flake with faceted striking-platform; quartzite, unrolled. Site H. 12.

7. Flake-blade core with prepared striking-platform; chert, slightly rolled. Site H. 12.

8. Single-edged scraper on flake with faceted striking-platform; quartzite, slightly rolled. Site H. 12.

PLATE 6

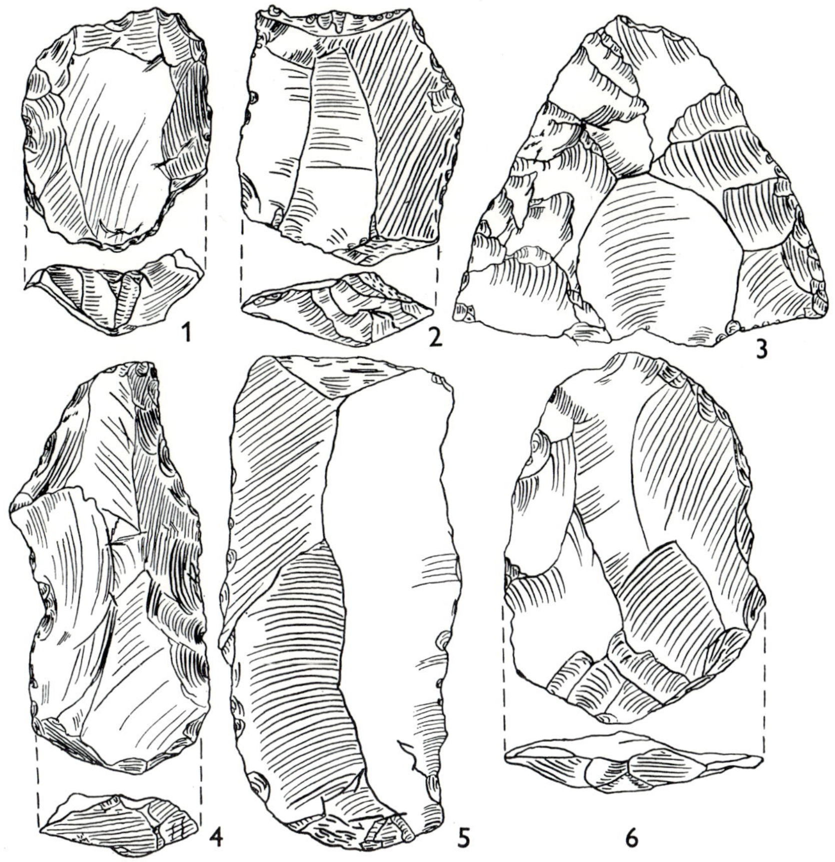

PLATE 7

Implements from Older Tug Gravels II at Hargeisa

Upper Acheulio-Levalloisian Culture

1. Small oval prepared core; quartzite, unrolled. Site H. 12.

2. Small prepared core approximating to the subtriangular in form; quartzite, unrolled. Site H. 12.

3. Small disc core; quartzite, slightly rolled. Site H. 12.

4. Primary flake, with crudely faceted striking-platform, struck from a prepared core; quartzite, unrolled. Site H. 12.

5. Discoidal chopper or core; quartzite, unrolled. Site H. 12.

6. Circular high-backed prepared core; quartzite, slightly rolled. Site H. 12.

PLATE 7

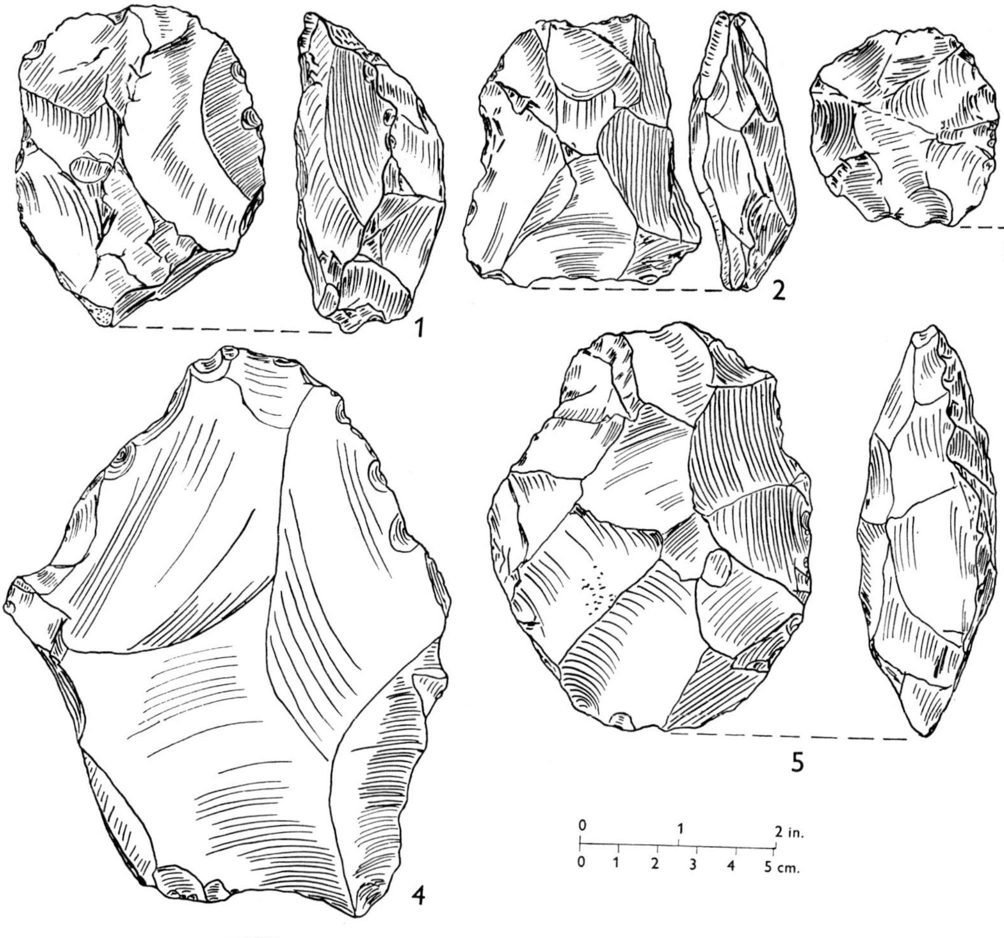

PLATE 8

Implements from Eil (Catalogue No. N.61. 1)

Acheulio-Levalloisian Culture (undifferentiated)

1. Thick, broad flake with crudely faceted striking-platform and all-round secondary retouch; ferruginous sandstone, weathered and with lime incrustation now being leached away by chemical weathering.

2. Thick, broad flake with inclined unfaceted striking-platform and all-round secondary retouch; ferruginous sandstone, weathered and similarly lime incrusted.

3. Small flake with faceted striking-platform worked to a single scraping edge; ferruginous sandstone, slightly weathered and with lime incrustation.

4. High-backed subtriangular prepared core; ferruginous sandstone, slightly weathered and lime incrusted.

5. Crude hand-axe; ferruginous sandstone, weathered and lime incrusted.

6. Elongated hand-axe on a side-flake, point missing; ferruginous sandstone, slightly weathered.

7. Flake-biade or long flake with unfaceted striking-platform worked to an end-scraper; ferruginous sandstone, weathered.

PLATE 8

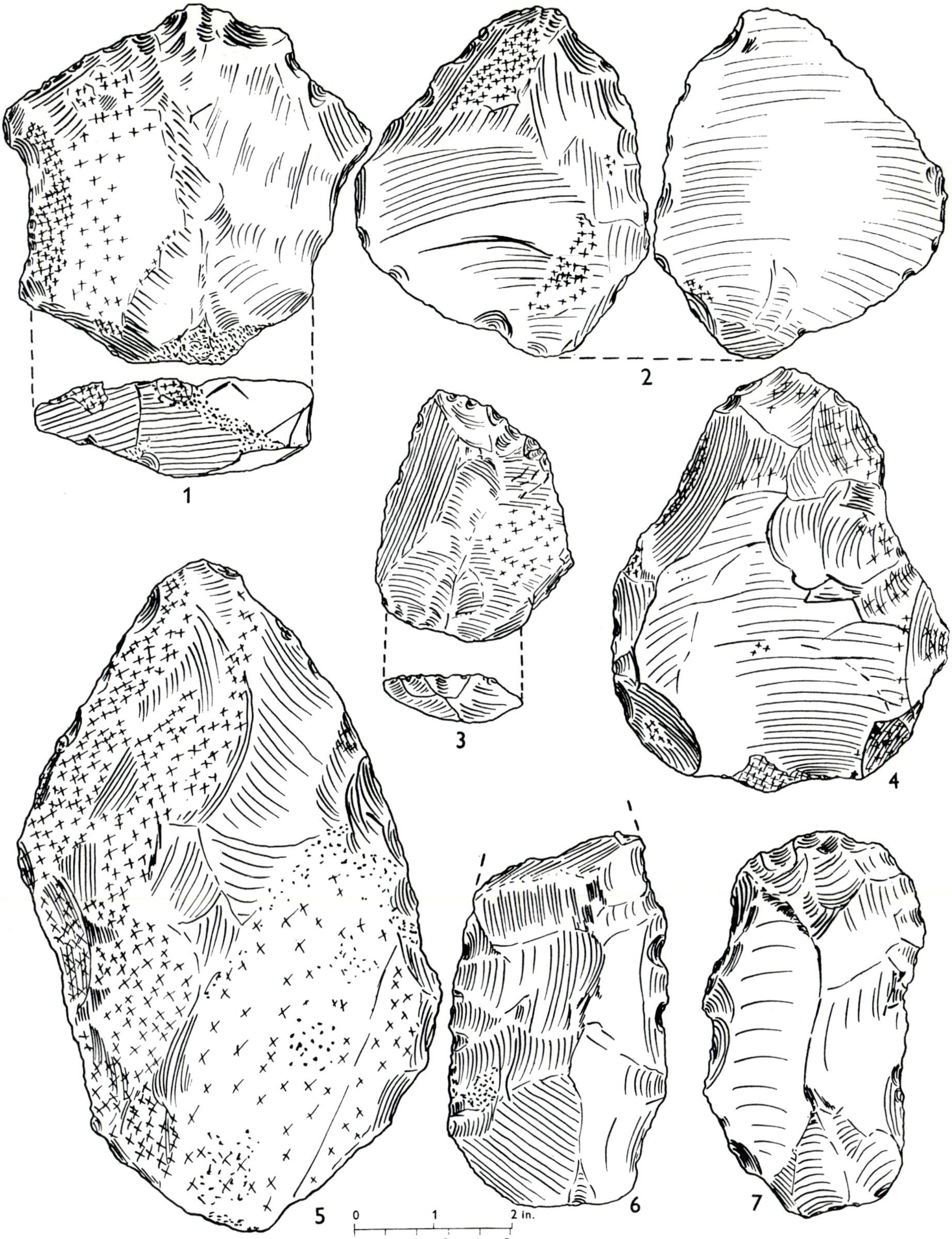

PLATE 9

Acheulio-Levalloisian (undifferentiated)

High-backed prepared core; ferruginous sandstone, slightly weathered and with lime incrustation. From Site N.61. 1 at Eil.

PLATE 9

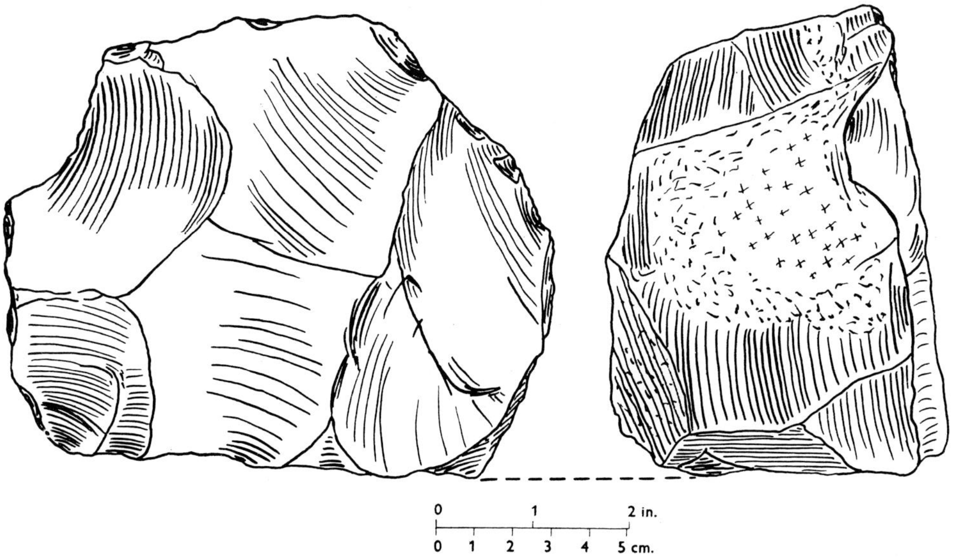

PLATE 10

Implements from Berbera (Dubar)

Acheulio-Levalloisian (undifferentiated)

Series I

1. Large and broad flake with crudely faceted striking-platform and all-round secondary retouch; quartzite, weathered.
2. Broad flake with unfaceted striking-platform and secondary retouch; quartzite, weathered.
3. Pear-shaped hand-axe; quartzite, weathered.
4. Large flattish prepared core; quartzite, weathered.
   Crude hand-axe, butt unworked; quartzite, weathered.

PLATE 10

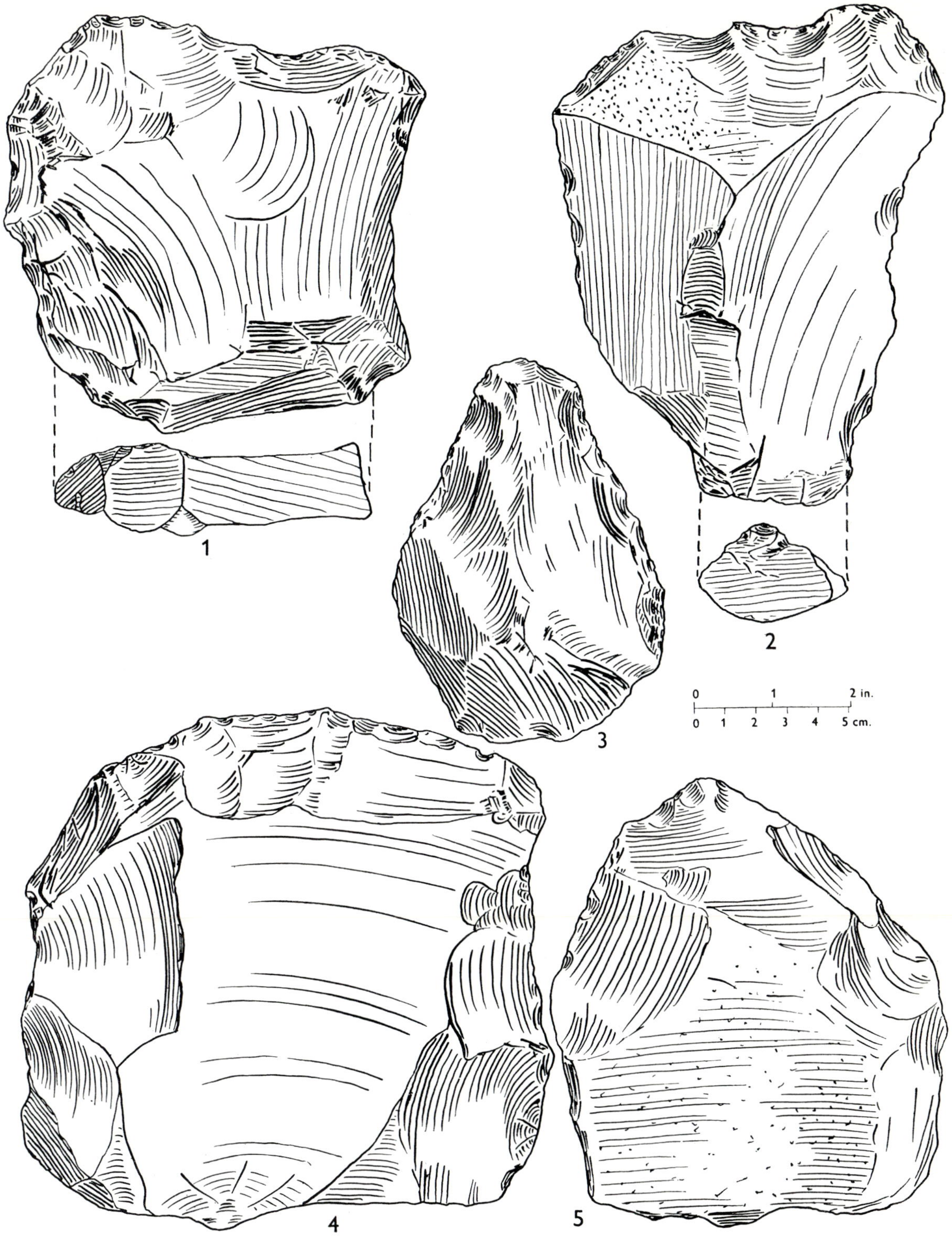

PLATE II

Implements from the Lower Tug Gravels at Garoe Tug

Lower Levalloisian

1. Prepared flake, bulb and platform removed, with notches and marginal retouch; chert, cream patina, rolled.

2. Broken faceted flake, with two opposed marginal notches forming a blunt *perçoir*; chert, cream patina, slightly rolled.

3. Short, broad, unfaceted flake with marginal retouch; chert, white patina, unrolled.

4. Flake-blade removed from double-ended core, showing marginal retouch; chert mottled patina, slightly rolled.

5. Notched flake; chert, buff patina, slightly rolled.

6. Double-ended, flat, subrectangular core; chert, white patina, rolled.

7. Subtriangular core with rounded back; chert, buff patina, slightly rolled.

8. Scraper with all-round marginal retouch; chert, buff patina, slightly rolled.

9. Unifaced flake-point on unprepared flake; chert, buff patina, slightly rolled.

10. Bifaced core chopper, or hand-axe; chert, white patina, slightly rolled.

11. Double-ended subrectangular core; chert, buff patina, slightly rolled.

PLATE II

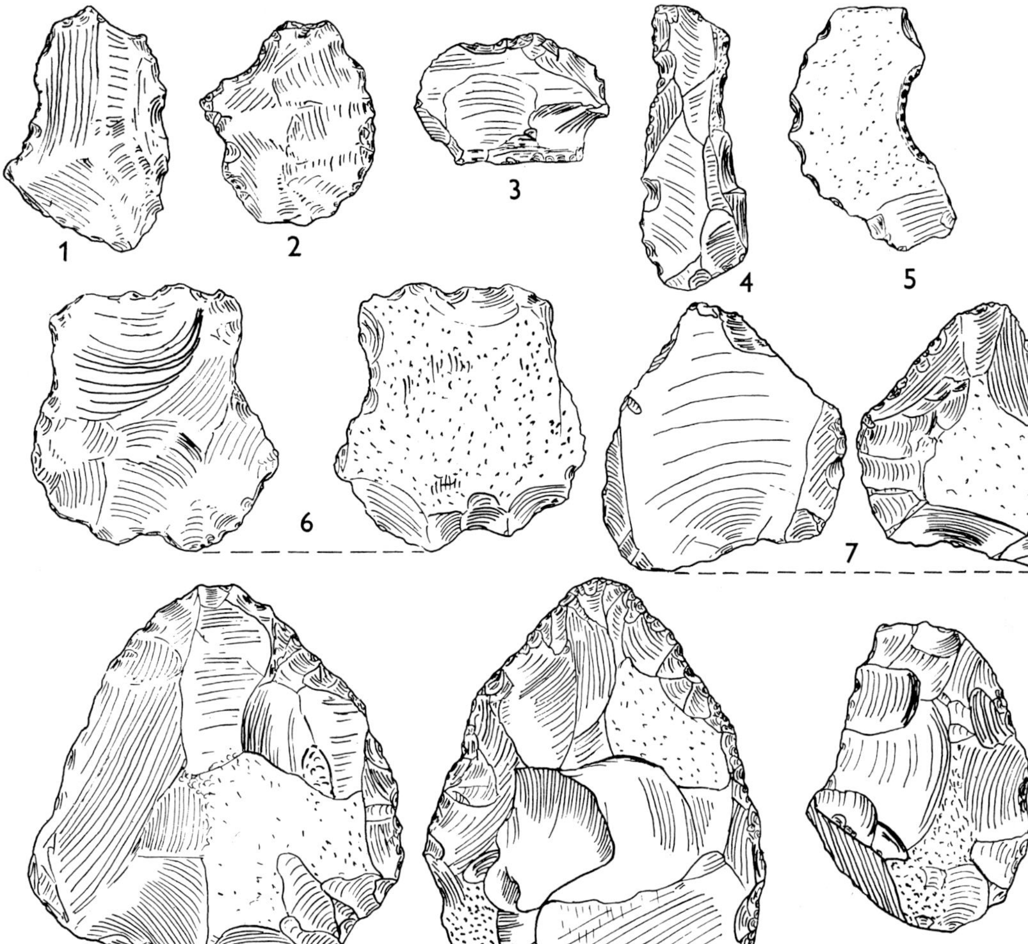

PLATE 12

Nos. 1–4. Implements from surface of the 50/100 ft. Terrace Gravels of the Webi Shebeli

Undifferentiated Levalloisian

1. Retouched fragment showing two opposed notches formed by steep retouch; chert, red patina, weathered and rolled. Sheik Gure.

2. Small single-ended subrectangular core on a pebble; chert, chestnut patina, weathered. Sheik Gure.

3. Flake detached from unprepared pebble core showing marginal retouch; chert?, red patina, probably rolled. Sheik Gure.

4. Small single-ended subrectangular core on a pebble; chert, brown patina, weathered. Cura Ba'ad.

Nos. 5, 9, 11–13. Implements from the Lower Ferruginous Sands at Jesomma, Somalia

Lower Levalloisian

5. Oval prepared core; siliceous sandstone, unrolled. Jesomma.

9. Subtriangular core with rounded back; siliceous sandstone, slightly abraded. Jesomma.

11. Trimming stone, or 'push-plane'; sandstone, weathered. Jesomma.

12. Crude unifaced point; siliceous sandstone, weathered. Jesomma.

13. Flake with plain striking-platform struck from prepared core; sandstone, ochreous patina. Jesomma.

Nos. 6–8, 10, 14. Implements from Lower Tug Gravels at Kabahan, Ogaden

Lower Levalloisian

6. Faceted flake-blade; chert, unrolled. Kabahan.

7. Faceted flake; siliceous sandstone, unrolled. Kabahan.

8. Faceted flake; siliceous sandstone, unrolled. Kabahan.

10. Double-ended unprepared core for the removal of flake-blades; chert, unrolled. Kabahan.

14. Crude core biface or chopper; ferruginous sandstone, rolled. Kabahan.

PLATE 12

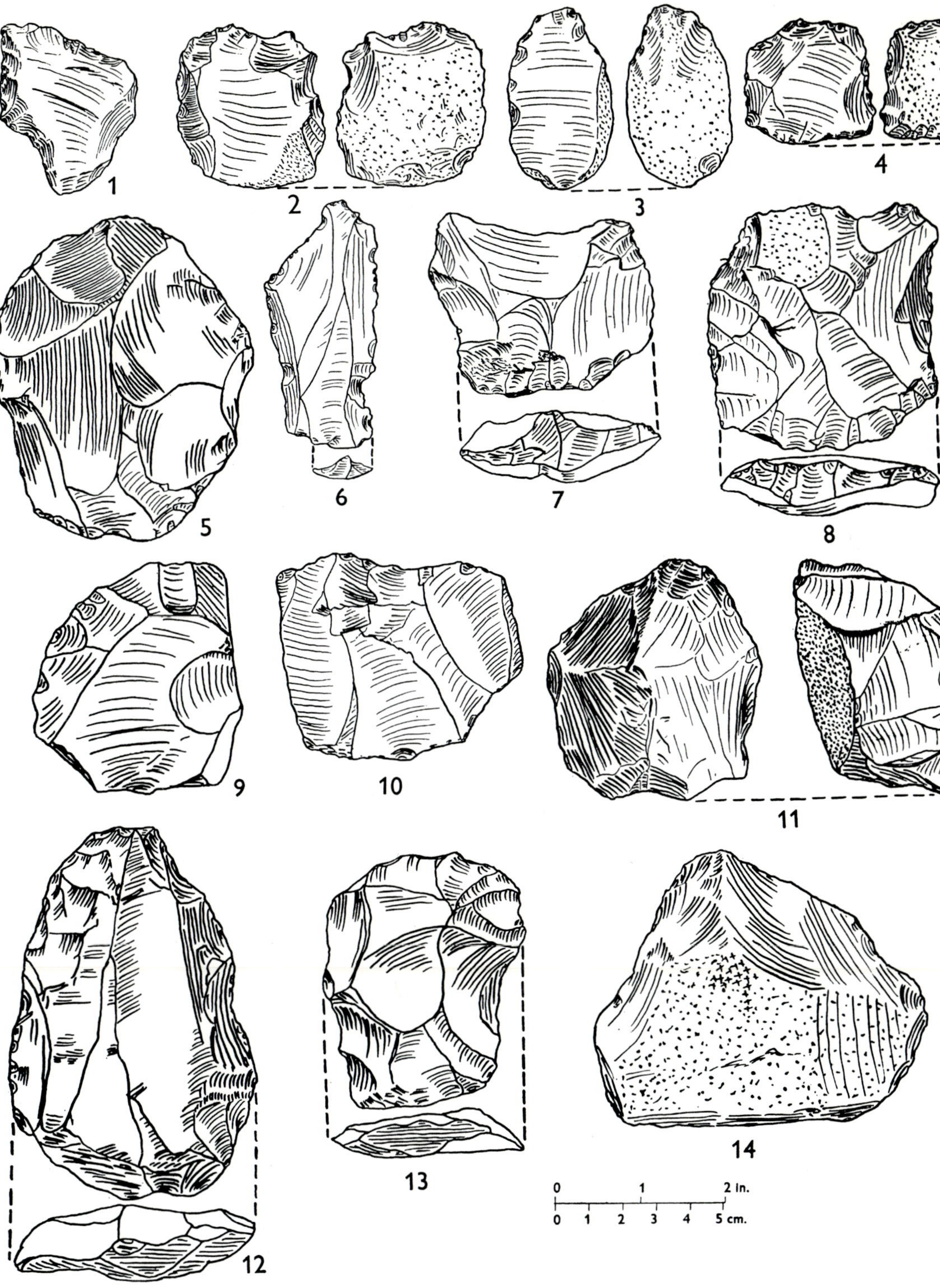

PLATE 13

Nos. 1–5, 7–14, 16–22. Implements from the Upper Tug Gravels at Garoe Tug, Somalia

Upper Levalloisian

1. Flake with right-angled striking-platform and converging edges at the butt; chert, buff patina, slightly rolled.

2. Faceted flake; chert, white patina, slightly rolled.

3. Flake with plain striking-platform, notched by marginal retouch; chert, white patina, slightly rolled.

4. Faceted flake-blade; chert, pale pink patina, unrolled.

5. Flake-blade with plain striking-platform and marginal retouch on the under-face; chert, buff patina, unrolled.

7. Side-scraper; chert, buff patina, rolled.

8. Pointed flake with plain striking-platform struck from prepared core; chert, buff patina, slightly rolled.

9. Faceted flake-blade (broken) with converging edges at butt; chert, buff patina, slightly rolled.

10. Flake (broken) showing converging edges at butt end and crude Tabelbalat reduction of the bulb and platform; chert, creamy patina, unrolled.

11. Faceted subrectangular flake; chert, cream patina, slightly rolled.

12. Proto-biconical pebble core; chert, buff patina, unrolled.

13. Flake (broken) worked to an end-scraper; chert, cream patina, slightly rolled.

14. Flake worked to an asymmetric unifaced point by steep marginal retouch; chert, buff patina, slightly rolled.

16. Subtriangular core with rounded back; chert, cream patina, slightly rolled.

17. Flat, subrectangular core; chert, cream patina, unrolled.

18. Subtriangular core with alternate flaking; chert, buff patina, unrolled.

19. Flat, double-ended, subrectangular core; chert, cream patina, slightly rolled.

20. Subtriangular prepared core; chert, pink patina, unrolled.

21. Proto-biconical pebble core with secondary marginal retouch; chert, buff patina, rolled.

22. Double-ended subrectangular core; chert, grey patina, unrolled.

Nos. 6, 15. Implements from below Upper Tug Alluvium at Donkukoq

Upper Levalloisian

6. Pointed faceted flake retouched at tip end to form a unifaced point; chert, unrolled, buff patina. Donkukoq.

15. Faceted flake; chert, buff patina, unrolled. Donkukoq.

PLATE 13

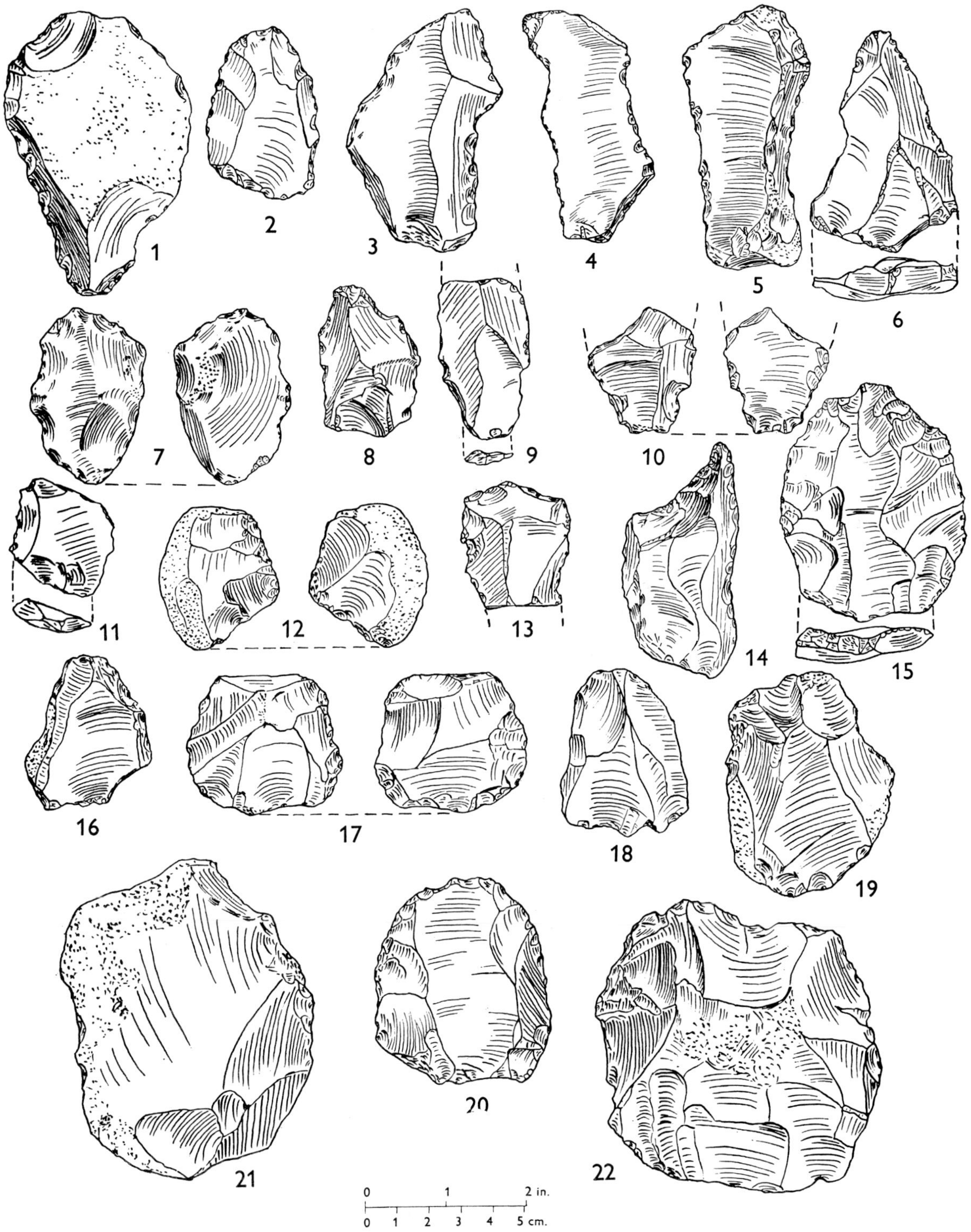

PLATE 14

Implements from the Upper Ferruginous Sands at Jesomma

Upper Levalloisian

1, 2. Unifaced, or partially bifaced foliate points; siliceous sandstone, No. 2 slightly weathered.

3. Faceted flake; siliceous sandstone, weathered.

4. End-scraper; sandstone, fresh.

5. Double-ended end-scraper; sandstone, fresh.

6. End-scraper on faceted flake-blade and marginal retouch on the underside of the right-hand edge; sandstone, fresh.

7. Faceted flake; sandstone, fresh.

8. Small subrectangular core; siliceous sandstone, fresh.

9. Disc core, or discoidal scraper; siliceous sandstone.

10. Subtriangular unifaced point; siliceous sandstone, fresh.

11. Side-scraper; silcrete, fresh.

12. Disc core; siliceous sandstone, fresh.

13. Broken, bifacially flaked, foliate point; siliceous sandstone, slightly weathered.

14. Disc core; ironstone, fresh.

15. Trimming stone, push-plane with two working edges at right angles to each other; sandstone, fresh.

16. Double-ended subrectangular prepared core; quartzite fresh.

17. Small high-backed disc core; siliceous sandstone, weathered.

PLATE 14

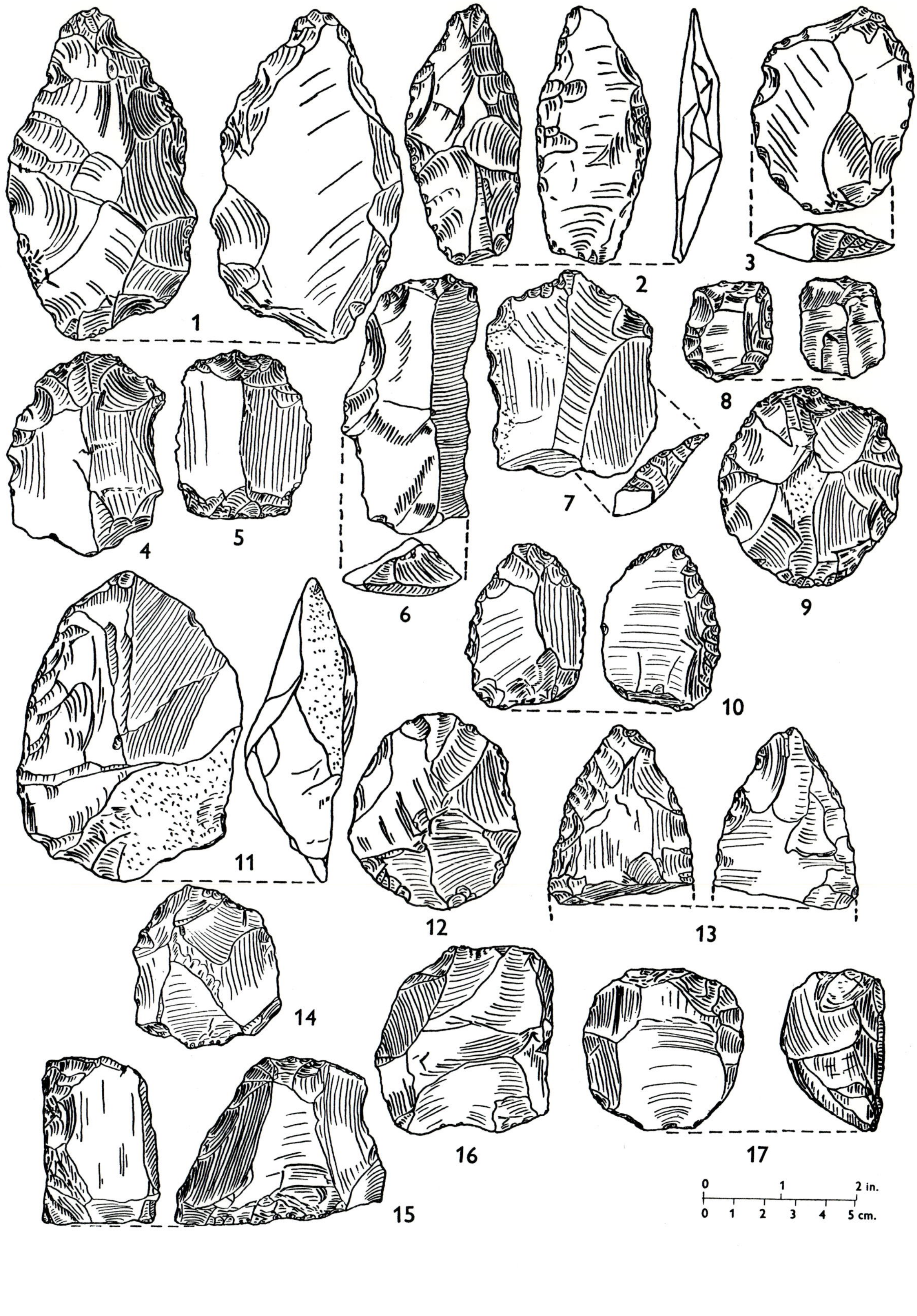

PLATE 15

Nos. 1–11. Implements from the surface of the 200 m. Conglomerates at Dubar, Berbera

Undifferentiated Levalloisian

1. Trimming stone, or push-plane; quartzite, weathered.

2. Bifaced core chopper or hand-axe; quartzite, weathered.

3. Flake with crude Tabelbalat retouch; quartzite, weathered.

4. Flat subrectangular prepared core; quartzite, weathered.

5. Retouched fragment with notches and steep marginal retouch; quartzite, weathered.

6. Ditto, with two opposed notches to form a blunt *perçoir*; quartzite, weathered.

7. Small high-backed disc core, unstruck; quartzite, weathered.

8–10. Faceted flakes; quartzite, weathered.

11. Diminutive retouched fragment with steep marginal retouch; quartzite, weathered.

Nos. 12–19. Implements from Eluvial Gravel at Garasleh Tug

Undifferentiated Levalloisian

12–13. Faceted flakes; lava, or indurated shale, weathered.

14. Side-scraper with steep marginal retouch; material unidentified, slightly weathered.

15. Flake-blade core; lava weathered.

16. Flake-blade with faceted striking-platform; lava?, weathered.

17. Faceted flake; lava?, weathered.

18. Prepared disc core; material unidentified, weathered.

19. Bifacially flaked chopper or hand-axe; baked shale?, slightly weathered.

PLATE 15

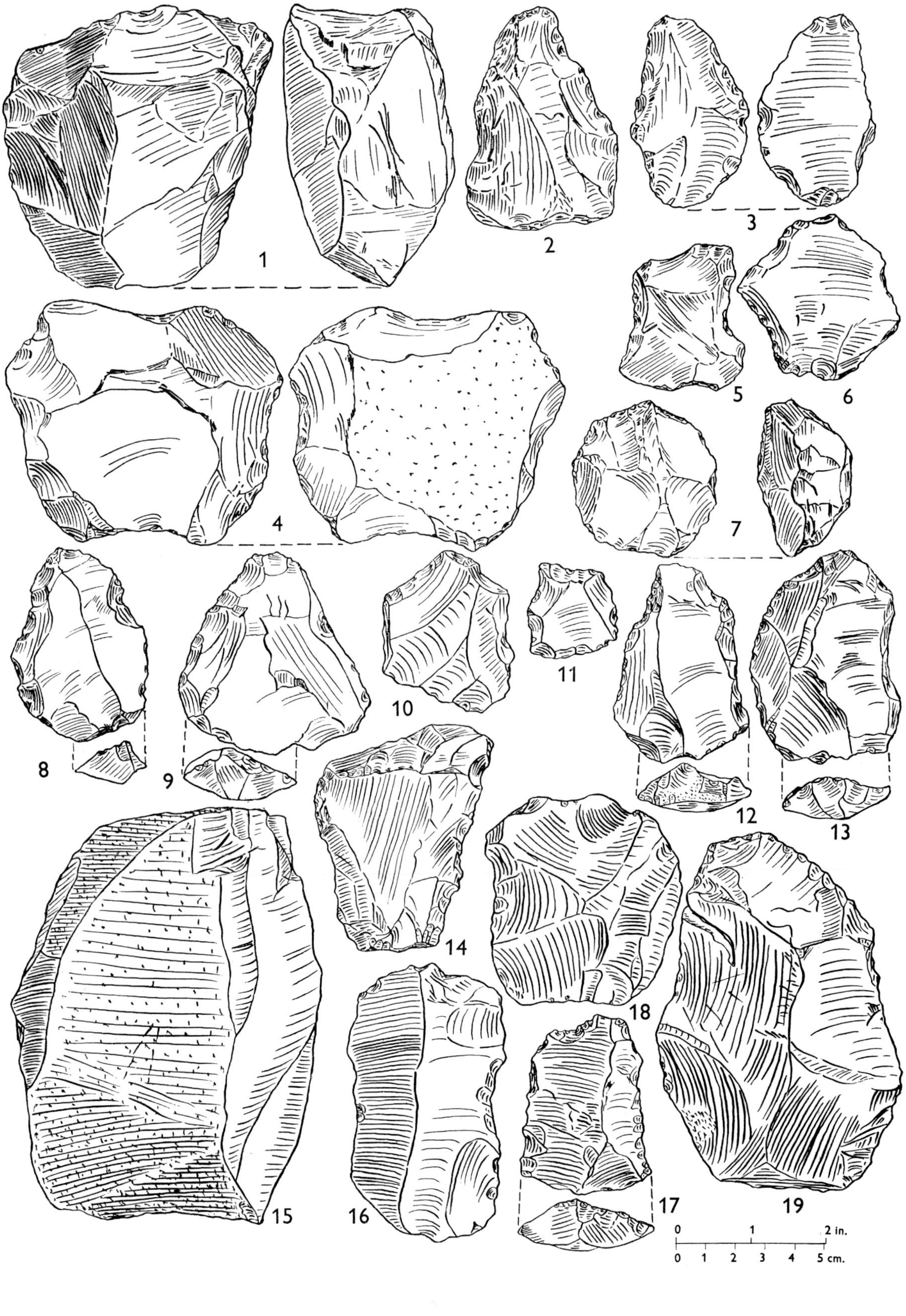

PLATE 16

Implements eroding from cemented scree deposits at Las Anod

Undifferentiated Levalloisian

1. Double-ended prepared core for the removal of flake-blades; chert, pink patina, fresh.

2. Unifaced point on a faceted flake; chert, red patina, weathered.

3. Bifacial core tool or hand-axe; chert, pinkish buff patina, fresh.

4. Bifacially flaked foliate point; chert, red patina, slightly weathered.

5. Unifaced point on a faceted flake and converging edges at butt end; chert, pinkish buff patina, slightly weathered.

6. Flake-blade with plain striking-platform; chert, pinkish buff patina, fresh.

7. Flake from prepared core showing Tabelbalat reduction and converging edges at the butt; chert, pinkish buff patina, slightly weathered.

PLATE 16

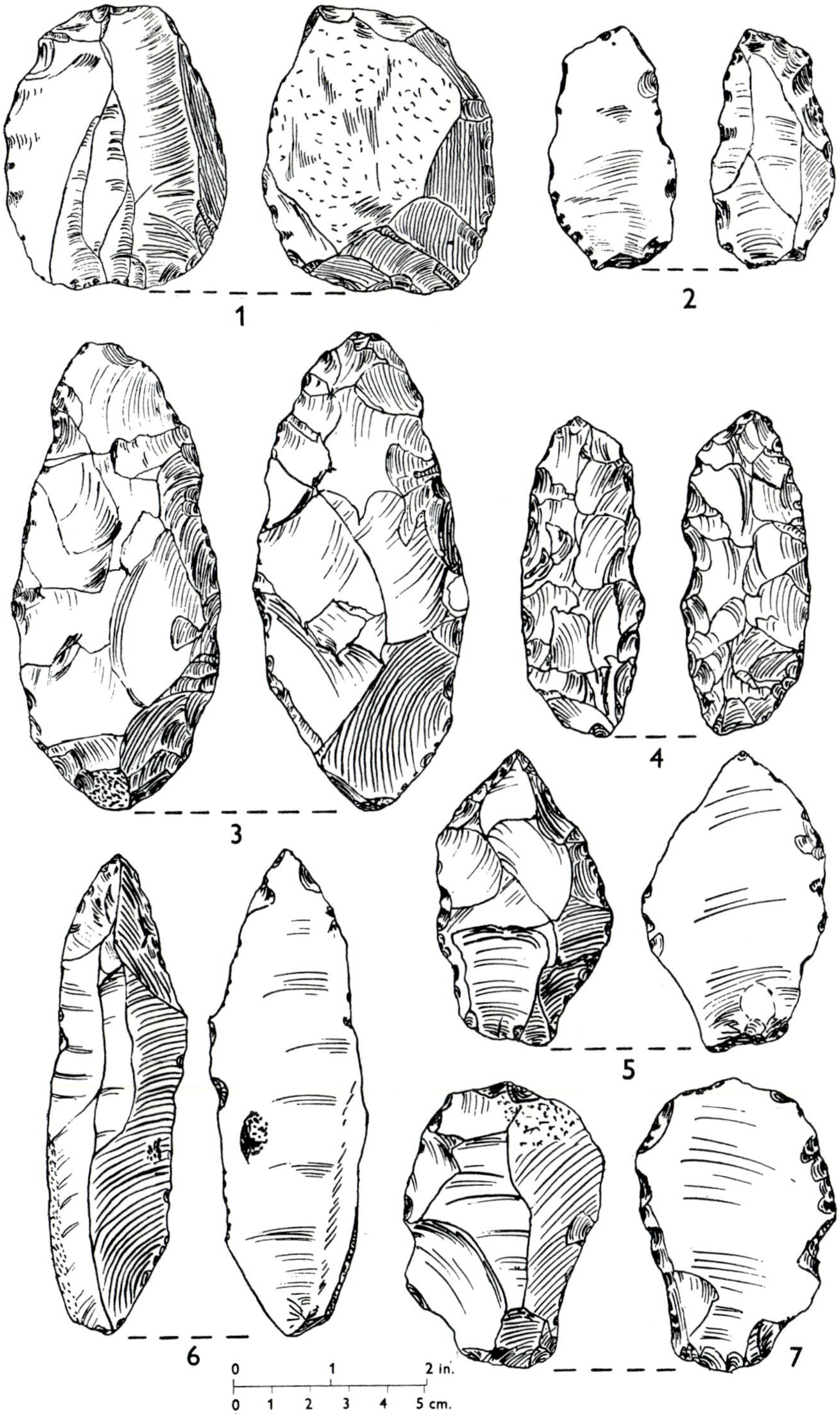

PLATE 17

Implements from Red Sandy Alluvium overlying Younger Tug Gravels at Site H.7 at Hargeisa

Lower Somaliland Stillbay; Hargeisa

1. Foliate bifaced point with plano-convex section.

2. Subtriangular parti-bifaced point with lenticular section.

3. Small foliate bifaced point with bi-convex section.

4. Subtriangular bifaced point with straight faceted butt.

5. Asymmetric and serrated unifaced point with centrally placed faceted butt.

6. Subtriangular unifaced point on faceted flake; secondary work is confined to the edges.

7. Unifaced foliate point with plain striking-platform and plano-convex section.

8. Parti-bifaced subtriangular to foliate point with thin and flat plano-convex section.

9. Crudely backed flake with straight serrated cutting edge; bulb and striking-platform removed.

10. Obliquely backed flake with straight, centrally placed, faceted striking-platform.

11. Small pyriform core biface flaked by percussion only; thickish bi-convex section.

12. Obliquely truncated flake, the point worked to an awl; centrally placed faceted striking-platform.

13. Small high-backed disc core; struck.

14. Flattish, radially flaked disc core; unstruck.

15. Small subtriangular prepared core; struck.

16. Subtriangular prepared core; (struck) with single-blow burin facet.

17. End-scraper on faceted flake.

18. Round scraper on faceted flake.

19. Side scraper on end-flake; convex working edge executed by resolved retouch.

All specimens are in chert (except No. 2 in silcrete), are unrolled, and show a matt, sometimes a glossy, white patina.

PLATE 17

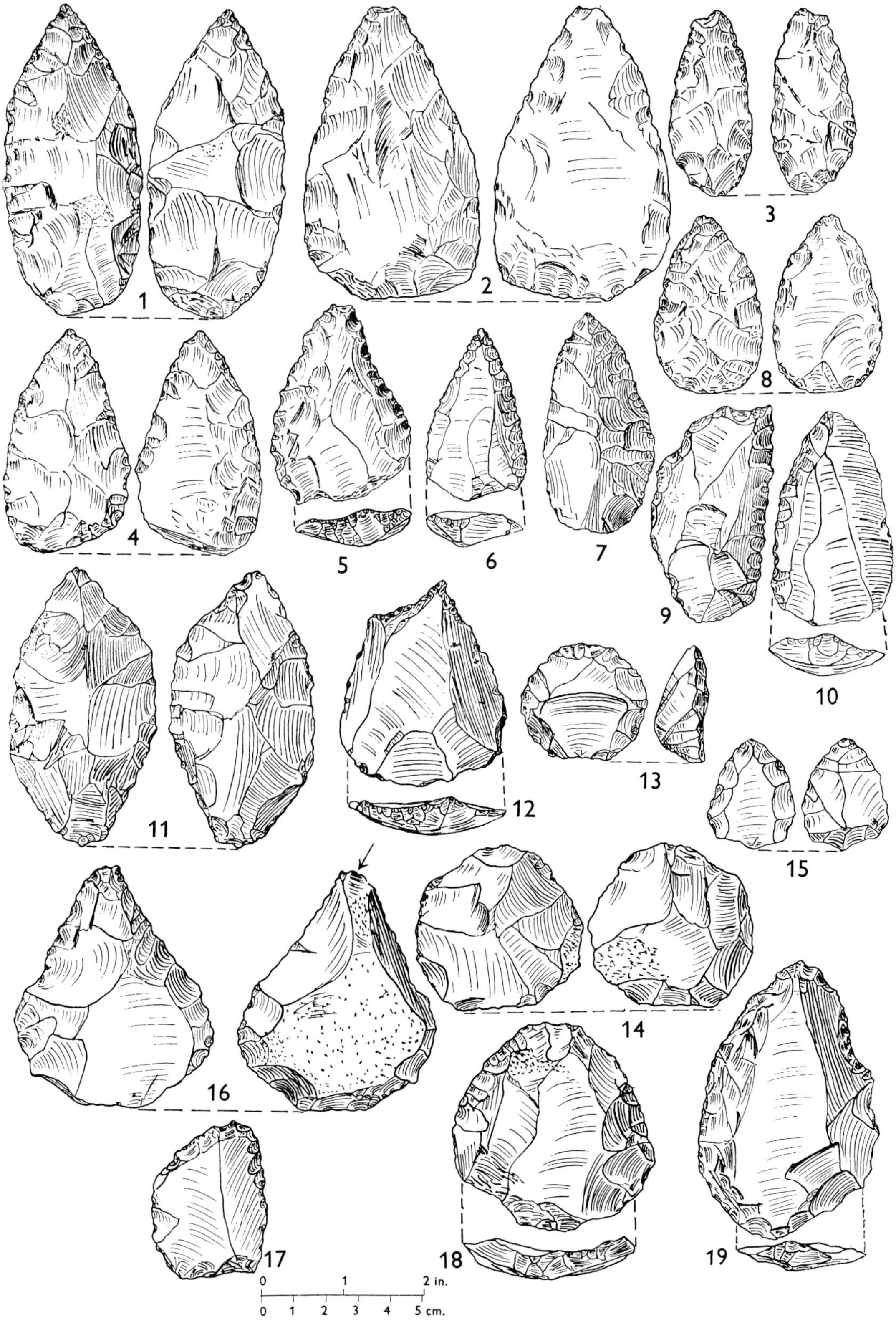

PLATE 18

A. Implements from lower levels of 'ironcrete' beds at Bur Yassin, Bur Eibe, Somalia

Somaliland Stillbay (Undifferentiated)

1. Subtriangular parti-bifaced point with faceted butt and thin lenticular section (4·5 mm. thick). Fresh, ochreous patina; chert.

3. Small subtriangular parti-bifaced point. Fresh, ochreous patina; chert.

4. Diminutive subtriangular core biface. Fresh, ochreous patina; chert.

5. Unifaced foliate, or limande, point. Weathered, ochreous patina; chert.

7. Backed flake with retouch down the cutting edge (broken). On a faceted flake with obliquely placed striking-platform. Fresh, ochreous patina; chert.

8. Small subtriangular unifaced point. Fresh, buff patina; chert.

9. Small asymmetric unifaced point. Fresh, buff patina; chert.

10. Oval unifaced point or disc. Fresh, buff patina; chert.

11. Small straight-backed blade. Fresh, buff patina; chert.

12. Redirecting flake used as a backed blade. Fresh, buff patina; chert.

13. Side-scraper on side-flake with faceted striking-platform. Fresh, cream patina; chert.

14. Small round scraper on faceted flake. Fresh, buff patina; chert.

15. Small disc core—struck. Fresh, buff patina; chert.

16. Double side-scraper on end-flake. Fresh, buff patina; chert.

17. Small end or round scraper on faceted flake. Fresh, buff patina; chert.

B. Implement from top of bedded yellow sands under 'ironcrete' at Jesomma Wells, near Bulo Burti, Somalia

2. Subtriangular bifaced point. Fresh, ochreous patina; silcrete.

C. Implement from eroded surface of Younger Calcretes at Ohole, near Ferfer, Somalia

6. Unifaced point with obliquely placed faceted striking-platform and plano-convex section. Fresh, unpatinated; silcrete.

D. Implements from Red Sandy Alluvium, Hargeisa, British Somaliland

Upper Somaliland Stillbay

18. Unifaced point with basal retouch and triangular section. Fresh, white patina; chert. Site H. 5.

19. Bifaced point with lenticular section. Fresh, white patina; chert. Site H. 4

20. Bifaced point with lenticular section. Fresh, unpatinated; silcrete. Site H. 5.

21. End- and side-scraper on end-flake. Fresh, white patina; chert. Site H. 5.

22. Backed flake with steep oblique retouch, trimmed on the opposite edge to form a unifaced point. Centrally placed faceted striking-platform. Fresh, white patina; chert. Site H. 1.

23. Flake-blade with oblique retouch. Fresh, white patina; chert. Site H. 3.

24. Unifaced point with centrally placed faceted striking-platform. Fresh, buff patina; chert. Site H. 5.

25. Small oval to subtriangular prepared core —struck. Fresh, white patina; chert. Site H. 5.

26. Prepared disc core—struck. Fresh, white to buff patina; chert. Site H. 1.

PLATE 18

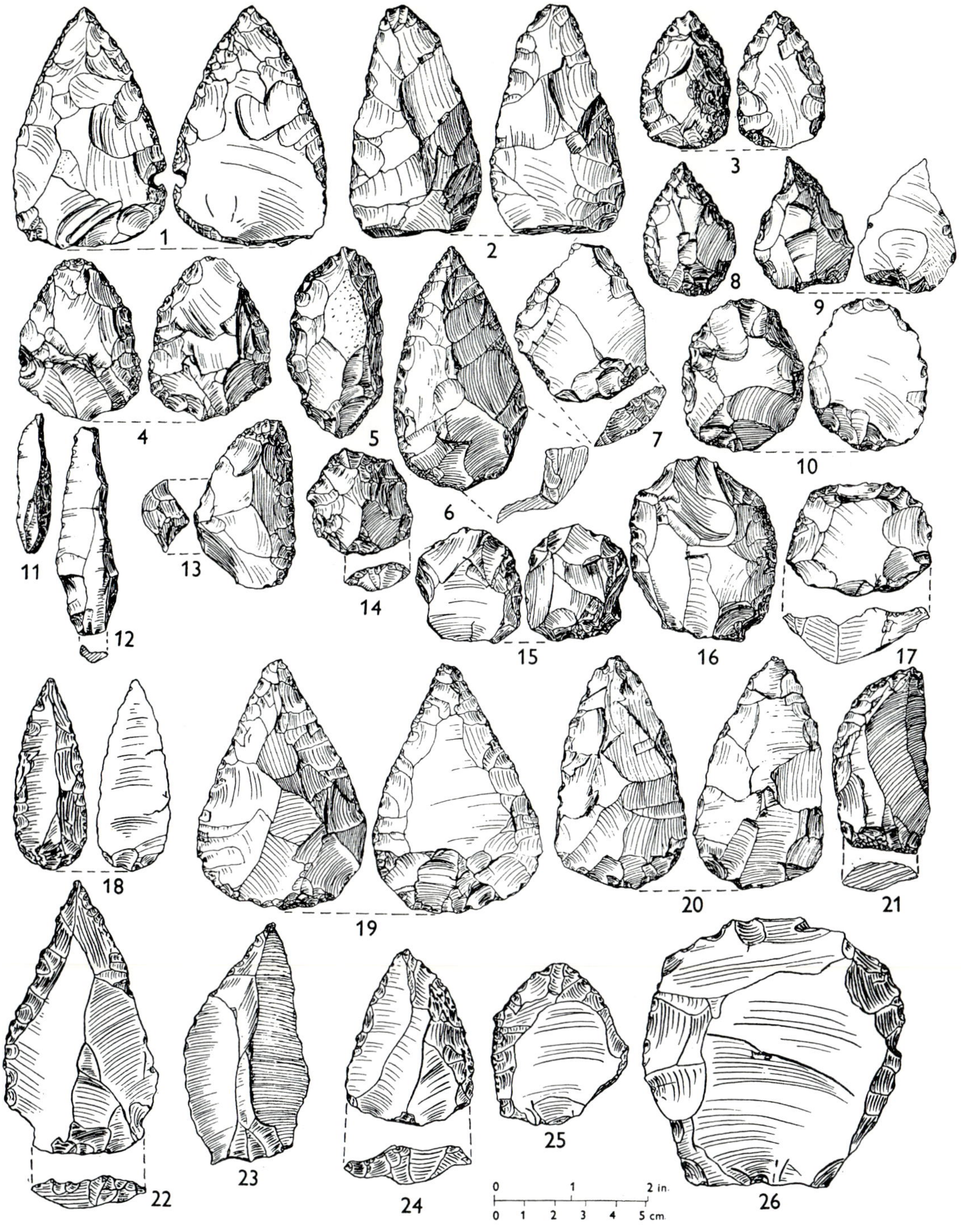
1
2
3
4
5
6
7
8
9
10
11
12
13
14
15
16
17
18
19
20
21
22
23
24
25
26
0 1 2 in.
0 1 2 3 4 5 cm

PLATE 19

Implements from Red Sandy Alluvium at Hargeisa, British Somaliland

Upper Somaliland Stillbay

1. Unifaced foliate point with plano-convex section. Fresh, white patina; chert. Site H. 5.

2, 3. Bifaced points with lenticular section. Fresh, white patina; silcrete. Site H. 4.

4. Bifaced point with notch at point. Fresh, buff patina; chert. Site H. 4.

5. Side-scraper on flake-blade. Fresh, white patina; chert. Site H. 5.

8. Flake-blade with serrated edge. Fresh, white patina; chert. Site H. 4.

Implements from Calcretes overlying Upper Tug Alluvium at Eil, Lower Nogal Valley

Somaliland Stillbay (Undifferentiated)

6. Side-scraper on side-flake. Fresh, white patina; chert.

7. End-scraper on end-flake. Fresh, white patina; chert.

PLATE 19

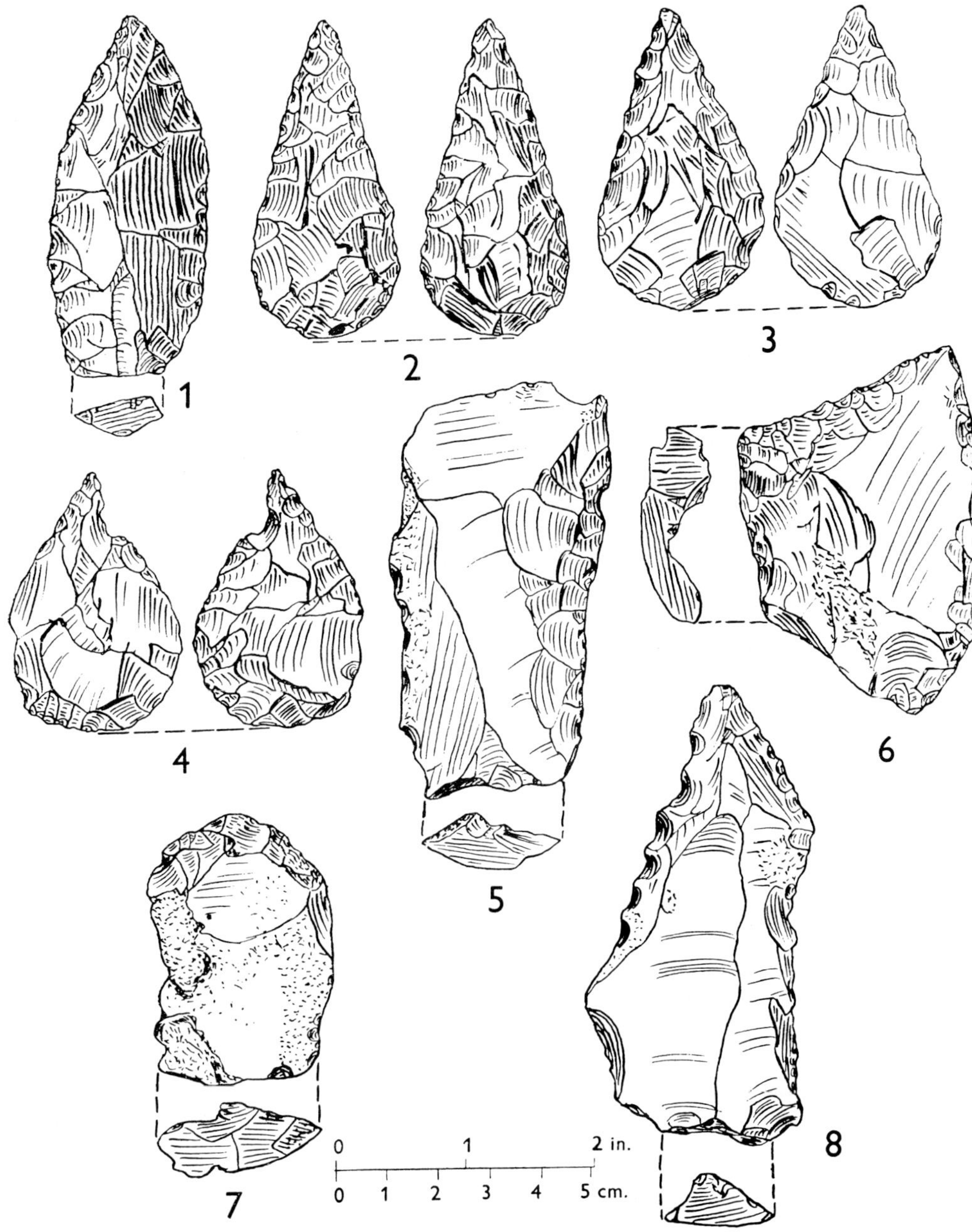

PLATE 20

Implements from Red Sandy Alluvium at Hargeisa

Hargeisan Culture

1–3, 7. Backed blades.

4, 5, 6. Backed blades with basal retouch on opposite edge.

8, 9. Double-backed blades.

10. Disc.

11. Lenticular bifaced point.

12. Transverse concave angle-burin.

13. Transverse straight angle-burin.

14. Double-ended polyhedral gouge-burin.

17. Unifaced point.

19. Bifaced triangle.

20. Blade core.

22. Oval scraper.

23. Double end-scraper on blade.

24, 25. End-scrapers.

All specimens in chert with white to pink patina, except No. 19 in silcrete. Nos. 1, 3–7, 9, 12, 20, 23–25 from Site H. 12R; No. 17 from Site H. 3; Nos. 2, 8, 10, 11, 13, 14, 18, 22 from Site H. 1.

Implements from Red Sandy Alluvium (Site N.u. 25) at Mandera North, British Somaliland

Hargeisan Culture

15. Utilized blade worked on both edges.

16, 18. Unifaced points.

21. Multi-platformed, or pseudo-disc, core.

All in chert with white patina.

Implements from Red Sandy Alluvium (Site H. 2) at Hargeisa

Hargeisan Derivative Industry

26. Triangular bifaced point.

27–29. Hollow-based bifaced points.

30. Unifaced point.

31, 32. Microlithic backed blades.

34. End-scraper.

35. Diminutive round scraper.

37. *Bec-de-flute* burin.

38. Rim sherd of light-brown well-fired pottery; 8 mm. thick at the rim. Edges are worn and the sherd is encrusted with a thin deposit of lime. Simple rolled rim.

All specimens in chert with white patina, except No. 26 in silcrete, and No. 30 in green/red mottled patina.

Implements from Eil, Lower Nogal Gorge

Hargeisan Derivative Industry

33. Blade utilized down both edges.

36. Burin on utilized blade with scraping edge.

Material chert with white patina.

Implements from sub-aerial deposits immediately overlying the ± 12 m. raised beach at Biede, Nogal Estuary Mouth, Somalia

Somaliland Stillbay

39. Diminutive hand-axe or bifaced point.

40. Flat disc core.

41. End-scraper on flake.

42. Utilized flake with centrally placed faceted striking-platform; probably a derived Levalloisian flake re-utilized in Stillbay times. Weathered.

All specimens in chert with white patina.

PLATE 20

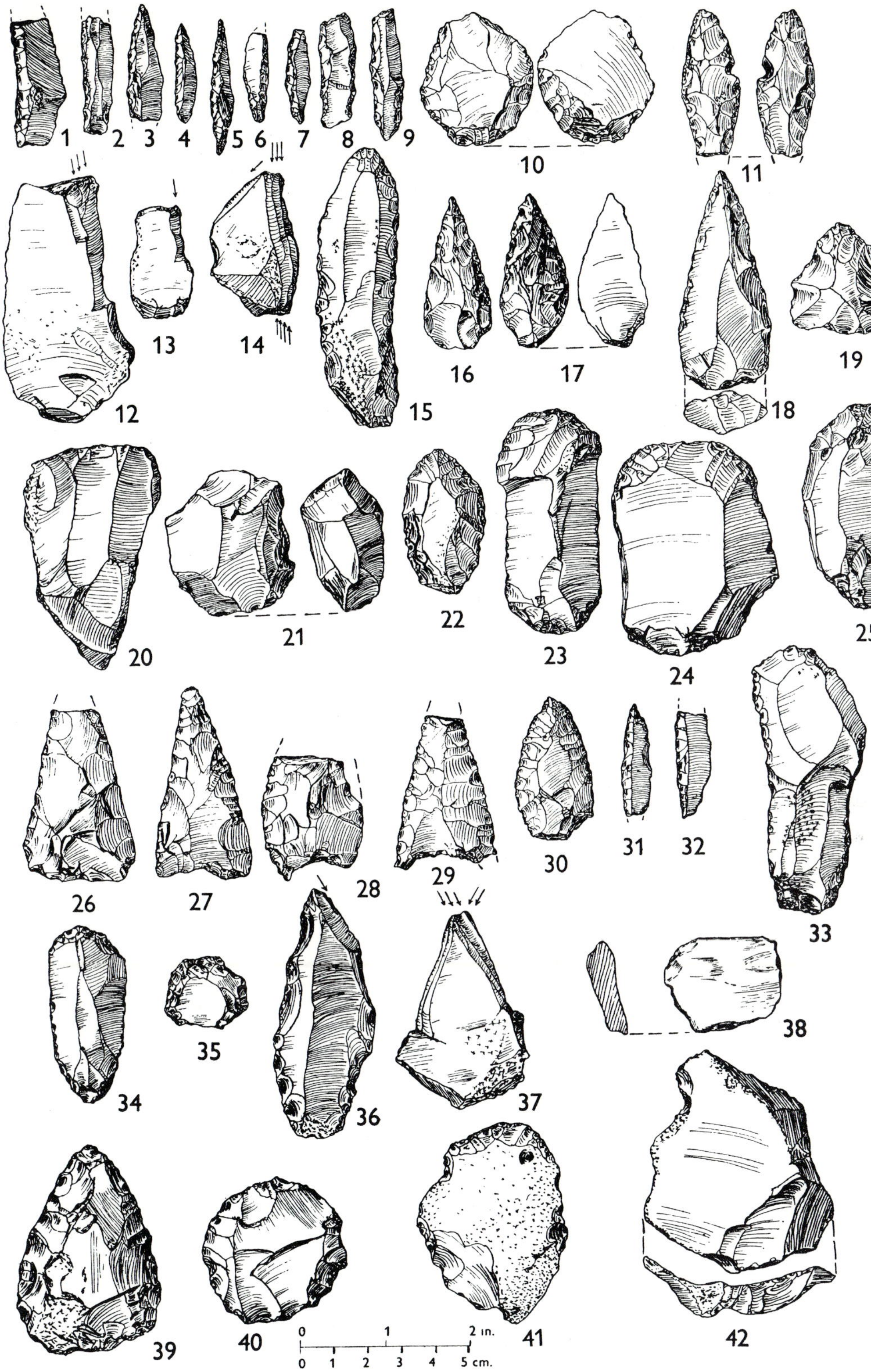

1
2
3
4
5
6
7
8
9
10
11
12
13
14
15
16
17
18
19
20
21
22
23
24
25
26
27
28
29
30
31
32
33
34
35
36
37
38
39
40
41
42
0 1 2 in.
0 1 2 3 4 5 cm.

PLATE 21

Implements from upper levels of superficial red sand at Alablah Balleh, Central Somalia. Glover Collection

Magosio/Doian Transitional Industry

1, 2. Bifaced points.

3. Lanceolate unifaced point.

4. Micro-blade core.

5. Diminutive flat disc-core.

6. Small disc-scraper.

7. Short end-scraper.

8. Double-ended polyhedral gouge-burin (?).

9–13. Backed bladelets.

14. Subtriangular prepared core.

All implements in chert. Fresh, sometimes with glossy patina.

Implements from eroded terrace alluvium at Hormo, north-eastern British Somaliland. Barrington Brown Collection

Hargeisan Culture

15–19. Straight-backed blades and bladelets.

20–21. Straight-backed blades with retouch on cutting-edge at one end.

22–24. Single end-scrapers on blades.

25. Alternate-ended blade core.

26–28. Double end-scrapers on blades.

29. Subtriangular core for the production of blades.

All implements in chert. White patina, some glossy.

Implements from Hobat, Hegligab District, north-eastern Somaliland. Barrington Brown Collection

30. Transverse-concave angle-burin.

31, 32. Utilized blades.

33. Blade core.

All implements in chert. Fresh, with reddish brown patina.

PLATE 21

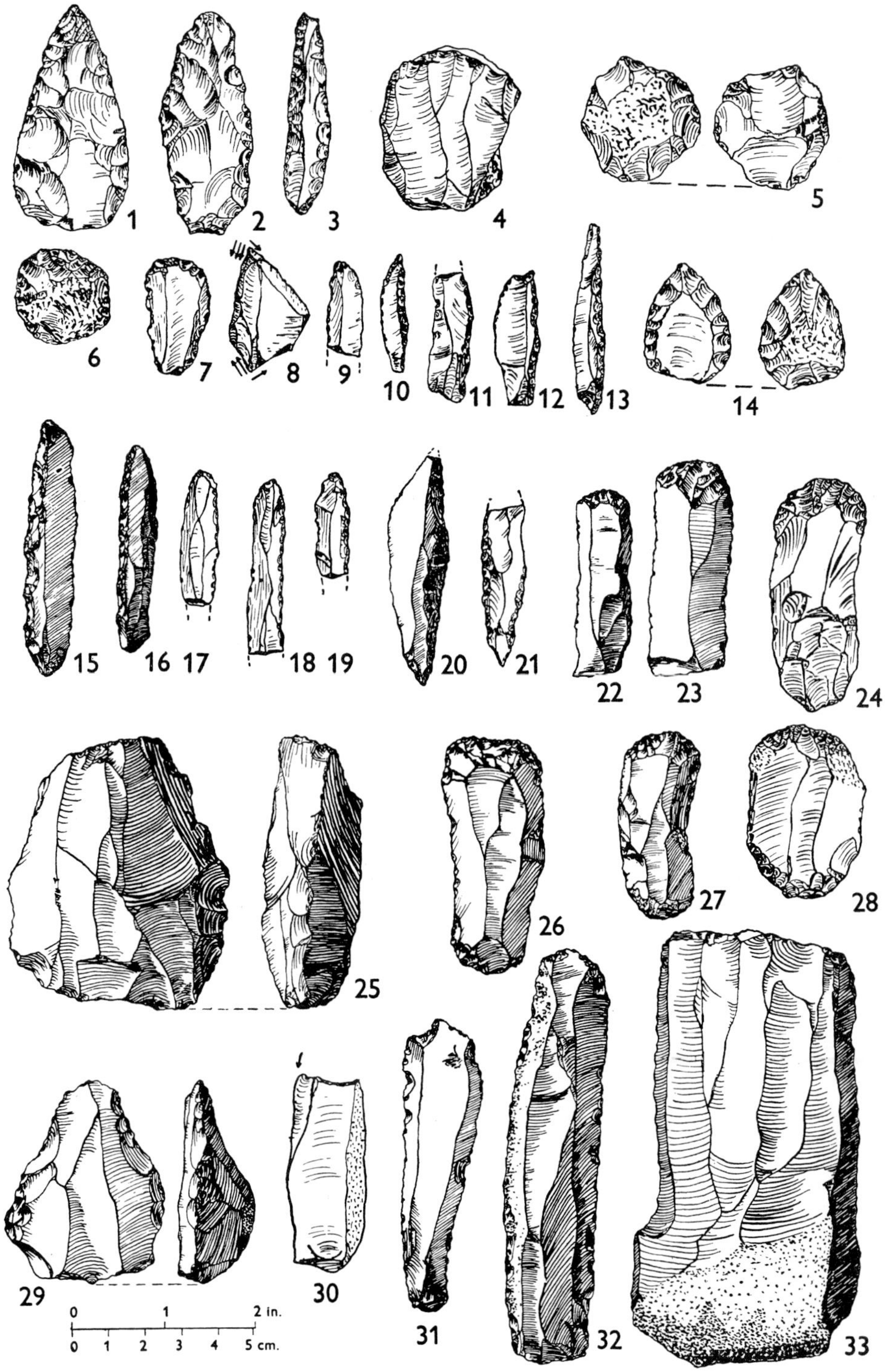
1
2
3
4
5
6
7
8
9
10
11
12
13
14
15
16
17
18
19
20
21
22
23
24
25
26
27
28
29
30
31
32
33
0 1 2 in.
0 1 2 3 4 5 cm.

PLATE 22

Implements from 'ironcrete' beds at Jesomma Wells, southern Somalia

Magosio/Doian Transitional Industry

1, 2, 5–10. Straight-backed microliths.

3, 4. Curved-backed microliths.

(Nos. 1, 5, 6, 8, 9, backing worked from both sides; 2, 3, 4, 7, 10, backing from under-face only; 2, 8, 10, cutting edge retouched at point.)

11–13. Crescents—cutting edge shows signs of utilization on Nos. 12 and 13.

14. Eared crescent (broken).

15. Bifaced foliate point with lenticular section.

16–21. Unifaced points—No. 20 shows reduction of bulb and platform, remainder square-butted.

22. Triangular bifaced point.

23, 24. Hollow-based arrowheads—bifacial and parti-bifacial.

25. Bifaced foliate point with serrated edges.

26. Unifaced point, or limace, with triangular section.

27. Blade, or rod, worked down the length of both edges.

28. Lanceolate bifaced point with triangular section.

29, 30. Flat disc cores, or discs.

31. Utilized flake-blade with plain striking-platform.

32. Side- and end-scraper on end-flake with faceted striking-platform.

33. High-backed 'slug' type of artifact.

Nos. 1, 3, 8, 17, 22, 23, 26 in silcrete; remainder chert, unpatinated.

Implements from 'ironcrete' beds at Bur Yassin, Bur Eibe, southern Somalia

Magosian Culture

34–41. Foliate and subtriangular bifaced points.

42. Triangular bifaced point.

43, 45. Asymmetric triangular bifaced points.

44. Unifaced point with basal retouch.

46, 47. Large tapered and eared crescents; backing is from under-face except at the 'ear', where it is directed from both sides.

48. Backed flake with faceted striking-platform.

49. Trapeze.

50. Curved backed microlith with rudimentary Heluan retouch.

51, 57, 58. Diminutive fan and thumbnail scrapers.

52–54. Backed microliths.

55. Utilized micro-blade.

56. Bifaced lanceolate point.

59, 60. End-scrapers.

61. Circular platformed micro-blade core.

62, 63. Disc cores, or discs.

64. Rubber and anvil-stone with centrally placed dimple scar.

Nos. 36, 38, 42, 43, 53, crystalline quartz; No. 64, quartz magnetite?. Remainder chert.

PLATE 22

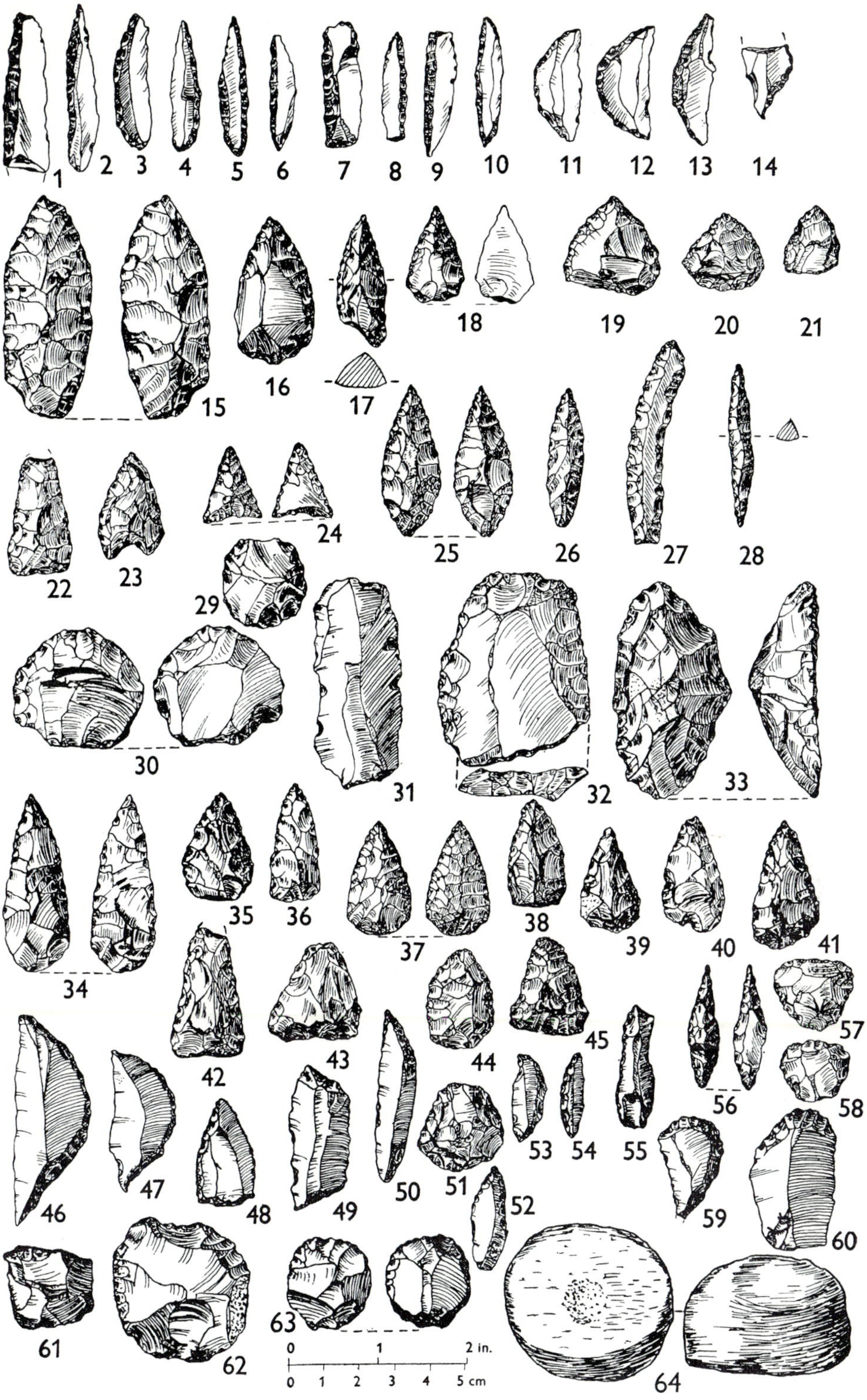

1 2 3 4 5 6 7 8 9 10 11 12 13 14
15 16 17 18 19 20 21
22 23 24 25 26 27 28
29 30 31 32 33
34 35 36 37 38 39 40 41
42 43 44 45 56 57 58
46 47 48 49 50 51 52 53 54 55 59 60
61 62 63 64
0 1 2 in.
0 1 2 3 4 5 cm

Implements from the Gure Warbei, Bur Eibe, southern Somalia (Nos. 17 and 21 from Sites 4 at Bur Eibe)

Nos. 1–27. Doian Culture from Layer B1

1–6. Lanceolate and foliate bifaced points, some with wavy and serrated edges.

7. Trihedral rod.

8, 9. Lanceolate unifaced and parti-bifaced points with serrated edges.

10. Subtriangular parti-bifaced point.

11, 12. Straight-backed microliths with basal retouch.

13. Double-backed microlith.

14. Crescent.

15. Eared crescent.

16. Bifaced tranchet(?) with protuberant boss.

17. Blade with Heluan-type retouch.

18, 19. Asymmetric bifaced triangles.

20. Ostrich egg-shell bead with black polish on one side.

21. Hollow-based arrowhead with serrated edges; unifacial.

22. Bipolar micro-blade core.

23. Conical, single-platformed, micro-blade core.

24. Small round core-scraper.

25, 26. Blades utilized down both edges.

27. Large blade with Kasouga retouch and eared ends.

All specimens in chert with exception of Nos. 6 and 19 in quartz. Nos. 6, 12, 15, 20, 26 from T section; remainder from M section.

Nos. 28–46. Doian Culture from Layer B2

28. Hammer and pestle stone.

29. Disc core.

30. Diminutive hand-axe or bifaced chopping tool.

31. End-scraper on broken flake.

32. Asymmetric (crescent) point with triangular section.

33, 34. Bifaced foliate points.

35. Unifaced subtriangular point.

36. Bifaced triangle.

37. Utilized blade worked over the whole of the upper face.

38, 40. Straight-backed microliths.

39. Curved-backed microlith.

41. Worked bone point.

42. Rim sherd from a thin-walled pot; well fired with fine dark brown paste. Tapered rim rounded on both sides.

43. Fragment of pierced ostrich egg-shell.

44. Ostrich egg-shell bead polished on both sides.

45. Ostrich egg-shell bead polished on one side only.

46. Diminutive fan-shaped scraper.

Nos. 33, 34, 36, in quartz, No. 30 in limestone; remainder in chert. Nos. 30, 32, 40–45 from T section; remainder from M section.

Nos. 47–79. Transitional Magosio/Doian (Lower Doian) from Layer C

47, 48. Large backed blades.

49. Subtriangular unifaced point.

50. Curved-backed microlith.

51, 54, 55. Straight-backed microliths with basal retouch.

52, 53. Straight-backed microliths without basal retouch.

56. Crescent.

57. Crescent with retouch of the cutting edge.

58. Blade with Heluan-type retouch.

59. Utilized blade worked over the whole of one face.

60. Utilized blade worked over the whole of both faces.

61. Asymmetric unifaced point.

62. Ostrich egg-shell bead polished only on one face.

63. Fragment of pierced ostrich egg-shell.

64. Asymmetric (crescent) unifaced point.

65, 67. Hollow-based arrowheads.

66. Asymmetric bifaced triangle.

68. Unifaced subtriangular point with partial retouch on under-face.

69–73. Limace and lanceolate unifaced points, some with serrated edges.

74. End- and side-scraper with notch on end-flake.

75. Hollow scraper with notch.

76. End-scraper on blade.

77. Prepared flake with faceted striking-platform.

78. Small unstruck disc core.

79. Single-platformed, conical, micro-blade core.

Nos. 60, 65, 66 in quartz; remainder in chert. Nos. 51, 52, 61–63 from T section; remainder from M section.

Nos. 80–90. Magosian Culture from Layer D

80. Asymmetric limace or unifaced point.

81–84. Subtriangular unifaced points.

85. Subtriangular bifaced point.

86. Side- and end-scraper on end-flake with faceted striking-platform.

87. Curved-backed microlith.

88. Flat disc core.

89. Small round or thumbnail scraper.

90. End-scraper on flake-blade.

All specimens in chert. All from T section.

PLATE 23

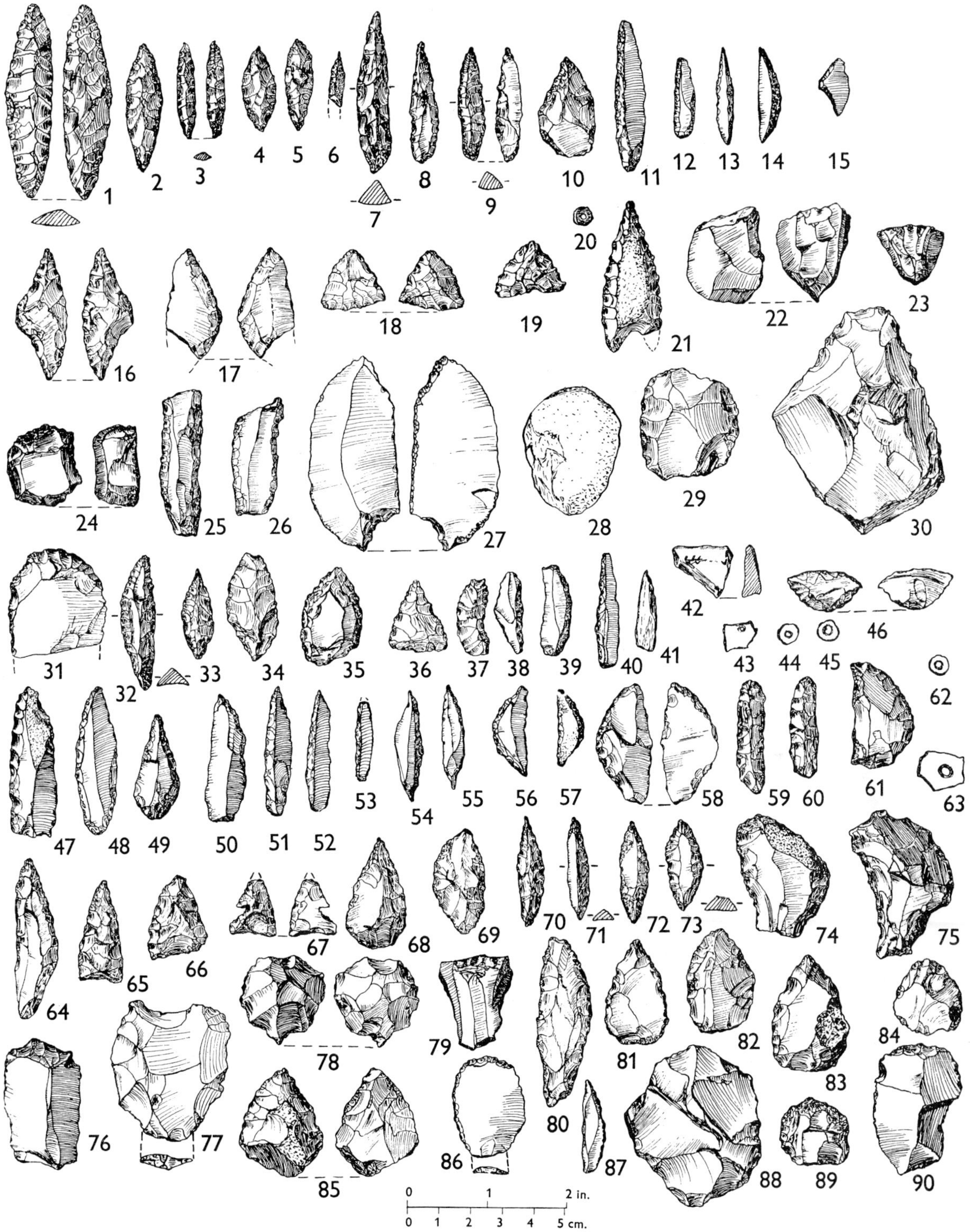
1 2 3 4 5 6 7 8 9 10 11 12 13 14 15
16 17 18 19 20 21 22 23
24 25 26 27 28 29 30
31 32 33 34 35 36 37 38 39 40 41 42 43 44 45 46
47 48 49 50 51 52 53 54 55 56 57 58 59 60 61 62 63
64 65 66 67 68 69 70 71 72 73 74 75
76 77 78 79 80 81 82 83 84 85 86 87 88 89 90
0 1 2 in.
0 1 2 3 4 5 cm.

PLATE 24

Implements from the Rifle Range site, Bur Hakaba, Somalia

Nos. 1–32. Doian Culture from Layer D

1, 2. Bifaced points—foliate.
3. Trihedral rod.
4–10. Blades and flakes showing Heluan-type retouch.
11. Blade utilized down one edge.
12. Broken bone point with oval section.
13, 14. Marine shell beads with fine direct perforation.
15. Fragment of perforated ostrich egg-shell.
16. Ostrich egg-shell bead.
17. Bifaced triangle.
18. Broken hollow-based arrowhead.
19–23. Crescents.
24. Double-backed blade.
25. Straight-backed blade.
26–30. Curved-backed blades.
31. Single-blow burin.
32. Double-ended *outil écaillé* fabricator on a bipolar core.
Nos. 2–4, 6–11, 19, 20, 22–26, 28, 30–32 in chert; remainder quartz.

Nos. 33–45. Magosian Culture from Layer F

33–35. Subtriangular bifaced points; No. 33 broken.
36–41. Straight-backed microliths.
42. Double-backed microlith.
43. Worked bone point.
44. Bipolar micro-blade core.
45. Diminutive disc core.
Nos. 33, 34, 35 in quartz; remainder in chert.

Implements from the Tug Fafan at Gabredarre, Ogaden

Nos. 46–56. Ogaden Variant of the Somaliland Wilton Culture

46. Single-platformed blade core on a pebble.
47, 48. Diminutive alternate-ended cores.
49, 50. Utilized blades.
51. Straight-backed microlith.
52. Curved-backed microlith.
53, 54. Crescents.
55. Triangle.
56. 'Shouldered' triangle.
All in chert.

Implements from sites near Giglei, Middle Webi Shebeli

Nos. 57–75. Derivative Industries of the Ogaden Variant of the Somaliland Wilton

57. Large primary flake of lava.
58. Crude unifaced point or hand-pick.
59. Hollow scraper.
60. *Bec-de-flute* burin.
61, 62. Large backed blades.
63–65. Crescents.
66. Crescent with Kasouga retouch.
67, 71 and 73. Straight-backed microliths.
68–70 and 72, 75. Curved-backed microliths.
74. Microlith with concave retouch at base.
All specimens in chert except No. 57 in lava.

PLATE 24

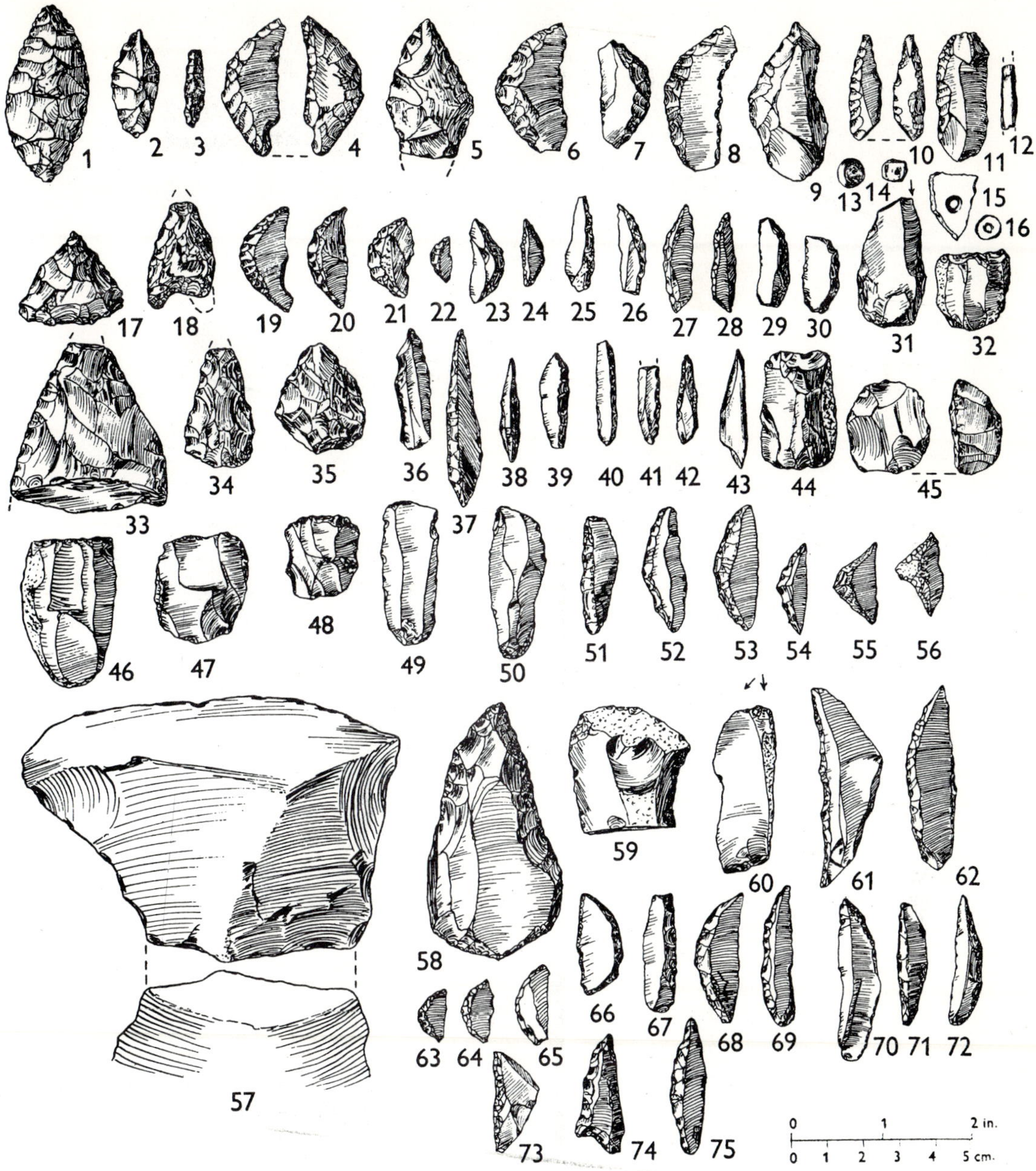
1
2
3
4
5
6
7
8
9
10
11
12
13
14
15
16
17
18
19
20
21
22
23
24
25
26
27
28
29
30
31
32
33
34
35
36
37
38
39
40
41
42
43
44
45
46
47
48
49
50
51
52
53
54
55
56
57
58
59
60
61
62
63
64
65
66
67
68
69
70
71
72
73
74
75
0 1 2 in.
0 1 2 3 4 5 cm.

PLATE 25

Nos. 1–28. Implements from the Gumbur Todoballa Rock-shelter, Mandera, British Somaliland

Somaliland Wilton Culture

1, 2. Straight-backed microliths with some opposite edge retouch at butt.

3–9. Curved-backed microliths.

10–15. Crescents.

16, 17. Double-backed microliths—broken.

18. Angle-burin.

19–20. Utilized blades.

21. Bipolar core.

22, 23. End-scrapers.

24. Double-ended thumbnail scraper.

25, 26. Single-ended thumbnail scrapers.

27. *Outil écaillé.*

28. Sherd of undecorated pottery; coarse brown paste, edges well worn. Thickness 6 mm.

Material, chert and some quartz.

Nos. 29–60. Implements from Afgudut Well, eastern Mudugh, Somalia

Somaliland Wilton Culture

29–36. Curved-backed microliths.

37. Shouldered backed blade.

38. Asymmetric double-backed blade.

39–43. Crescents.

44. Microlithic triangle.

45–52. Straight-backed microliths.

53. Bipolar core.

54. Double transverse concave angle-burin.

55. Transverse concave angle-burin.

56–58. Short-end or round scrapers.

59–60. Thumbnail scrapers.

Material, chert.

Nos. 61–97. Implements from Rakon Wells, eastern Mudugh, Somalia

Somaliland Wilton Industry of 'Neolithic' Facies

61. Unfinished parti-bifaced foliate point.

62. Limace or slug point.

63. Asymmetric (crescent) point.

64. Broken bifaced lanceolate point.

65. Blade worked over the whole of the upper face.

66. Lanceolate unifaced point.

67, 68. Diminutive limace points.

69, 70, 73, 75, 77. Straight-backed microliths.

71, 72, 74, 76. Curved-backed microliths.

78–81. Ostrich egg-shell beads and disc.

82. Blade with concave retouch at base.

83. Crescent with concave retouch at base.

84–86. Crescents.

87–91. Hollow-based bifacial arrowheads.

92. Oblique straight angle-burin.

93. Crude large foliate point—broken.

94, 95. End-scrapers.

96. Thumbnail scraper.

97. Rhomboidal thumbnail scraper worked over entire upper surface.

Material, chert.

Nos. 98–115. Implements from Mirsale Wells, eastern Mudugh, Somalia

Doian Industry of 'Neolithic' Facies

98–100, 108. Large and crude bifaced foliate points.

101. Small slug or limace.

102, 103. Asymmetric (crescent) point worked over one face; No. 103 with obverse retouch at base.

104. Lanceolate unifaced point with triangular section.

105–107, 109. Foliate and lanceolate bifaced points.

110, 111. Backed blades.

112, 113. Short end- or round scrapers; No. 113 with double scraping edge.

114, 115. Double-ended thumbnail scrapers.

All specimens in chert and showing varying degrees of weathering due to wind action.

PLATE 25

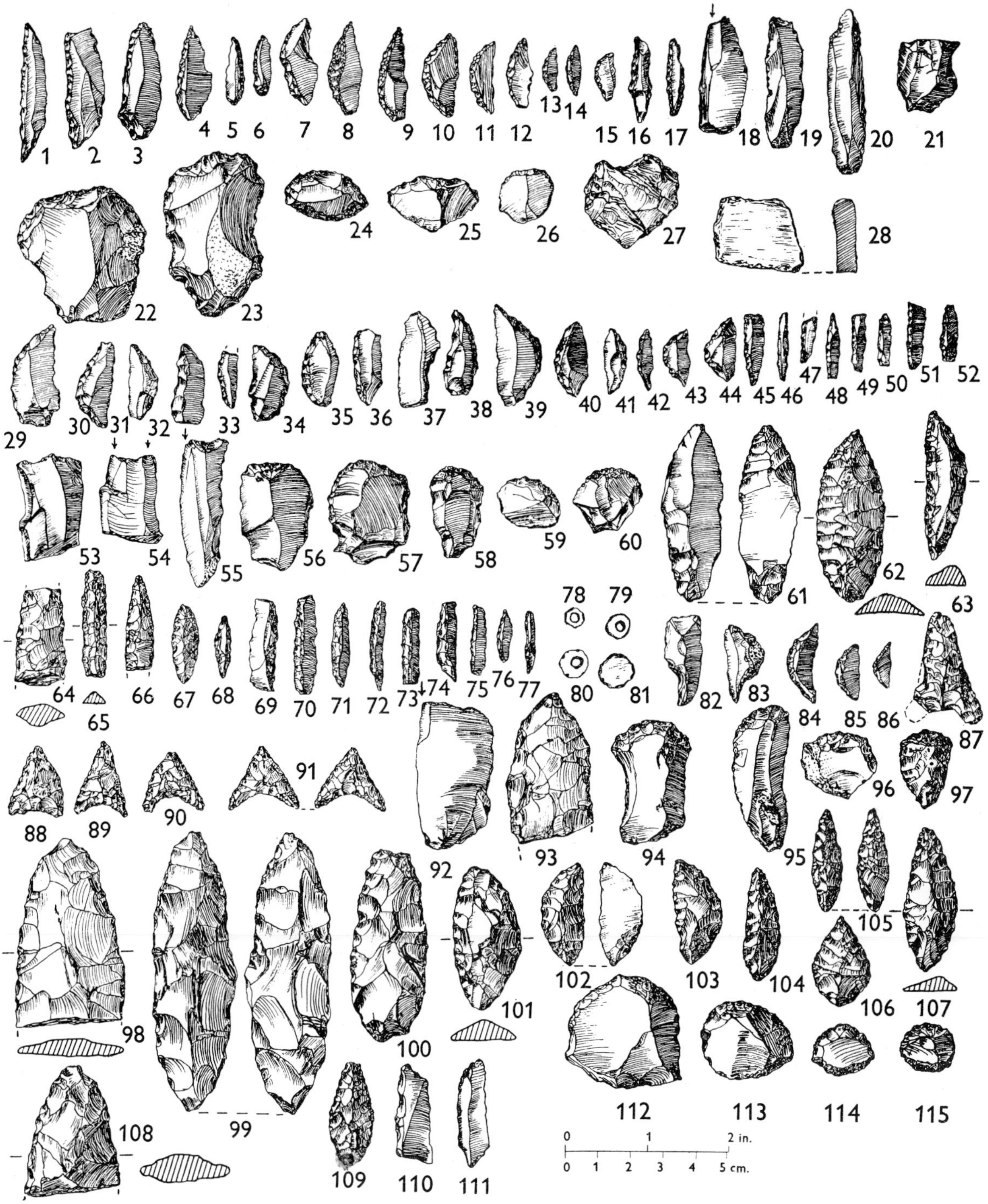
1
2
3
4
5
6
7
8
9
10
11
12
13
14
15
16
17
18
19
20
21
22
23
24
25
26
27
28
29
30
31
32
33
34
35
36
37
38
39
40
41
42
43
44
45
46
47
48
49
50
51
52
53
54
55
56
57
58
59
60
61
62
63
64
65
66
67
68
69
70
71
72
73
74
75
76
77
78
79
80
81
82
83
84
85
86
87
88
89
90
91
92
93
94
95
96
97
98
99
100
101
102
103
104
105
106
107
108
109
110
111
112
113
114
115
0
1
2 in.
0
1
2
3
4
5 cm.

PLATE 26

Strandlooping Industry from Biede, Nogal Estuary Mouth, Somalia
Nos. 1–16. From the top levels of detrital deposit overlying the ± 12 m. raised beach

1. End-chopper on limestone pebble.
2. Pestle or rubbing stone.
3. Side-scraper on large limestone flake.
4. Bifacial pebble chopper.
5. Pebble with opposing notches.
6. Crude flake point with basal retouch.
7. Crudely backed blade worked for boring.
8. Diminutive blade core.
9. Microlithic double-backed blade or borer.
10. Microlithic backed blade.
11. Core-scraper.
12. Backed blade.
13. Backed flake.
14. Utilized blade from double-ended blade core.
15. Crude end-scraper on an end-flake.
16. Unifaced point.

Nos. 1–5 in limestone; remainder in chert.

Nos. 17–23. From the Upper Midden, Biede Rock-shelter

17. Sherd of thick hand-made pottery in fine grey paste, buff to red external finish, black interior. One edge shows section of an oblique hole or intentional perforation, made before firing. Outer surface shows scratches, perhaps an attempt at decoration—made after firing. Thickness 8 mm.

18. Two rim sherds of fine grey paste, probably hand-made; buff to red interior and exterior finish. Simple turned-over rim with incised line running round the base of the rim inside.

19. Sherd of wheel-turned red 'stone-ware' pottery, with yellow external slip, and ribbed interior.

20. Mother-of-pearl shell disc.
21. Perforated fragment of mussel shell.
22. Sherd of fine black paste from a thin-walled pot.
23. Crude bone point.

Nos. 24–35. Implements from the Lower Midden, Biede Rock-shelter

24. Utilized flake.

25. End-scraper on flake.

26. Pebble side-scraper.

27. Small core-scraper.

28–30. Microlithic backed blades.

31. Microlithic utilized blade.

32. Perforated shell disc.

33, 34. Perforated cowrie shells.

35. Portion of long bone showing wear; the cancellous bone has also been removed.

All stone tools in chert.

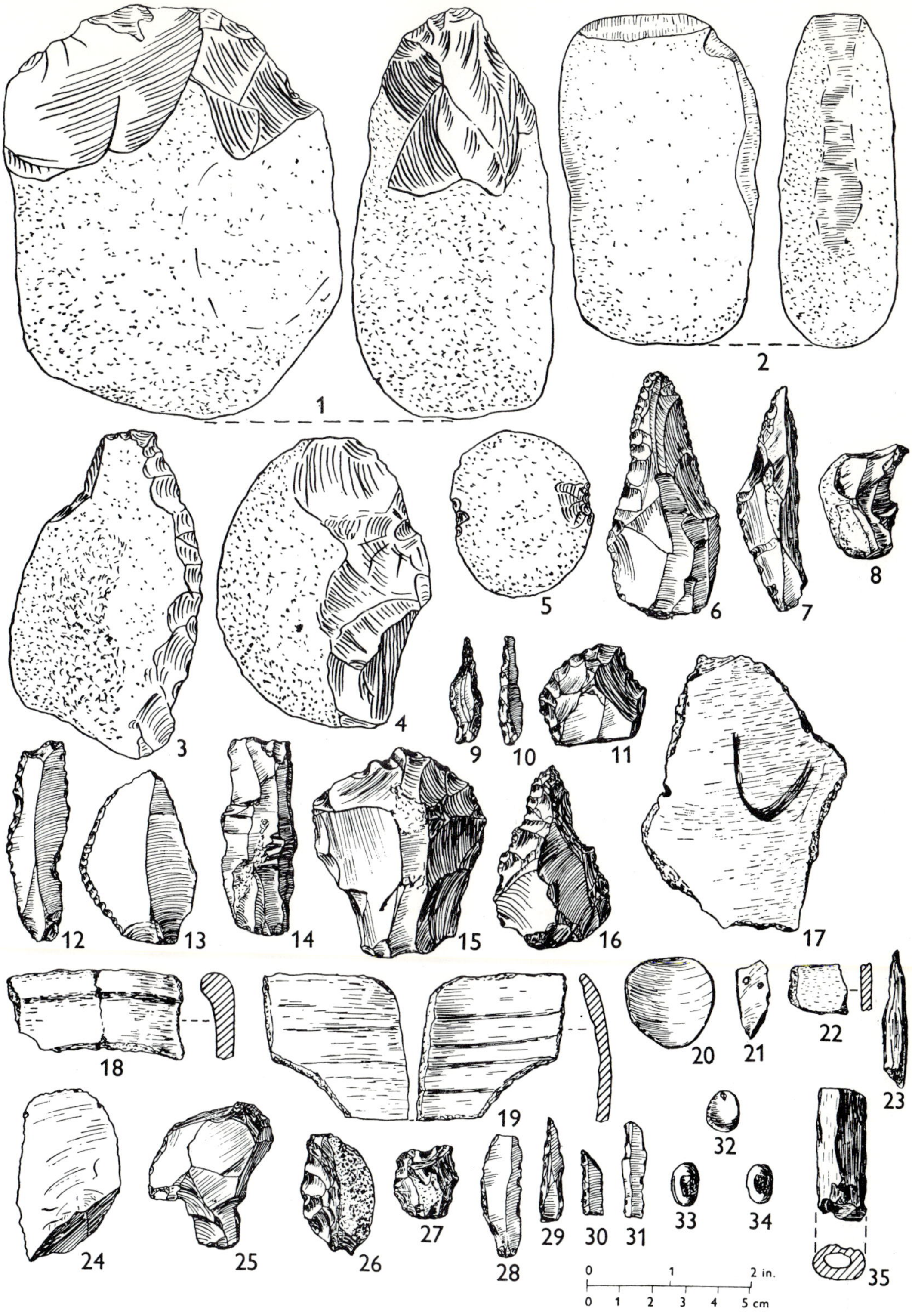
1
2
3
4
5
6
7
8
9
10
11
12
13
14
15
16
17
18
19
20
21
22
23
24
25
26
27
28
29
30
31
32
33
34
35
0 1 2 in.
0 1 2 3 4 5 cm

PLATE 27

Engravings from a rock-shelter at Bur Dahir, Middle Webi Shebeli, Somalia

Series 1

1. Main group on flat upper surface of larger rock. Showing cattle (both *Bos africanus* and zebu), camels, goats?, rhinoceros, and filiform humans armed with shield and spear.

2. Inset. Isolated figures from upper surface of smaller rock. (1) Indecipherable figure, but see similar motif from a rock-shelter at Yavello, southern Abyssinia (Clark, J. D., 1945, Fig. 2, No. 1). (2) and (3) Human figures with arms raised. Fig. 3 has an engraving of what is probably a camel above the human figure. Human figures in a similar position occur on the monoliths discovered by P. Azais at Silte, Maska and Thya, north of Lake Zwai (see Kammerer, A., 1926).

Series 2

1. Comparatively recent Amharic (?) engravings. The inscription is translated as 'quelelett', meaning 'melting'.

2. Crude pecked engraving of an elephant.

Series 3

Finely incised engravings done with a metal knife. Somali (?).

PLATE 27

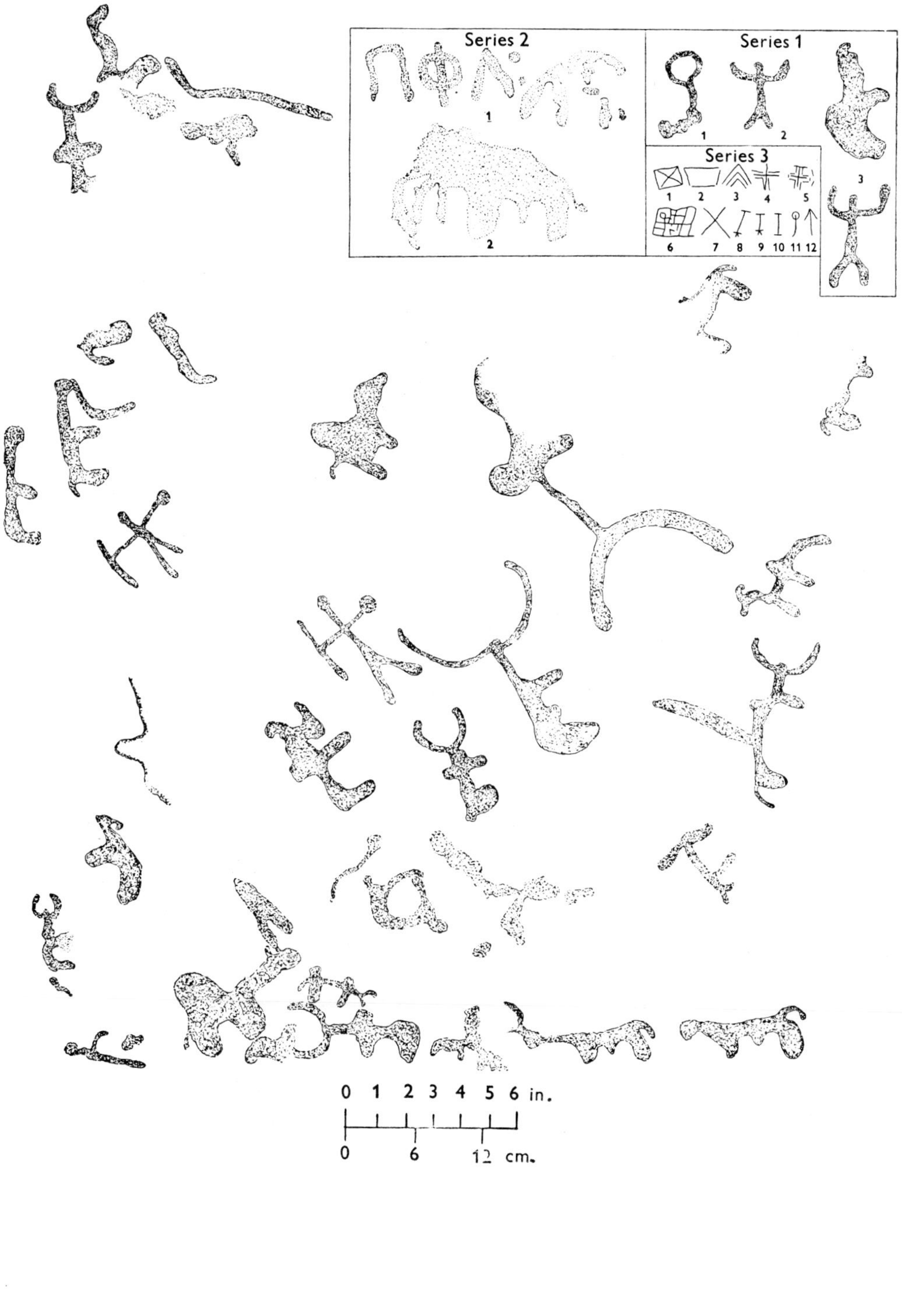
Series 2
1
2
Series 1
1
2
3
Series 3
1 2 3 4 5
6 7 8 9 10 11 12
0 1 2 3 4 5 6 in.
0 6 12 cm.

PLATE 28

Pecked Engravings from rock-shelter above upper water-hole at El Goran, Middle Webi Shebeli, Somalia

Series 1. Weathered to a chestnut patina identical with the rock on which they are engraved.

1 and 4. Camels.

2 and 5. Cattle (cf. Bur Dahir engravings).

3. Elephant.

6. Partially weathered human figure (cf. Bur Dahir engravings and illustrations of motifs on monoliths discovered by P. Azais at Silte, Maska and Thya, north of Lake Zwai. Kammerer, A. (1926)).

7. Camel (?) and indeterminate motifs.

8. Tectiforms and curvilinear motifs.

9. Curvilinear motif (cf. engraved monoliths from the upper Maganamo Valley, Abyssinia. *L'Africa Orientale* (1936)).

Series 2. Greatly weathered and polished with much overworking thus rendering interpretation difficult and obscuring the underlying engravings.

1. Curvilinear motif.

2. This sign has been adopted today as a tribal mark by the Hawiya Galjaal and other tribes.

3, 4. Inverted U's or half-circles.

5. Possibly a degenerate representation of similar signs on the Lake Zwai monoliths. The same sign occurs on grave stones on the old Zeila/Boramo road in British Somaliland.

6, 7. A sign now adopted by the Eile and Shekal tribes.

8, 9. Circles and twinned circles (cf. Maganamo monoliths and Jid Banan engravings).

10. Arabic letter 'min' adopted as a tribal mark by the Rahanwein Harau tribe.

11, 12. *Garre* boards (No. 12 incomplete). The two left-hand holes of No. 11 recently repecked.

13. Tectiform motifs. The engraving above may be a representation of the *garre* board known in Abyssinia as *gabata.* It occurs on the Lake Zwai monoliths.

PLATE 28

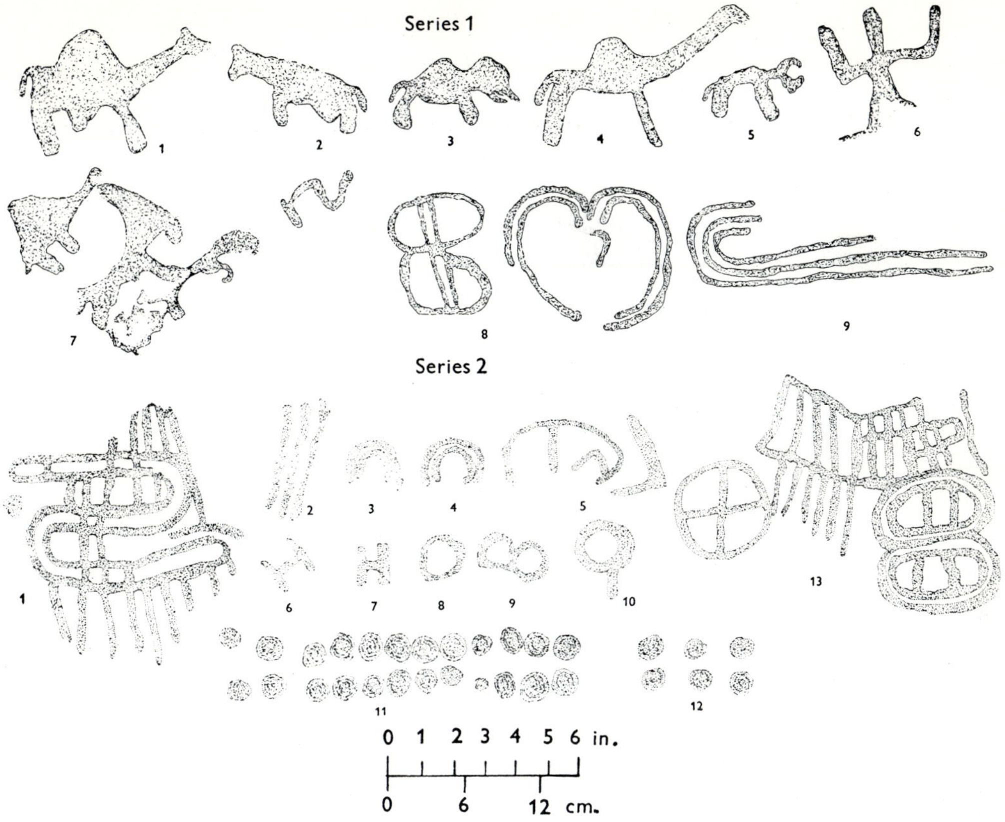
Series 1
1
2
3
4
5
6
7
8
9
Series 2
1
2
3
4
5
6
7
8
9
10
11
12
13
0 1 2 3 4 5 6 in.
0 6 12 cm.

PLATE 29

Distribution Map of Prehistoric Sites in the Horn of Africa

1. Berbera Rock-shelter.
2. Laferug.
3. Hargeisa.
4. Jigjiga.
5. Borama.
6. Aisha.
7. Adele.
8. Garasleh Tug.
9. Adjin.
10. Haud Surface: Aware.
11. Mandera.
12. Balleh Bio Mudo.
13. Khansaha Au Plain.
14. Dubar.
15. Guban Plain: Berbera.
16. Harnas.
17. Bur Degis.
18. Bur Eibe.
19. Bur Mahamudo.
20. Bur Hakaba.
21. Warishe.
22. Gawan wein.
23. Deileb.
24. Wanei.
25. Isha Baidoa.
26. Manaas.
27. Hoddur.
28. Wejit.
29. Legullei.
30. Madeida.
31. Morogavi.
32. Alamo.
33. Far Abou.
34. Bohol Washaga.
35. God'dere.
36. Ato.
37. Saha Gurral.
38. Ted.
39. Burdo.
40. El Dubbo.
41. Sijjia.
42. Ohole.
43. Goffado.
44. Rakon.
45. Mirsale.
46. Daba Gallo.
47. Afgudut.
48. Gowein: mile 27 from Obbia.
49. Obbia.
50. Kourou.
51. Dusa Mareb.
52. Donkukoq.
53. Laba Higlo: mile 14 from Garoe.
54. Garoe Tug.
55. Hararile: mile 44 from Eil.
56. Eil.
57. Biede.
58. Got Ad.
59. Babani Tug.
60. Bihen.
61. Jid Banan.
62. Domo.
63. Wardere.
64. Gherlogubi.
65. Gabredarre: Tug Fafan.
66. Kaptinag.
67. Damburweini.
68. Durualeh.
69. Kabahan.
70. Melka Taka: Bama'asi Cave.
71. Melka Taka: Webi Shebeli.
72. Cura Ba'ad.
73. Darchaua Tug.
74. Bur Eighei.
75. Mile 32 Callafo/Mustahil Road.
76. Durdere: mile 18 Callafo/Mustahil Road.
77. El Goran.
78. Bohol Bandoi.
79. Daban.
80. Habr Fordisso.
81. Durdur Heime (Dondere).
82. Fossil Dune: 7·6 miles from Duldir.
83. Sheik Gure.
84. Bugda Acable.
85. Jesomma.
86. Bulo Burti.
87. Gesira.
88. Danane.
89. Mogadishu.
90. Bardera.
91. El Colcolo.
92. Berdale.
93. Mile 75: Wajir Road.
94. Salagle.
95. Bun Mess.
96. Bur Mun.
97. Bur Dinsor.
98. Allengo.
99. Guretca.
100. Lugh Ferrandi.
101. Dolo.
102. Washaga Guran.
103. El Ure.
104. Mahaddeiwein.
105. Between Beletwein and Tijieglo.
106. Mustahil.
107. Adad.
108. Ilbehla.
109. Durgale.
110. Adani.
111. Warandi.
112. Hill 49 km. from Gorrahei.
113. Ubitale.
114. Walwal.
115. Bio Ado.
116. Dagahbur.
117. Galkayu/Garoe Road: about 60 km. from Galkayu.
118. Between Galkayu and Garoe.
119. Upper Nogal.
120. Deavani.
121. Trogo Dait.
122. Gulule Tug.
123. DolDol Plateau.
124. Between Bakbah and Bio Ado.
125. Between Bihen and Ariri.
126. Bihen.
127. Between Bender Beila and Bihen.
128. Between Warassimoghe and Bender Beila.
129. Badi Ahmud.
130. Cubo.
131. Between Martissor Dunti and Scerer Hor.
132. Ras Hafun.
133. Lower Darror (Daror).
134. Upper Darror (Daror).
135. Ras Filuc.
136. Taleh.
137. Buran.
138. Hobat.
139. Hormo.
140. Erigavo.
141. Sheik (Sheikh).
142. El Anod.
143. Tug Issutugan.
144. Between Zeila and Berbera.
145. Ras Kiro.
146. Between Gulf of Tajura and Lake Assal.
147. Gulf of Tajura.
148. Kor Ali.
149. Mount Mabla.
150. Obok.
151. Djebel Djinn.
152. Porc-Epic Cave: Dire Dawa.
153. Segag (Sagak).
154. Tug Fidadaedi (Fidadedih)
155. Orahaut.
156. Dallehallé (Delahalleh)
157. Tug Ensiral.
158. Habaoule.
159. Dakata/Webi confluence.
160. Left Bank of the Webi Shebeli.
161. Left Bank of the Webi Shebeli.
162. Left Bank of the Webi Shebeli.
163. Dagahbur/Webi confluence.
164. Left Bank of the Webi Shebeli.
165. Souri.
166. Oksa.
167. Chekifton.
168. Iddi.
169. Dabrin.
170. Karsaalek (Carsalec).
171. Djilohanno.
172. Rahaïtou.
173. Robabouta (Roba Butta).
174. Abbé.
175. Ghimir (Guiner).
176. Kadjoua.
177. Goba (Gobbà).
178. Sheik Hussein.
179. Valley of the Tug Sullul (Salout).
180. Bed of the Tug Dakata: 20 km. above confluence with Daure Tug.
181. Las Anod.
182. Kirit.
183. Bohotleh: Merkanwein (Marqanweina) Balleh.
184. El Dur Elan.
185. Raguda Tug.
186. 5 miles from Hudin: Hudin/Ainabo Road.
187. Bur Khansa: Bihen (Hubera Valley).
188. Halin.
189. Haded: Balad Ad.
190. Shimbir Beris.
191. 50 miles from Las Anod: Las Anod/Halin Road.
192. 12 miles from Hudin: Hudin/Halin Road.
193. Sunto.
194. Al Hills.
195. Teisa Tug.
196. Arabsiyo.
197. Gebile Tug.
198. Ijara Tug.
199. Sirrau (Sirau) Balleh.
200. 45 miles from Hargeisa: Hargeisa/Wardere Road.
201. Dudubo Coffee Shop.
202. Ged Dobo.
203. Haleya.
204. Skutar.
205. Jifa Ure (Jifo Uri).
206. Alablah Balleh.
207. Genda-Biftou: Sourré.
208. Harar (Saka Sharifa, Bake Khallo and Errer Kimiet).
209. Gerbakele Shelter.
210. Tug Khaboba Shelter.
211. Dombosleh Shelter.
212. Balleh Shelters, Skutar.
213. Dukokolol yero.
214. Bur Dahir.
215. Bur Eschunle.
216. Bur Dauleh.
217. Tug Argan.
218. Darbruk (Darburuk).
219. Roukatcha (Ruchessa) Wells.
220. Omerdin.
221. Lago Oda.
222. Taffe.
223. Araue.
224. Warka.
225. Tchalinko (Ciallanco).
226. Kouloubï (Collubi).
227. Kourkourron (Curgurru).
228. Dubasso (Dobasso).
229. Lake Adili.
230. Dubato.
231. Karin Heganah.
232. Gedeis.
233. Manjasseh (Manja Asseh).
234. Gala-Ad Shelters

Note. The boundaries of Somalia are shown as including the Ogaden and Odo Provinces. These were returned to Abyssinia on 23 September 1948. Abyssinia also took over the administration of the Reserved Areas of Dire Dawa and Jigjiga in May 1946 and May 1948 respectively.

Fig. 1. General view of the Hargeisa valley, British Somaliland, looking towards the town from the top of 'Fly Hill' on the southern scarp. Main Jigjiga/Berbera road in the foreground. Gully site H. 12 in centre left.

Fig. 2. View of site H. 2, Hargeisa, British Somaliland. Eroded aeolian sands, yielding an Hargeisan Derivative industry, overlie eluvial gravels on the surface of which are found Somaliland Stillbay tools.

Fig. 1. Site H.12, Hargeisa, British Somaliland. Stratified gravels and grits with Acheulio-Levalloisian home site exposed in side of tributary tug (Older Tug Gravels II). The occupation horizon can be seen on a level with the excavating tool. Two hand-axes can be seen at right centre.

Fig. 2. Site H.8, Hargeisa, British Somaliland. Older Tug Gravels and Alluvium, containing Acheulio-Levalloisian tools, overlain by Younger Tug Gravels with a Levalloisian industry. Shows the torrent nature of the deposits exposed in this tributary tug.

Fig. 1. General view of the Gumbur Todoballa, an isolated granite hill at the foot of the Gulf of Aden Fault Scarp at Mandera, British Somaliland. Type site for the Somaliland Wilton Culture.

Fig. 2. View from the Gumbur Todoballa shelter, showing the surrounding peneplain and the section excavated through the deposits in this shelter.

Fig. 3. Aeolian/alluvial Sands at Site H.7, Hargeisa, British Somaliland. Containing flaking floors (one partially exposed) of the Hargeisan Culture.

Fig. 4. Erosion gully at Mandera, looking from the camp towards the bakery, showing gravels (Older Tug Gravels?) overlain by the Younger Alluvium and land rubble. Somaliland Stillbay tools in the top levels are overlain by aeolian red sands with Hargeisan chipping floors.

Fig. 1. General view of the southern scarp of the Nogal valley from the Boundary Road, near Donkukoq. Showing the structural nature of the scarp terraces in this area. A residual gravel with Levalloisian tools is seen in the foreground.

Fig. 2. General view of the Central Nogal plain looking north-east from Callis towards the northern scarp. A monotonous gypsum plain with little or no water or vegetation.

Fig. 3. Two deposits of Tug Gravels and Alluvium (the lower gravel only just exposed) overlain by gritty gravels and alluvium seen in the right bank of the tug about 1 mile upstream from Garoe Fort. The Tug Gravels yielded Levalloisian tools and are of Upper Pleistocene date.

Fig. 4. Section exposed at Got Ad, near Eil, in the left bank of the tug within the Nogal Gorge. Shows the cross-bedded nature of the Estuarine Beds, capped by massive tufa and surface limestone, and resting on sandstones of Miocene (?) age. The seated figures (middle centre) give the scale.

Fig. 1. General view of the Lower Nogal Gorge at Got Ad, near Eil. The 22–30 m. terrace of the Estuarine Beds can be seen on the right. On the left is the aggradation terrace of the Tug Gravels and Alluvium containing Levalloisian tools. The low-level terrace of torrent gravels, dating to the First Post-Pluvial Wet Phase, can be seen immediately below the Tug Gravels.

Fig. 2. Two denudation terraces on the right side of the Lower Nogal Gorge between Eil and Biede. The upper terrace at 24 m. corresponds to the level of the top of the Estuarine Beds on the left side of the gorge. The lower denudation terrace at 12 m. is thought to represent the height to which the Tug Gravels were accumulated in this part of the gorge. Below these two denudation terraces is an aggradation terrace at approx. 3 m. above the tug bed, representing the deposits of the First Post-Pluvial Wet Phase.

Fig. 3. General view of tufa deposits from springs at Eil, Lower Nogal Gorge. These tufas yielded a wide range of tools.

Fig. 4. Torrent Gravels and Alluvium of the First Post-Pluvial Wet Phase resting unconformably on grey Tug Alluvium. Seen in a tributary gully at Eil, Lower Nogal Gorge. Some Stillbay tools come from the top of the Tug Alluvium while a Doian living site overlies the upper alluvium.

Fig. 1. View of the scarp at Isha Baidoa where the southern plateau joins the coastal plain. The cliff at the left centre is composed entirely of tufa deposited by the 'isha' (spring) during Pleistocene and Recent times.

Fig. 2. General view of the wide flat valley of the middle Webi Shebeli between Bulo Burti and Beletwein. The northern scarp can be seen in the left background.

Fig. 1. The 15 m. terrace gravels of the middle Webi Shebeli at Cura Ba'ad, Ogaden. The line of trees (back centre) marks the line of the Webi. An early erosion level cutting the southern scarp is seen in the right background.

Fig. 2. The eroded 30 m. terrace gravels of the middle Webi Shebeli at Sheik Gure, near Bugda Acable. The gravels rest on beds of Cretaceous limestones and marls.

Fig. 1. The Upper Tug Gravels and calcareous alluvium exposed in the Tug Fafan at Gabredarre, Ogaden. The figure is standing at the junction of the upper and lower calcretes, and points to the implementiferous horizon at the base of the upper beds which yielded diminutive Levalloisian tools of developed form. The upper calcretes are overlain by a fine powdery alluvium, and at the junction of the two can be seen the land surface yielding living floors of an Ogaden Wilton industry.

Fig. 2. General view of Bur Eibe, southern Somalia, a granite inselberg rising over 1000 ft. out of the coastal plain. Photo taken looking north-east from the Bur Hakaba/Bur Eibe road. The scrub in the foreground is typical of that supported by the red and orange sands of the 'doi' country.

Fig. 1. View of Bur Eibe from the top of Bur Degis, southern Somalia. Looking east showing the distribution on the plain of the light-coloured 'adable' (grey and black agricultural soil), and (dark patches) the red and orange 'doi' sands of aeolian and eluvial origin.

Fig. 2. Part of the Bur Yassin Site at Bur Eibe, southern Somalia. Showing eroded 'ironcrete' beds containing Somaliland Stillbay and Magosian industries, and overlain towards the foot of the bur by sands and grits with a Doian industry.

Fig. 1. Section exposed in the right bank of the Garasleh Tug 30 miles north-east of Dire Dawa in the southern Danakil Rift. Showing Upper Pleistocene (?) fan deposits of gravels and grits and alluvium overlain by lavas.

Fig. 2. General view, looking south-west, of the eroded freshwater lake and fan deposits of the Aisha Plain, southern Danakil Rift.

Fig. 1. View looking south-west across the Aisha Plain, southern Danakil Rift. Showing the hills of the fault scarp in the background, and lavas (black line centre) intruded through the sedimentary deposits of the plain. The eroded area of these sedimentary beds commences to the left of the line of lavas.

Fig. 2. View looking towards Adele from the Aisha/Jibuti road, showing faulted lavas overlying clays and fine gravels of early Upper Pleistocene age (?). The Duanle Horst can be seen in the centre background.

Fig. 1. General view of the Berbera coastal plain looking towards the barrier hills at Dubar, British Somaliland.

Fig. 2. View of the 8 m. raised beach at the hospital at Berbera. Photo taken from the 2 m. coastal flat.

Fig. 1. General view of Biede, Nogal Estuary mouth, northern Somalia. The surface of the Estuarine Beds can be seen in the left centre. The isolated building in the centre (Customs House) stands on the 12 m. raised beach.

Fig. 2. View of the highest part of the 12 m. raised beach at the back of the Customs House at Biede. The figure in the centre is seated just below the base of the beach, which rests on the eroded Estuarine Beds.

Fig. 3. View looking towards the cliffs from unconsolidated dunes which block the mouth of the Nogal estuary. The low-level 0 m. beach (consolidated) is seen outcropping in the right centre. At the foot of the cliffs (left centre) are eroded relicts of fossil dunes. The figure (centre) is seated near the foot of the 12 m. raised beach.

Fig. 4. Low-level 0 m. beach at Obbia, Central Somalia, overlying current-bedded marine sands.

Fig. 1. Low-level 0 m. beach at Mogadishu Lido; resting on marine coral sands and overlain by fossil dunes.

Fig. 2. General view showing the false-bedded nature of the fossil dunes overlying the 0 m. beach at Mogadishu Lido. The figure stands on the level of the beach. The surface of the dunes is a red bed.

Fig. 3. Section through the upper part of the coastal dunes exposed in a gully at Danane, southern Somalia coast. The figure at the left bottom corner gives the scale. At the top are orange sands overlying an irregular land surface with Late Stone Age strand-looper sites. This surface rests on compact red sands, underlain by compact grey to yellow calcareous sands, underlain by consolidated fossil dune sands (top just seen at base of section).

Fig. 4. Typical differential weathering of the coastal dunes at Danane, southern Somalia coast.

PLATE 44

Fig. 1. General view of excavations in the Gure Warbei rock-shelter at Bur Eibe, southern Somalia, seen from the left rear side of the shelter.

Fig. 2. Gure Warbei excavations, Bur Eibe. Showing T trench excavated to bed rock in the rear sections. The light-coloured cave earth contains the Early Eile and Doian industries, while the dark earth contains the Magosio/Doian Transition and the Magosian industries.

Fig. 3. View of M trench, Gure Warbei excavations, Bur Eibe. The first four sections have been excavated to the base of Layer B 2. The fifth section is excavated to the base of Layer B 1.

Fig. 1. General view of the Rifle Range site, Bur Hakaba, southern Somalia. Showing the bank and ditch of the butts and the excavated area towards the farther end of the ditch.

Fig. 2. Rifle Range site excavations, Bur Hakaba. The entrenching tool rests on the granite rock, while the brush lies on the Doian occupation layer.

Fig. 1. General view of Biede rock-shelter, Nogal Estuary mouth; northern Somalia coast.

Fig. 2. Biede rock-shelter excavations. Showing the near part of the excavation carried down to bed rock. The remaining sections show the Lower Midden with hearths and a Late Stone Age strandlooping industry overlain by sterile yellow sand. The Upper Midden can be seen in section in the end wall of the excavation.

Fig. 3. Biede rock-shelter excavations. View of excavation taken from the rear of the shelter showing the Lower Midden overlain by sterile yellow sand overlain by the Upper Midden.

PLATE 47

Fig. 1. General view of Rakon Wells, central Somalia, from the Late Stone Age occupation site. Eroded artifacts can be seen in the foreground. The flat and bare nature of the eastern Mudugh can be clearly seen.

Fig. 2. Mirsale Wells excavation, central Somalia. Showing the unbedded nature of the red sands which fill the pan and contain the Late Stone Age industry.

Fig. 3. View of the upper surface of the larger engraved rock in shelter No. 2 at Bur Dahir, middle Webi Shebeli valley. Showing the main engraved panel.

Fig. 4. Side of the larger engraved rock in shelter No. 2 at Bur Dahir. Showing engravings of camels, cattle and one human of Series 1 overlain by camels and an Amharic inscription of Series 2.

Fig. 1. Engraved rock-shelter at El Goran, on the southern edge of the middle Webi Shebeli valley. The area where the Series 1 engravings occur is in the part of the shelter where the sloping rear wall meets the roof. Series 2 engravings are found on the sloping rear wall and are overlain by recent engravings of hands and 'garre' boards.

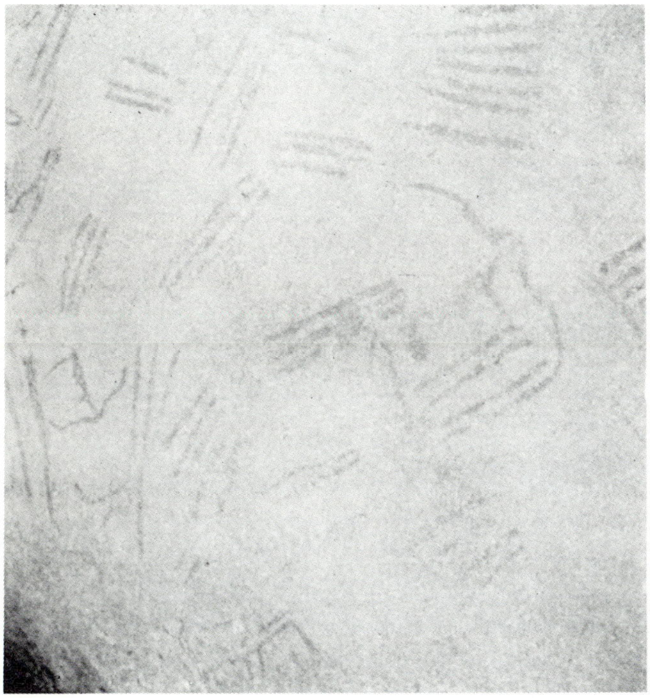

Fig. 2. Paintings in faded red paint on the roof of a small rock-shelter at Bur Dauleh, Bur Eibe. Note the possibly conventionalized 'cattle' motif.

PLATE 49

Figs. 1–4. Pecked engravings on limestone rocks in the pass at Jid Banan, north-eastern Haud, British Somaliland. These engravings include tribal marks of local Somali tribes. Some of the engravings are very recent (white in photograph).

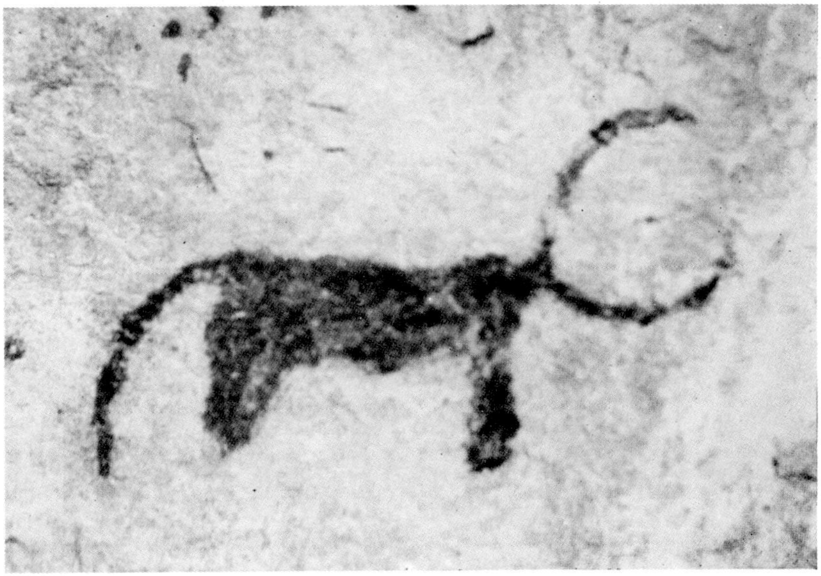

Fig. 1. Flat-wash painting of large-horned domestic ox (? *Bos africanus*) in reddish brown paint. From Saka Sharifa, Harar. Reproduced by permission of Count Björn von Rosen.

Fig. 2. Flat-wash human figure and outline drawings of (?) cattle, one with branding mark on the rump, in reddish brown paint. From Saka Sharifa, Harar. Reproduced by permission of Count Björn von Rosen.

Fig. 1. Probably a representation of a hunter shooting an ostrich; in reddish brown paint. From Saka Sharifa, Harar. Reproduced by permission of Count Björn von Rosen.

Fig. 2. Human figure in reddish brown paint with circular headdress or bundle, overlain by painting in white pigment. From Errer Kimiet I, Harar. Reproduced by permission of Count Björn von Rosen.

Fig. 1. Domestic cattle scene in reddish brown pigment. Note angular profile and saddle-backs of some of the animals. From Errer Kimiet I, Harar. Reproduced by permission of Count Björn von Rosen.

Fig. 2. White paintings of a (?) cheetah, ox and stylized human figures in the form of a capital H. From Errer Kimiet II, Harar. Reproduced by permission of Count Björn von Rosen.

INDEX